Soil Biology

Volume 60

Series Editor

Ajit Varma, Amity Institute of Microbial Technology,
Amity University, Noida, Uttar Pradesh, India

The Springer series *Soil Biology* publishes topical volumes in the fields of microbiology, environmental sciences, plant sciences, biotechnology, biochemistry, microbial ecology, mycology and agricultural sciences.

Special emphasis is placed on methodological chapters or volumes. This includes coverage of new molecular techniques relevant to soil biology research or to monitoring and assessing soil quality as well as advanced biotechnological applications. Leading international authorities with a background in academia, industry or government will contribute to the series as authors or editors.

Key Topics: microbial-plant interactions; microbial communities; root symbiosis, mycorrhiza; rhizosphere environment; soil fauna, e.g. termites etc.; biochemical processes, soil proteins, enzymes, nucleic acids; degradation of biomaterials, decomposition, nutrient cycles; soil genesis, mineralization; bioremediation of contaminated sites; biotechnological applications of soil microorganisms.

More information about this series at http://www.springer.com/series/5138

Neeraj Shrivastava • Shubhangi Mahajan •
Ajit Varma
Editors

Symbiotic Soil Microorganisms

Biology and Applications

Springer

Editors
Neeraj Shrivastava
Amity Institute of Microbial Technology
Amity University Uttar Pradesh (AUUP)
Noida, Uttar Pradesh, India

Shubhangi Mahajan
Amity Institute of Microbial Technology
Amity University Uttar Pradesh (AUUP)
Noida, Uttar Pradesh, India

Ajit Varma
Amity Institute of Microbial Technology
Amity University Uttar Pradesh (AUUP)
Noida, Uttar Pradesh, India

ISSN 1613-3382 ISSN 2196-4831 (electronic)
Soil Biology
ISBN 978-3-030-51915-5 ISBN 978-3-030-51916-2 (eBook)
https://doi.org/10.1007/978-3-030-51916-2

This Springer imprint is published by the registered company Springer Nature Switzerland AG.
The registered company address is: Gewerbestrasse 11, 6330 Cham, Switzerland

Contents

Part I
Fungal Symbiosis

Chapter 1
Current Status–Enlightens in Its Biology and Omics Approach on Arbuscular Mycorrhizal Community

Tulasikorra, O. Siva Devika, K. Mounika, I. Sudhir Kumar, Suman Kumar, G. Sabina Mary, Uday Kumar, and Manoj Kumar

Abstract Symbiotic association has been subject to great heed in part because of the prime of the disease to world agriculture, but also because both host and arbuscular mycorrhizal fungi are persuadable to proceed investigational approaches. The goal of the review is to furnish an overview of the microbial association system and designate concurrent remarkable studies that amend our comprehension of the biology and Omics of VAM fungi. The genomic studies have been organized to long-term well-established areas of investigation, including disease development and the characterization of proteins in relation to host. VAM fungi act as biological control in global sustainable development under varied agroecological regions. VAM fungi illustrate the plant–root/microbial interactions which serves gaining knowledge and a wide out-look of issues in crop production/protection.

Keywords AMF · Genes · Omic approach · Spores · Enzyme activity · Biocontrol · Symbiosis · Ecology

Tulasikorra · I. S. Kumar · S. Kumar · U. Kumar
Department of Mycology and Plant Pathology, Institute of Agricultural Sciences, Banaras Hindu University, Varanasi, Uttar Pradesh, India

O. Siva Devika
Department of Soil Science and Agricultural Chemistry, Institute of Agricultural Sciences, Banaras Hindu University, Varanasi, Uttar Pradesh, India

K. Mounika
Department of Genetics and Plant Breeding, Institute of Agricultural Sciences, Banaras Hindu University, Varanasi, Uttar Pradesh, India

G. Sabina Mary
Department of Plant Pathology, Odisha University of Agriculture & Technology, Bhubaneswar, India

M. Kumar (✉)
Department of Plant Pathology, Acharya N.G. Ranga Agricultural University, Agricultural College, Bapatla, Andhra Pradesh, India

N. Shrivastava et al. (eds.), *Symbiotic Soil Microorganisms*, Soil Biology 60,
https://doi.org/10.1007/978-3-030-51916-2_1

1.1 Introduction

Arbuscular mycorrhizal fungi (AMF) is one of the most widely circulated, ecologically and economically significant fungal groups and are ubiquitously associated with the vast majority of plant families in numerous habitats around the world, extending from the tropics (Zhao et al. 2001) or arctic-alpine habitats (Haselwandter 1987) to mesic (Muthukumar and Udaiyan 2000) and arid habitats (O'Connor et al. 2002). AMF is included in the order of Glomales (Morton and Benny 1990) classified in the phylum of the Zygomycota, defined as compulsively symbiotic, asexual organisms, and form symbioses with most land plants in almost any terrestrial environment (Smith and Read 2008). AMF uplifts crop development through enhancement of growth, intake of nutrients and water, confronts the plant under diverse stress situations, and also releases antioxidants. Omic approach is a crucial aspect to perceive a deep insight into appropriate mechanisms of the mentioned activities of AMF. Several SSU-targeting PCR primers (Lee et al. 2008) that amplify fragments of 500–800 bp have been widely applied in ecological studies (Zhang et al. 2010). AMF encompasses a wide range of species that performs variety of functions. Phylogenetic analyzes, based on three rDNA markers, offer accurate and stable resolution from phylum to species level. In comparison, 109 known species and 27 populations comprising as yet unknown organisms have been studied (Kruger et al. 2012). Besides these, AMF release several proteins and enzymes. Proteins implicate disparate signaling pathways and also plays crucial role in arbuscule formation. Enzymes catalyze in the decomposition of complex unavailable organic substrates to simple utilizable forms through which plant uptake can be enhanced. However, by keeping in view regarding change in climate and varied ecology, current study emphasized the role of AMF in modern agriculture and sustainability.

1.2 Goal for Studying Its Biology and Omics Approaches

Mycorrhizal fungi have a broad range of roles and can also lead as nectrotroph lifestyle, endophytic, antagonists among host and nonhost plants, with roles which vary in association of life. It is a long-term goal to unravel and cope mycorrhizal functioning in our world. We must review full spectrum of varied factors, i.e., deciphering the intricacies of plant–microbe interactions, gene function, trade-off in mutualistic symbiosis, capacity to adapt the changing soils or environment or how to described genus and species.

Aim to Study

- How the AMF symbiosis form and arise?
- Broaden our perspective to include all functional and phylogeny studies.
- The most enigmatic groups are Glomales.
- Furthermore, characterization studies on wide range of plant and fungal species could provide hints for understanding the unexplained relationship mutualistic association in VAM fungi.
- Owing to their role, VAM can be classified as essential linkage species of plant community.
- Elaborating research network pools that span from genetics to ecosystem levels.
- Essential tools for phytoremediation practices.
- It shed light on various pathways regulating the development of VAM fungi.
- In addition, the potential role of AM as biocontrol agents needs to be considered and utilized in plant breeding programs for the selection of pathogen-resistant cultivars.
- AMF as potential bicontrol agent due to low cost, wide availability, and environmental friendliness.
- Applications in the field of agroecology.

1.3 Phylogeny of New Speices of AMF

Diversispora *Diversisporaceae* family (order *Diversisporales*, subphylum Glomeromycotina) are described by species producing spores at the tip of a funnel-shaped hypha or cylindrical hyphal structure (Błaszkowski and Chwat 2013). Diversisporales comprises five families of the Diversisporaceae that has five genera (Spatafora et al. 2016a, b). Symanczik et al. (2018) reconstructed the true phylogeny of the "*omaniana*" fungus and to describe a new species in the genus Diversispora.

Glomus Phylogenetic analyzes of rDNA sequences were performed in the Glomus group A sensu (Schwarzott et al. 2001) which verified their uniqueness relative to other recognized Glomus species. Thus the fungi are described as *G. africanum* sp. nov. and *G. iranicum* sp. nov. Błaszkowski (2010) was the first person to report a phylogenetic sequencing of the two new species *Glomus africanum* and *G. iranicum.*

Dominikia and Kamienskia All *Dominikia* and *Kamienskia* spp. form colorless or very pale colored, small glomoid hypogeous spores in more or less compact clusters. Błaszkowski et al. (2015) included seven group of fungi *G. achrum, G. bistratum, G. iranicum, G. minutum, G. perpusillum*, and two undescribed fungi with glomoid.

Lactarius It is vital species of ectomycorrhizal communities in different habitat types ranging from low elevation to higher elevation rural and urban areas. Twenty genus identified with three different species along with one unresolved species (Barge and Cripps 2016). Eight taxa *Lactarius alpinus v. mitis* with *Alnus incana, L. barrowsii* with *Pinus flexilis, L. aff. brunneoviolaceus* with *Salix reticulata,*

L. luculentus v. laetus under mixed conifers, *L. aff. Olivinus* with *Picea engelmannii, L. pseudodelicatus* with *Populus tremuloides*, and L. *aff. tuomikoskii* with *Picea engelmannii* were identified and, respectively.

Laccaria *Laccaria* species has long been recognized as important ectomycorrhizal associates of ectotrophic plants worldwide. They are known to form associations with members of the Pinaceae, Dipterocarpaceae, Fagaceae, Myrtaceae, Betulaceae, Tiliaceae and Salicaceae (Wilson et al. 2013). Ramos et al. (2017) during their study, they found two species *Fagus grandifolia* var. *mexicana* and *Laccaria trichodermophora*.

Acaulospora *Acaulospora ignota* and *Claroideoglomus hanlinii*, two new species of arbuscular mycorrhizal fungi (Glomeromycota) from Brazil and Cuba. On basis of morphological studies *A. ignota*, is distinguished by yellow colored, relatively small (65–80 μm diam when globose) spores, whose upper surface of the structural laminate spore wall layer is ornamented with flattened elevations or inconspicuous warts these are usually deteriorating and disappearing with age. The most distinguishing features of the second species, *C. hanlinii*, are its relatively small (45–90 μm) and dark-colored (olive-yellow) spores, with a simple two-layered spore wall. *Acaulospora koreana* isolated from forest soils in South Korea (Lee et al. 2018). New species listed of VAM fungi (Table 1.1).

1.4 Genomics

Genomics plays an great role to investigate genes in association with plant root and VAM fungi Early decades scientists used limited technologies about genetically to ecology level. We recently focus on AM fungus interaction and physiology of the AM symbiosis and also is turning point to study their biology that yet remain to be elucidated would now became accessible for investigation. Genomic sequencing would provide information on genetic variants that may occur disease or reduction of threat of disease development, even in asymptomatic plant.

To date, full genome sequences are available for only two AM fungi from a single genus, *Rhizophagus* (Chen et al. 2018; Kobayashi et al. 2018). The isolate DAOM197198 of *Rhizophagus irregularis* was used for genome sequencing as (1) it is widely used for laboratory studies (2) it is highly sporulating, providing large quantities of biological material and (3) a smaller genome size than many other AM fungal species. The sequencing program was announced in 2004 and was supported by the Joint Genome Initiative and founded by the US Department of Energy. Glomus Genome Consortium was formed with several contributors, participated to provide biological material and sequences. Furthermore, the genome size was estimated to be 153 Mb, in accordance with the value of 154.8_6.2 Mb measured by flow cytometry. The first AM fungal genome was released in May 2013 (http://genome.jgi.doe.gov/Gloin1/Gloin1.home.html) before publication (Tisserant et al. 2013). It was found that the genome of *R. irregularis* is haploid

Table 1.1 New species listed of VAM fungi

VAM fungi	Host	Location	Accession no.	References
Acaulospora koreana	*Lindera obtusiloba* and *Styrax obassia*	Forest soils in South Korea	MN792884.1 (SSU)	Lee et al. (2018)
Acaulospora saccata *A. fragilissima* *Scutellospora ovalis* *Rhizophagus neocaledonicus*	*Sorghum vulgare*	Soils in New Caledonia (South Pacific)	KY362430 (ITS)	Crossay et al. (2018)
Acaulospora ignota	Rhizosphere soils of plant species of the Poaceae	Natal, Brazil	KP191472 (ITS)	Błaszkowski et al. (2015)
Claroideoglomus hanlinii	Roots of *Phoenix dactylifera*	Cuba	KP191482.1 (its)	Błaszkowski et al. (2015)
Septoglomus mexicanum	Rhizospheric soil of two endemic Mexican legumes: *Prosopis laevigata* and *Mimosa luisana*	Mexico	MK570915.1 (its)	Chimal-Sánchez et al. (2020)
Acaulospora papillosa	*Zea mays.*	Northeastern Brazil	LN881565.1 (its)	Pereira et al. (2015)
Septoglomus fuscum and *S. furcatum*	Roots of *Arctotheca populifolia* and *Cordia oncocalyx* in dry forest	South Africa and Brazil	KF060313.1 (its) MG183939.1 (its)	Błaszkowski (2017), Zu et al. (2019)
Rhizoglomus dalpeae	Savanna zone	Benin, West Africa	MN130954.1 (its)	Błaszkowski et al. (2019)
R. Silesianum.	Coal mine spoil heap in	Poland	MN130954.1 (its)	Błaszkowski et al. (2019)
R. maiae	Shrubs growing in a tropical humid reserve.	Brazil	MN130957.1 (its)	Błaszkowski et al. (2019)
Diversispora omaniana, *Septoglomus nakheelum*, and *Rhizophagus arabicus* spp	Date palm plantation.	Oman	MG459210.1,	Symanczik et al. (2018)
Septoglomus nigrum	*Lolium perenne*, *Trifolium pratense*, *Plantago lanceolata* and *Hieracium pilosella*.	France, Germany and Switzerland	LR723644.1 (its) LR723643.1	Oehl et al. (2019)
Paraglomus occidentale	Inka nut (*Plukenetia volubilis*)	Peru	MN081581.1	Corazon-Guivin et al. (2020)

and does not show any evidence of recent duplication. Transposable elements (TE) are strongly represented (up to 55 Mb of the genome is formed by repeated TE) as expected from a large fungal genome. The mitochondrial (mt) genome of the AM fungi is also considered during the sequencing, because it played a great role to investigate the fungal ecology unlike normal fungi Malbreil et al., 2014 .

These genomic data has opened a new era in the study of AM fungi. The advent of large scale sequencing approaches the studies on AM fungal taxonomy and systematics had risen to a new level (Spatafora et al. 2016a, b). New operational taxonomic units (OTUs) at the species, genus, and higher taxonomic levels were identified with the help of metagenomic approaches based on next-generation sequencing methodologies (Opik et al. 2013). Analysis of the sequenced genomes revealed that the capacity for plant cell wall degradation and secondary metabolite production was less even in AM fungi like many other biotrophic and some ectomycorrhizal fungi. But they contain a large repertoire of small secreted proteins of unknown functions, potentially candidate effectors for modulating the interaction with their host (Lin et al. 2014; Zeng et al. 2018). The first indication of the AM fungi being fatty acid auxotrophs was provided by genome sequence (Wewer et al. 2014). However, the symbiotic physiology of AM fungi has been widely investigated, and it shows that phosphorus is not only a nutrient recruited by extraradical hyphae in soil and translocated to the host plant (Graham et al. 1981), but it is also a major regulator for the establishment and function of AM symbiosis (Javot et al. 2007). A large number of transcriptomic studies have given a good knowledge of physiology, metabolism, and cell regulation modified in the host plants after the symbiotic relationship.

1.5 Fungal Metabolism During Symbiotic Life

The role of AM fungi in mineral nutrient supply for the plants is the most widely studied aspect. This supply of nutrients requires different metabolic machinery and the coenocytic nature of AM fungal hyphae allows easier, faster, and low-energy-demanding transport. Genomics has helped in the identification of a number of genes and transcripts responsible for the effective symbiotic relationship and are explained below.

- Most importantly for Phosphate transport and metabolism, *R. irregularis* contains at least four different kinds of putative phosphatases that are expressed in intraradical mycelium (IRM) that are able to cleave a broad range of substrates to release Pi (Tisserant et al. 2012). Alkaline phosphatase activity and candidate genes have been identified in AM fungi (Liu et al. 2013).
- Coming to the Nitrogen transport and metabolism, which is often underestimated, AM fungi supply significant amounts of the total N taken up by plants (Govindarajulu et al. 2005). Two high-affinity N transporters have been partially characterized in *R. irregularis* namely, GinAMT1 and GinAMT2 (Pérez-Tienda

et al. 2011). Both the transporters are expressed in extraradical mycelium (ERM), but GinAMT2 transcript levels are higher in intraradical mycelium (IRM) and GinAMT1 is induced in ERM at low N concentrations. Tian et al 2010 has identified a putative high-affinity nitrate transporter and a transcriptomic approach showed that this gene was expressed more in IRM (Tisserant et al. 2012).

- Total 445 gene models involved in carbohydrate metabolism including transport of sugars from the plant to the fungus have been identified in *R. irregularis* and AM fungal monosaccharide transporters (MST) and sucrose transporters were also isolated further (Helber et al. 2011).
- A set of 432 lipid-related genes were identified for either transport or metabolism in Gloin1 with respect to lipid metabolism. An unexpected and specific regulation mechanism has been observed for lipid metabolism, as shown by numerous experiments (Trépanier et al. 2005). All the genes involved in fatty acid synthesis are present in *R. irregularis* has given by the biochemical studies showing that the fungus did not obtain its fatty acid from the plant but was able to synthesize them (Tisserant et al. 2012).

1.6 Proteomics

Limited research work has been reported in the field of protein biology. Scientists had a hard task to differentiate the proteins of AM Fungi from plants. Moreover, Post-translation modification assigns the covalent and usually enzymatic protein modification after the biosynthesis of proteins. Moreover, proteins were controlled by post-translation mechanism more than 200 different types which have been reported it includes phosphorylation, ubiquitination, and proteolysis (Packer et al. 1997). Furthermore, polyprotiens separated by 2DE Electrophoresis on the basis of isoelectric points and analysis of polypeptides by mass spectrometry methods either de novo sequencing or peptide mass fingerprinting might outcome in identification of fungal proteins. Antibodies corresponding to fungal proteins were used to identify specific fungal species (Göbel et al. 1995). Early proteomics work on AM interactions made it clear that proteins aware of AM symbiosis in the early stages of colonization are not quite the same caused in the late stages when the symbiosis is formed and functional (Bestel-Corre et al. 2004). Della proteins are core for several signaling pathways of gibberellic acid (GA), are required for arbuscule formation. GA signaling also influences the arbuscule formation in monocots. (Floss et al. 2013). Glomalin Soil Protein (GRSP), a widespread glycoprotein developed by arbuscular mycorrhizal fungi (AMF), plays a critical role in the functioning of the ecological regeneration and the environment (Jiang et al. 2017). The subsequent proteomics revealed that *E.augustifolia* seedlings inoculated with AMF improved the secondary metabolism level of the phenylpropane pathway, increased the signal transduction capacity of Ca^{2+} and ROS scavenging, encouraged protein biosynthesis, acceleration protein folding and prevented protein degradation under salt stress

(Jia-Dong et al. 2019). We listed few concurrent proteomic studies on rhziobial symbioses (Table 1.2).

1.7 Symbiotic Root–Microbe Interactions

The Symbiotic Plant-Mycorrhizal Symbioses

Earlier decades, two model legumes were used as forefont of proteomic research into root association symbiosis, *Medicago truncatula and Lotus japonicas* (Cook 1999). Proteins secreted by the general use of secretion systems. Primarily, Gram-negative bacteria classifies four types of secretion systems (Types I, II, III, IV, V, VI (T1SS or TOSS, T2SS, T3SS or TTSS, T4SS or TFSS, T5SS, and T6SS, respectively). Type V to Type VI Animal secretion system and remaining plant secretion system. Nodo is the first secreted rhizobial protein for which symbiosis played a role could be seen in *Rhizobium leguminosarum* (de Maagd et al. 1989). The secretion pathways discussed so far are also important for processes that are not related to host interactions. Several secretion systems, by contrast, appear to have specialized in mediating such interactions, with ability to translocate effector proteins into the cytoplasm of the host cell as a defining element. These include the T3SS, T4SS, T6SS Frame works. Viprey et al. (1998) provided the first definite evidence of involvement of T3SS in the rhizobium-legume symbiosis. T3SS substrates come in two different forms: effectors and proteins for helpers. Effectors are translocated into eukaryotic host cells whereas helpers proteins help to translocate. Proteins secreted by rhizobial T3SS are called outer proteins or Nops nodulations. So far three proteins have been identified as helpers viz., NopA, NopB, and NopX (Saad et al. 2008). HrpW has recently been shown to promote translocation of the T3SS effector into plant leaves, making it an aid protein rather than a true effector (Kvitko et al. 2007). HopG1 cuts down basal tobacco resistance (Oh and Collmer 2005). These homology-based findings expand significantly the pool of potential T3SS rhizobial effectors.

The Symbiotic Plant-Mycorrhizal Symbioses

In AM symbioses, which are the most common associations and have existed for about the last 450 M years (Remy et al. 1994). Arbuscular mycorrhiza (AM) originated from the Latin word "arbusculum" and the Greek words "mycos" and "rhiza," respectively, meaning small tree, fungus, and root. Predominantly two types classified into ectomycorrhiza and endomycorrhiza. Ectomycorrhiza tends to form symbiotic mutual relationships with woody plants, including birch, beech, willow, pine, oak, spruce, and fir. An intracellular surface, known as the Hartig Net is distinguished by ectomycorrhizal connections. The Hartig net is a highly branched hyphae which connect the epidermal and cortical root cells. In addition, ectomycorrhiza can be identified by forming a dense hyphal sheath that surrounds the surface of the root so-called mantle. In other words, ectomycorrhiza has just survived off the root. Together, ectomycorrhiza has only 5–10% of terrestrial plant

Table 1.2 Proteomic studies on rhziobial symbioses

Sl. No	Species	Identified proteins	Methods	References
1	*S. meliloti*	60 deficient strain altered proteins in QSS, 52 identified	MALDI-TOF 2DE	Gao et al. (2007)
2	*Rhizobium leguminosarum* bv. *trifolii*	Identification of two inducible flavonoids and 10 constituent proteins	2DE, N-terminal Sequencing	Guerreiro et al. (1997)
3	*Frankia alni*	126 proteins identified as N-replete, relative to N-limiting circumstances	MALDI-TOF,2DE	Alloisio et al. (2007)
4	*B. japonicum*	17 proteins identified	2DE	Panter et al. (2000)
5	*M. truncatula*	7 proteins identified	2DE	Bestel-Corre et al. (2004)
6	*Rhizobium leguminosarum* bv. *trifolii*	10 proteins identified in association with subterranean clover cultivar	2DE	Morris and Djordjevic (2001)
7	*Funneliformis mosseae* and *Rhizophagus irregularis*	3 genes (Fm201, Ri14–3-3 and RiBMH$_2$) encoding 14–3-3-like proteins for abiotic stress responses and arbuscule formation during AM Symbiosis	Starch gel-electrophoresis	Sun et al. (2018)
8	*R. Leguminosarumbv. viciae*	Identification of protein abundance variations as part of the local reactions of pea nodules grown under split-root conditions and subjected to water stress.	2DE	Irar et al. (2014)
9	*S. meliloti*	377 plant protein identification in nodules,	2DE	Larrainzar et al. (2007)
10	*Mesorhizobium loti*	SNARE proteins (LjVAMP72a and LjVAMP72b) for root symbiosis and root growth in *Lotus Japonicas* seeds	SDS-PAGE	Sogawa et al. (2019)
11	*P. hybrida*, *N. benthamiana*	VAPYRIN protein interacts with a symbiotic R-SNARE of the VAMP721 family *A. thaliana* as model plant	Semiautomated image analysis pipeline	Bapaume et al. (2019)
12	*Glomus intraradices*	450 proteins in perennial plants.	2DE	Lingua et al. (2002)
13	*R. Leguminosarum*	111 differentially expressed proteins seed profiling I pea seed	SDS-PAGE	Mamontova et al. (2019)

(continued)

Table 1.2 (continued)

Sl. No	Species	Identified proteins	Methods	References
14	*Rhizophagus irregularis*	75 proteins identified in *E. angustifolia* seedlings	LC-MS/MS analysis	Jia et al. (2019)
15	*R. irregularis*	529 different peptides that were confidently mapped to 474 *R. irregularis* proteins in chicory roots	LC-MS/MS analysis	Murphy et al. (2019)
16	*Pinus pinaster–Hebeloma cylindrosporum*	869 proteins constituting the exoproteome of *Hebeloma cylindrosporum.*	LC-MS/MS analysis	Doré et al. (2015)
17	*M. truncatula–R. irregularis, G. intraradices*	Root membrane proteome, cadmium tolerance	2-DE, 2D-DIGE, LC–MS/MS	Zhan et al. (2018)
18	Tree root–*Laccaria bicolor*	224 proteins identified in involved in the cell wall remodeling linked to hyphal growth	2-DELC-MS/MS	Vincent et al. (2012)
19	*Glomus mosseae (GM)* and *G. intraradices,*	3473 proteins were identified in the roots of *Amorpha fruticosa*	LC-MS/MS	Song et al. (2015)
20	*Funneliformis mosseae*	131 differentially expressed proteins (DEPs) were identified in *F. mosseae* treated samples of soyabean.	iTRAQ, LC-MS/MS	Bai et al. (2019)

species. On the other hand, endomycorrhizae are present in more than 80% of established plant species, including grains and greenhouse plants such as most fruits, grasses, flowers, and fruit trees. Endomycorrhizal relationships are characterized by fungal penetration of the cortical cells, and the formation of abscesses and vesicles by the fungi. Endomycorrhiza, in other words, has an attachment process inside the root, with the hyphae of the fungus spreading outside the root (Fig. 1.1). Compared with the ectomycorrhiza it is a more invasive relationship. Endomycorrhiza is further subdivided into specific types: Arbuscular Mycorrhizae, Ericaceous, Arbutoid, and Orchidaceous. Bestel and Co-workers (2004) characterized on *Medicago truncatula* infected with *Glomus mosseae* find 14 protein changes in the preinfection stage and 23 protein changes in 14 days of arbuscular development. Analysis of proteins shows that mycorrhizal fungi induced improvements in the redox or stress response (peroxides and glutathione-transferases), respiration, and alteration of the cell wall. Velvet and colleagues identified that membrane-associated *M. truncatula* proteins with the fungus *G. intraradices.* Due to a lack of sequence knowledge and difficulties in growing AM fungi in pure culture, the detection of proteins from the fungal partners of AM symbioses has been much more challenging. A proteome map of mycorrhizal fungal proteins from *Glomus intraradices* has been founded and eight

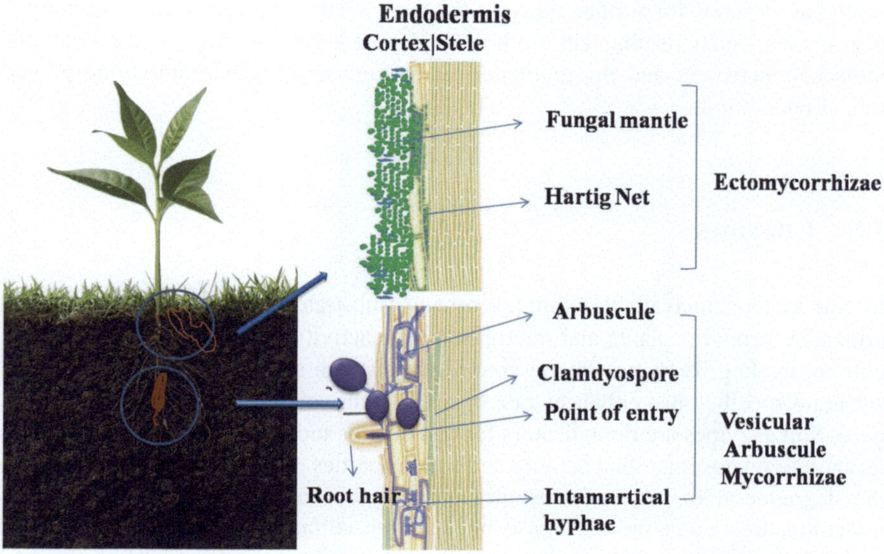

Fig. 1.1 A schematic representation diagram on Ectomycorrhiza and Endomycorrhiza

of the 438 solved proteins were detected using tandem mass spectrometry (Dumas-Gaudot et al. 2004).

Other Beneficial Interactions

Mostly grasses may enter into symbiotic relationships with nitrogen-fixing bacteria promoting plant growth, which does not form a particular root structure like a nodule (Reinhold-Hurek and Hurek 1998). The Trichoderma genus "free-living" fungi are considered to be beneficial to plants, mainly through their defense against pathogenic soil fungi, but also through systemic resistance caused in plant roots and shoots (Harman et al. 2004). A study of the *T.harzianum* secretome identified an aspartic protease that could target fungal cell walls containing parasitized fungus proteins. (Suarez et al. 2005). Gomes et al., 2017 analyzed gene regulation changes in *T. harzianum* Epl-1 protein involved botrydial biosynthesis (BcBOT genes) in the course of mycoparasitism association. Even, Brassicaceae includes plants that are not hosted by arbuscular mycorrhizal fungi (AMF) such as used plant model *Arabidopsis thaliana* (Poveda et al. 2019). However, one of the main challenges to be addressed is how *Trichoderma* modulates the plant's immune response to establish beneficial interactions (Ramírez-Valdespino et al. 2019). In building a beneficial relationship between *Trichoderma* and plants, the effectors can play key roles, as illustrated in mycorrhizal systems like those of *Laccaria bicolor* and *Glomus intraradices*. (Ramírez-Valdespino et al. 2019). Exploring these protein functions, our study provides a detailed understanding of the architectural alteration of the root, cell remodeling, and upkeep of cellular homeostasis during the mutualistic and development stage of plant life. Hopefully, our review provided valuable candidate proteins for further study of this beneficial combination of plant–fungal interactions. Such results will further allow researchers to fully understand the molecular networks and the regulation of biological pathways underlying plant–fungal interactions.

1.8 Enzymes

In soil system, unavailable complex organic substrates can be decomposed into utilizable forms for plants and microbes by the activity of enzymes which govern nutrient cycling. Enzymes in soil commence through soil flora and fauna and are substrate specific, also enhances rate of reaction at which organic residues decompose. Soil enzymes are bioindicators for soil quality and soil health (Killham 2002), its activity relates microbial activity and soil properties. Soil enzymes associate with the degradation of organic substrates (e.g., hydrolase, glucosidase), in nutrient mineralization (e.g., amidase, urease, phosphatase, sulfates) besides further activities such as pesticide degradation. In practice, enzymes have ample utilization in agriculture and bio-degradation which is eco-friendly. In the attempt to enhance soil enzyme activity, pertinent research reveals the notably varying effect of AMF on enzyme activity.

AMF forms mutual association with the roots of majority of land plant species (Wang and Qiu 2006) and are mostly familiar to help phosphorus uptake by host plant (Smith and Read 2008). Various soil enzymes and its activity alteration as a result of AMF application is shown (Table 1.3.) Favorable growth conditions for AMF i.e., generally neutral pH enhance the enzyme activity. Various enzymes were identified and classified into different releasing groups based on nutrient transformation (Table.1.3) indicates that AMF positively affects the enzyme activity. In wide range of crops great increase in the enzyme activity was observed by the application of AMF.

1.9 Biology of AMF on Different Crops: Insight and Impact

Biology Literal meaning of the Latin word Arbusculum, mycos, and rhiza-meaning tree, fungus, and root, respectively. The primary characteristic feature of arbuscular mycorrhizal, the evolutionary precursor of most other mutualistic root–microbe associations (belonging to phylum—Glomeromycota) fungi being the growth of arbuscles and vesicles in the cortical cells of plant roots. AMF consists of 9–55% of soil microbes and benefits around 90% of the total agricultural plants.

Insights of the Mycorrhizal Biology

The plants predominantly used to study the plant–fungal partnership are *Medicago truncatula* and *Lotus japonicus* (Hause & Schaarschmidt (2009) Fukai et al. 2012).

Table 1.3 Response of soil enzymes activity to the application of AMF

S. No.	Enzyme	Releasing type	% increase in enzyme activity	Crop	References
1	Esterase	C-releasing	256%	Maize	Vázquez et al. (2000)
2	Chitinase	C-releasing	197%		
3	Trehalase	C-releasing	444%		
4	Phosphatase	P-releasing	166%		
5	Endo-xyloglucanase	Plant enzymes	78–320%	Soybean	Garcia-Garrido et al. (2000)
6	Endoglucanase		33–145%		
7	Endopolymethylgalacturonase		73–266%		
8	Endopolygalacturonase		148–416%		
9	Alkaline phosphatase	P-releasing	8–51.6%	Cucumber (soil incorporated with clover leaves)	Joner and Jakobsen (1995)
10	Phosphatase	P-releasing	39–58%	Red clover	Ming-Yuan et al. (2007)
11	Urease	N-releasing	34–123%		
12	Alkaline phosphatase	P-releasing	193%	Berseem clover	Raiesi and Ghollarata (2006)
13	Acid phosphatase	P-releasing	143%		

The development of AM fungi is based on the induction of specific gene recognition cites and the elicitors such as strigolactones (Akiyama et al. 2010). The first interaction results into reciprocation of diffusible signals and once the fungi senses the vicinity of plant root system the signals are triggered toward the development of host fungal symbiotic relationship (Bonfante and Requena 2011; Gough and Cullimore 2011). The root released phytohormones belonging to the class Strigolactones plays the next major role and the highly sensitive (up to a minimal concentration of 10 nM) relevance in the establishment of symbiosis (Besserer et al. 2006).

The Strigolactone (SL) biosynthesis pathway and its regulation by the NSP1 and NSP2 transcription factors. NSP1 and NSP2 (yellow) promote expression of DWARF27 (D27) (Liu et al. 2003). *nsp2* mutants additionally accumulate orobanchol, but the misregulated gene responsible for this accumulation is unknown. All compounds with SL activity on plants and fungi are marked in orange. PDR1 is required for SL transport within the plant tissue and for root exudation. There is evidence for plant SL perception via a complex between the α/β-fold hydrolase D14/DAD2 and the F-box leucine-rich repeat (LRR) E3 ligase D3/RMS4/MAX2 (Hamiaux et al. 2012). The mechanism of SL perception by AM fungi is elusive. Products of deviating biosynthesis pathways (green) also play a role in AM (Herrera-Medina et al. 2007). (Fig. 1.2).

The hypopodium of the fungi develops at the root surface based on the physico-chemical features of root cell walls. The hyphopodia were noticed by *G. gigantea* on

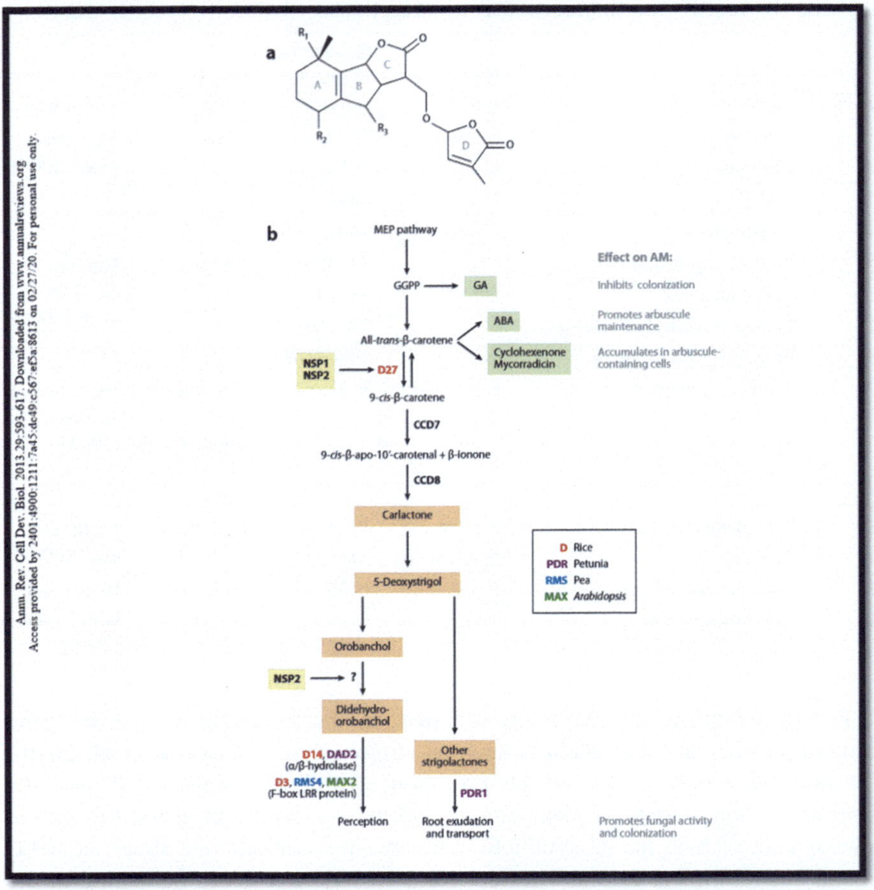

Fig. 1.2 The SL biosynthesis pathway and its regulation

isolated cell wall fragments of roots from the host plant carrot but not on the nonhost *Beta vulgaris*. Further studies were performed based on the *M. truncatula* mutants (ram) 1 and 2 (Gobbato et al. 2012). The fungal hyphopodium once attached releases plant inducing signals, which induce expression of a suite of plant genes, calcium spiking in rhizodermal cells, starch accumulation in roots, and lateral root formation prior to colonization (Chabaud et al. 2011) Table 1.4.

Table 1.4 Host plants for various arbuscular mycorrhizal fungi

Fungus	Host
Glomus mosseae, G. intraradices	Pepper
G. Mosseae	Tomato *(Lycopersicon esculentum* cv Zhongzha105)
Glomus mosseae	Peanut
G. Mosseae, G. versiforme	Trifoliate orange
Glomus etunicatum, G. mosseae, G. intraradices	Tabasi (mutated)
G. Intraradices	*Zucchini squash* (*Cucurbita pepo* L.)
G. Mosseae	Cotton
Glomus intraradices, G. aggregatum Glomus sp.	Onion (*Allium cepa* L) lettuce (*Lactuca sativa* L.)
Glomus mosseae	Maize, alfa
G. clarum	Mungbean (*V. radiata*)
Acaulospora, Glomus	*Parthenium hysterophorus* (L.)
Acaulospora, Glomus	*Cynodon dactylon* (L.) Pers
Glomus	*Lantana camara* (L.)
Acaulospora	*Digitaria sanguinalis* (L.) Scop
Acaulospora	*Cassia tora* (L.)
A. scrobiculata, G. aggregatum	*Andrographis paniculata*
A. laevis, G. dimorphicum	*Catharanthus roseus*
A. scrobiculata, G. fasciculatum, Gi. albida, S. calospora	*Azadirachta indica*
G. multicaule, G. clarum, G. fasciculatum, A. delicata, S. scutata	*Centella asiatica*
G. maculosum, G. glomerulatum, A. scrobiculata	*Hibiscus rosa- sinensis*

1.10 AMF Applications in Different Crops Under Varied Agro-Ecology

Agroecology can be described as the practice of applying ecological concepts and principles to maintain agriculture sustainable (Gliessman 1992). India encompasses huge diversity in landforms with differing environmental conditions produced wide range of soils. Under these fluctuating conditions to obtain ensured yield, application of chemical fertilizers, herbicides, and pesticides became mandatory which encumber the farmer economically. Because of changing climate and stress conditions, benefit from the products shall be decreased. Promptly increasing population places heavy strain on the natural resources for fulfilling requirements. These critical situations lead the researchers toward inventing environmental friendly, economical, and efficient techniques like adoption of microorganisms which can be beneficial in varying environmental conditions to maintain sustainable agriculture. In this study, we focused on the vesicular arbuscular mycorrhiza fungi and its role in agriculture (Table 1.5) which can form association with plant roots, enhances nutrient uptake especially phosphorus, thus helps in better growth of the crop and also resist the plants toward biotic and abiotic stress conditions.

Table 1.5 Role of AMF in crop performance under varied agroecology

S. No.	Agroecology	Crop	AMF	Impact	Reference
1	Hot summer Mediterranean to cold semiarid	*Persea Americana* (avocado)	*Glomus fasciculatum*	Improved the development of root system, shoot: Root ratio in plants inoculated with AMF. Inoculation also enhanced the uptake of N, P, and K by plant tissue as compared to the rest of the treatments.	Vidal et al. (1992)
2	Humid subtropical	Transgenic tobacco hairy roots	*Glomus intraradices*	Results reported that anti-oxidation enzyme activities were higher in the plants inoculated with AMF, through which plants showed tolerance against toxic oxygen species and phenol level up to the concentration 25 mg L^{-1}	Ibáñez et al. (2011)
3	Humid subtropical	*Solanum tuberosum* (potato)	*Glomus intraradices Glomus mosseae*	Chlorophyll content, fresh and dry weight of plants at harvest, growth of above and below ground parts were enhanced and maximum in the mycorrhizal infested potato plants	Lone et al. (2015)
4	Humid subtropical	Cowpea, Pigeonpea, groundnut	*Glomus etunicatum, Gigaspora margarita*	Growth responses of these three legume plants inoculated with mycorrhiza were significant. Shoot dry matter production and shoot phosphorus content were enhanced.	Ahiabor and Hirata (1994)
5	Hot desert climate	*Panicum turgidum*	*Glomus etunicatum Glomus intraradices Glomus mosseae*	Anti-oxidation enzyme activities viz., superoxide dismutase, peroxidase, glutathione reductase, and compatible solutes were enhanced in AMF inoculated plants under salt stress and tolerance to salt stress was increased significantly.	Hashem et al. (2015)

(continued)

Table 1.5 (continued)

S. No.	Agroecology	Crop	AMF	Impact	Reference
6	Cold semiarid	*Solanum lycopersicum*	*Glomus intraradices*	In salt stress conditions less H_2O_2, lipid peroxidation, and higher proline content were noticed in AMF inoculated plants. AMF also reduced the impact of salt stress on P, Ca, and K uptake by plants.	Hajiboland et al. (2010)
7	Humid continental	*Zea mays*	*Glomus intraradices*	Plants cultivated in clay brick granules with the colonization of *Glomus* showed higher colonization, spore production and root length as compared to rest of the treatments	Gaur and Adholeya (2000)
8	Semiarid	*Triticum aestivum*	*Glomus mosseae* *Glomus fasciculatum*	Growth parameters such as plant height, root length, shoot, and root biomass production, grain size, and photosynthetic pigment were found maximum in Glomus inoculated treatments.	Pal and Pandey (2017)
9	Humid subtropical	*Scutellaria integrifolia*	*Gigaspora margarita* *Glomus etunicatum* *Glomus intradicas*	AMF inoculation to plant roots resulted in positive effect on growth parameters especially plant height, fresh weights of root, shoot, and seed.	Joshee et al. (2007)
10	Hot and dry climate	*Zea mays*	*Glomus constrictum*	Plant biomass production on dry weight basis was significantly superior in plants inoculated with AMF to that of the treatments of non-inoculated.	Omar (1995)
11	Island	*Malus pumila* *Allium porrum* *Tagetes patula*	*Glornus epigaeum* *Glomus monosporum*	Positively influenced the growth of the plants. Endomycorrhizal colonization decreased with increase in phosphorus levels.	Plenchette et al. (1983)

(continued)

Table 1.5 (continued)

S. No.	Agroecology	Crop	AMF	Impact	Reference
12	Semiarid	*Sorghum bicolor*	*Glomus spp.*	In response to CO_2 hyphal length of AMF, protein globulin and aggregate water stability were increased. Hyphal length, globulin were positively correlated with aggregate stability	Rillig et al. (2001)
13	Humid	Maize, *Sorghum*, chickpea	*Glomus clarum*	Enhancement of plant growth, dry weight, and root length were recorded in the treatments inoculated with AMF	Simpson and Daft (1990)
14	Hot summer Mediterranean to cold semiarid	*Trifolium repens*	*Glomus mosseae*	Root growth, nodule formation, dehydrogenase activity, phosphatase, β-glucosidase activities, and soil fertility were improved with inoculation in cd polluted soils.	Vivas et al. (2005)
15	Humid subtropical	*Poncirus trifoliata* (citrus)	*Glomus mosseae* *Glomus versiforme* *Glomus diaphanum*	Mycorrhizal fungus improved plant biomass, soil protein concentration, soil hyphal length densities, and peroxidase activities. It enhanced water stable aggregates and also plant growth under drought stress conditions. Reduced the hot water extractable and hydrolyzable carbohydrates concentrations in soil.	Wu et al. (2008)

1.11 Opportunities and Challenges: AMF

We believe that significant progress will be made through the development and use of genetic model systems in the early diverging host and nonhost lineages. Powerful genetic tools are available already in *Physcomitrella patens* (moss) and Marchantia (liverwort) (Zimmer et al. 2013; Hiss et al. 2014). Use of these and the Charophyte green algae *Penium margaritaceum* as research models will lead to greater understanding of the host mechanisms that gave rise to efficient AM symbioses (Sørensen et al. 2014).

Significant progress in the analysis of plant systems modulating symbiotic interactions in AM has been made through the application of genetic, ultra-cytological, biochemical, and molecular biology techniques. Mycorrhiza-defective legume mutants provide direct evidence for specific symbiosis-related gene control and afford the potential for analyzing some host genes and gene functions involved in early and late events. However, the range of mutant phenotypes available is still extremely limited. Expansion of the genetic approach to additional mutations is essential to understand more fully the molecular mechanisms leading to morphological integration and reciprocal functional compatibility between AM symbionts. Moreover, mutagenesis programs, including mycorrhiza-forming non-legumes such as maize or tomato, should indicate to what extent AM development is determined by processes other than those common to nodulation. Despite obvious similarities between the infection processes of AM fungi and rhizobia in legumes, important differences also exist. A number of non-nodulating legume genotypes are able to form mycorrhiza (Duc et al. 1989; Wyss et al. 1990; Gianinazzi-Pearson et al. 1991b), and some nodule infection thread components are not synthesized around the fungal symbionts in mycorrhizal roots (Gianinazzi-Pearson et al. 1990; Perotto et al. 1994). Evidence for AM symbioses appears in the fossil record before the evolution of legume species. This raises the interesting possibility that during evolution, rhizobia may have exploited plant processes involved in symbiotic interactions with arbuscular mycorrhizal fungi and modified them to a completely different purpose, that of nodule development (Gollotte et al. 1995b). Use of heterologous probes for genes that mediate other plant–microbe interactions has provided much information on cell-cell interactions in AM. This approach has been particularly effective in evaluating the restricted expression of defense responses that is essential for compatibility between plant and fungal symbionts. Clearly, plant–fungal compatibility is an active phenomenon in AM, because in activation of a single plant gene, as in the case of the Myc-1 pea mutants, results in resistance. Moreover, the fact that AM symbioses are widespread in the plant kingdom makes this symbiosis an interesting model system for dissecting the genetic and molecular bases of biotrophic plant–microbe interactions in general. Although transport processes between plant and fungal cells are crucial for the proper functioning of AM, they represent one of the most poorly understood aspects of the symbiosis. The use of tools such as antibody and nucleic acid probes and promoter-GUS fusions in transgenic plants will accelerate the investigation of the ill-defined area of interface function. The complex cellular relationship between roots and AM fungi necessitates continuous recognition and signal exchange between both partners. These exchanges affect the regulation of genes whose products participate in the metabolic and structural changes leading to the symbiosis (Gianinazzi et al. 1995). Signal or receptor molecules involved in this dialog have not been identified. Polysaccharides, hormones, and polyamines are among the candidates for signaling, because they are important components of various developmental processes in plants and are associated with AM development (Dannenberg et al. 1992; E1 Ghachtouli 1995). The search for plant genes mediating symbiotic events in defined root tissues at specific times has begun. The next few years should see exciting advances toward identifying these plant determinants, the molecules they encode, and the ways they are regulated in AM, the most common root symbiosis in the plant kingdom.

The idea of mycorrhizae and their use in the agronomic sector has taken hold in many scientific experiments around the world. Most of the research has focused on the host plant benefits attributed to arbuscular fungi from the points of view of yield and resistance to biotic and abiotic stresses. Great efforts have been made in order to study the processes and metabolic pathways involved in the fungus, aimed at the greater absorption of nutrients and water and greater resistance to pathogens, salinity, and heavy metals. Despite the numerous studies on horticultural and forestry plants, wheat, which is one of the most important food crops worldwide, has been the subject of many tests on mycorrhizal inoculation. With a view to greater environmental sustainability, the selection and cultivation of cereals in agricultural systems with a low environmental impact could be based on the selection of wheat varieties with highly effective mycorrhizal symbiosis. In the last few years, the research has recognized notable differences in plant susceptibility and/or responsiveness to AMF among wheat genotypes that differ in ploidy number or geographic origin (DeVita et al. 2018). Significant genotypic differences were detected in the ability to form mycorrhizal symbiosis, and some significant markers, representing a Quantitative Trait Locus (QTL), were detected on wheat chromosomes (Lehnert et al. 2017).

Future research should therefore not focus only on the AM fungus colonization capacity, but it could take into account the ability of single grain accessions to form a mycorrhiza, based on the results obtained by previous genetic characterization. The identification of molecular markers closely associated with a mycorrhiza could be a very effective tool for selecting highly effective plants for symbiosis and developing wheat varieties suitable for low-environmental impact agricultural systems. At the same time, the identification and selection of the most infectious and efficient mycorrhizal fungi in combination with wheat will facilitate their use as biofertilizers to overcome the loss of soil biological fertility, reduce chemical inputs, and alleviate the effects of biotic and abiotic stress.

1.12 Conclusion

VAM fungi studies offer shot-term snapshots of inner depth behavior on physiological changes. However, it is to be expected that the rapid technological advances in the field of molecular biology would fill the still current scientific and functional differences between arbuscular mycorrhizas and other more symbiotic species whatever experimentally high tractable. Thus, it was concluded that the discovery of new genera and species provides a means of recognizing and investigating the practical importance of AM and provides a new potential for habitat protection, revegetation, and sustainable growth. Mycorrhizal fungus could demonstrate its beneficial effect in a diverse range of environmental conditions, be aggressive against pathogens, and be able to colonize roots of even disease-resistant varieties. Genomic sequence databases responsible for the symbiosis for estimation of fungal taxa number and its ecology. Unpublished work on additional plant mutants in the mycorrhizal colonization pathway in several laboratories and studies to achieve the transformation of mycorrhizal fungi would provide innovative input into process control mechanisms in the near future. Therefore, we encourage future studies to investigate both root and fungal characteristics as separate entities concurrently.

References

Ahiabor BD, Hirata H (1994) Characteristic responses of three tropical legumes to the inoculation of two species of VAM fungi in andosol soils with different fertilities. Mycorrhiza 5(1):63–70

Akiyama K, Ogasawara S, Ito S, Hayashi H (2010) Structural requirements of strigolactones for hyphal branching in AM fungi. Plant Cell Physiol 51:1104–1117

Alloisio N, Félix S, Maréchal J et al (2007) *Frankia alni* proteome under nitrogen-fixing and nitrogen-replete conditions. Physiol Plantarium 130:440–453

Bai L, Sun HB, Liang RT, Cai BY (2019) iTRAQ proteomic analysis of continuously cropped soybean root InoculatedWithFunneliformismosseae. Front Microbiol 10:61. https://doi.org/10.3389/fmicb.2019.00061

Bapaume L, Laukamm S, Darbon G, Monney C, Meyenhofer F, Feddermann N, Reinhardt D (2019) VAPYRIN marks an endosomal trafficking compartment involved in arbuscular mycorrhizal symbiosis. Front Plant Sci 10:666

Barge EG, Cripps CL (2016) New reports, phylogenetic analysis, and a key to Lactarius Pers. in the greater Yellowstone ecosystem informed by molecular data. Mycokeys 15:1–58

Besserer A, Puech-Pages V, Kiefer P, Gomez-Roldan V, Jauneau A et al (2006) Strigolactones stimulate arbuscular mycorrhizal fungi by activating mitochondria. PLoS Biol 4:e226. https://doi.org/10.1371/journal.pbio.0040226

Bestel-Corre G, Dumas-Gaudot E, Gianinazzi S (2004) Proteomics as a tool to monitor plant-microbe endosymbioses in the rhizosphere. Mycorrhiza 14(1):1–10

Błaszkowski J (2010) Two new *Glomus* spp. *Glomus africanum* and *G. iranicum*, two new species of arbuscular mycorrhizal fungi (Glomeromycota). Mycol Soc America 102(6):1450–1462

Błaszkowski J (2017) Taxonomy of arbuscular fungi. http://www.zor.zut.edu.pl/Glomeromycota/Taxonomy.html (consultado diciembre de 2017)

Błaszkowski J, Chwat G (2013) *Septoglomus deserticola* emended and new combinations in the emended definition of the family *Diversisporaceae*. Acta Mycol 48(1):89–103. https://doi.org/10.5586/am.2013.011

Błaszkowski J, Chwat G, Góralska A, Bobrowska-Chwat A (2014) *Glomus tetrastratosum*, a new species of arbuscular mycorrhizal fungi (Glomeromycota). Mycoscience 30:1–7. https://doi.org/10.1016/j.myc.2015.08.003

Błaszkowski J, Chwat G, Góralska A, Ryszka P, Kovács GM (2015) Two new genera, Dominikia and Kamienskia, and *D. disticha* sp. nov. in Glomeromycota. Nova Hedw 100:225–238. https://doi.org/10.1127/nova_hedwigia/2014/0216

Błaszkowski J, Niezgoda P, Piątek M, Magurno F, Malicka M, Zubek S et al (2019) Rhizoglomus dalpeae, R. maiae, and R. silesianum, new species. Mycologia 111(6):965–980

Bonfante P, Requena N (2011) Dating in the dark: how roots respond to fungal signals to establish arbuscular mycorrhizal symbiosis. Curr Opin Plant Biol 14:451–457

Chabaud M, Genre A, Sieberer BJ, Faccio A, Fournier J et al (2011) Arbuscular mycorrhizal hyphopodia and germinated spore exudates trigger Ca^{2+} spiking in the legume and nonlegume root epidermis. New Phytol 189:347–355

Chen EC, Morin E, Beaudet D, Noel J, Yildirir G, Ndikumana S et al (2018) High intraspecific genome diversity in the model arbuscular mycorrhizal symbiont Rhizophagus irregularis. New Phytol 220(4):1161–1171

Chimal-Sánchez E, Senés-Guerrero C, Varela L, Montaño NM, García-Sánchez R, Pacheco A et al (2020) Septoglomus mexicanum, a new species of arbuscular mycorrhizal fungi from semiarid regions in Mexico. Mycologia 112(1):121–132

Cook DR (1999) *Medicago truncatula* — a model in the making! Commentary. Curr Opin Plant Biol 2:301–304

Corazon-Guivin MA, Cerna-Mendoza A, Guerrero-Abad JC, Vallejos-Tapullima A, Ríos-Ramírez O, Vallejos-Torres G et al (2020) Paraglomus occidentale, a new arbuscular mycorrhizal fungus from the sources of the Amazon river in Peru, with a key to the Paraglomeromycetes species. Sydowia 72:85

Crossay T, Cilia A, Cavaloc Y, Amir H, Redecker D (2018) Four new species of arbuscular mycorrhizal fungi (Glomeromycota) associated with endemic plants from ultramafic soils of New Caledonia. Mycol Prog 17(6):729–744

de Maagd RA, Wijfjes AH, Spaink HP, Ruiz-Sainz JE, Wijffelman CA, Okker RJ, Lugtenberg BJ (1989) nodO, a new nod gene of the rhizobium leguminosarum biovar viciae sym plasmid pRL1JI, encodes a secreted protein. J Bacteriol 171:6764–6770

Doré J, Perraud M, Dieryckx C, Kohler A, Morin E, Henrissat B, Lindquist E, Zimmermann SD, Girard V, Kuo A (2015) Comparative genomics, proteomics and transcriptomics give new insight into the exoproteome of the basidiomycete hebeloma cylindrosporum and its involvement in ectomycorrhizal symbiosis. New Phytol 208:1169

Dumas-Gaudot E, Valot B, Bestel-Corre G, Recorbet G, St-Arnaud M, Fontaine B et al (2004) Proteomics as a way to identify extra-radicular fungal proteins from Glomus intraradices — RiT-DNA carrot root mycorrhizas. FEMS Microbiol Ecol 48:401–411

Floss DS, Levy JG, Lévesque-Tremblay V, Pumplin N, Harrison MJ (2013) DELLA proteins regulate arbuscule formation in arbuscular mycorrhizal symbiosis. Proc Natl Acad Sci 110(51): E5025–E5034

Fukai E, Soyano T, Umehara Y, Nakayama S, Hirakawa H et al (2012) EstablishmentofaLotusjaponicusgene tagging population using the exon-targeting endogenous retrotransposon LORE1. Plant J 69:720–730

Gao M, Chen H, Eberhard A, Gronquist MR, Robinson JB, Connolly M et al (2007) Effects of AiiA-mediated quorum quenching in Sinorhizobium meliloti on quorum-sensing signals, proteome patterns, and symbiotic interactions. Mol Plant-Microbe Interact 20(7):843–856

Garcia-Garrido JM, Tribak M, Rejon-Palomares A, Ocampo JA, Garcia-Romera I (2000) Hydrolytic enzymes and ability of arbuscular mycorrhizal fungi to colonize roots. J Exp Bot 51 (349):1443–1448

Gaur A, Adholeya A (2000) Effects of the particle size of soil-less substrates upon AM fungus inoculum production. Mycorrhiza 10(1):43–48

Gliessman SR (1992) Agroecology in the tropics: achieving a balance between land use and preservation. Environ Mngt 16:681–689

Gobbato E, Marsh J, Vernie T, Wang E, Maillet F et al (2012) A GRAS-type transcription factor with a specific function in mycorrhizal signaling. Curr Biol 22:2236–2241

Göbel C, Hahn A, Hock B (1995) Production of polyclonal and monoclonal antibodies against hyphae from arbuscular mycorrhizal fungi. Crit Rev Biotechnol 15(3–4):293–304

Gomes EV, Ulhoa CJ, Cardoza RE, Silva RN, Gutiérrez S (2017) Involvement of Trichoderma harzianum Epl-1 protein in the regulation of Botrytis virulence-and tomato defense-related genes. Front Plant Sci 8:880

Gough C, Cullimore J (2011) Lipo-chitooligosaccharide signaling in endosymbiotic plant-microbe interactions. Mol Plant-Microbe Interact 24:867–878

Govindarajulu M, Pfeffer PE, Jin H, Abubaker J, Douds DD, Allen JW et al (2005) Nitrogen transfer in the arbuscular mycorrhizal symbiosis. Nature 435:819–823

Graham JH, Leonard RT, Menge JA (1981) Membrane-mediated decrease in root exudation responsible for phosphorus inhibition of vesicular-arbuscular mycorrhiza formation. Plant Physiol 68:548–552

Guerreiro N, Redmond JW, Rolfe BG, Djordjevic MA (1997) New rhizobium leguminosarum flavonoid-induced proteins revealed by proteome analysis of differentially displayed proteins. Mol Plant-Microb Interact 10:506–516

Hajiboland R, Aliasgharzadeh N, Laiegh SF, Poschenrieder C (2010) Colonization with arbuscular mycorrhizal fungi improves salinity tolerance of tomato (Solanum lycopersicum L.) plants. Plant Soil 331(1–2):313–327

Hamiaux C, RSM D, Janssen BJ, Ledger SE, Cooney JM et al (2012) DAD2 is an α/βhydrolase likely to be involved in the perception of the plant branching hormone, strigolactone. Curr Biol 22:2032–2036

Harman GE, Howell CR, Viterbo A, Chet I, Lorito M (2004) Trichoderma species — opportunistic, avirulent plant symbionts. Nature Rev Microbiol 2:43–56

Haselwandter K (1987) Mycorrhizal infection and its possible ecological significance in climatically and nutritionally stressed alpine plant communities. Angew Bot 61:107–114

Hashem A, Abd_Allah EF, Alqarawi AA, Aldubise A, Egamberdieva D (2015) Arbuscular mycorrhizal fungi enhances salinity tolerance of Panicum turgidum Forssk by altering photosynthetic and antioxidant pathways. J Plant Interact 10(1):230–242

Hause B, Schaarschmidt S (2009) The role of jasmonates in mutualistic symbioses between plants and soil-born microorganisms. Phytochemistry 70:1589–1599

Helber N, Wippel K, Sauer N, Schaarschmidt S, Hause B, Requena N (2011) A versatile monosaccharide transporter that operates in the arbuscular mycorrhizal fungus Glomus sp is crucial for the symbiotic relationship with plants. Plant Cell 23:3812–3823

Herrera-Medina M, Steinkellner S, Vierheilig H, Ocampo Bote J, García Garrido J. (2007) Absicic acid determines arbuscule development and functionality in the tomato arbuscular mycorrhiza. New Phytol 175:554–564

Ibáñez SG, Medina MI, Agostini E (2011) Phenol tolerance, changes of antioxidative enzymes and cellular damage in transgenic tobacco hairy roots colonized by arbuscular mycorrhizal fungi. Chemosphere 83(5):700–705

Irar S, González EM, Arrese-Igor C, Marino D (2014) A proteomic approach reveals new actors of nodule response to drought in split-root grown pea plants. Physiol Plant 152(4):634–645

Javot H, Penmetsa RV, Terzaghi N, Cook DR, Harrison MJ (2007) A *Medicago truncatula* phosphate transporter indispensable for the arbuscular mycorrhizal symbiosis. Proc Nat Acad Sci 104:1720–1725

Jia T, Wang J, Chang W, Fan X, Sui X, Song F (2019) Proteomics analysis of E. angustifolia seedlings inoculated with arbuscular mycorrhizal fungi under salt stress. Int J Mol Sci 20(3):788

Jia-Dong H, Tao D, Hui-Hui W, Ying-Ning Z, Qiang-Sheng W, Kamil K (2019) Mycorrhizas induce diverse responses of root TIP aquaporin gene expression to drought stress in trifoliate orange. Sci Hortic 243:64–69

Jiang Y, Wang W, Xie Q, Liu N, Liu L, Wang D et al (2017) Plants transfer lipids to sustain colonization by mutualistic mycorrhizal and parasitic fungi. Science 356(6343):1172–1175

Joner EJ, Jakobsen I (1995) Growth and extracellular phosphatase activity of arbuscular mycorrhizal hyphae as influenced by soil organic matter. Soil Biol Biochem 27(9):1153–1159

Joshee N, Mentreddy SR, Yadav AK (2007) Mycorrhizal fungi and growth and development of micropropagated Scutellaria integrifolia plants. Ind Crop Prod 25(2):169–177

Killham K (2002) Bioindicators and sensors of soil health and the application of geostatistics. In: Burns RG, Dick RP (eds) Enzymes in the environment: activity, ecology and applications. Marcel Dekker, Inc., New York, pp 391–405

Kobayashi Y, Maeda T, Yamaguchi K, Kameoka H, Tanaka S, Ezawa T, Kawaguchi M (2018) The genome of Rhizophagus clarus HR1 reveals a common genetic basis for auxotrophy among arbuscular mycorrhizal fungi. BMC Genomics 19(1):465

Kruger M, Kruger C, Walker C, Stockinger H, Schußler A (2012) Phylogenetic reference data for systematics and phylotaxonomy of arbuscular mycorrhizal fungi from phylum to species level. New Phytol 193:970–984. https://doi.org/10.1111/j.1469-8137.2011.03962.x

Kvitko BH, Ramos AR, Morello JE, Oh HS, Collmer A (2007) Identification of harpins in Pseudomonas syringae pv. Tomato DC3000, which are functionally similar to HrpK1 in promoting translocation of type III secretion system effectors. J Bact 189:8059–8072

Larrainzar E, Wienkoop S, Weckwerth W, Ladrera R, Arrese-Igor C, González EM (2007) *Medicago truncatula* root nodule proteome analysis reveals differential plant and bacteroid responses to drought stress. Plant Physiol 144(3):1495–1507

Lee J, Lee S, Young JPW (2008) Improved PCR primers for the detection and identification of arbuscular mycorrhizal fungi. FEMS Microbiol Ecol 65:339–349

Lee E-H, Park S-H, Eo J-K (2018) Kang-Hyeon Ka, and Ahn-Heum Eom *Acaulosproa koreana*, a new species of arbuscular mycorrhizal Fungi (Glomeromycota) associated with roots of Woody plants in Korea. Mycobiology 46(4):341–348

Lin K, Limpens E, Zhang Z, Ivanov S, Saunders DG, Mu D et al (2014) Single nucleus genome sequencing reveals high similarity among nuclei of an endomycorrhizal fungus. PLoS Genet 10 (1):e1004078

Lingua G, Agostino GD, Massa N, Antosiano M, Berta G (2002) Mycorrhiza-induced differential response to a yellows disease in tomato. Mycorrhiza 12:191–198

Liu J, Blaylock L, Endre G, Cho J, Town C et al (2003) Transcript profiling coupled with spatial expression analysis reveals genes involved in distinct developmental stages of the arbuscular mycorrhizal symbiosis. Plant Cell 15:2106–2123

Liu Q, Parsons AJ, Xue H, Jones CS, Rasmussen S (2013) Functional characterization and transcript analysis of an alkaline phosphatase from the arbuscular mycorrhizalfungus Funneliformis mosseae. Fungal Genet Biol 54:52–59

Lone R, Shuab R, Sharma V, Kumar V, Mir R, Koul KK (2015) Effect of arbuscular mycorrhizal fungi on growth and development of potato (Solanum tuberosum) plant. Asian J Crop Sci 7 (3):233–243

Mamontova T, Afonin AM, Ihling C, Soboleva A, Lukasheva E, Sulima AS, Frolov A (2019) Profiling of seed proteome in pea (*Pisum sativum* L.) lines characterized with high and low responsivity to combined inoculation with nodule bacteria and arbuscular mycorrhizal fungi. Molecules *24*(8):1603

Malbreil, M., Tisserant, E., Martin, F., & Roux, C. (2014). Genomics of arbuscular mycorrhizal fungi: out of the shadows. In: Advances in Botanical Research, Vol. 70 Academic Press, pp 259–290

Ming-Yuan W, Ren-Xue X, Qiang-Sheng W, Ji-Hong L, Li-Ming H (2007) Influence of arbuscular mycorrhizal fungi on microbes and enzymes of soils from different cultivated densities of red clover. Ann Microbiol 57(1):1–7

Morris AC, Djordjevic MA (2001) Proteome analysis of cultivar-specific interactions between rhizobium leguminosarum biovar trifolii and subterranean clover cultivar Woogenellup. Electrophoresis 22(3):586–598

Morton JB, Benny GL (1990) Revised classification of arbuscular mycorrhizal fungi (Zygomycetes): a new order, Glomales, two new suborders, Glomineae and Gigasporineae, and two new families, *Acaulosporaceae* and *Gigasporaceae,* with an emendation of Glomaceae. Mycotaxon 37:471–491

Murphy CL et al (2019) The extraradical proteins of Rhizophagus irregularis: a shotgun proteomics approach. Fungal Biol 124:91. https://doi.org/10.1016/j.funbio.2019.12.001

Muthukumar T, Udaiyan K (2000) Arbuscular mycorrhizas of plants growing in the Western Ghats region, southern India. Mycorrhiza 9:297–313

O'Connor PJ, Smith SE, Smith AF (2002) Arbuscular mycorrhizas influence plant diversity and community structure in a semiarid herbland. New Phytol 154:209–218

Oehl M, Brandenburg S, Huemer AK (2019) German bike messengers' experiences and expressions of cycling anger. Traffic Inj Prev 20(7):753–758

Oh HS, Collmer A (2005) Basal resistance against bacteria in Nicotiana benthamiana leaves is accompanied by reduced vascular staining and suppressed by multiple Pseudomonas syringae type III secretion system effector proteins. Plant J 44:348–359

Omar SA (1995) Growth effects of the vesicular-arbuscular mycorrhizal fungus Glomus constrictum on maize plants in pot trials. Folia Microbiol 40(5):503–507

Opik M, Zobel M, Cantero JJ, Davison J, Facelli JM, Hiiesalu I et al (2013) Global sampling of plant roots expands the described molecular diversity of arbuscular mycorrhizal fungi. Mycorrhiza 23:411–430

Packer NH, Pawlak A, Kett WC, Gooley AA, Redmond JW, Williams KL (1997) Proteome analysis of glycoforms: A review of strategies for the microcharacterisation of glycoproteins separated by two-dimensional polyacrylamide gel electrophoresis. Electrophoresis 18 (3–4):452–460

Pal A, Pandey S (2017) Role of arbuscular mycorrhizal fungi on plant growth and reclamation of barren soil with wheat (Triticum aestivum L.) crop. Int J Soil Sci 12:25–31

Panter S, Thomson R, De Bruxelles G, Laver D, Trevaskis B, Udvardi M (2000) Identification with proteomics of novel proteins associated with the peribacteroid membrane of soybean root nodules. Mol Plant-Microbe Interact 13(3):325–333

Pereira CMR, Goto BT, da Silva DKA, de Ferreira ACA, de Souza FA, da Silva GA, Maia LC, Oehl F (2015) Acaulospora reducta sp. nov. and A. excavata-two glomeromycotan fungi with pitted spores from Brazil. Mycotaxon 130(4):983–995. https://doi.org/10.5248/130.983

Pérez-Tienda J, Testillano PS, Balestrini R, Fiorilli V, Azcón-Aguilar C, Ferrol N (2011) GintAMT2, a new member of the ammonium transporter family in the arbuscular mycorrhizal fungus Glomus intraradices. Fungal Genet Biol 48:1044–1055

Plenchette C, Furlan V, Fortin JA (1983) Responses of endomycorrhizal plants grown in a calcined montmorillonite clay to different levels of soluble phosphorus. I Effect on growth and mycorrhizal development. Can J Bot 61(5):1377–1383

Poveda J, Hermosa R, Monte E, Nicolás C (2019) Trichoderma harzianum favours the access of arbuscular mycorrhizal fungi to non-host Brassicaceae roots and increases plant productivity. Sci Rep 9(1):1–11

Raiesi F, Ghollarata M (2006) Interactions between phosphorus availability and an AM fungus (Glomus intraradices) and their effects on soil microbial respiration, biomass and enzyme activities in a calcareous soil. Pedobiologia 50(5):413–425

Ramírez-Valdespino CA, Casas-Flores S, Olmedo-Monfil V (2019) Trichoderma as a model to study effector-like molecules. Front Microbiol 10:1030

Ramos A, Bandala VM, Montoya L (2017) A new species and a new record of *Laccaria* (Fungi, Basidiomycota) found in a relict forest of the endangered *Fagus grandifolia* var. Mexicana Myco Keys 27:77–94

Reinhold-Hurek B, Hurek T (1998) Interactions of gramineous plants with Azoarcus spp. and other diazotrophs: identification, localization, and perspectives to study their function. Crit Rev Plant Sci 17:29–54

Remy W, Taylor TN, Hass H, Kerp H (1994) Four-hundred million-year-old vesicular–arbuscular mycorrhizae. Proc Natl Acad Sci U S A 91:11841–11843

Rillig MC, Wright SF, Kimball BA, Pinter PJ, Wall GW, Ottman MJ, Leavitt SW (2001) Elevated carbon dioxide and irrigation effects on water stable aggregates in a sorghum field: a possible role for arbuscular mycorrhizal fungi. Glob Chang Biol 7(3):333–337

Saad MM, Staehelin C, Broughton WJ, Deakin WJ (2008) Protein-protein interactions within type III secretion systemdependent pili of rhizobium sp. strain NGR234. J Bacteriol 190:750–754

Schwarzott D, Walker C, Schüßler A (2001) Glomus, the largest genus of the arbuscular mycorrhizal fungi (*Glomales*) is nonmonophyletic. Mol Phyl Evol 21:190–197

Simpson D, Daft MJ (1990) Spore production and mycorrhizal development in various tropical crop hosts infected with Glomus clarum. Plant Soil 121(2):171–178

Smith SE, Read DJ (2008) Mycorrhizal symbiosis. Academic Press, London

Sogawa A, Yamazaki A, Yamasaki H, Komi M, Manabe T, Tajima S, Nomura M (2019) SNARE proteins LjVAMP72a and LjVAMP72b are required for root symbiosis and root hair formation in *Lotus japonicus*. Front Plant Sci 9:1992

Song F, Qi D, Liu X, Kong X, Gao Y, Zhou Z et al (2015) Proteomic analysis of symbiotic proteins of glomus mosseae and Amorpha fruticosa. Sci Rep 5:18031

Spatafora JW, Chang Y, Benny GL, Lazarus K, Smith ME, Berbee ML, Bonito G, Corradi N, Grigoriev I, Gryganskyi A, James TY, O'Donnell K, Roberson RW, Taylor TN, Uehling J, Vilgalys R, White MM, Stajich JE (2016a) A phylum-level phylogenetic classification of zygomycete fungi based on genome-scale data. Mycologia 108:1028–1046

Spatafora JW, Chang Y, Benny GL, Lazarus K, Smith ME, Berbee ML et al (2016b) A phylum-level phylogenetic classification of zygomycete fungi based on genome-scale data. Mycologia 108:1028–1046

Suarez MB, Sanz L, Chamorro MI, Rey M, Gonzalez FJ, Llobell A et al (2005) Proteomic analysis of secreted proteins from *Trichoderma harzianum* — identification of a fungal cell wall-induced aspartic protease. Fungal Genet Biol 42:924–934

Sun Z, Song J, Xin XA, Xie X, Zhao B (2018) Arbuscular mycorrhizal fungal 14-3-3 proteins are involved in arbuscule formation and responses to abiotic stresses during AM symbiosis. Front Microbiol 9:91

Symanczik S, Al-Yahya'ei MN, Kozłowska A, Ryszka P, Błaszkowski J (2018) A new genus, Desertispora, and a new species, Diversispora sabulosa, in the family Diversisporaceae (order Diversisporales, subphylum Glomeromycotina). Mycol Prog 17(4):437–449

Tian C, Kasiborski B, Koul R, Lammers PJ, Bücking H, Shachar-Hill Y (2010) Regulation of the nitrogen transfer pathway in the arbuscular mycorrhizal symbiosis, gene characterization and the coordination of expression with nitrogen flux. Plant Physiol 153:1175–1187

Tisserant E, Malbreil M, Kuo A, Kohler A, Symeonidi A, Balestrini R et al (2013) Genome of an arbuscular mycorrhizal fungus provides insight into the oldest plant symbiosis. Proc Nat Acad Sci 110:20117–20122

Trépanier M, Bécard G, Moutoglis P, Willemot C, Gagné S, Avis TJ et al (2005) Dependence of arbuscular-mycorrhizal fungi on their plant host for palmitic acid synthesis. Appl Environ Microbiol 71:5341–5347

Vázquez MM, César S, Azcón R, Barea JM (2000) Interactions between arbuscular mycorrhizal fungi and other microbial inoculants (Azospirillum, Pseudomonas, Trichoderma) and their effects on microbial population and enzyme activities in the rhizosphere of maize plants. Appl Soil Ecol 15(3):261–272

Vidal MT, Azcón-Aguilar C, Barea JM, Pliego-Alfaro F (1992) Mycorrhizal inoculation enhances growth and development of micropropagated plants of avocado. HortScience 27(7):785–787

Vincent D, Kohler A, Claverol S, Solier E, Joets J, Gibon J, Lebrun MH, Plomion C, Martin F (2012) Secretome of the free-living mycelium from the ectomycorrhizal basidiomycete laccaria bicolor. J Proteome Res 11:157

Viprey V, Del Greco A, Golinowski W, Broughton WJ, Perret X (1998) Symbiotic implications of type III protein secretion machinery in rhizobium. Mol Microbiol 28:1381–1389

Vivas A, Barea JM, Azcón R (2005) Interactive effect of Brevibacillus brevis and Glomus mosseae, both isolated from cd contaminated soil, on plant growth, physiological mycorrhizal fungal characteristics and soil enzymatic activities in cd polluted soil. Environ Pollut 134(2):257–266

Wang B, Qiu YL (2006) Phylogenetic distribution and evolution of mycorrhizas in land plants. Mycorrhiza 16:299–363

Wewer V, Brands M, Dörmann P (2014) Fatty acid synthesis and lipid metabolism in the obligate biotrophic fungus R hizophagus irregularis during mycorrhization of L otus japonicus. Plant J 79(3):398–412

Wilson AW, Hosaka K, Perry BA, Mueller GM (2013) *Laccaria* (Agaricomycetes, Basidiomycota) from Tibet (Xizang autonomous region, China). Myco Science 54(6):406–419. https://doi.org/10.1016/j.myc.2013.01.006

Wu QS, Xia RX, Zou YN (2008) Improved soil structure and citrus growth after inoculation with three arbuscular mycorrhizal fungi under drought stress. Eur J Soil Biol 44(1):122–128

Zeng T, Holmer R, Hontelez J, te Lintel-Hekkert B, Marufu L, de Zeeuw T et al (2018) Host-and stage-dependent secretome of the arbuscular mycorrhizal fungus Rhizophagus irregularis. Plant J 94(3):411–425

Zhang H, Tang M, Chen H, Tian Z, Xue Y, Feng Y (2010) Communities of arbuscular mycorrhizal fungi and bacteria in the rhizosphere of Caragana korshinkii and *Hippophae rhamnoides* in Zhifanggou watershed. Plant Soil 326:415–424

Zhao ZW, Xia YM, Qin XZ, Li XW, Cheng LZ, Sha T, Wang GH (2001) Arbuscular mycorrhizal status of plants and the spore density of arbuscular mycorrhizal fungi in the tropical rain forest of Xishuangbanna, Southwest China. Mycorrhiza 11(3):159–162

Zu Y, Ping Y, Mu L, Yang T (2019) The diversity of arbuscular mycorrhizal fungi of *Rosa acicularis* 'Luhe'in saline areas. J For Res 30(4):1507–1512

Chapter 2
An Insight through Root-Endophytic-Mutualistic Association in Improving Crop Productivity and Sustainability

Nitika Thakur

Abstract The Endophytic Symbiotic Association (ESA) has proven as an important tool in uplifting various areas related to crop productivity, nutritional profile, issues related to sustainability, food quality, and security. These beneficial associations have highlighted the burning scenario that is generally equipped with the conventional and chemical cultivation patterns, but moving toward a safe area which includes the use of EPHs (Enhanced Plant Holobionts) consortiums is a best choice toward a sustainable agriculture. The use of these types of associates with mutualistic associations have benefited the host in terms of increased uptake of nitrogen, efficient utilization of nutrients, increased photosynthetic activity and levels, introducing a systemic resistance toward various plant pathogens, pests, and diseases. Secretion of osmoprotectant, decreasing the availability of ROS (Reactive Oxygen Species) and finally leading to enhanced plant growth, crop productivity, and yields. Further, in this direction many integrated systems (SRI-System of Rice Intensification) have been incorporated with these endophytic symbionts, which have yielded spectacular results in terms of pest resistance and crop yields. Thus, it can be concluded that integrating the endophytic symbiotic associations with the different cropping patterns can easily solve the issues related to crop quality, yields, and sustainability.

Keywords Endophytes · Enhanced Plant Holobionts (EPHs) · Agricultural Sustainability · System of Rice Cultivation (SRI) · Reactive Oxygen Species · Crop quality and productivity

---

N. Thakur (✉)
Department of Biotechnology, Shoolini University of Biotechnology and Management Sciences, Solan, Himachal Pradesh, India

N. Shrivastava et al. (eds.), *Symbiotic Soil Microorganisms*, Soil Biology 60,
https://doi.org/10.1007/978-3-030-51916-2_2

2.1 Introduction

The endophytes have been known to colonize the plant roots for decades, but the fact that these mutualistic associations can be directed toward crop sustainability and improvement came into limelight since recent years. Initial researches took a new turn when the symbiotic associations between plants and microorganisms were identified in the form of root galls. Furthermore, when these studies were expanded these root galls were identified as “Nodules” composed of both bacterial and plant proteins, which can help in fixing nitrogen (Hirsch 2009) and other essential nutrients to the leguminous plants. Similarly, these symbiotic associators which helped in increasing plant productivity were recognized as “Mutualistic fungi’s” or the arbuscular mycorrhizal fungi (AMF) (Harrison 1997). In the 1920s and 1930s some strains of fungi “*Trichoderma*” were reported with an antifungal activity which was working against the pathogenic fungi (Weindling 1932, 1934) rendering multiple benefits and enhanced productivity to plants (Berch et al. 2005). A similar group of fungi *“Piriformospora”* with multiple benefits (Harman 2000; Harman et al. 2004; Varma et al. 1999) was observed which working efficiently against various plant diseases. So, the endophytes can be broadly defined as the beneficial microorganisms that can be extracted or isolated from the surface disinfected plant tissues and do not visibly harm the plant (Hallmann et al. 1997). Through penetration and colonization process, these microorganisms become part of regular symbiotic plant–microbe system representing a composite entity (Shrivastava et al. 2018).

Plants existing in association with the microbial entities constitutes the “Holobionts” which refers to the combination or assemblage of different ecological units which symbiotically works together (Margulis 1990; Rogers 2016; Malley 2017) not only restricting itself to some microorganisms but covering a wide era of all microbial symbiont communities.

The growth curve and performance rate of holobionts can be successfully altered positively by introducing screened microbial stains of specified microbial communities that basically enhances or modifies the root microflora. These types of alterations which are intervened purposely result in modification of simple holobionts to EPH’s (Enhanced Plant Holobionts). The EPH’s can be induced more numerously for agricultural purposes by inculcating the selected microbial strains which when given the opportunity to colonize the root zone, can raise crop yields by root colonization resulting in growth enhancement of both roots and shoots, enhancing efficient nutritional uptake, inducing resistance to pests, diseases, and developing tolerance toward biotic/abiotic stress. The EPHs have an altered root zone having larger and deeper root system, enhanced assured crop yields, efficiency in nitrogen supply and its utilization, enhanced photosynthesis levels, etc.

2.2 Endophyte Distribution Pattern, Grouping, and Transmission Mode

A diverse microbial community (Archaeal, fungal, bacterial, and parodistic communities) are widely distributed within plant groups (Hardoim et al. 2015a, b).

The identification and estimation of these fungal associates range from traditional 1.5 million species (Hawksworth 2001) to the updated newer data by sequence analysis which highlights 5.1 million fungal species, with one million involved in mutualistic associations (Strobel and Daisy 2003). These fungal associates predominately belong to class "Ascomycetes" that have been recovered ubiquitously from regions ranging from extreme hot temperatures to the extreme cold one's (Arnold 2007, 2008). Similarly, these also colonize different regions within a plant, that may vary from leaves, stems to the extreme roots, playing an active role in synthesizing the secondary metabolites (Porras-Alfaro and Bayman 2011; Thakur 2018a). However, the positive increase in the use of endophytes suggests an expanding recognition by the scientific communities to exploit various potential benefits of these endophytes for crop improvement and to attain sustainability in yields (Hardoim et al. 2015a, b; Card et al. 2016).

The classification criteria to differentiate various fungal endophytes were based on unique symbiotic and ecological framework, phylogeny, and life history traits.

The endophytes are classified into Class 1 Endophytes (Clavicipitaceous), which are known to form (Rodriguez et al. 2009) most economic associations with the aboveground tissues (Johnson et al. 2013) of grasses. The non-clavicipitaceous are further categorized into 2, 3, and 4 subclasses. The Class 2 endophytes can effectively colonize roots, stems, and leaves, generally belonging to ascomycota and few basidiomycota.

The plant-endophyte association in case of tropical, non-vascular, and vascular plants can be observed in case of Endophytes-3. However, Class 4 endophytes or DSE (Dark Septate Endophytes) are biotrophic facultative fungi which has a special feature of producing dark melanized septate hyphae while colonizing plant roots (Vidal and Jaber 2015).

The Endophytic-transmission routes can vary from Horizontal to vertical transmission. A typical example that satisfies both transmission routes is the Epichloe species. The horizontal transmission depends on the reproductive structures of endophytes which include the spores that can be either (Gao and Mendgen 2006) dispersed by wind or by a vector from plant to plant transmission. While the vertical transmission can be easily mobilized in laboratories enabling the artificial inoculation of grass-endophytic association for commercial purposes (Johnson et al. 2013).

2.3 The Beneficial Endophytic Secretions: Bioactive Natural Products

The endophytes have been bestowed with beneficial secretions (Fig. 2.1) which includes bioactive natural products that can work as antimicrobials, anti-insects, and as plant beneficial addictive.

2.3.1 The Antimicrobial Bioactive Metabolites

There are many promising secondary metabolites which have been identified as potential antimicrobials (230 secondary metabolites) against pathogens and pests in various horticultural and agricultural operations. For example, ambuic acid (Li et al. 2001) which is found to be an antifungal and anti-oomycete agent isolated from *Pestalotiopsis microspora* which is found to be active against fusarium and Pythium species. Similarly, a compound showing antifungal, antibacterial, and antialgal properties known as "Colletonoic acid" in Colletotrichum sp. isolated from (Hussain et al. 2014) *Artemisia annua*. A strong bioactive compound called as "Scandenin" isolated from *Derris scandens* has a potential antibacterial activity against *Bacillus megaterium.* These types of bioactive metabolites are known to be produced by the fungal associates and none have been reported to be produced by bacterial strains.

Twenty-eight volatile compounds secreted by an endophytic fungus (*Muscodor albus*) have been used for inhibiting and killing certain fungi and bacteria. Interestingly, 50 different volatiles have been reported to inhibit the growth of Gram-positive and Gram-negative bacteria, and some plant pathogenic fungi. Furthermore, artificial volatile mixtures are being processed and hold great potential and

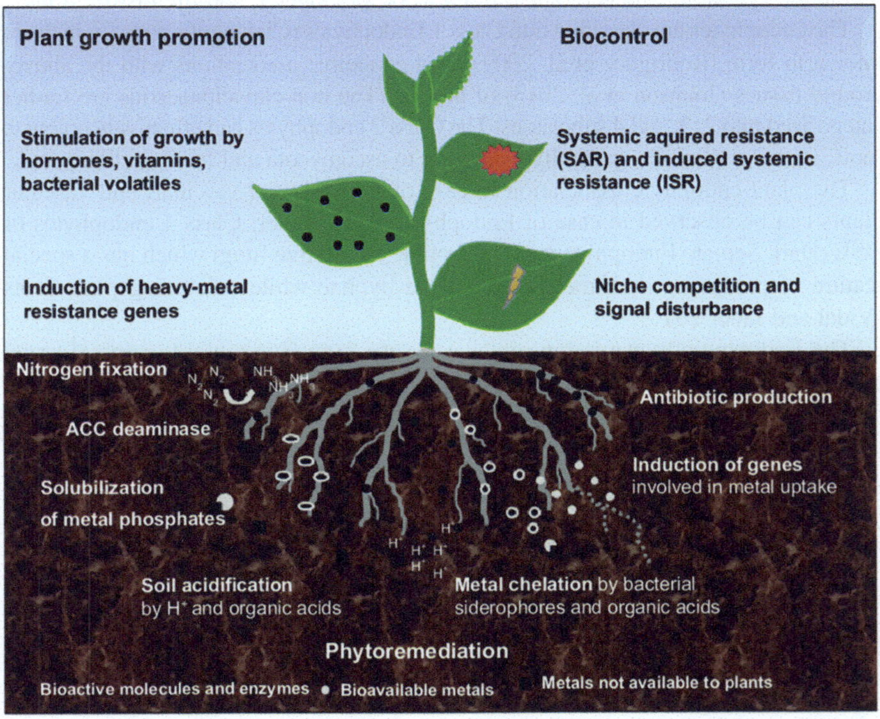

Fig. 2.1 Potential beneficial effects of plant-associated symbionts on plant growth and health (Mitter et al. 2013)

application in food industry, household applications, industrial uses, and agricultural implementation.

2.3.2 Anti-Insect Bioactive Potential

The drastic equation of chemical insecticides has led to devastating decrease in crop yields and crop quality. The use of safe alternatives like biopesticides is a boon to agricultural sustainability. Additionally, the secretion of these types of anti-insect metabolites by the endophytes has added an organic texture to the agricultural practices. The Nodulisporic acids (Indole diterpenes), displays anti-insecticidal activity against blowfly by activating the insect's glutamate-induced chloride pathways. Similarly, an insect repellent "naphthalene" produced by *Muscodor vitigenus* which is effective against mothballs (Daisy et al. 2002).

The endophytic secondary metabolites (Ola et al. 2013) that may work efficiently as antimicrobials, anti-insect, etc., may get triggered by the limitation of food supplies, presence of different cell components, activation and deactivation procedures of cell signaling and increasing competition rates between the different organisms (Thakur 2017b, c).

2.4 The Quadra-Brigade of Endophytic Microbes: The Components of EPH's

The most important symbiotic microbes which have been recognized as backbone in terms of multiple benefit potential that may range from deriving nutrients and enhancing plant–endophyte association.

The four important categories have been designed which serves as components of EPHs:

2.4.1 Rhizobiaceae

This class generally introduces a wide variety of bacteria belonging to family Rhizobiaceae, which shows (Jones et al. 2007; Boogerd and Van Rossum 1997) complex interactions in the form of root nodules with the leguminous plants. An infection thread gets initiated which confronts the bacterial path toward plant root cells, where final structures known as "bacteroid" can be formed. The complex nodulated structure is hence initiated by the leguminous plants which contains leghemoglobin (Iron containing protein), thus proving a hypoxic environment (Schmidt et al. 1994; Chi et al. 2005) for nitrogen fixation and finally the free nitrogen gets reduced to ammonium ions that can be easily utilized by the plant. This class has highly specific interactions that generated ammonium pool for efficient plant growth (Thakur 2017a, b).

2.4.2 Piriformaspora indica

This class generally penetrates the root cortical cells and colonize the area by establishing root interactions accompanying programmed plant cell death (Varma et al. 1999; Samuels and Hebbar 2015).

2.4.3 Trichodermal strains

Trichoderma has been recognized as the most widespread species (Saprophytic associations) which is probably found in abundance in soil, tree stems, and shoots. Their association can be short lived via direct penetration into the plant cell walls or can become perfect endophytes persisting throughout the lifetime (Klein and Eveleigh 1998; Neumann and Laing 2006; Gill et al. 2016).

2.4.4 AMF: Arbuscular Mycorrhizal Fungi

The obligate association is required as they cannot be cultured or grown unless or until they get a plant host. The process includes a chemical signaling pathway which initiates the infection establishment and colonization process, followed by the formation of "Arbuscules" that are located between the host cell wall and plant cell membranes (Strack et al. 2003a, b; Parniske 2008a, b).

2.5 EPHs: A Mutualistic Boon for Sustainable Agriculture

The successful development and spread of EPHs highlight a strong potential (Table 2.1) for increasing the crop productivity, promising better crop yields, and ensuring agricultural sustainability in terms of cost to benefit ratio. The following parameters are listed which focusses on EPHs role in various plant protection mechanisms and enhancing crop productivity.

Table 2.1 Examples of the abilities of endophytic symbiotic microorganisms to increase plants' productivity and yield (Harnam and Uphoff 2019)

Interactions of symbiotic microbes	Associated crop	Effects of endophytic association with the host plant
Rhizobiaceae	Soyabean	Meta-analysis showed 6–176% increase in soybean yields across 28 studies. On farmer fields in Michigan, yields were increased by 23–45% where inoculants had not been used previously. Average yield increased 2–3% where inoculants had previously been used. In Indiana, yield, increases were ~1.5–2%.
Rhizobium	Common bean	Increases of 2–3.5 t/ha under dry conditions
R. leguminosarum bv. *trifolii*	Rice, wheat, and corn	Increases in yield were seen under field conditions. With corn, not all plant genotype-microbial combinations increased yield
AMF (*Glomus versiforme*)	All types of crops	Across numerous studies in the literature, AMF inoculation has resulted in increases in yield but not statistically different from zero. In grasses, the combination of aerially applied endophytic fungi and AMF gave greater than expected results than from either alone.
Piriformaspora indica	Over 150 plant species	Increased shoot and root growth is seen compared to untreated controls in drought but not well-watered conditions. Inactivation of reactive oxygen species (ROS) by gene expression change was required. Various studies have identified plant growth-promoting activities of plants whose roots were colonized by *P. indica*, as reviewed. Improvements in plant performance include better seed germination under temperature, improved resistance of plantlets during micropropagation, and stress resistance.
P. indica	Barley	*P. indica* reduced effects of stresses and pathogens, inducing reprogramming of plant gene expression, which resulted in increased plant biomass and resistance to abiotic stresses. Include the upregulation of enzymes that inactivate toxic levels of reactive oxygen species (ROS) that are formed in plants under stress.
Trichoderma afroharzianum, *T. virens*, *T. viride*, and other species	Numerous crops	Inoculation with the organism induced increased growth responses in numerous vegetable species, greenhouse ornamental plants, and cereal crops
T. afroharzianum	Tomato, corn	Seed treatments applied to corn or tomato resulted in endophytic colonization of plant roots. Colonization is associated with increased resistance to stresses and is causally associated with higher levels of expression of enzymes that inactivate ROS

(continued)

Table 2.1 (continued)

Interactions of symbiotic microbes	Associated crop	Effects of endophytic association with the host plant
Chitooligosaccharides, lipochitooligosaccharides	Rhizobiaceae and AMF	Application, even to the soil, increased fruit yield, and increased total amount of polyphenols. Increased seedling growth of roots; increased yields of corn and other crops including leaf area, shoot mass, and root mass; root branching; increased photosynthesis; changes in plant gene expression; induced resistance to plant diseases. LCOs are produced by the bacteria, but COs may elicit similar plant responses. Compounds added to plants of many kinds result in season-long disease resistance and plant yield increases.
6-Pentyl-α-pyrone (6PP), 1-Octen-3-ol (1o3), Harzianic acid (HzA)	*T. afroharzianum*	Application of this volatile unsaturated lactone molecule, even to the soil, increased fruit yield and increased the total amount of polyphenols as effectively as did treatments with the organism. Seed treatments with picoliter quantities of this volatile metabolite resulted in season-long improvements to shoot and root growth in corn as effectively as did treatments with the fungus itself.
Hydrophobins and other hydrophobin-like proteins (Hp)	*Trichoderma* spp.	Induced plant Défense responses and are inhibitory to soil microflora. Hydrophobic proteins induce plant resistance and increase plant growth is great variability between these proteins, and only a few have beneficial activity. Induces immunity to a virus, a fungus, a bacterium, and an oomycete plant pathogen

2.5.1 *Promoting Plant Growth and Increasing Market Value*

The mutualistic associations have led to an increase in nitrogen-fixing and nitrogen utilization abilities, which affects the overall plant growth. The various seed treatments and colonization of legumes with various strains of Rhizobiaceae family have successfully led to increase in crop yields (Thilakarathna and Raizada 2017). The current market value of such products corresponds to $ 230 million respectively (Mordor 2018). The proliferation strategies in the roots of rice, wheat, and other cereals have led to effective colonization and increasing growth and crop yields.

The need of the hour is to increase the photosynthetic efficiencies, which can be easily met by "reprogramming of plant genes." This technique includes reframing or redesigning chloroplasts (For example, redesigning tobacco chloroplasts) which has shown to enhance crop yield by increasing the photosynthetic abilities (Chi et al. 2010; Wu et al. 2018a, b; Long et al. 2018).

AMF stands as strong foundation for many plant species for survival purposes, especially those living under adverse conditions. It has been observed that AMF has been found to offer limited benefits in terms of crop productivity. (Markmann and Parniske 2009), but these organisms can increase the uptake of nutrients from the soil especially the uptake of phosphorus (Strack et al. 2003a, b). Additionally, they are beneficial in terms of inducing resistance (Tahat et al. 2014; Akhtar et al. 2015; Alaux et al. 2018) against plant pathogens, reducing abiotic stresses, and providing maximum benefit in agriculture where the soil disruption is minimum and carbon-farming systems are incorporated. On the other hand, *Trichodermal* strains have been known to set up long associations with the host enhancing the root and shoot growth. The selected known strains (*P. indica*) of *Trichoderma* induce crops resistant to pests and pathogens, reduce abiotic stress, and finally activate the internal systems which refill the plant operating system with more photosynthetic capabilities and adding on more chlorophyll for efficient plant growth (Thakur 2018a). The *Trichodermal* strains and AMF can be easily propagated for successful implementation in agricultural practices (Woo et al. 2014).

2.5.2 Integrated Pest and Disease Management

The endophytes possess an ability to systemically induce the potential to reduce or to completely eliminate the plant diseases and pests by introducing "systemic resistance." The best example of endophytic association is seen in Rhizobia-legume interaction through mycoparasitic colonization, production of bacteriocins which can fight against number of plant pathogens, production of siderophores which can lead to sequestration of metal ions preferably iron ions required by pathogens for their growth. AMF are also categorized in terms of producing an internal systemic resistance which enables the host to fight against multiple diseases (Mustafa et al. 2017; Kashyap and Thakur 2017). The Solanaceae family when colonized by endophyte *Rhizophagus*, is observed with lesser incidence of *Phytopthora.* The plant resistance against viruses can also be induced where the symptoms can be found to a lesser degree than usual cases (Example: tomato leaf curl virus) (Maffei et al. 2014).

The root colonization process through various endophytic associations has led to successful reduction in diseases, including root rot pathogens (*Verticillium, Rhizoctonia,* etc.). It has also led to reduction in various types of mycotoxins, as in case of *P. indica* colonization with wheat roots for the reduction of wheat head blight, and activation of ROS (Reactive Oxygen Species), thus surviving the biotic and abiotic stresses (Narayan et al. 2017).

Trichoderma possesses various abilities to control the harmful organisms (*Fusarium, Colletotrichum, Phytopthora,* and nematodes) (Thakur 2018a, b). Various strains like *T. harzianum have* been successfully employed to control infections initiated by *Pythium* (Harman et al. 2018).

2.6 Successful Trials: Integrating Crop Management Practices with Endophytic Symbionts

2.6.1 The SRI Trial: System of Rice Intensification

The effects of integrating fungal endophyte (*Trichoderma asperellum*) with the crop management systems recommended for rice cultivation, were found to be beneficial and had more impact when grown in SRI, when compared to the conventional cultivation systems (Uphoff et al. 2010). It was found that the SRI Systems could modify the environments for rice cultivation and resulted in increased microbial activities around the root zone, decreased susceptibility to sheath blight, reduction in methane emissions, and net reduction in greenhouse gases, increased biomass, increased water use efficiency and higher rates of photosynthesis. This system also promotes the utilization of organic biofertilizers and amendments in preference to chemical pesticides and fertilizers.

2.7 Improved Agricultural Prospective of EPHs

EPHs are the promising consortium for improving different aspects of cultivation systems, providing different cellular and molecular benefits

2.7.1 Developing Tolerance to Stress by Optimizing the Redox Environment

The symbiotic association of endophytes with their respective hosts results in altering the genetic makeup by changing the gene expression that ultimately leads to scavenging or detoxifying the active ROS (Reactive oxygen species), which may lead to development of toxins. The enzymes involved in detoxification of ROS include APX (Ascorbate Peroxidases) and CAT (Catalase) etc. (Zuccaro 2002). The antioxidants which are released by the symbionts may result in lowering the ROS stress and keeping the internal environment at low ROS levels, so that the internal system can work efficiently and neutralize the adverse effects. Another advantage is that these associates can secrete osmoprotectants (proline, betaines, and sugars) which can help the host partner to survive under drought conditions (Amato et al. 2008; Thakur 2018a, b). This in turn could help in raising photosynthetic levels and result in better plant growth and productivity.

2.7.2 *Sequestering the Future Agricultural Stress*

- Developing an efficient mechanism through symbiotic association to reduce harmful forms of nitrates and to efficiently increase the nitrogen efficiency rates.
- To reduce the stress (abiotic/biotic) to be faced in near future due to climatic variation and global warming.
- Enhancing the carbon sequestration so that it can be fixed during plant growth and reducing methane emissions.
- Increasing soil productivity through the use of organic amendments and symbiotic associations, so that food quality and security issues do not intervene in agricultural sustainability.
- Creating awareness among the farmer folk to successfully implement the organic strategies and make more use of these symbiotic associates rather than moving toward a chemical cultivation mode (Thakur 2018a, b).

2.8 Conclusion

The use of symbiotic endophytes (EPHs) in agricultural applications can enhance nutritional quality, increased resistance to pest and diseases, increased uptake of nutrients and water utilization, thus introducing systemic resistance for scavenging the ROS (Reactive Oxygen Species) which can finally detach or decrease the levels of toxin production. The utilization of such symbiotic associates can increase soil productivity by increasing the microbial communities (PGPR's) around the rhizospheric area, making it healthier and more fertile for plant growth and finally boosting up the crop productivity and agricultural sustainability. Therefore, it is recommended in near future to utilize more EPHs consortium to solve issues related to carbon sequestering and micro-farming situations.

References

Akhtar MS, Panwar J, Abdullah NA, Siddiqui Y, Swamy MK, Ashkani S (2015) Biocontrol of plant parasitic nematodes by fungi: efficacy and control strategies. In: Meghvansi MK, Varma A (eds) Organic amendments and soil suppressiveness in plant disease management, vol 46, pp 219–247

Alaux PL, Cesar V, Naveau F, Cranenbrouck S, Declerck S (2018) Impact of *Rhizophagus irregularis* MUCL 41833 on disease symptoms caused by *Phytophthora infestans* in potato grown under field conditions. Crop Prot 107:26–33

Amato M, Bass B, Celano G, Bitella G, Morelli G, Rossi R (2008) In situ detection of tree root distribution and biomass by multi-electrode resistivity imaging. Tree Physiol 28:1441–1448

Arnold AE (2007) Understanding the diversity of foliar endophytic fungi: progress, challenges, and frontiers. Fungal Biol Rev 21:51–66

Arnold A (2008) Hidden within our botanical richness, a treasure trove of fungal endophytes. Plant Press 32:13–15

Berch SM, Massicotte HB, Tackaberry LE (2005) Re-publication of a translation of "vegetative organs of *Monotropa hypopitys* L." published by F. Kamienski in 1882, with an update on Monotropa mycorrhizas. Mycorrhiza 15:323–332

Boogerd F, Van Rossum D (1997) Nodulation of groundnut by *Bradyrhizobium*: a simple infection process by crack entry. FEMS Microbiol Rev 21:5–27

Card SD, Johnson LJ, Teasdale S (2016) Deciphering microbial behaviour—the link between endophyte biology and efficacious biological control agents. FEMS Microbiol Ecol 92: fiw114. https://doi.org/10.1093/femsec/fiw114

Chi F, Shen SH, Cheng HP, Jing YX, Yanni YG, Dazzo FB (2005) Ascending migration of endophytic rhizobia, from roots to leaves, inside rice plants and assessment of benefits to rice growth physiology. Appl Environ Microbiol 71:7271–7278

Chi F, Yang P, Han F, Jing Y, Shen S (2010) Proteomic analysis of rice seedlings infected by *Sinorhizobium meliloti* 1021. Proteomics 10:1861–1874

Daisy BH, Strobel GA, Castillo U et al (2002) Naphthalene, an insect repellent, is produced by Muscodor vitigenus, a novel endophytic fungus. Microbiology 148:3737–3741

Gao K, Mendgen K (2006) Seed-transmitted beneficial endophytic Stagonospora sp. can penetrate the walls of the root epidermis, but does not proliferate in the cortex, of *Phragmites australis*. Can J Bot 84:981–988

Gill SS, Gill R, Trivedi DK et al (2016) *Piriformospora indica*: potential and significance in plant stress tolerance. Front Microbiol 7:332

Hallmann J, Quadt-Hallmann A, Mahaffee WF (1997) Bacterial endophytes in agricultural crops. Canadian Journal of Microbiol 43:895–914

Hardoim PR, van Overbeek LS, Berg G (2015a) The hidden world within plants: ecological and evolutionary considerations for defining functioning of microbial endophytes. Microbiol Mol Biol R 79:293–320

Hardoim PR, Van Overbeek LS, Berg G (2015b) The hidden world within plants: ecological and evolutionary considerations for defining functioning of microbial endophytes. Microbiol Mol Biol R 79:293–320

Harman GE (2000) Myths and dogmas of biocontrol changes in perceptions derived from research on *Trichoderma harzianum* T-22. Plant Dis 84:377–393

Harman GE, Howell CR, Viterbo A, Chet I, Lorito M (2004) Trichoderma species—opportunistic, avirulent plant symbionts. Nat Rev Microbiol 2:43–56

Harman GE, Cadle-Davidson M, Nosir W (2018) Patent application WO2017192117A1: highly effective and multi-functional microbial compositions and uses. USPTO, Alexandria, VA

Harnam GE, Uphoff N (2019) A review article on symbiotic Root-endophytic soil microbes improve crop productivity and provide environmental benefits. Hindawi Scientifica 2019:1–25. https://doi.org/10.1155/2019/9106395

Harrison MJ (1997) Arbuscular mycorrhizal symbiosis: an underground association. Trends Plant Sci 2:54–60

Hawksworth DL (2001) The magnitude of fungal diversity: the 1.5 million species estimate revisited. Mycol Res 105:1422–1432

Hirsch AM (2009) Brief history of the discovery of nitrogen-fixing organisms. https://www.mcdbuclaedu/Research/Hirsch/imagesb/HistoryDiscoveryN2fixingOrganismspdf

Hussain H, Root N, Jabeen F et al (2014) Seimatoric acid and colletonoic acid: two new compounds from the endophytic fungi, Seimatosporium sp. and Colletotrichum sp. Chin Chem Lett 25:1577–1579

Johnson LJ, De Bonth ACM, Briggs LR et al (2013) The exploitation of epichloae endophytes for agricultural benefit. Fungal Divers 60:171–188

Jones KM, Kobayashi H, Davies BW, Taga ME, Walker GC (2007) How rhizobial symbionts invade plants: the *Sinorhizobium-Medicago* model. Nat Rev Microbiol 5:619–633

Kashyap AS, Thakur N (2017) Problems and Prospects of Lychee Cultivation in India. In: Kumar M, Kumar V, Bhalla-Sarin N, Varma A (eds) Lychee Disease Management. Springer, Singapore

Klein D, Eveleigh DE (1998) Ecology of *trichoderma*. In: Kubicek CP, Harman GE (eds) Trichoderma and Gliocladium, vol 1. Taylor & Francis, London, pp 57–74

Li JY, Harper JK, Grant DM et al (2001) Ambuic acid, a highly functionalized cyclohexenone with antifungal activity from Pestalotiopsis spp. and Monochaetia sp. Phytochemistry 56:463–468

Long BM, Hee WY, Sharwood RE et al (2018) Carboxysome encapsulation of the CO_2-fixing enzyme rubisco in tobacco chloroplasts. Nat Commun 9:3570

Maffei G, Miozzi L, Fiorilli NVM, Lanfranco L, Accotto GP (2014) Arbuscular mycorrhizal symbiosis attenuates symptom severity and reduces virus concentration in tomato infected by tomato yellow leaf curl Sardinia virus (TYLCSV). Mycorrhiza 24:179–186

Malley MAO (2017) From endosymbiosis to holobionts: evaluating a conceptual legacy. J Biol 434:34–41

Margulis L (1990) Words as battle cries: symbiogenesis and the new field of endocytobiology. Bio Science 40:673–677

Markmann K, Parniske M (2009) Evolution of root endo-symbiosis with bacteria: how novel are nodules. Trends Plant Sci 14:77–86

Mitter B, Muhamad A, Stephane C, Muhammad N (2013) Advances in elucidating beneficial interactions between plant, soil and bacteria, beneficial interactions between plants, soil, and bacteria. In: Sparks DL (ed) Advances in agronomy, vol 121. Academic Press, Burlington, pp 381–445. https://doi.org/10.1016/B978-0-12-407685-3.00007-4

Mordor IntellIgence (2018) *Rhizobium*-based biofertilizer market-segmented by geography-growth, trends, and forecast (2018–2023). https://mordorintelligence.com/industry-reports/rhizobium-based-biofertilizer-market

Mustafa G, Khong NG, Tisserant B et al (2017) Defence mechanisms associated with mycorrhiza-induced resistance in wheat against powdery mildew. Funct Plant Biol 44:443–454

Narayan OP, Verma N, Singh AK et al (2017) Antioxidant enzymes in chickpea colonized by *Piriformospora indica* participate in defense against the pathogen *Botrytis cinereal*. Sci Rep 7:135–153

Neumann B, Laing M (2006) *Trichoderma*: an ally in the quest for soil system sustainability. In: Uphoff N et al (eds) Biological approaches to sustainable soil systems. CRC Press, Boca Raton, FL, pp 491–500

Ola ARB, Thomy D, Lai D et al (2013) Inducing secondary metabolite production by the endophytic fungus Fusarium tricinc-tum through coculture with Bacillus subtilis. J Nat Prod 76:2094–2099

Parniske M (2008a) Arbuscular mycorrhiza: the mother of plant root endosymbiosis. Nat Rev Microbiol 6:763–775

Parniske M (2008b) Arbuscular mycorrhiza: the mother of plant root endosymbioses. Nat Rev Microbiol 6:763–775

Porras-Alfaro A, Bayman P (2011) Hidden fungi, emergent proper-ties: endophytes and microbiomes. Annu Rev Phytopathol 49:291–315

Rodriguez RJ, White JF, Arnold AE (2009) Fungal endophytes: diversity and functional roles. New Phytol 182:314–330

Rogers S (2016) Human microbiome. Encyclopædia Britannica, Chicago, IL. https://www.britannica.com/science/human-microbiome

Samuels GJ, Hebbar PK (2015) Trichoderma: identification and agricultural properties, vol 32. American Phytopathological Society, St. Paul, MN

Schmidt D, James EK, Ellis GJ, Shaw JE, Sprent JI (1994) Interaction between rhizobia and potato tis-sues. J Exp Bot 45:1475–1482

Shrivastava N, Li J, Pan L, Sharma AK, Sanling W, Pandey R, Gao QK, Lou BG (2018) Proteomic approach to understand molecular physiology of symbiotic interaction between *Piriformospora indica* and *Brassica napus*. Sci Rep 8(1):5773

Strack D, Feste T, Hause B, Schliemann W, Walter MH (2003a) Arbuscular mycorrhiza: biological, chemical, and molecular aspects. J Chem Ecol 29(9):1955–1979

Strack D, Fester T, Hause B, Schliemann W, Walter H (2003b) Arbuscular mycorrhiza: biological, chemical, and molecular aspects. Journal of Chemical Ecology 29:1955–1979

Strobel G, Daisy B (2003) Bioprospecting for microbial endophytes and their natural products bioprospecting for microbial endophytes and their natural products. Microbiol Mol Biol R 67:491–502

Tahat MM, Nejat N, Sijam K (2014) *Glomus mosseae* bioprotection against aster yellows phytoplasma (16srI-B) and *Spiroplasma citri* infection in Madagascar periwinkle. Physiol Mol Plant Pathol 88:1–9

Thakur N (2017a) Organic farming, food quality, and human health: a trisection of sustainability and a move from pesticides to eco-friendly biofertilizers. In: Kumar V, Kumar M, Sharma S, Prasad R (eds) Probiotics in agroecosystem. Springer, Singapore

Thakur N (2017b) Integrated approach for the management of differential patterns of diseases and pest incidence in lychee. In: Kumar M, Kumar V, Bhalla-Sarin N, Varma A (eds) Lychee disease management. Springer, Singapore

Thakur N (2017c) Increased Soil-Microbial-Eco-Physiological Interactions and Microbial Food Safety in Tomato Under Organic Strategies. In: Kumar V, Kumar M, Sharma S, Prasad R (eds) Probiotics and Plant Health. Springer, Singapore

Thakur N (2018a) Heat stability and antioxidant potential of beta-carotene isolated from a fungal isolate. Bulgarian J Agr Sci 24(5):891–896

Thakur N (2018b) In silico modulation techniques for upgrading sustainability and competitiveness in agri-food sector. In: Choudhary D, Kumar M, Prasad R, Kumar V (eds) In silico approach for sustainable agriculture. Springer, Singapore

Thilakarathna MS, Raizada MN (2017) A meta-analysis of the effectiveness of diverse rhizobia inoculants on soybean traits under field conditions. Soil Biol Biochem 105:177–196

Uphoff N, Anas I, Rupela OP, Thakur AK, Thiyagarajan TM (2010) Learning about positive plant-microbial interactions from the system of rice intensification (SRI). Annals of Applied Biology 98:29–53

Varma A, Verma S, Sudha SN, Butehorn B, Franken P (1999) *Piriformospora indica*, a cultivable plant-growth-promoting root endophyte. Appl Environ Microbiol 65:2741–2744

Vidal S, Jaber LR (2015) Entomopathogenic fungi as endophytes: plant–endophyte–herbivore interactions and prospects for use in biological control. Current Scidence 109:46–54

Weindling R (1932) *Trichoderma lignorum* as a parasite of other soil fungi. Phytopathology 22:837–845

Weindling R (1934) Studies on a lethal principle effective in the parasitic action of *Trichoderma lignorum* on *Rhizoctonia solani* and other soil fungi. Phytopathology 24:1153–1179

Woo SJ, Ruoccco M, Vinale F et al (2014) *Trichoderma*-based products and their widespread use in agriculture. Open Mycology Journal 8:71–126

Wu Q, Peng X, Yang M et al (2018a) Rhizobia promote the growth of rice shoots by targeting cell signaling, division, and expansion. Plant Molecular Biology 97:507–523

Wu Q, Peng X, Yang M et al (2018b) Rhizobia promote the growth of rice shoots by targeting cell signaling, division, and expansion. Plant Mol Biol 97:507–523

Zuccaro R (2002) Oxidative stress, antioxidants and stress Tol-erance. Trends in Plant Science 7:05–410

Chapter 3
Interaction Between Root Endophytes and Plants: Their Bioactive Products and Significant Functions

Dhriti Kapoor and Nitika Kapoor

Abstract Root endophytes exist in almost every identified plant species. Their capability of entering and flourishing in the plant tissues make them distinctive, which indicates their multidimensional connections within the host plants. Root endophytes influence various significant processes of the host. They trigger the growth of the plants, stimulate defensive response against pathogens, and mitigate abiotic stresses. Several studies have confirmed that endophytes interaction with the host plant is the same as that of the plant growth-promoting (PGP) microbes reside in the rhizosphere. Besides, these endophytes play crucial role in stimulating the productivity of plants and in boosting up their resistance against temperature, salinity, heavy metals, drought, and biotic stresses. Furthermore, root endophytes have potential to synthesize myriad of bioactive compounds including antioxidants, antibiotics, anticancer and antimicrobial agents, immune-suppressive compounds, and insecticides. Such compounds comprise a broad range of organic molecules such as peptides, carbohydrates, aromatics hydrocarbons, etc. and these compounds play vital role in host–microbe relationship. This chapter aimed to give background knowledge of the developments in endophyte biology, functions of root endophytes, and bioactive compounds synthesized by them.

Keywords Root endophytes · Bioactive products · Genotypic and phenotypic expression · Stress resistance

D. Kapoor
Department of Botany, School of Bioengineering and Biosciences, Lovely Professional University, Phagwara, Punjab, India

N. Kapoor (✉)
PG Department of Botany, Hans Raj Mahila Maha Vidyalaya, Jalandhar, Punjab, India

N. Shrivastava et al. (eds.), *Symbiotic Soil Microorganisms*, Soil Biology 60,
https://doi.org/10.1007/978-3-030-51916-2_3

3.1 Introduction

The term endophyte, coined by De Bary (1866) is defined as any micro-organism colonizes healthy plant tissues without causing any symptoms and noticeable injury to the host plant. Most common microbes existing as endophytes in plants are fungi and bacteria (Padhi et al. 2013; Khare et al. 2018). Endophytes are microorganisms that live in plant tissue partly or in all of their life cycle. They can be beneficial, neutral, or disadvantageous depending on their interaction with host plant. For example, rhizobia and mycorrhizal fungi are regarded as beneficial whereas *Fusarium* sp. that causes wilt disease in many economically important food crops is taken as disadvantageous (Anyasi and Atagana 2019; Rodriguez et al. 2009). An array of symbiotic lifestyle has been defined depending on fitness benefits or impacts to host and symbionts. Plant ecology, fitness level, and biological activities of plants are profoundly affected by microbial symbionts. The association of plants with microbial symbionts was found to more than 400 million years old. This relationship starts the moment a land is inhabited by plants (Kozyrovska 2013; Anyasi and Atagana 2016). Endophytes behave the same way in their colonization ability as phytopathogens. So they can be used as biocontrol agents for various pathogens of plants. Some of the microorganisms help to assist the plant growth under adverse conditions, whereas other endophytes have been involved in the synthesis of some novel compounds and antifungal metabolites (Kumar et al. 2014; Afzal et al. 2014).

After colonizing the plant roots, these microorganisms become part of a symbiotic plant-microbe system. Now, these plants should not be regarded as independent entities instead they should be regarded as composite organisms. Such plants that exist in association with their microbial colonizers constitute holobionts. The term holobionts was originally proposed by Margulis (1990) to describe diverse microorganisms that combine asexually to create new integrated hereditary symbionts. But this concept was later expanded to describe more generally host organisms and all of their microbial symbionts (Rogers 2016).

The performance and growth of plant holobionts can be improved by purposely altering the root's microflora. Such plant holobionts show better growth and performance and they are called as EPHs, i.e., enhanced plant holobionts (EPHs). EPHs can be induced within agricultural systems by the selective addition of microbes. EPHs are better than conventional plants as associated microorganisms can be capitalized on both for the human race and for the sustainability of our environment. Several benefits associated with EPHs include enhanced plant growth, larger and deeper root system, higher yield, efficient uptake and use of nutrients, enhanced rate of photosynthesis and enhanced immunity against various biotic and abiotic stresses (Harman and Uphoff 2019).

The present chapter focuses on the concept of root endophytes, their lifestyle and relationship with plants, and beneficial interactions with host plant particularly in relation to bioactive products synthesized by endophytes.

3.2 Lifestyle of Root Endophytes

An inclusive definition of endophytes does not specify their functional relationship with host plant. Apart from commensalistic symbionts, they can exist from dormant pathogens or saprotrophs to mutualistic associates. The mutualistic association by colonizing plant tissues both intercellularly or intracellularly is a well-versed component of their lifestyle. Most of the recent studies clearly show that survival and health of plants are very much dependent upon these microorganisms (Fesel and Zuccaro 2016; Potshangbam et al. 2017). This concept is well explained with rhizobia-legume symbiosis, which is also considered as one of the best-described endophytic relationships (Santoyo et al. 2016).

It has been suggested that endophytes have instigated from the rhizosphere microbes or seed-borne microbial communities, but genomic studies and their correlation with them show that these microbes are more versatile and may contain genes for novel traits which are beneficial to the host plant (Ali et al. 2014). Endophytes also manufacture or induce the host plant to produce metabolites that improve plant growth and assist them to adapt the different environmental conditions (Vega et al. 2008; Lugtenberg et al. 2016; Lata et al. 2018; Khare et al. 2018).

There is another concept of "plant microbiome" which explains the coevolution of plants and their symbionts and tracked the benefits out of the relationship (Turner et al. 2013). The expression of plant genes in the presence of the endophytes provides evidences about their effects on the host plant (Berendsen et al. 2015). The contemporary "omics" based approaches like genome sequencing, comparative genomics, next-generation sequencing (NGS), microarray, and metagenomics may provide an in-depth detail on endophytic lifestyle (Kaul et al. 2016). Endophytes play an imperious role to maintain the health of plants. They help in enhancing growth and yields of plants by protecting or preparing them against various abiotic and biotic stresses (Khare et al. 2018).

Harman and Uphoff (2019) divided the plant root endophytes into four major groups viz. (a) bacteria of the family Rhizobiaceae; (b) arbuscular mycorrhizal fungi (AMF) of the phylum Glomeromycota; (c) certain strains of fungi of Ascomycetous genus *Trichoderma*; and (d) fungi of the order Sebicales, *Piriformaspora indica*. Although these organisms are phylogenetically distinct, each group has independently evolved and colonizes the plant roots, thus becoming resident plant root endophytes. These four groups are regarded as true plant symbionts as they confer advantages to host plant and at same time they drive nutrients, shelter, and other benefits from the host plant. Although their mode of infection and lifestyle within plant roots differ they provide similar advantages to plants (O'Malley 2017).

Bacteria of family Rhizobiaceae infect the leguminous plants through plant root hairs in conjunction with complex plant-microbe chemical signaling. The host plant produces infection threads that guide the bacteria to cortical cells of root where they morph into nitrogen-fixing bacteroids. Later on, the host plant produces a complex structure called nodule for nitrogen-fixing bacteroids. The cells of nodule contain a specific protein called leghemoglobin which provide anaerobic environment to bacteroids necessary for nitrogen fixation. (Andrews and Andrews 2017; Jones et al. 2007). Some of the members of family Rhizobiaceae are also colonize the roots of nonleguminous plants like cereals and potatoes (Schmidt et al. 1994; Chi et al. 2005). The second group of symbionts, arbuscular mycorrhizal fungi (AMF) are obligate plant symbionts. AMF form association with roots of terrestrial plant species which involve complex process of chemical signaling between fungus and host plant (Markmann and Parniske 2009). After the penetration of fungi in host plant, a prepenetration apparatus (PPA) analogous to infection thread of rhizobium bacteria helps the fungi to reach a specific cell of host. After penetration AMF forms large lobed structures called arbuscules that are located between cell wall and cell membrane of host plant cell. Large surface area of arbuscules provides greater area for exchange of nutrients and other metabolites between fungus and host plant cells (Parniske 2008).

Trichoderma species have many diverse lifestyles including saprophytic growth in soils where they degrade complex substrates like cellulose, lignin, chitin, etc. In addition to saprophytic lifestyle they also exhibit symbiotic behavior by colonizing tree's branches and roots. *Trichoderma* species directly penetrate the plant cell wall and persist as endophyte for entire life of the plant (Harman et al. 2004; Zachow et al. 2016). *Piriformaspora indica,* the fourth group of symbionts also directly penetrate the root cells and establish colonies within them. The penetrated fungus further strengthens their colonies by programmed death of root's cortical cells (Samuels and Hebbar 2015).

All the four groups of symbionts produce signaling molecules called SAMPs (*symbiont-associated molecular patterns*) which were previously designated as MAMPs (microbial-associated molecular patterns). SAMPs interact with receptors located on plasmalemma of plant cell. Such interactions are essential to permit infection and plant receptivity leading to nodulation in case of rhizobium-legume interactions and to root infections with AMF (Jones et al. 2007; Parniske 2008). SAMPs include diverse lipochitooligosaccharides (also called as *Nod* factors) that are produced by Rhizobiaceae and AMF; cellotriose produced by *P. indica*; and peptides and proteins produced by *Trichoderma* strains (Johnson et al. 2018; Das et al. 2015; Ruocco et al. 2015).

SAMP transduced MAPK (mitogen-activated protein kinases) mediated signals in host plants which in consequence reprogrammed the gene expression of plants. Reprogramming occurs by alterations in the cell's chromatin; by DNA methylation in the upstream regulatory portions of genes or by modification of the histone proteins. These changes result in "gene priming," a process where gene products are not expressed until these products are needed. In a "primed" state, genes are expressed more rapidly and more fully in response to various abiotic and biotic stimulating factors (Ramirez-Prado et al. 2018; Conrath et al. 2015). Various SAMPS produced by different symbionts and their role in plant productivity and yield are summed up in Table 3.1.

Table 3.1 Role of symbiont produced SAMPs in plant productivity and yield

Symbionts	SAMPs	Effect on plants	References
Rhizobiaceae and AMF	Chitooligosaccharides and lipochitooligosaccharides	Improved seedling growth, enhanced yield of corn and other crops, and boosted photosynthetic efficiency.	Janczarek et al. (2015), Das et al. (2015)
Trichoderma spp.	1-Octen-3-ol	Seed treatment with this metabolite results in season-long improvement in shoot and root growth in corn	Harman et al. (2018)
Trichoderma spp.	Harzianic acid	Show antifungal and growth-promoting activities; help to chelate iron.	Vinale et al. (2013)
Trichoderma spp.	Hydrophobins and other hydrophobin like proteins	Increased plant growth and induce resistance against various plant diseases.	Ruocco et al. (2015)
T. afroharzianum	6-phenyl-α-pyrone	Increased fruit yield and the total amount of polyphenols.	Pascale et al. (2017)
Trichoderma formosa	Plant response like proteins	Improve yield of plants by inducing immunity against a wide spectrum of pathogens	Cheng et al. (2018)

3.3 Plant–Endophyte Relationship

There are multidimensional interactions between endophytes and their plant hosts, particularly, in relation to maintaining the health of the plant (Fig. 3.1). Fungal endophytes change chemical and physical characteristics of the leaf such as high-cellulose content and lamina density, which provide toughness resulting in reduced herbivory, specifically by leaf-cutting ants. Endophytes also prime the host plant's defensive responses against phytopathogens (Table 3.2). Early detection of the phytopathogen is possible by cell surface receptor kinases (RK) and cytoplasmic kinases (CK) mediated activation of ethylene/jasmonic acid transduction pathway. Endophytes have the ability to control many plant diseases and pests through induction of systematic resistance (Harman and Uphoff 2019; Khare et al. 2018; Chadha et al. 2014). Endophytes also help to remediate the several abiotic and biotic stresses by enhancing the expression of various stress-responsive genes and by regulating the level of certain phytohormones like abscisic acids and gibberellins. Plants colonized by endophytes show enhanced expression of genes that detoxify reactive oxygen species (ROS). ROS are neutralized by antioxidants and antioxidative enzymes such as superoxide dismutase (SOD), catalase (CAT) ascorbate peroxidase (APOX), glutathione reductase (GR), dehydroascorbate reductase (DHAR) and monodehydroascorbate reductase (MDHAR), etc. Antioxidants like ascorbic acid and glutathione are only effective if they are present in reduced form. Endophytes enhanced the expression of genes encoding for antioxidative enzymes that maintain the level of reduced forms of antioxidants and detoxify the ROS (Moore et al. 2016; Tyagi et al. 2017). Endophytic associations also alleviate

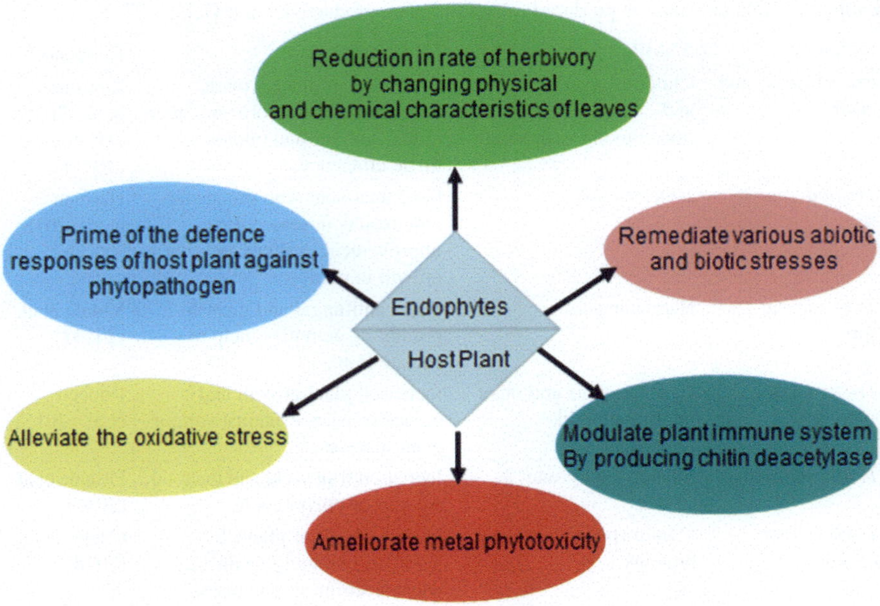

Fig. 3.1 Interaction of host plant with endophytes

metal phytotoxicity by various methods which include sequestration, extracellular precipitation, intracellular accumulation, and biotransformation of toxic metals to less toxic or nontoxic forms (Mishra et al. 2017).

3.4 Bioactive Product Synthesized by Endophytes

Isolation of the secondary products from the endophytes started in 1980s, where the main concern was the toxins which are associated with the disease outbreak in livestock. The information regarding their beneficial effects for human beings and plants were not apparent till far long. Endophytes can be used as a novel and beneficial source of drug specifically used as taxol, an anticancer drug, which came into existence with the recent discovery of *Taxomyces andreanae* in Montana in a yew tree (Strobel et al. 1993). Therefore, extraction of the endophytes and their bioactive products can be useful for the living organisms including human beings as it further leads to improved vitality in plants and further contributes to the product formation in plants. According to the reports, taxol acts as a persuasive compound which is having a strong capacity to fight against oomycetes and the mechanism it carries against the oomycetes is almost similar to the strategies against hastily dividing cancer cells (Young et al. 1992). Plants can be protected from the water molds by the use of taxol and similar compounds obtained from *Taxus* sp.

Table 3.2 Abilities of root endophytic symbionts to increase host resistance to pathogens

Symbiont	Host	Pathogen	Response of host plant	References
Rhizobia	Legumes like soybean, lentil, chickpea, and lupine	*Fusarium, Sclerotinia, Cylindrocladium,* and *Rhizoctonia*	Pathogen controlled by mycoparasitism of bacteria within fungal pathogen	Das et al. (2017)
Mesorhizobium loti	Indian mustard	*Sclerotinia sclerotiorum*	Significant control of white rot disease	Chandra et al. (2007)
AMF (*Funneliformis mosseae*)	Barley, tomato	*Blumeria graminis* (powdery mildew of barley), Tomato leaf curl virus	Systematic resistance reduced level of disease by almost 80%	Mustafa et al. (2017), Maffei et al. (2014)
AMF (*Rhizophagus irregularis*)	Potato	*Phytophthora infestans* (late blight of potato)	Reduction in symptoms but not under high disease pressure	Alaux et al. (2018)
Piriformaspora indica	Wheat	*Fusarium* head blight	Significant reduction in symptoms and also reduced the production of mycotoxins by *Fusarium*	Rabiey and Shaw (2016)
Piriformaspora indica	Chick pea, barley	*Alternaria brassicae, Botrytis cinerea* (fungal disease of leaves and powdery mildew)	Disease controlled by stimulation of antioxidative system of plants	Waller et al. (2005), Narayan et al. (2017)
Trichoderma spp.	Tomato	*Meloidogyne hapla* (root nematodes)	Controlled severity of disease by inducing systematic resistance and production of phytoalexins	Harman et al. (2004, 2018)

Certain techniques of fermentation are required for the characterization of bioactive secondary products synthesized by the living organisms, thus endophytic organisms are extracted from the pure culture for their application in this methodology. For this method, certain medium like M-I-D medium rather than a complex medium is used (Strobel and Daisy 2003). In the process of purification, certain defined medium is applied, which further removes certain other products. For the duration of 2–3 weeks, the fermentation process is carried out in big containers. Certain compounds like butanol, ethyl acetic acid or methyl chloride can be used initially for the extraction. After extraction, flash chromatography, TLC, and HPLC are performed. Each step is followed by certain bioassay, where bioactive compounds are taken. Therefore, in the end, results are obtained in the form of crystals, which is the most required form and finally can be analyzed by using X-ray crystallography. Eventually, the data obtained from spectroscopic analysis must be in correlation with the X-ray crystallographic analysis (Jones et al. 2008).

Globally, different endophytic isolates of *Pestalotiopsis* microspora are collected from several plants which are present at different sites. This fungal species is one of the most commonly isolated endophytic fungi specifically from temperate and tropical regions. Many times it has been observed and allocated for various species, mainly depends upon the host plant from where it is extracted.

3.4.1 Ambuic Acid

An isolate of *Pestalotiopsis microspora* gives rise to the formation of extremely active cyclohexenone, which acts as an endophytic organism of *Fagraea bodenii* and exists in plateau of Papua New Guinea (Li et al. 2001). This antifungal compound has slight antifungal activities. Through solid-state NMR techniques, it was found out that this compound is a first natural compound with the appropriate structure which allows the spatial arrangement of hydroxyl group on the seventh carbon atom (Harper et al. 2003a, b). Ambuic acid was considered as the most important compound having anti-quorum sensing activity specifically in the Gram-negative bacteria and it was estimated from the primary work done on its extraction and resolving the structure (Nakayama et al. 2009). Production of ambuic acid causes inhibition of the quormones of *Staphylococcus aureus* and *Listeria innocua*, which are cyclic peptide. Ambuic acid acts as a principal product, which further leads to the formation of anti-pathogenic drugs targeting quorum sensing-facilitating virulence expression in Gram-positive bacteria.

3.4.2 Cryptocin

One of the Asiatic plants namely *Tripterygium wilfordii* is having resilient immunosuppressive activities. Nature of endophytes of any living organisms must be studied completely whether all of them are likely to synthesize the same or different bioactive compounds. Apart from this, an endophytic fungus was extracted from one of the key species of *Cryptosporiopsis*. Its conidiospores were found to be segmented, which is different from other members of the group (Li et al. 2000). Certain unique antifungal properties were contained in the compounds namely cryptocandin, which is a lipopeptide and cryptocin, that is tetrameric acid, of this particular organism. Spectroscopic techniques and X-ray crystallography are used to extract out and characterize this acid (Strobel 2014a). This particular compound plays a significant role against fungi, which is specifically plant pathogen in comparison to those which cause disease in human beings. *Pyricularia oryzae* is considered as the most susceptible fungus, which acts as a pathogen for causing rice blast disease all over the world, hence affects productivity.

3.4.3 *Colutellin A*

In the tropical forest region of *Costa Rica,* an endophytic fungus namely *Colletotrichum dematium* was extracted out (Strobel 2014b). While dealing with the different species of tropical regions, only this endophyte was isolated under infrequent situations from that particular plant species outlandishly. Against *Botrytis cinerea* and *Sclerotinia sclerotiorum*, respectively, this fungus leads to the synthesis of a unique peptide antimycotic namely Colutellin A, which has least inhibitory concentrations of 3.6 μg/mL at 48 h.

3.4.4 *Pesatcin*

Pesatcin is a unique benzofuran, which was obtained from a culture of an endophytic fungus namely *Pestalotiopsis microspora*, extracted from *Taxus wallichiana* native to the Himalayan foothills. It was very interesting fact that these endophytes also lead to the formation of taxol along with the two unique antioxidants. 1,3-dihydro-isobenzofuran a new compound is synthesized by the endophyte which is having strong antioxidant potential superior to the trolox, a derivative of tocopherol. Mild antifungal properties are also possessed by this unique compound (Harper et al. 2003a, b). With the help of methylene blue, extraction of culture fluid is done, which is subsequently followed by the silica gel chromatography and hence leads to the isolation of pestacin. Its structure and composition were estimated by X-ray diffraction method and ^{13}C and ^{1}H NMR spectroscopic techniques. According to the X-ray analysis, it was found that pestacin exits in nature as a racemic mixture. These analyses also include certain other features of the compound like its antioxidant potential and post-biosynthetic racemization. Another compound namely isopestacin has been synthesized by this specific endophyte and besides it contains identical bioactivities as in pestacin (Strobel et al. 2002).

3.4.5 *Torreyanic Acid*

Certain specific endophytes existence is estimated in *Florida Torreya* plant as it is considered as a rare and threatened species. From *Torreya taxifolia* plant, torreyanic acid was extracted out, which is native to Northern Florida. Torreyanic acid is a dimeric quinone acquired from the endophyte namely *Pestalotiopsis microspora* (Lee et al. 1996). This specific compound exhibits the feature of cytotoxicity against various human cancer cell lines.

3.5 Influence of Endophytes on Genetic and Phenotypic Expression of Plants

In plants, endophytes play a significant role in stimulating induced systemic resistance against the pathogens (Kloepper and Ryu 2006). Host gene expression is regulated by the endophytes which are foliar in their existence and further influence the physiology and defensive responses of plants (Salam et al. 2017). Against the attack of pathogens, certain plant growth regulators like salicylic acid and jasmonic acid play significant roles to combat it (Khare et al. 2016). Resistance against pathogens including insects are improved by gibberellin-synthesizing endophytes by following jasmonic acid and salicylic acid pathway of ameliorating their toxic effects (Waqas et al. 2015). According to Kavroulakis et al. (2007), induced systemic resistance is elicited by *Fusarium solani* against *Septoria lycopersici*, acts as a pathogen for tomato by inducing the expression of pathogenesis-related genes in the root tissues. Inoculation of *Theobroma cacao* was done by foliar endophytic fungi, *Colletotrichum tropicale*, which further led to decrease in the infection caused by *Phytophthora sp* (Mejia et al. 2008). In several plants like *Arabidopsis*, *Theobroma*, etc. resistance against diseases is provided by inoculating *Candida tropicalis*, which results in elicitating different constituents of the ethylene defense strategy and also activates certain defensive genes (Mejía et al. 2014).

Certain volatile organic compounds are possessed by some endophytic bacteria, which are also having antibacterial, antifungal properties and apart from this, they can also combat the attack of other plant pathogens like nematodes. It was reported that the endophyte present in black pepper namely *Pseudomonas putida* BP25 contributed in combating the diseases with the help of volatile organic compounds, caused by various pathogens like *Pythium myriotylum*, *Gibberella moniliformis*, *Phytophthora capsici*, *Colletotrichum gloeosporioides*, *Radopholus similis* (a parasitic nematode), etc. (Sheoran et al. 2015). *Macrophomina phaseolina*, a plant pathogen caused charcoal rot is controlled by siderophore contributing to rhizobium formation in various crops (Arora et al. 2001). *Pseudomonas fluorescens*, endophytic fungi were isolated from the olive roots, which is having antagonistic property against different plant pathogens (Mercado-Blanco et al. 2004). Certain endophytes have also been reported by Etminani and Harighi in 2018, which were found to be having antagonistic properties against *Bacillus*, *Pseudomonas*, *Serratia*, and *Stenotrophomonas* from *Pistacia atlantica* which further showed the regulation of *Pseudomonas syringae* and *Pseudomonas tolaasii*.

Stimulation and expression of stress genes, production of secondary metabolites, quenching of free radicals like hydroxyl radical, singlet oxygen, etc. are certain defensive strategies activated by the endophytes (Lata et al. 2018). Osmotic stress and other abiotic stresses are ameliorated by a plant growth regulator namely abscisic acid, which further regulate the closing of stomata and plant growth and survival (Waqas et al. 2012). Synthesis and signaling pathway of ABA get restrained in the presence of significantly important microorganisms in the plant endosphere, hence helpful in improving the development and productivity of plants in salt stress.

Lately, ABA signaling is modulated by salt-tolerant *Dietzia natronolimnaea*, which is blamable for salt resistance in *Triticum* plants, was authenticated by the upregulation of ABA responsive gene namely TaABARE and TaOPR1 genes (Ilangumaran and Smith 2017). For the mutual symbiotic association of roots of *Arabidopsis* and beneficial fungus *Piriformospora indica*, ABA is required (Peskan-Berghofer et al. 2015). For the amelioration of abiotic stress from *Oryza sativus* plant, upregulation of certain genes like dehydrin, aquaporin, and malonialdehyde have been observed by inoculating endophytic *Trichoderma harzianum* (Pandey et al. 2016).

According to a report, it was analyzed that the endophyte namely *Enterobacter* sp. SA187 can form colonies on the surface as well as the inner tissues of roots and shoots of *Arabidopsis* and caused salinity resistance by stimulating the ethylene signaling strategies with the use of synthesized 2-keto-4-methylthiobutyric acid. This strategy can be used to improve the salinity stress resistance and hence growth and productivity by using the endophyte *Enterobacter* sp. Another endophyte namely *Curvularia protuberata*, a fungal species leads to improving the survival of grass *Dichanthelium lanuginosum* even at stressed conditions like increased soil temperatures, specifically in the grass exist in Yellowstone National Park (Márquez et al. 2007).

3.6 Significant Role of Root Endophytes

3.6.1 Endophytes Are Saprobic Decomposers

Certain reports suggested the role of endophytes as saprobes and their interactions with each other (Tan et al. 2003). The thing is incidental; though, some of the saprobes are obtained from the endophytes (James et al. 2002). There are various reports which suggest that endophytes are helpful in regulating the components of the host, endophytes residing inside host plants, and host specificity of saprobes (Frank et al. 2017). There are various endophytes of wood and leaf tissues which are specific to the host family, colony formation, and the matters within leaf and wood (Lo Piccolo et al. 2010). Certain endophytes are pathogens whereas some possess mutualism (White et al. 2014).

3.6.2 Endophytes as Producer of Antibiotics

Attack of various plant pathogens like fungus, bacteria, nematodes, etc. is overcome by endophytes. With the hardwood species of Europe, a fungus species called *Pezicula cinnamomea* is associated and *Cryptosporiopsis quercina* is considered as its imperfect stage. It was extract out as an endophyte from a medicinal plant *Tripterygium wilfordii*, which is native to Eurasia (Castanheira et al. 2017).

Outstanding antifungal property was reported in *C. quercina* against *Candida albicans* and *Trichophyton sp*., human pathogens, when experiment was performed on petri plates. From *C. Quercina*, antifungal compound cryptocandin was extracted out and characterized, which acts as a distinctive type of peptide (Castanheira et al. 2017).

The compound cryptocandin and the related compounds are also used against different fungi, which leads to skin and nail diseases. Another compound namely, Cryptocin, which is an exclusive tetramic acid, also synthesized by *C. Quercina* (Taghavi et al. 2009). This rare compound has persuasive action against *Pyricularia oryzae* along with a number of phytopathogenic fungi. *P. viridiflava* is one of the subordinate of an assembly of plant-related fluorescent bacteria. In the leaves of grass species, it is positioned on and inside the tissues (Perrine-Walker et al. 2007).

Pseudomycins is alternative group of antifungal compounds which is synthesized by a plants related pseudomonad (Paungfoo-Lonhienne et al. 2010). From the *Artemisia annua*, another endophyte *Colletotrichum* sp. is isolated, which leads to the formation of biologically active metabolites that possess diverse antimicrobial activity. *Artemisia annua* is an old Chinese herb that is very familiar for the production of an anti-malarial drug called artemisinin which is having potential to reside over various geographically regions. In *A. annua*, *Colletotrichum* sp. is present, which not only contributes in the synthesis of metabolites protecting against human pathogen but also from bacteria, fungi etc., which cause diseases in human beings and some of the fungus species in plants also (Baker and Orlandi 1993).

3.6.3 Antiviral Compounds

Antibiotic compounds obtained from endophytic fungi can be used for the reduction of survival rate of viruses. Solid-state fermentation of endophytes specifically fungus *Cytonaema* sp. leads to the isolation of protease inhibitors namely Cytonic acids A and B which are two novel human cytomegalovirus protease inhibitors. With the help of NMR and mass spectrometry techniques, their structures isomers were illuminated (Ali et al. 2014). Because of the absence of relevant screening system, compound discoveries program may face certain confines.

3.6.4 Endophytic Fungal Products as Anticancer Agents

Certain anticancerous compounds can be isolated from the endophytes like Paclitaxel and its derivatives, which is considered as the foremost group. Paclitaxel is usually observed in almost in all species of taxus globally and it is an active diterpenoid. This compound contributes to excluding tubulin molecules by depolymerization at the time of cell division (Doty 2017). This anticancerous compound is considered to make world's first anticancerous medicine, which is used in the treatment of human tissue-proliferating diseases.

3.6.5 Antidiabetic Agents from Endophytes

From the endophytes namely *Pseudomassaria* sp., a nonpeptidal fungal metabolite [L-783] was extracted out, which is gathered from an African rainforest near Kinshasa in the Democratic Republic of the Congo (Germaine et al. 2006). This is used as a substitute for the insulin without causing any damaging effects on the digestive tract. A drastic inhibition in the blood sugar level was observed in two mice models, when they have been orally given L-783,281. Such significant findings may cause new development in therapies for diabetes (Germaine et al. 2006).

3.7 Conclusion and Future Insights

Endophytes are a rich and consistent source from which certain novel compounds are extracted out. They are very diverse in nature, can lead to the synthesis of various pharmaceutically important compounds which are highly bioactive in nature which further seek attention from the scientific societies in extraction and examination of their biotechnological potential. Comparatively uncharted ecological source is represented by them, and they possess strong secondary metabolism as they have metabolic interactions with the hosts. Plants are associated with the endophytes closely. Therefore, a consistency of the research is required to avoid the vanishing of different plant species that would further leads to disappearance of endophytic potential. Collection, arrangement, and proper utilization of endophytes all over the globe the world may lead to development in agricultural practices, industries, and medicines. Management of microbial communities is a great challenge, which further needs colonization of significantly important endophytes.

Conflicts of Interest The authors declare that they have no conflicts of interest.

References

Afzal M, Khan QM, Sessitsch A (2014) Endophytic bacteria: prospects and applications for the phytoremediation of organic pollutants. Chemosphere 117:232–242

Alaux PL, Cesar V, Naveau F, Cranenbrouck S, Declerck S (2018) Impact of *Rhizophagus irregularis* MUCL 41833 on disease symptoms caused by *Phytophthora infestans* in potato grown under field conditions. Crop Prot 107:26–33

Ali S, Duan J, Charles TC, Glick BR (2014) A bioinformatics approach to the determination of genes involved in endophytic behavior in Burkholderia spp. J Theor Biol 343:193–198

Andrews M, Andrews ME (2017) Specificity in legume- rhizobia symbioses. Int J Mol Sci 18 (4):705

Anyasi RO, Atagana HI (2016) Endophytes: an indicator for improved phytoremediation of industrial waste. In: Proceedings of the 23rd waste con conference. Emperors Palace, Johannesburg, pp 140–150

Anyasi RO, Atagana HI (2019) Endophyte: understanding the microbes and its applications. Pak J Biol Sci 22:154–167

Arora NK, Kang SC, Maheshwari DK (2001) Isolation of siderophore producing strains of Rhizobium meliloti and their biocontrol potential against Macrophomina phaseolina that causes charcoal rot of groundnut. Curr Sci 81:673–677

Baker CJ, Orlandi EW (1993) Active oxygen in plant pathogenesis. Annu Rev Phytopathol 33:299–321

Berendsen RL, van Verk MC, Stringlis IA, Zamioudis C, Tommassen J, Pieterse CM et al (2015) Unearthing the genomes of plant-beneficial Pseudomonas model strains WCS358, WCS374 and WCS417. BMC Genomics 16:539. https://doi.org/10.1186/s12864-015-1632-z

Castanheira NL, Dourado AC, Pais I, Semedo J, Scotti-Campos P, Borges N, Carvalho G, Barreto Crespo MT, Fareleira P (2017) Colonization and beneficial effects on annual ryegrass by mixed inoculation with plant growth promoting bacteria. Microbiol Res 198:47–55

Chadha N, Mishra M, Prasad R, Varma A (2014) A root endophytic fungi: research update. J Biol Life Sci USA 5(2):135–158

Chandra S, Choure K, Dubey RC, Maheshwari DK (2007) Rhizosphere competent *Mesorhizobium loti* MP6 induces root hair curling, inhibits *Sclerotinia sclerotiorum* and enhances growth of Indian mustard (*Brassica campestris*). Brazilian J Microbiol 38(1):124–130

Cheng CH, Shen BN, Shang QW et al (2018) Gene-to-gene network analysis of the mediation of plant immunity by the eliciting plant response-like 1 (EPL1) elicitor of *Trichoderma formosa*. Mol Plant-Microbe Interact 31(7):683–691

Chi F, Shen SH, Cheng HP, Jing YX, Yanni YG, Dazzo FB (2005) Ascending migration of endophytic rhizobia, from roots to leaves, inside rice plants and assessment of benefits to rice growth physiology. Appl Environ Microbiol 71(11):7271–7278

Conrath U, Beckers GJM, Langenbach CJG, Jaskiewicz MR (2015) Priming for enhanced defense. Annu Rev Phytopathol 53(1):97–119

Das SN, Madhuprakash J, Sarma PVSRN et al (2015) Biotechnological approaches for field applications of chitooligosaccharides (COS) to induce innate immunity in plants. Crit Rev Biotechnol 35(1):29–43

Das K, Prasanna R, Saxena AK (2017) Rhizobia: a potential biocontrol agent for soilborne fungal pathogens. Folia Microbiol 62(5):425–435

De Bary A (1866) Morphologie und Physiologie der Pilze, Flechten und Myxomyceten Hofmeister's handbook of physiological botany. Wilhelm Engelmann, Leipzig. https://doi.org/10.5962/bhl.title.120970

Doty SL (2017) Functional importance of the plant endophytic microbiome: implications for agriculture, forestry, and bioenergy. In: Doty SL (ed) Functional importance of the plant microbiome. Springer International Publishing, New York, NY, pp 1–5

Etminani F, Harighi B (2018) Isolation and identification of endophytic bacteria with plant growth promoting activity and biocontrol potential from wild pistachio trees. Plant Pathol J 34:208–217

Fesel PH, Zuccaro A (2016) Dissecting endophytic lifestyle along the parasitism/mutualism continuum in Arabidopsis. Curr Opin Microbiol 32:103–112. https://doi.org/10.1016/j.mib.2016.05.008

Frank A, Saldierna Guzmán J, Shay J (2017) Transmission of bacterial endophytes. Microorganisms 5:70

Germaine KJ, Liu X, Cabellos GG, Hogan JP, Ryan D, Dowling DN (2006) Bacterial endophyte-enhanced phytoremediation of the organochlorine herbicide. 2,4-dichlorophenoxyacetic acid. FEMS Microbiol Ecol 57:302–310

Harman GE, Uphoff N (2019) Symbiotic root-endophytic soil microbes improve crop productivity and provide environmental benefits. Scientifica 2019:1–25. https://doi.org/10.1155/2019/9106395

Harman GE, Howell CR, Viterbo A, Chet I, Lorito M (2004) Trichoderma species-opportunistic, avirulent plant symbionts. Nat Rev Microbiol 2(1):43–56

Harman GE, Cadle-Davidson M, Nosir W (2018) Patent application WO2017192117A1: highly effective and multifunctional microbial compositions and uses. USPTO, Alexandria, VA

Harper JK, Barich DH, Hu JZ, Strobel GA, Grant DM (2003a) Stereochemical analysis by solid-state NMR: structural predictions in Ambuic acid. J Org Chem 68:4609–4614

Harper JK, Ford EJ, Strobel GA, Arif A, Grant DM, Porco J, Tomer DP, Oneill K, Pestacin A (2003b) 1,3-dihydro isobenzofuran from Pestalotiopsis microspora possessing antioxidant and antimycotic activities. Tetrahedron 59:2471–2476

Ilangumaran G, Smith DL (2017) Plant growth promoting rhizobacteria in amelioration of salinity stress: a systems biology perspective. Front Plant Sci 8:1768

James EK, Gyaneshwar P, Mathan N, Barraquio WL, Reddy PM, Iannetta PPM, Olivares FL, Ladha JK (2002) Infection and colonization of rice seedlings by the plant growth-promoting bacterium Herbaspirillum seropedicae Z67. Mol Plant-Microbe Interact 15:894–906

Janczarek M, Rachwał K, Marzec A, Grzadziel J, Palusińska-Szysz M (2015) Signal molecules and cell-surface components involved in early stages of the legume-rhizobium interactions. Appl Soil Ecol 85:94–113

Johnson JM, Thurich J, Petutschnig EK et al (2018) A poly(a) ribonuclease controls the cellotriose-based interaction between *Piriformospora indica* and its host *Arabidopsis*. Plant Physiol 176 (3):2496–2514

Jones KM, Kobayashi H, Davies BW, Taga ME, Walker GC (2007) How rhizobial symbionts invade plants: the *Sinorhizobium-Medicago* model. Nature Rev Microbiol 5(8):619–633

Jones EBG, Stanley SJ, Pinruan U (2008) Marine endophytes sources of new chemical natural products: a review. Bot Mar 51:163–170

Kaul S, Sharma T, Dhar MK (2016) "Omics" tools for better understanding the plant–endophyte interactions. Front Plant Sci 7:955. https://doi.org/10.3389/fpls.2016.00955

Kavroulakis NS, Zervakis GI, Ehaliotis C, Haralampidis K, Papadopoulou KK (2007) Role of ethylene in the protection of tomato plants against soil-borne fungal pathogens conferred by an endophytic *Fusarium solani* strain. J Exp Bot 58:3853–3864

Khare E, Kim KM, Lee KJ (2016) Rice OsPBL1 (*Oryza sativa* Arabidopsis PBS1-LIKE 1) enhanced defense of Arabidopsis against Pseudomonas syringae DC3000. Eur J Plant Pathol 146:901–910

Khare E, Mishra J, Arora NK (2018) Multifaceted interactions between endophytes and plant: developments and prospects. Front Microbiol 9:2732. https://doi.org/10.3389/fmicb.2018.02732

Kloepper JW, Ryu CM (2006) Bacterial endophytes as elicitors of induced systemic resistance. In: Schulz BJE, Boyle CJC, Sieber TN (eds) Microbial root endophytes. Springer, Berlin, pp 33–52

Kozyrovska NO (2013) Crosstalk between endophytes and a plant host within information-processing networks. Biopolymers and Cell 29:234–243

Kumar U, Aggarwal C, Paul S, Annapurna K (2014) Endophytes as biocontrol agents of plant pathogens and insects. Kavaka 41:92–95

Lata R, Chowdhury S, Gond SK, White JF Jr (2018) Induction of abiotic stress tolerance in plants by endophytic microbes. Lett Appl Microbiol 66:268–276. https://doi.org/10.1111/lam.12855

Lee JC, Strobel GA, Lobkovsky E, Clardy JC (1996) Torreyanic acid: a selectively cytotoxic quinone dimer from the endophytic fungus Pestalotiopsis microspora. J Org Chem 61:3232–3233

Li J, Strobel GA, Harper J, Lobkovsky E, Clardy J (2000) Cryptocin, a potent tetramic acid antimycotic from the endophytic fungus *Cryptosporiopsis quercina*. Org Lett 2:767–770

Li JY, Harper JK, Grant DM, Tombe BO, Bashyal B, Hess WM, Strobel GA (2001) Ambuic acid, a highly functionalized cyclohexenone with bioactivity from *Pestalotiopsis* spp. and *Monochaetia* sp. Phytochemistry 56:463–468

Lo Piccolo S, Ferraro V, Alfonzo A, Settanni L, Ercolini D, Burruano S, Moschetti G (2010) Presence of endophytic bacteria in Vitis vinifera leaves as detected by fluorescence in situ hybridization. Ann Microbiol 60:161–167

Lugtenberg BJJ, Caradus JR, Johnson LJ (2016) Fungal endophytes for sustainable crop production. FEMS Microbiol Ecol 92:fiw194. https://doi.org/10.1093/femsec/fiw194

Maffei G, Miozzi L, Fiorilli V, Novero M, Lanfranco L, Accotto GP (2014) The arbuscular mycorrhizal symbiosis attenuates symptom severity and reduces virus concentration in tomato infected by tomato yellow leaf curl Sardinia virus (TYLCSV). Mycorrhiza 24(3):179–186

Margulis L (1990) Words as battle cries: symbiogenesis and the new field of endocytobiology. Bio Sci 40(9):673–677

Markmann K, Parniske M (2009) Evolution of root endosymbiosis with bacteria: how novel are nodules? Trends in Plant Sci 14(2):77–86

Márquez LM, Redman RS, Rodriguez RJ, Roossinck MJ (2007) A virus in a fungus in a plant: three-way symbiosis required for thermal tolerance. Science 315:513–515

Mejia LC, Rojas EI, Maynard Z, Van Bael SA, Arnold EA, Hebbar P et al (2008) Endophytic fungi as biocontrol agents of *Theobroma cacao* pathogens. Biol Control 46:4–14

Mejía LC, Herrel EA, Sparks JP, Winter K, García MN, Van Bael SA et al (2014) Pervasive effects of a dominant foliar endophytic fungus on host genetic and phenotypic expression in a tropical tree. Front Microbiol 5:479

Mercado-Blanco J, Rodríguez-Jurado D, Hervás A, Jiménez-Díaz RM (2004) Suppression of Verticillium wilt in olive planting stocks by root associated fluorescent Pseudomonas spp. Biol Control 30:474–486

Mishra J, Singh R, Arora NK (2017) Alleviation of heavy metal stress in plants and remediation of soil by rhizosphere microorganisms. Front Microbiol 8:1706. https://doi.org/10.3389/fmicb.2017.01706

Moore Y, Wang Y, Yang R et al (2016) Regulation of plant growth, photosynthesis, antioxidation and osmosis by an arbuscular mycorrhizal fungus in watermelon seedlings under well-watered and drought conditions. Front Plant Sci 7:644

Mustafa G, Khong NG, Tisserant B et al (2017) Defence mechanisms associated with mycorrhiza-induced resistance in wheat against powdery mildew. Funct Plant Biol 44(4):443–454

Nakayama J, Uemura Y, Nishiguchi K, Yoshimura N, Igarashi Y, Sonomoto K (2009) Ambuic acid inhibits the biosynthesis of cyclic peptide Quormones in gram-positive Bacteria. Antimicrob Agents Chemother 53:580–586

Narayan OP, Verma N, Singh AK et al (2017) Antioxidant enzymes in chickpea colonized by *Piriformospora indica* participate in defense against the pathogen *Botrytis cinerea*. Sci Rep 7 (1):13553

O'Malley MA (2017) From endosymbiosis to holobionts: evaluating a conceptual legacy. J Theo Biol 434:34–41

Padhi L, Mohanta YK, Panda SK (2013) Endophytic fungi with great promises: a review. J Adv Pharm Edu and Res 3(3):152–170

Pandey V, Ansari MW, Tula S, Yadav S, Sahoo RK, Shukla N et al (2016) Dose-dependent response of Trichoderma harzianum in improving drought tolerance in rice genotypes. Planta 243:1251–1264

Parniske M (2008) Arbuscular mycorrhiza: the mother of plant root endosymbiosis. Nature Rev Microbiol 6(10):763–775

Pascale F, Vinale G, Manganiello G et al (2017) *Trichoderma* and its secondary metabolites improve yield and quality of grapes. Crop Prot 92:176–181

Paungfoo-Lonhienne C, Rentsch D, Robatzek S, Webb RI, Sagulenko E, Näsholm T, Schmidt S, Lonhienne TGA (2010) Turning the table: plants consume microbes as a source of nutrients. PLoS One 5:11915

Perrine-Walker FM, Prayitno J, Rolfe BG, Weinman JJ, Hocart CH (2007) Infection process and the interaction of rice roots with rhizobia. J Exp Bot 58:3343–3350

Peskan-Berghofer T, Vilches-Barro A, Muller TM, Glawischnig E, Reichelt M, Gershenzon J et al (2015) Sustained exposure to abscisic acid enhances the colonization potential of the mutualist fungus Piriformospora indica on *Arabidopsis thaliana* roots. New Phytol 208:873–886

Potshangbam M, Devi SI, Sahoo D, Strobel GA (2017) Functional characterization of endophytic fungal community associated with *Oryza sativa* L. and *Zea mays* L. Front Microbiol 8:325. https://doi.org/10.3389/fmicb.2017.00325

Rabiey M, Shaw MW (2016) *Piriformospora indica* reduces fusarium head blight disease severity and mycotoxin DON contamination in wheat under UK weather conditions. Plant Pathol 65 (6):940–952

Ramirez-Prado JS, Piquerez SJM, Bendahmane A, Hirt H, Rayaud C, Bendamed M (2018) Modify the histone to win the battle: chromatin dynamics in plant-pathogen interactions. Front in Plant Sci 9:355

Rodriguez RJ, White JF Jr, Arnold AE, Redman ARA (2009) Fungal endophytes: diversity and functional roles. New Phytol 182(2):314–330

Rogers S (2016) Human microbiome. Encyclopædia Britannica, Chicago, IL. https://www.britannica.com/science/human-microbiome

Ruocco M, Lanzuise S, Lombardi N et al (2015) Multiple roles and effects of a novel Trichoderma hydrophobin. Mol Plant-Microbe Interact 28(2):167–179

Salam N, Khieu TN, Liu MJ, Vu TT, Chu-Ky S, Quach NT, Phi QT, Narsing Rao MP, Fontana A, Sarter S, Li WJ (2017) Endophytic actinobacteria associated with Dracaena cochinchinensis Lour.: isolation, diversity, and their cytotoxic activities. Biomed Res Int

Samuels GJ, Hebbar PK (2015) *Trichoderma:* identification and agricultural properties. In: The American phytopathological society, vol 32. The American Phytopathological Society, St.Paul, MN

Santoyo G, Moreno-Hagelsieb G, Orozco-Mosqueda M, Glick BR (2016) Plant growth-promoting bacterial endophytes. Microbiol Res 183:92–99. https://doi.org/10.1016/j.micres.2015.11.008

Schmidt D, James EK, Ellis GJ, Shaw JE, Sprent JI (1994) Interaction between rhizobia and potato tissues. J Exp Bot 45(279):1475–1482

Sheoran N, Valiya Nadakkakath A, Munjal V, Kundu A, Subaharan K, Venugopal V et al (2015) Genetic analysis of plant endophytic Pseudomonas putida BP25 and chemo-profiling of its antimicrobial volatile organic compounds. Microbiol Res 173:66–78

Strobel GA (2014a) The story of mycodiesel. Curr Opin Microbiol 19:52–58

Strobel GA (2014b) Methods of discovery and techniques to study endophytic fungi producing fuel-related hydrocarbons. Nat Prod Rep 39:259–272

Strobel GA, Daisy B (2003) Bioprospecting for microbial endophytes and their natural products. Microbiol Mol Biol Rev 67:491–502

Strobel GA, Stierle A, Stierle D, Hess WM (1993) Taxomyces andreanae, a proposed new taxon for a bulbilliferous hyphomycete associated with Pacific yew. Mycotaxon 47:71–78

Strobel GA, Ford EJ, Worapong J, Harper JK, Arif AM, Grant DM, Chau RMW (2002) Isopestacin, a unique isobenzofuranone from Pestalotiopsis microspora possessing antifungal and antioxidant properties. Phytochemistry 60:179–183

Taghavi S, Garafola C, Monchy S, Newman L, Hoffman A, Weyens N, Barac T, Vangronsveld J, Van Der Lelie D (2009) Genome survey and characterization of endophytic bacteria exhibiting a beneficial effect on growth and development of poplar trees. Appl Environ Microbiol 75:748–757

Tan Z, Hurek T, Reinhold-Hurek B (2003) Effect of N-fertilization, plant genotype and environmental conditions on nifH gene pools in roots of rice. Environ Microbiol 5:1009–1015

Turner TR, James EK, Poole PS (2013) The plant microbiome. Gen Biol 14:209. https://doi.org/10.1186/gb-2013-14-6-209

Tyagi J, Varma A, Pudake RN (2017) Evaluation of comparative effects of arbuscular mycorrhiza (*Rhizophagus intraradices*) and endophyte (*Piriformospora indica*) association with finger millet (*Eleusine coracana*) under drought stress. European J Soil Biol 8:1–10

Vega FE, Posada F, Aime MC, Pava-Ripoll M, Infante F, Rehner SA (2008) Entomopathogenic fungal endophytes. Biol Control 46:72–82. https://doi.org/10.1016/j.biocontrol.2008.01.008

Vinale F, Nigro M, Sivasithamparam K et al (2013) Harzianic acid: a novel siderophore from *Trichoderma harzianum*. FEMS Microbiol Lett 347(2):123–129

Waller F, Achatz B, Baltruschat H et al (2005) Endophytic fungus *Piriformospora indica* reprograms barley to salt-stress tolerance, disease resistance, and higher yield. Proc Natl Acad Sci 102(38):13386–13391

Waqas M, Khan AL, Kamran M, Hamayun M, Kang SM, Kim YH et al (2012) Endophytic fungi produce gibberellins and indole acetic acid and promotes host-plant growth during stress. Molecules 17:10754–10773

Waqas M, Khan AL, Hamayun M, Shahzad R, Kang SM, Kim JG et al (2015) Endophytic fungi promote plant growth and mitigate the adverse effects of stem rot: an example of Penicillium citrinum and Aspergillus terreus. J Plant Interact 10:280–287

White JF, Torres MS, Somu MP, Johnson H, Irizarry I, Chen Q, Zhang N, Walsh E, Tadych M, Bergen M (2014) Hydrogen peroxide staining to visualize intracellular bacterial infections of seedling root cells. Microsc Res Tech 77:566–573

Young DH, Michelotti EJ, Sivendell CS, Krauss NE (1992) Antifungal properties of taxol and various analogues. Experientia 48:882–885

Zachow C, Berg C, Müller H, Monk J, Berg G (2016) Endemic plants harbour specific Trichoderma communities with an exceptional potential for biocontrol of phytopathogens. J Biotech 235:162–170

Chapter 4
Unravelling the Role of Endophytes in Micronutrient Uptake and Enhanced Crop Productivity

Kanchan Vishwakarma, Nitin Kumar, Chitrakshi Shandilya, and Ajit Varma

Abstract Endophytes belong to the domain of bacterial and fungal species that inhabit the host plant and associates symbiotically with plant tissue. The most common forms of endophytes are found to be associated with the species of *Enterobacter, Colletotrichum, Phyllosticta*, etc. Some of the potent methods of emerging sustainable agriculture to make sure enhancement in crop production having a minimal disturbance caused to environment are exploration of beneficial microbial interactions with plants. The endophytic interactions with plants are advantageous to both the host plant as well as for endophytic microorganism. The ecological and environmental conditions of the host plant significantly influence the associated endophytic population. Endophytes functions as plant growth promoters; enhance uptake of minerals and nutrients; restrict the survival of phytopathogens and increase plant tolerance to environmental stress. The study of functions and host-endophytic relations is important to focus on endophytic significance. Over recent years, substantial studies have been carried out based on their potential to produce valuable substances responsible to enhance plant growth, biocontrol potential as well as their sustenance in adverse environment. The present article focuses on understanding the taxonomy of endophytes and their functions in nutrient uptake and sustainable crop production.

Keywords Endophytes · Micronutrients · Crop productivity · Biocontrol · Biofertilization

---

K. Vishwakarma (✉) · C. Shandilya · A. Varma
Amity Institute of Microbial Technology, Amity University Uttar Pradesh, Noida, India
e-mail: kvishwakarma@amity.edu

N. Kumar
Department of Biotechnology, Periyar Maniammai Institute of Science and Technology, Thanjavur, Tamil Nadu, India

N. Shrivastava et al. (eds.), *Symbiotic Soil Microorganisms*, Soil Biology 60,
https://doi.org/10.1007/978-3-030-51916-2_4

4.1 Introduction

Endophyte is the mixture of two words taken together, i.e., endon meaning "within the plant" and phyton meaning 'plant'. Therefore, this term is a topological expression that includes all organisms collectively which are capable of colonization of living internal tissues of plant hosts without giving any symptom along the uneven period of their life cycle (Stone et al. 2000). This term has broad-spectrum literal meaning with respect to its specificity to host plants and inhabitants (Kobayashi and Palumbo 2000; Stone et al. 2000; Marler et al. 1999; Feller 1995). The host-endophyte interactions may have originated from the initial appearance of higher plants on earth. This is supported by the manifestation of plant-related microbes that were found to present on fossilized stem and leaf tissues (Strobel 2003; Andrzej 2002).

Increase in the plant growth occurs with the help of acquisition of enhanced nutrients and/or stimulating the hormones (Dobbelaere et al. 2003). Plant growth-promoting rhizobacteria, i.e., PGPR releases the substances that enhance plant growth and also important in macro- and micronutrient cycling. Micronutrients important for plants, animals, humans include iron (Fe), zinc (Zn), and copper (Cu) that helps in increased micronutrient density in staple crops such as wheat. (Dobbelaere et al. 2003).

4.2 Major Groups of Endophytic Microbes

There are various distinguishable groups of endophytic microorganisms on the basis of their source of plant organ, which is presented in Table 4.1.

Table 4.1 Different types of endophytes with respect to their host plant

S. No.	Types of endophytes
1.	Endophytic *Clavicipitaceae*
2.	Endophytic fungi of dicots
3.	Systemic fungal endophytes
4.	Fungal endophytes of lichens
5.	Endophytic fungi of bryophytes and ferns
6.	Endophytic fungi of tree bark;
7.	Fungal endophytes of xylem
8.	Fungal endophytes of root
9.	Fungal endophytes of galls and cysts
10.	Endophytic prokaryotes of plants (involves bacterial endophytes and actinomycetes)

4.2.1 Fungal Endophytes

The fungal symbionts pour extreme impact on ecological significance, fitness, and evolution of plants (Brundrett 2006), forming communal plants (Clay and Holah 1999) and indicating profound impact on the biodiversity of integrated organisms which may include insects and bacteria (Omacini et al. 2001). Fungal endophytes inhabit completely within the plants and spread within roots, stems/leaves and emerge to form spores at plant tissue senescence and are not similar to the mycorrhizal fungi that establish their colony in plant roots and spread to rhizosphere (Sherwood and Carroll 1974; Stone et al. 2004). In effect, there are two broad categories of endophytic fungi reported earlier which reflects the differences in evolution studies, taxonomic studies, host plants and ecological properties: clavicipitaceous endophytes (C-endophytes), infecting some grasses; and non-clavicipitaceous endophytes (NC-endophytes), retrieved from asymptomatic tissues of non-vascular plant hosts.

4.2.2 Class 1 Clavicipitaceous Endophytes

Majority of *Clavicipitaceous* endophytes impart resistance to insect attack on hosts (Clay 1990; Patterson et al. 1991; Riedell et al. 1991). Kimmons et al. (1990) carried out a study that enlists anti-nematodal activity of Class 1 endophytes. Since there are multiple studies that do not support the defensive roles of C-endophytes to hosts, some researchers put forward the questions on classifying them as defense-mutualists (Faeth 2002). Many Class 1 endophytes release some substances that inhibit the in vitro spread of other fungi. In a research conducted by Yue et al. (2000) various indole derivatives, i.e., sesquiterpene and diacetamide were obtained from *Epichloë festucae*. It is still undiscovered whether the disease-resistant mechanism is credited to antifungal substances released by endophytes, substances secreted by plants with respect to endophytes, tropical competition between fungi or any physico-exclusion mechanism.

In a study carried out by Malinowski and Belesky (2000), extensive root systems were developed after the infection of *N. coenophialum*. This further enabled the acquiring capacity of plants to better capture water content in the form of moisture and nutrients thereby resulting in resistance of abiotic stress such as drought resistance and faster recovery from water-stress.

Non-clavicipitaceous Endophytes They are majorly found to be present in asymptomatic leaf tissues. Majority of research was focused on isolating and analyzing the bioactive substances, using NC-endophytes as biocontrol agents, and executing phylogenetic identification (Schulz 2006; Arnold et al. 2007; Kithsiri Wijeratne et al. 2008). Despite the lack of functional understanding, some of the NC-endophytes have been shown to display mutualistic behavior toward plants, taking part in tolerating biotic and abiotic stress thereby giving fitness benefits, enhanced growth, productivity, and yield (Redman et al. 2002; Mucciarelli et al. 2003; Waller et al. 2005; Rodriguez et al. 2008).

4.2.3 Class 2 Endophytes

Class 2 endophytes involves variety of species, which are members of Dikarya (*Ascomycota* or *Basidiomycota*). Majority of them are considered under *Ascomycota* while only few come under *Basidiomycota*. It is observed that almost all Class 2 endophytes increase host root and/or shoot biomass. The reason may be dedicated to the fact that plant hormones are induced by the host or fungi that biologically synthesize plant hormones (Tudzynski and Sharon 2002). They are also thought to give shelter to host plants to some extent against fungal attack by secreting secondary metabolites or inducing systemic resistance (Narisawa et al. 2002; Vu et al. 2006; Campanile et al. 2007).

4.2.4 Class 3 Endophytes

Class 3 endophytes are differentiated according to their occurrence

1. Primarily or exclusively in the tissues above ground.
2. Transmissions made horizontally.
3. Occurrence of highly localized infections.
4. Potential to confer benefits or costs on hosts that are not necessarily habitat-specific.
5. Extremely increased in planta biodiversity.

The environmental roles of Class 3 endophytes are not much explored because of their much-noted diverse species range within individual host or its tissues (Rodriguez et al. 2009).

4.2.5 Class 4 Endophytes

Class 4 endophytes are also known as dark septate endophytes (DSE). These are differently labeled on the basis of the functional group which characterizes dark melanized septa and restrict to plant roots. They are shown to have very little host specificity. Interactive relation of DSE with 600 plants has been reported, which includes non-mycorrhizal plants and the plants which belong to Antarctic, Arctic, and temperate zones and also African coastal plains and low-lands, and certain tropical ecological systems (Jumpponen and Trappe 1998; Jumpponen 2001).

4.2.6 *Bacterial Endophytes*

Endophytic bacteria are observed to colonize internal host tissues. Their colonizing number may be high, but it does not cause any damage to host and elicit strong defense responses. These are not like endosymbionts that reside in living plant cells and covered by a membranous chamber (Hurek and Hurek 2011). Endophytic bacteria that are present inside the plants have been noticed for more than 120 years. The occurrence of bacterial endophytes within plants is thought to be variable and sometimes transient but they are also seemed to elicit physiological changes helping in modulating growth and productivity of plants (Zhang et al. 2006; Hardoim et al. 2008; Santoyo et al. 2016).

In the endosphere, a strategic step has been taken by modulating plant physiology by changing levels of plant ethylene. This is because any changes in the levels of ethylene signal will impart major changes in bacterial diversity (Rouws et al. 2010). Therefore, how bacteria respond to plant ethylene concentration signals is key to their ecological success or competence as endophytes.

The phrase of "competent endophytes" is brought into context in order to dictate path in characterizing those bacteria which have the required generic machinery to colonize in the endosphere and persist in it. Whereas the "opportunistic endophytes," unlike the competent ones, have the competency to colonize the rhizosphere that might turn into endosphere colonizer by entering the root tissues; they also lack genes that play a key role in defining their ecological role inside the host plant.

4.2.7 *The Ecology of Competent Endophytes*

The stochastic events govern the diversity and composition of bacterial community in endosphere, and these events are further affected by deterministic procedure of colonization (Taghavi et al. 2009). The colonization of endophytes and their community structures are driven by several parameters that include genotype of plants, several stages of growth and state of physiology, plant tissue type, soil environment conditions, and agricultural practices (Chi et al. 2005; Ikeda et al. 2010). In addition to characteristics that confer competence in the rhizosphere, numbers of other characteristics are proposed to make competent endophytes successful in the plant endosphere.

The comparison of observed endophytic bacterial communities and soil bacterial communities revealed that bacterial endophytes were relatively simple than soil-inhabiting communities, giving up to hundreds of bacterial types that differentiate each other. Hence, it is observed that plants are capable of performing the function of true "filters" for microorganisms of the rhizosphere, selecting those that are successful, competent endophytes (Shah 2009).

4.2.8 Host Range

The plants that grow mainly in temperate, tropical and boreal forests are known to carry the various classes of endophytes. Their host range includes herbaceous plants with their habitats in extreme arctic, alpine, and extremely dry environment (Mushin et al. 1989); moist temperate zones and tropical forests. Moss, ferns, angiosperms, and gymnosperms along with palms, trees having broad leaves, estuary plantations, and deciduous and evergreen forest trees are the carriers of endophytic fungi (Ligrone et al. 1993; Schmid and Oberwinkler 1993). Although, from past studies, it is evident that there are only 100,000 fungal species (approximately) described together with endophytic fungus and around 900,000 fungi are still unknown (Hawksworth and Rossman 1997). Since the number of new endophyte species might exist in plants, they are considered to have the significant contribution to microbial biodiversity (Clay 1992; Kandel et al. 2017).

In a study, 21 cacti species that occur in various zones within Arizona were found to have the presence of 900 endophytic isolates that belonged to 22 fungal species. *Cylindropuntia fulgida* possessed the highest endophyte species diversity, while C. *ramosissima* harbored the most endophyte isolates (Suryanarayanan et al. 2005). It has been broadly accepted that there are very few plants that are free from endophytes, and these include particularly herbs and shrubs (Gennaro et al. 2003). It is also reported that most frequent endophytes comprise bacteria and fungi (Petrini 1991). Generally, fungal endophytes exhibit specificity to single host with respect to plant. However, environment conditions do affect their specificity toward host (Susan 2004). Only exception is *Epichloë typhina* with wide host range (Caruso et al. 2000).

4.2.9 Physiological Role

It is known that endophytes perform potential and advantageous role in physiology of plant hosts. It is also accepted that plants containing endophytes are healthier than those deprived of endophytes (Waller et al. 2005). The reason may be attributed partially to the fact that endophytes produce various phytohormones, for e.g., indole acetic acid, cytokines, and plant growth-promoting compounds such as vitamins; and partially owe to the property of endophytes to increase the absorption of nutrients like N (Reis et al. 2000; Lyons 1990), P (Guo et al. 2000; Malinowski et al. 1999) by host and regulating C: N ratio (Raps and Vidal 1998). For instance, Kaldorf et al. (2005) inoculated *Populus* Esch5 roots with *Piriformospora indica* and observed an increase in biomass of root with significant numbers of second-order roots.

The plants infected with endophytes offer a range of advantages, i.e., increased resistance to pests (Siegel and Schardl 1991; Breen 1994), acclimatization in drought areas (Cheplick et al. 2000; Eerens et al. 1998), enhanced competition (Hill et al.

1991), improved tolerance to stress parameters like heavy metals (Monnet et al. 2001), acidity (Lewis 2004), high pH levels (Waller et al. 2005) and microbe-level infections (Bacilio-Jimenez et al. 2001).

4.2.10 Ecological Role

Endophytic microbes have key ecological roles in systems which involves shaping of plant host communities and governing their associations (Ganley et al. 2004). Endophytic fungi colonized in few plants carry out novel ecological operations such as studying thermal tolerance of plants that grow in geothermal soils (Redman et al. 2002). The ability of endophytes to impact communal biodiversity and interactions between microbes has been considered to be useful determinants of biodiversity of plants (Clay and Holah 1999). Toxic alkaloids are produced by the endophytic microbial communities predominant in grass and herbs in order to prevent or hinder herbivores (Wilkinson et al. 2000). They also play defensive roles to diminish or prevent the damage caused by pathogens in woody plants (Miller et al. 2002; Arnold et al. 2003). Complete ecological properties of endophytes residing in woody plants are still not clearly understood but it can be associated with their species biodiversity (Ganley et al. 2004). Clay and Holah (1999) carried out a study in eastern USA observed that as the interaction between tall fescue and *Neotyphodium coenophialum* was enhanced, the plant biodiversity was declined.

4.3 Micronutrients and their Role in Plant Growth

Elements necessary for plant growth and required in trace amount are called micronutrients. In comparison to macro-minerals, their allowed dose is not more than 100 milligrams/day. The following minerals are considered as micronutrients: iron (Fe), cobalt (Co), chromium (Cr), copper (Cu), iodine (I), manganese (Mn), selenium (Se), zinc (Zn), and molybdenum (Mo) (Fig. 4.1). Organic compounds like vitamins and phytochemicals can also be included as micronutrient as they are also required by organism in trace amount. Excellent source of micronutrients to growing plants includes recycled organic matters consist of grass and tree leaves (White and Brown 2010).

4.3.1 Boron (B)

In soil, both inorganic and organic forms of boron are found and are available to plants in one of the two formats, i.e., boron composed minerals dissolved in soil and soil organic matter. Around 10–300 mg/kg of boron is generally found in the upper

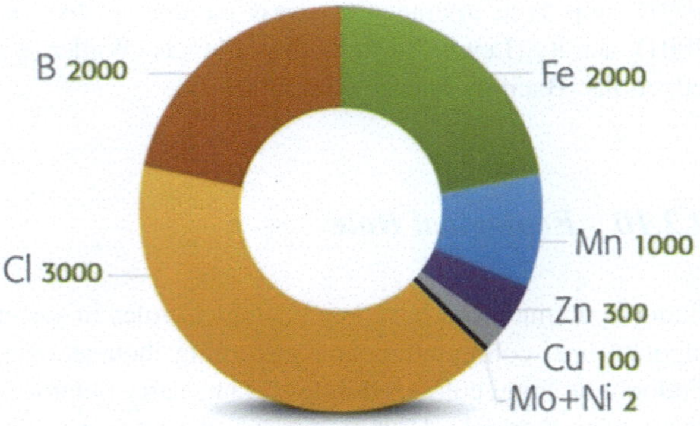

Fig. 4.1 Relative concentration of micronutrients required for plant growth

surface of soils of which only a small quantity of boron is available to plants (Howe 1998).

Boron has several functions in plant cell physiology. Primarily, it is involved in structuring the cell and maintaining properties of plasma membrane (Brown et al. 2002; Wimmer et al. 2009; Atique-ur-Rehman et al. 2018). Boron deficiency causes a reduction in plant growth by blocking the cell wall elongation at growth stage and cause an increase in cellulose, uronic acid and hemicellulose and decrease in pectin (Zehirov and Georgiev 2003). Symptoms for deficiency may include low flowering rate, thickening, curling of leaves, and spots on fruits and leaves (Reid et al. 2004).

4.3.2 Zinc (Zn)

Zinc plays a bigger role in the regulation of early growth of plants by involving in the enzymatic system and is considered important for root, seed, and fruit maturation, stress alleviation, expression of plant growth regulators, and photosynthesis. Further zinc help in many plant developmental processes by acting together with other minerals like potassium, nitrogen, and phosphorus. Zinc is also required in the synthesis of proteins and regulation of growth stages. Delay in maturity is shown in zinc-deficient plants.

4.3.3 Manganese (Mn)

The application of manganese can be in the form of foliar or soil inclusions to eradicate manganese deficiency. It is majorly involved in plants for being used as a cofactor in the functioning of enzymes. It helps directly in photosynthesis mainly in chlorophyll synthesis and activating many metabolic functions. It accelerates maturation and germination of plants by increasing the availability of calcium and phosphate (Millaleo et al. 2010).

4.3.4 Iron (Fe)

Generally, plants show deficiency of iron in soil due to the condition which makes iron unavailable to the plant. The reason for the unavailability of iron to the plants is basic pH of soil, oxygen scarce soil, high soil moisture, high temperatures and high soil manganese, zinc, copper, and phosphorus levels. Unavailable iron is made available by plants by secreting acidic compounds or phytosiderophores in rhizospheric soil so that it complexes with iron at rhizosphere and make iron in available form for the plant itself (Vishwakarma et al. 2018). Iron helps in synthesis of chlorophyll and enzymatic metabolic pathways associated with the transfer of energy, fixation and reduction of nitrogen, and formation of lignin in plants.

4.3.5 Copper (Cu)

Copper is necessary for carbon and nitrogen metabolism. The deficiency of copper causes stunted growth of plants. It is also required in the process of lignin synthesis, as it is needed for preventing wilt and strengthening cell wall. Deficiency symptoms are pale greenish leaves, short growth, and yellow leaves. Generally young plants show these symptoms.

4.3.6 Molybdenum (Mo)

Molybdenum is required in synthesis and activity of nitrate reductase and it is important for symbiotic fixation of nitrogen by rhizobium species. It is also having direct role in converting inorganic to organic phosphorus. pH of soil is the major factor for deficiency of molybdenum, i.e., pH below 6.0 shows the major cause of available molybdenum in soil. Molybdate (MnO_4^{2-}) is the form in which molybdenum is found in soil.

4.3.7 Chlorine (Cl)

Chlorine comes in essential nutrient elements and it exists in the soil in the form of chloride anion (Cl^-) (White and Broadley 2001). Chloride content in fertilizers can affect the quality of crops. For example, tomato, potato, and tobacco need potassium nitrate and potassium sulfate fertilizers rather than potash. Common root rot, leaf, and head fungal infections shown by small grains can be cured by the use of chloride.

4.4 Mechanism of Micronutrient Uptake by Endophytes

Rhizospheric microbes provide plant nutrition by two universal mechanism. They either transform the unavailable forms of nutrients to available form, for e.g., nitrogen fixation or phosphate solubilization or by increasing the transport of mineral and uptake of nutrients by the plant (Hayat et al. 2010; Ahemad and Kibret 2014). It has been reported that both the bacterial as well as fungal endophytes have the potential to colonize their host plants for their plant-inhabiting strategies, promotion of plant growth and protection at the time of biotic and abiotic stress and uptake of soil nutrients and minerals (Johnston-Monje and Raizada 2011). As marine plants absorb vitamins and minerals from water by the process of diffusion, the primitive land plants (e.g., *Aglaophyton major*) had to develop the potential to absorb mineral nutrients present in the soil. The actual root system is not found in the premature bryophyte-like plants (e.g., mosses) therefore it may be possible that earlier mycorrhizal hyphae served this function in early land plant evolution (Brundrett 2002). Endophytes are microbes that colonize the endosphere, a zone within the plants establish them within the cells (Izumi et al. 2008; Izumi 2011). The most eminent kind of symbiosis observed in plants has been observed by mycorrhizae. Some properties indicate their availability to plants, and these are chemical properties, mobility of macronutrients. Specific parameters for example soil microbes and their consequences on the production of root exudates can substantially have an impact on nutrient availability in soil (Vishwakarma et al. 2017a, 2017b). Extensive research proved that mycorrhizal fungi are capable of increasing the uptake of various micronutrients in association with their host plant which correlates the mobilization of minerals and nutrients by the fungi. This becomes possible as a result of (1) the secretion of various enzymes by fungi, (2) the association of the fungi with other soil microorganisms, (3) the impact of fungi upon plant rhizosphere, and (4) the effect of fungi on physiological and morphological (root growth) parameters of host plant that influence the root exudation (Fig. 4.2). The extraradical mycelium extended from the plant root to the encompassing soil is able to enlarge the surface area of plant roots and enhance the nutrient uptake (Landeweert et al. 2001; Kariman et al. 2018).

The ancient fungal partner mainly belongs to the phylum of *Glomeromycota* as the 90% of the land plants forming partnership with arbuscular mycorrhizal fungi (AMF) belongs to *Glomus* species. More than 200 species of this AMF have functions like transmission through soil spores, obligate biotrophy, and inability to grow independently without a compatible host root. The AMF are capable of elevating the supply of nutrients such as N, P, Cu, Zn, Ni, S, Fe, Ca, K, B, and Mn to plant (Clark and Zeto 2000), mainly because the faster hyphae growth explored soil volume more proficiently than roots. The area for nutrient absorption

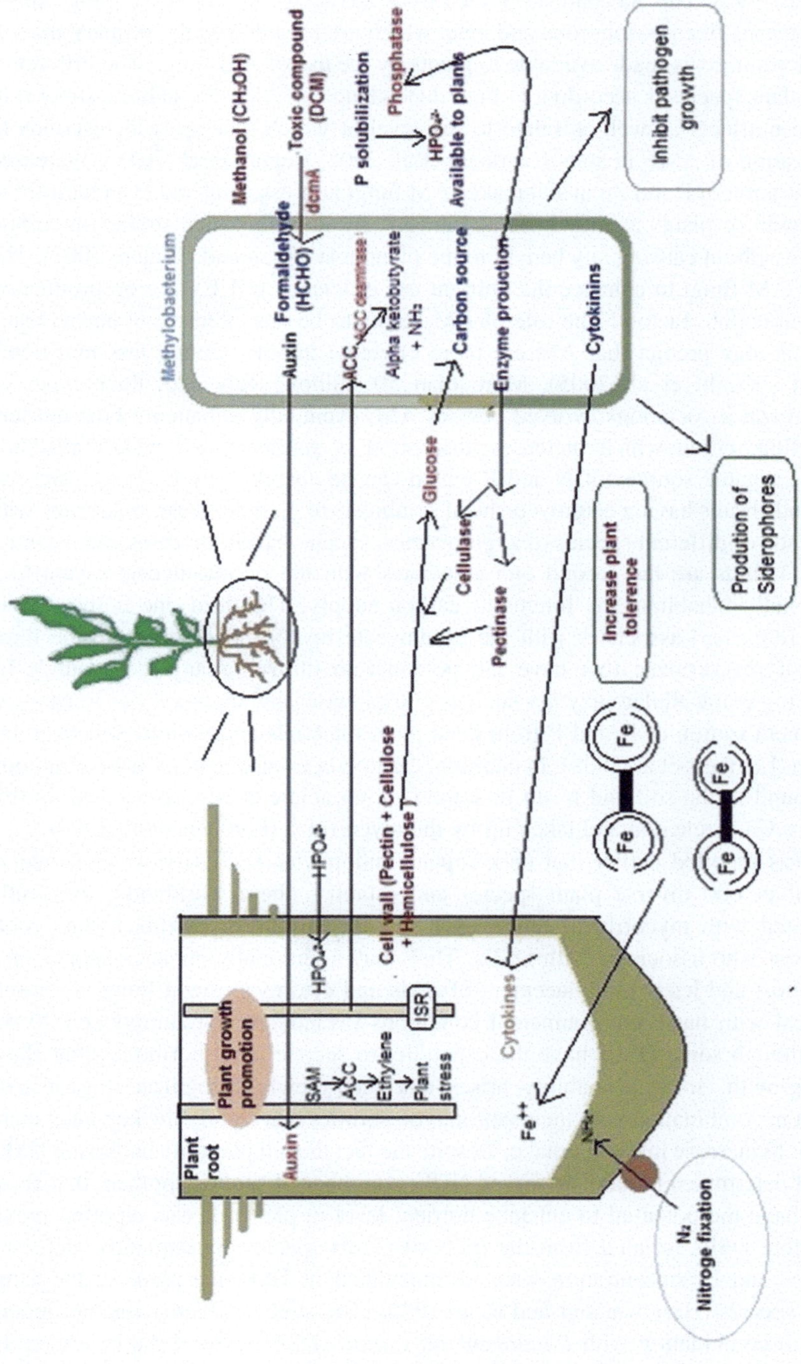

Fig. 4.2 Various mechanisms of increasing root growth by endophytic bacterium

increased by 100 times relative to root length by the association found in AM fungi (Hetrick 1991). This mechanism proved to be advantageous for scavenging immobile nutrients like phosphorous and zinc, which are regarded as the primary macro- and micronutrients made available to plants by the use of AM fungi. The difference in the plant species is according to their dependence on AMF for gaining support to acquire nutrients as well as relate to the level at which their growth responds to colonization of AMF in soils (Lambers et al. 2008; Begum et al. 2019). Increased concentration of P and Zn in soil makes AM fungi non-essential and even harmful to the growth of plants as they become parasitic by imposing their drains on carbon sources without causing any benefit to the plant host (Ryan and Graham 2002). The role of AM fungi to enhance the nutrient use efficiency (NUE) in crop production relies on certain factors. The role of AM seems to be not effective in nutrient-rich soil, this may predict that AM are more effective in soils having less nutritional content (Wright et al. 2005). More than 50 million years ago, three types of mycorrhizal associations evolved besides AM, eventually enhancing host nutrient interacting region which increases absorption of nutrients and makes available various organic sources of N and P which can be absorbed by the fungi. Several trees and shrubs having ectomycorrhizal symbiosis (e.g., pine trees), associates with thousands of different species of zygomycetes, septate basidiomycetes and ascomycetes. Orchids are the second one associates with the basidiomycete mycorrhiza additionally inhabitant the fungus as carbon supply. The third one is the ericoid plants (e.g., tea) associates with the ascomycete mycorrhiza. These are the three types of mycorrhizae that have the potential to utilize organic compounds by producing extracellular enzymes such as phosphatases and carboxylases to catalyze the mineralization of N and P from dead plant materials and various soil microorganism (Lambers et al. 2008). In addition, carboxylases also interact with aluminum (Al) found in the soil and result in some Ca-rich acidic complexes; hence, in this manner, Ca is released and taken up by the mycorrhiza (Lambers et al. 2008).

It was reported earlier that dark septate endophytes (DSEs) were observed in more than 600 diverse plant species and establish them worldwide, frequently associated with mycorrhizal fungi, even though various mycorrhizal plant roots also exist who associates with DSEs. They may additionally enhance supply of P to the host and leads to replacement of AMs and ectomycorrhizal fungi at regions that deal with harsh environmental conditions (Mandyam and Jumpponen 2005). Even though some DSEs have the capability to secrete plant hormones that show plant growth, some probability arises that their newly recognized mycorrhizal inhabit in combination with their saprophytic abilities that signify the key machinery of DSE to increase nutrient uptake. Despite the fact that all plant life harboring DSEs are not determined to have improved NUE, a number of studies are there that show DSEs have the potential to enhance nutrient level in plants. It was reported previously that DSEs isolated from the roots of *Carex* species consequently increased biomass and P concentration when re-inoculated in DSE-free plant of the same *Carex* species (Haselwandter and Read 1982). The study of *Pinus contorta* inoculation in combination with *Phialocephala fortinii* (DSE) showed the increment of leaf P and increase nutrient uptake from soil and increase biomass of plant

(Jumpponen et al. 1998). Nearly 90% of soil phosphorus is immobilized in natural and organic sources, and DSEs may be capable of mineralizing P for their uptake and utilization by plants, but their ability to try this has no longer but been explicitly proven. On the other hand, DSEs have been proven to secrete a quantity of enzymes such as cellulases, pectinases, lipases, laccases, pectinases, and polyphenol oxidases capable of degrading organic materials and make nutrient available to plant host (Mandyam and Jumpponen 2005).

4.5 Role of Endophytes in Plant Growth Promotion

A number of studies have been conducted on the plant growth-promoting potentials of diverse endophytic microbes. The growth promotion mechanism adapted by them involves both the direct method by enhancing the cycling of nutrients such as nitrogen, phosphate, producing valuable metabolites as well as indirectly as biocontrol strains by inhibiting the growth plant pathogens (Garbaye 1994; Cardoso et al. 2011). The attributes of root endophytes as plant inoculants for promoting the growth of plant has been significantly studied (Thakore 2006), and the present research is moving toward the beneficial application to envisage the advantages of endophytes in promoting plant growth and development at field conditions.

The mechanism adapted by endophytes to promote plant growth is very similar to rhizobacteria which include the supply of nutrients by producing siderophore (Costa and Loper 1994), solubilizing the insoluble phosphate (Verma et al. 2001; Wakelin et al. 2004) and the production of growth regulators such as IAA (Lee et al. 2004) as well as by providing essential mineral and vitamins to plants (Pirttilä et al. 2004). The indirect method of promoting plant growth by endophytic bacterial was also reported. The potential of bacterial endophytes to colonize plant niche in the same manner as plant pathogens make them an appropriate biocontrol agent (Berg et al. 2005). The diverse role of endophytes on plant growth was well studied which includes modification of stomata, regulating osmotic stress, altering root morphology, improving mineral uptake, and regulating nitrogen immobilization and mineralization (He et al. 2017; Chandrasekaran et al. 2019). The most promising field which utilizes the potential of these growth-promoting endophytes comprises rhizoremediation as well as revegetation.

The plant-inhabiting life strategies of endophytes follow three main categorizations, i.e., obligate, facultative, and passive endophytes. Obligate endophytes are host-dependent endophytes that are not able to survive outside the plant and the possible way of its transmittance is by means of seed (Hardoim et al. 2008; Gaiero et al. 2013). Facultative endophytes are free-living endophytes that reside dominantly in soil and colonize plant roots as a result of infection. Passive endophytes are the third class of endophytes that colonize plant roots because of any coincidental damage to the root hairs (Hardoim et al. 2008, Gaiero et al. 2013).

The majority of the endophytes responsible for plant growth promotion belong to the category of facultative endophytes due to their potential of colonizing plant roots.

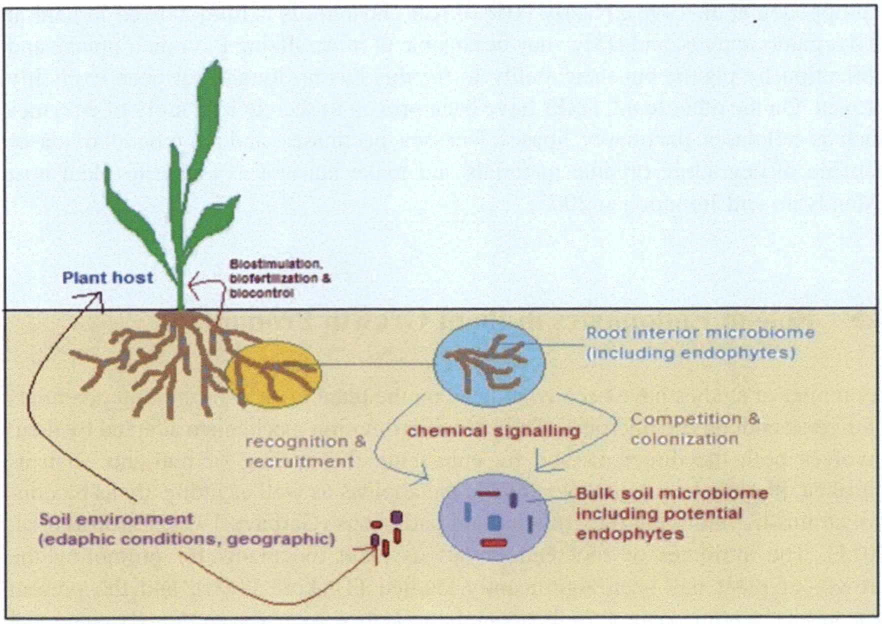

Fig. 4.3 Mechanistic illustration of biostimulation, biofertilization, and biocontrol

The most ordinary process adapted by root endophytes to colonize the epidermis and infiltrate includes the sites of emerging lateral roots, beneath the zone of root hair, and root cracks (Dong et al. 2003; Compant et al. 2005; Zakria et al. 2007) as well as follows intercellular and intracellular establishment (Hurek et al. 1994; Zakria et al. 2007). Some endophytes can pass to different regions of the plant after preliminary colonization via penetrating into the vascular tissues and spreading systemically (Compant et al. 2005; Zakria et al. 2007; Johnston-Monje and Raizada 2011). Three different as well as interconnected mode of action was involved for facilitating plant growth promotion by bacterial endophytes comprises phytostimulation, biofertilization, and biocontrol (Bloemberg and Lugtenberg 2001) (Fig. 4.3).

4.5.1 Phytostimulation

Phytostimulation involves direct method of plant growth enhancement via secreting growth-regulating plant hormones (Bloemberg and Lugtenberg 2001). Probably the most incredibly studied instance of phytostimulation includes the reduction of ethylene levels (a plant hormone) by the enzyme 1-aminocyclopropane-1-carboxylate (ACC) deaminase. A number of endophytes that produce ACC deaminase were observed to display plant growth and development, together with *Arthrobacter* spp. and *Bacillus* spp. in *Capsicum annuum* (pepper) (Sziderics et al.

2007), in addition to *Pseudomonas putida* and *Rhodococcus* spp. in *Pisum sativum* (peas) (Belimov et al. 2001). The mechanism of enhancing plant growth is not well identified although ACC deaminase activity might also lower the abiotic stress by means of balancing levels of plant ethylene production, because high ethylene concentration inhibits replication, cell cycle as well as plant growth (Burg 1973). Several reports are available that show the production of other phytohormones such as abscisic acid, jasmonates, and indole-3-acetic acid by endophytes which enhance plant growth (Patten and Glick 2002; Forchetti et al. 2007a, 2007b).

4.5.2 Biofertilization

Biofertilization is the process of enhancing the growth of plants and yield by stimulating nutrient supply to plant required for its growth (Bashan 1998). Phosphate solubilization, i.e., solubilization of insoluble phosphate and fixation of atmospheric nitrogen, i.e., conversion of atm. Nitrogen to ammonia are the most common mechanism of plant growth promotion which can categorize as a process of biofertilization (Bargaz et al. 2018; Gouda et al. 2018). A few plant growth-promoting bacterial endophytes (PGPBEs) have been studied largely because of their ability of fixing nitrogen inclusive of *Azospirillum* spp. (Hill and Crossman 1983), *Pantoea agglomerans* (Verma et al. 2001), and *Azoarcus* spp. (Hurek et al. 2002). To overcome the phosphorus stress in plants, endophytes start secreting organic acids which solubilize the phosphate and enhances phosphorus availability to the plant. These low molecular weight organic acids chelate the metal associated with phosphorus and aid the plants to utilize the phosphorus (Kpomblekou-A and Tabatabai 2003). A study of Forchetti et al. (2007a, 2007b) on the isolation and characterization of phosphate solubilizing endophytes of sunflower (*Helianthus annuus*) shows that *Achromobacter xiloxidans* and *Bacillus pumilus* were having the phosphate solubilizing potential. Yazdani and Bahmanyar (2009) confirmed that the addition of PGPBEs in fertilizer for corn (*Zea mays*) treatments decreased the need for phosphorus and enhanced crop yield.

4.5.3 Biocontrol

The potential plant growth-promoting endophytic bacteria can inhibit the growth of phytopathogenic organisms and indirectly increase the plant growth; this is referred as biocontrol activity. Biocontrol mechanism includes the production of antibiotics, antifungal compounds, siderophore production, and production of hydrogen cyanide. Siderophores, for instance, pyochelin and salicylic acid are iron-chelating compounds therefore contribute to disease suppression indirectly by blocking phytopathogens for trace metals (Duffy and Defago 1999). Endophytes follow the similar growth-promoting mechanism as explained for plant growth-promoting rhizobacteria which have been studied previously by several authors such as Kloepper et al. (1999), Gray and Smith (2005) and Compant et al. (2010). Some examples of antimicrobial compounds produced by endophytic bacteria are 2, 4-diacetylphloroglucinol (DAPG) that inhibits the growth of phytopathogens. It was previously reported by Ramesh et al. (2008) that the chances of causing wilt by *Ralstonia solanacearum* infection are reduced to 70% in eggplant. It was also reported previously that endophytic microorganism reduced the chance of infection or damage caused by nematodes (Hallmann et al. 1998), insects (Azevedo et al. 2000) or bacterial, viral and fungal species (Kerry 2000; Sturz et al. 2000; Ping and Boland 2004; Berg and Hallmann 2006). It was also described in certain studies that endophytes also promote plant growth in a stressed environment (Bent and Chanway 1998) and additionally support the acceleration of seedling emergence (Chanway 1997). A mechanism referred to as induced systemic resistance (ISR) is adapted by several endophytes and is analogous on the basis of phenotype to systemic-acquired resistance (SAR) (Harman and Uphoff 2019). SAR gets activated when plants effectively initiate their defense mechanism against any primary infection caused by a pathogen (van Loon et al. 1998). ISR proved to be successful for various pathogenic organisms but differs from SAR in which the associated bacteria do not produce symptoms which is visible on the host plant (van Loon et al. 1998). Kloepper and Ryu (2006) studied broadly the role of bacterial endophytes in ISR. Numerous studies are available on exploring the ability of PGPBEs as a potential plant growth promoter and used as bioinoculant for agriculturally important crops (Kloepper and Schroth 1978; Kuklinsky-Sobral et al. 2004).

4.6 Conclusion and Future Prospects

Ecologically, the association of host plants, pathogens, and herbivores with each other is mediated by endophytes. The compositional and distributional species of endophytes and their associations s within and among hosts, the reciprocal of effective ecology of endophytic colonization on fitness of host and composition of communal plants, ecological effects of endophyte colonization on host fitness and the composition of plant communities, are some of general problems of interest to endophytologists and plant ecologist. The upgradation of colonization of bacteria stimulated by explicit carbonaceous exudates by plant roots and the limit of specific microorganisms for modulation of metabolism of plants are key issues for future investigation, as these can be major studies for understanding relationships of potentially mutualistic plant–endophytes. Specific endophytes have significance in the enhancement of growth and development of plants. If the transmission of such endophytes is not on vertical base, (for example, by means of the seed), at that point the development of productive physiological systems that empower their selectivity from oil could have been the key fitness improving attributes that have upgraded the transformative accomplishment of these plant species.

References

Ahemad M, Kibret M (2014) Mechanisms and applications of plant growth promoting rhizobacteria: current perspective. Journal of King Saud University-science 26(1):1–20

Andrzej C (2002) Coexistence and coevolution: various levels of interactions I. Internal fungi (endophytes) and their biological significance in coevolution. Wiad Bot 46:35–44

Arnold AE, Mejia LC, Kyllo D, Rojas EI, Maynard Z, Robins N, Herre EA (2003) Fungal endophytes limit pathogen damage in a tropical tree. Proc Natl Acad Sci USA 100:15649–15654

Arnold AE, Henk DA, Eells RA, Lutzoni F, Vilgalys R (2007) Diversity and phylogenetic affinities of foliar fungal endophytes in loblolly pine inferred by culturing and environmental pcr. Mycologia 99:185–206

Atique-ur-Rehman FM, Rashid A, Nadeem F, Stuerz S, Asch F, Bell RW, Siddique KH (2018) Boron nutrition of rice in different production systems. A review. Agron Sustain Dev 38(3):25

Azevedo JL, Maccheroni Junior W, Pereira JO, Araújo WL (2000) Endophytic microrganisms: a review on insect control and recent advances on tropical plants. Electron J Biotechnol 3:40–65

Bacilio-Jimenez M, Aguilar-Flores S, del Valle MV, Perez A, Zepeda A, Zenteno E (2001) Endophytic bacteria in rice seeds inhibit early colonization of roots by *Azospirillum brasilense*. Soil Biol Biochem 33:167–172

Bargaz A, Lyamlouli K, Chtouki M, Zeroual Y, Dhiba D (2018) Soil microbial resources for improving fertilizers efficiency in an integrated plant nutrient management system. Front Microbiol 9:1606

Bashan Y (1998) Inoculants of plant growth-promoting bacteria for use in agriculture. Science 16:729–770

Begum N, Qin C, Ahanger MA, Raza S, Khan MI, Ahmed N, Ashraf M, Zhang L (2019) Role of Arbuscular Mycorrhizal Fungi in plant growth regulation: implications in abiotic stress tolerance. Front Plant Sci 10:1068

Belimov AA, Safronova VI, Sergeyeva TA, Egorova TN, Matveyeva VA, Tsyganov VE, Borisov AY, Tikhonovich IA, Kluge C, Preisfeld A, Dietz KJ (2001) Characterization of plant growth promoting rhizobacteria isolated from polluted soils and containing 1-aminocyclopropane-1-carboxylate deaminase. Can J Microbiol 47:642–652

Bent E, Chanway C (1998) The growth-promoting effects of a bacterial endophyte on lodgepole pine are partially inhibited by the presence of other rhizobacteria. Can J Microbiol 44:980–988

Berg G, Hallmann J (2006) Control of plant pathogenic fungi with bacterial endophytes. In: Schulz B, Boyle C, Sieber TN (eds) Microbial root endophytes. Springer, Berlin, pp 53–67

Berg G, Eberl L, Hartmann A (2005) The rhizosphere as a reservoir for opportunistic human pathogenic bacteria. Environ Microbiol 7:1673–1685

Bloemberg GV, Lugtenberg BJJ (2001) Molecular basis of plant growth promotion and biocontrol by rhizobacteria. Curr Opin Plant Biol 4:343–350

Breen JP (1994) *Acremonium* Endophyte interactions with enhanced plant resistance to insects. Annu Rev Entomol 39:401–423

Brown PH, Bellaloui N, Wimmer MA, Bassil ES, Ruiz J, Hu H, Pfeffer H, Dannel F, Römheld V (2002) Boron in plant biology. Plant Biol 4:211–229

Brundrett MC (2002) Coevolution of roots and mycorrhizas of land plants. New Phytol 154 (2):275–304

Brundrett MC (2006) Understanding the roles of multifunctional mycorrhizal and endophytic fungi. In: Schulz BJE, Boyle CJC, Sieber TN (eds) Microbial root endophytes. Springer, Berlin, pp 281–293

Burg SP (1973) Ethylene in plant growth. PNAS 70(2):591–597

Campanile G, Ruscelli A, Luisi N (2007) Antagonistic activity of endophytic fungi towards *Diplodiacorticola* assessed by in vitro and in planta tests. Eur J Plant Pathol 117:237–246

Cardoso EJBN, Vasconcellos RLF, Ribeiro CM, Miyauchi MYH (2011) PGPR in coniferous trees. In: Maheshwari DK (ed) Bacteria in agrobiology: crop ecosystems. Springer, Berlin, pp 345–359

Caruso M, Colombo AL, Crespi-Perellino N, Fedeli L, Malyszko J, Pavesi A, Quaroni S, Saracchi M, Ventrella G (2000) Studies on a strain of *Kitasatospora* sp. paclitaxel producer. Ann Microbiol 50:89–102

Chandrasekaran M, Chanratana M, Kim K, Seshadri S, Sa T (2019) Impact of arbuscular mycorrhizal fungi on photosynthesis, water status, and gas exchange of plants under salt stress—a meta-analysis. Front Plant Sci 10:457

Chanway CP (1997) Inoculation of tree roots with plant growth promoting soil bacteria: an emerging technology for reforestation. For Sci 43:99–112

Cheplick GP, Perera A, Koulouris K (2000) Effect of drought on the growth of *Lolium perenne* genotypes with and without fungal endophytes. Funct Ecol 14:657–667

Chi F, Shen SH, Cheng HP, Jing YX, Yanni YG, Dazzo FB (2005) Ascending migration of endophytic rhizobia, from roots to leaves, inside rice plants and assessment of benefits to rice growth physiology. Appl Environ Microbiol 71:7271–7278

Clark RB, Zeto SK (2000) Mineral acquisition by arbuscular mycorrhizal plants. J Plant Nutr 23 (7):867–902

Clay K (1990) Fungal endophytes of grasses. Annu Rev Ecol Syst 21:275–297

Clay K (1992) Fungal endophytes of plants: biological and chemical diversity. Nat Toxins 1:147–149

Clay K, Holah J (1999) Fungal endophyte symbiosis and plant diversity in successional fields. Science 285:1742–1745

Compant S, Reiter B, Sessitsch A, Nowak J, Clément C, Ait BE (2005) Endophytic colonization of *Vitis vinifera* L. by plant growth-promoting bacterium Burkholderia sp. strain PsJN. Appl Environ Microbiol 71:1685–1693

Compant S, Clément C, Sessitsch A (2010) Plant growth-promoting bacteria in the rhizo- and endosphere of plants: their role, colonization, mechanisms involved and prospects for utilization. Soil Biol Biochem 42:669–678

Costa JM, Loper JE (1994) Characterization of siderophore production by the biological control agent *Enterobacter cloacae*. Mol Plant-Microbe Interact 7:440–448

Dobbelaere S, Vanderleyden J, Okon Y (2003) Plant growth promoting effects of diazotrophs in the rhizosphere. Crit Rev Plant Sci 22:107–149

Dong Y, Iniguez AL, Triplett EW (2003) Quantitative assessments of the host range and strain specifi city of endophytic colonization by *Klebsiella pneumonia* 342. Plant Soil 257:49–59

Duffy BK, Defago G (1999) Environmental factors modulating antibiotic and siderophore biosynthesis by *Pseudomonas fluorescens* biocontrol strains. Appl Environ Microbiol 65:2429–2438

Eerens JPJ, Lucas RJ, Easton S, White JGH (1998) Influence of the ryegrass endophyte (*Neotyphodium lolii*) in a cool-moist environment II. Sheep production. NZ J Agric Res 41:191–199

Faeth SH (2002) Are endophytic fungi defensive plant mutualists? Oikos 98:25–36

Feller IC (1995) Effects of nutrient enrichment on growth and herbivory of dwarf red mangrove (*Rhizophora mangle*). Ecol Monogr 65:477–505

Forchetti G, Masciarelli O, Alemano S, Alvarez D, Abdala G (2007a) Endophytic bacteria in sunflower (*Helianthus annuus* L.): isolation, characterization, and production of jasmonates and abscisic acid in culture medium. Appl Microbiol Biotechnol 76:1145–1152

Forchetti G, Masciarelli O, Alemano S, Alvarez D, Abdala G (2007b) Endophytic bacteria in sunflower (*Helianthus annuus* L.): isolation, characterization, and production of jasmonates and abscisic acid in culture medium. Appl Microbiol Biotechnol 76:1145–1152

Gaiero JR, McCall CA, Thompson KA, Day NJ, Best AS, Dunfield KE (2013) Inside the root microbiome: bacterial root endophytes and plant growth promotion. Am J Bot 100:1738–1750

Ganley RJ, Brunsfeld SJ, Newcombe G (2004) A community of unknown, endophytic fungi in western white pine. Proc Natl Acad Sci U S A 101:10107–10112

Garbaye J (1994) Tansley review no. 76. Helper bacteria: a new dimension to the mycorrhizal symbiosis. New Phytol 128:197–210

Gennaro M, Gonthier P, Nicolotti G (2003) Fungal Endophytic communities in healthy and declining *Quercus robur* L. and *Q. cerris* L. trees in northern Italy. J Phytopathol 151:529–534

Gouda S, Kerry RG, Das G, Paramithiotis S, Shin HS, Patra JK (2018) Revitalization of plant growth promoting rhizobacteria for sustainable development in agriculture. Microbiol Res 206:131–140

Gray EJ, Smith DL (2005) Intracellular and extracellular PGPR: commonalities and distinctions in the plant-bacterium signalling processes. Soil Biol Biochem 37:395–412

Guo LD, Hyde KD, Liew ECY (2000) Identification of endophytic fungi from *Livistona chinensis* based on morphology and rDNA sequences. New Phytol 147:617–630

Hallmann J, Quadt-Hallmann A, Rodríguez-Kábana R. and Kloepper JW. (1998) Interactions between *Meloidogyne incognita* and endophytic bacteria in cotton and cucumber. Soil Biol Biochem 30:925–937

Hardoim PR, Overbeek LS, Elsas JD (2008) Properties of bacterial endophytes and their proposed role in plant growth. Trends Microbiol 16(10):463–471

Harman GE, Uphoff N (2019) Symbiotic root-endophytic soil microbes improve crop productivity and provide environmental benefits. Scientifica 2019:1

Haselwandter K, Read DJ (1982) The significance of a root-fungus association in two Carex species of high-alpine plant communities. Oecologia 53(3):352–354

Hawksworth DL, Rossman AY (1997) Where are all the undescribed fungi? Phytopathology 87:888–891

Hayat R, Ali S, Amara U, Khalid R, Ahmed I (2010) Soil beneficial bacteria and their role in plant growth promotion: a review. Annals Microbiol 60(4):579–598

He F, Sheng M, Tang M (2017) Effects of *Rhizophagus irregularis* on photosynthesis and antioxidative enzymatic system in *Robinia pseudoacacia* L under drought Stress. Front Plant Sci 8:183

Hetrick BAD (1991) Mycorrhizas and root architecture. Cell Mol Life Sci 47(4):355–362

Hill A, Crossman SM (1983) Characterization of N2-fixing bacteria associated with sweet potato roots. Can J Microbiol 29:860–862

Hill NS, Belesky DP, Stringer WC (1991) Competitiveness of tall fescue as influenced by Acremonium coenophialum. Crop Sci 31:185–190

Howe PD (1998) A review of boron effects in the environment. Biol Trace Elem Res 66:153–166

Hurek BR, Hurek T (2011) Living inside plants: bacterial endophytes. Curr Opin Plant Biol 14:435–443

Hurek T, Reinhold-Hurek B, Montagu MV, Kellenberger E (1994) Root colonization and systemic spreading of *Azoarcus* strain BH72 in grasses. J Bacteriol 176:1913–1923

Hurek T, Handley LL, Reinhold-Hurek B, Piche Y (2002) Azoarcus grass endophytes contribute fixed nitrogen to the plant in an un-culturable state. Mol Plant-Microbe Interact 15:233–242

Ikeda S, Okubo T, Anda M, Nakashita H, Yasuda M, Sato S, Kaneko T, Tabata S, Eda S, Momiyama A et al (2010) Community- and genome-based views of plant-associated bacteria: plant–bacterial interactions in soybean and rice. Plant Cell Physiol 51:1398–1410

Izumi H (2011) Diversity of endophytic bacteria in forest trees. In: Pirttilä AM, Frank AC (eds) Endophytes of forest trees, biology and applications. Springer, New York, pp 95–105

Izumi H, Anderson IC, Killham K, Moore ERB (2008) Diversity of predominant endophytic bacteria in European deciduous and coniferous trees. Can J Microbiol 54:173–179

Johnston-Monje D, Raizada MN (2011) 4.58 Plant and Endophyte relationships: nutrient management. In: Comprehensive biotechnology, 2nd edn. Academic, Burlington, pp 713–727

Jumpponen A (2001) Dark septate endophytes – are they mycorrhizal? Mycorrhiza 11:207–211

Jumpponen A, Trappe JM (1998) Dark septate endophytes: a review of facultative biotrophic root colonizing fungi. New Phytol 140:295–310

Jumpponen A, Mattson KG, Trappe JM (1998) Mycorrhizal functioning of Phialocephala fortinii with Pinus contorta on glacier forefront soil: interactions with soil nitrogen and organic matter. Mycorrhiza 7(5):261–265

Kaldorf M, Koch B, Rexer KH, Kost G, Varma A (2005) Patterns of interaction between Populus Esch5 and *Piriformospora indica*: a transition from mutualism to antagonism. Plant Biol 7 (2):210–218

Kandel SL, Joubert PM, Doty SL (2017) Bacterial endophyte colonization and distribution within plants. Microorganisms 5(4):77

Kariman K, Barker SJ, Tibbett M (2018) Structural plasticity in root-fungal symbioses: diverse interactions lead to improved plant fitness. PeerJ 6:e6030

Kerry BR (2000) Rhizosphere interactions and the exploitation of microbial agents for the biological control of plant-parasitic nematodes. Annu Rev Phytopathol 38:423–441

Kimmons CA, Gwinn KD, Bernard EC (1990) Nematode reproduction on endophyte-infected and endophyte-free tall fescue. Plant Dis 74:757–761

Kithsiri Wijeratne EM, Paranagama PA, Marron MT, Gunatilaka MK, Arnold AE, Gunatilaka AAL (2008) Sesquiterpene quinines and related metabolites from *Phyllosticta spinarum*, a fungal strain endophytic in *Platycladus orientalis* in the Sonoran Desert. J Nat Prod 71:218–222

Kloepper JW, Ryu CM (2006) Bacterial endophytes as elicitors of induced systemic resistance. Microbial root endophytes. Springer, Berlin, pp 33–52

Kloepper J, Schroth MN (1978) Plant growth-promoting rhizobacteria on radishes. In: Fourth international conference on plant pathogenic Bacteria, vol II. Gibert Clarey, Tours, pp 879–882

Kloepper JW, Hallmann J, Rodriguez-Kabana R (1999) Chitin-mediated changes in bacterial communities of the soil, rhizosphere and within roots of cotton in relation to nematode control. Soil Biol Biochem 31(4):551–560

Kobayashi DY, Palumbo JD (2000) Bacterial endophytes and their effects on plants and uses in agriculture. In: Bacon CW, White JF (eds) Microbial endophytes. Dekker, New York, pp 199–236

Kpomblekou-A K, Tabatabai MA (2003) Effect of low-molecular weight organic acids on phosphorus release and phytoavailability of phosphorus in phosphate rocks added to soils. Agric Ecosyst Environ 100:275–284

Kuklinsky-Sobral J, Araújo WL, Mendes R, Geraldi IO, Pizzirani-Kleiner AA, Azevedo JL (2004) Isolation and characterization of soybean-associated bacteria and their potential for plant growth promotion. Environ Microbiol 6:1244–1251

Lambers H, Chapin FS, Pons TL (2008) Biotic influences. In: Plant physiological ecology, 2nd edn. Springer, New York, NY, pp 403–443

Landeweert R, Hoffland E, Finlay RD, Kuyper TW, van Breemen N (2001) Linking plants to rocks: ectomycorrhizal fungi mobilize nutrients from minerals. Trends Ecol Evol 16(5):248–254

Lee WC, Yoshihara M, Littleton JT (2004) Cytoplasmic aggregates trap polyglutamine-containing proteins and block axonal transport in a Drosophila model of Huntington's disease. Proc Natl Acad Sci U S A 101(9):3224–3229

Lewis GC (2004) Effects of biotic and abiotic stress on the growth of three genotypes of *Lolium perenne* with and without infection by the fungal endophyte *Neotyphodium lolii*. Ann Appl Biol 144:53–63

Ligrone R, Pocock K, Duckett JG (1993) A comparative ultrastructural study of endophytic basidiomycetes in the parasitic hepatic *Cryptothallus mirabilis* and the closely allied photosynthetic species *Aneura pinguis* (Metzgeriales). Can J Bot 71:666–679

Lyons PC (1990) Effects of the fungal Endophyte *Acremonium coenophialum* on nitrogen accumulation and metabolism in tall fescue. Plant Physiol 92:726–732

Malinowski DP, Belesky DP (2000) Adaptations of endophtye-infected cool-season grasses to environmental stresses: mechanisms of drought and mineral stress tolerance. Crop Sci 40:923–940

Malinowski DP, Brauer DK, Belesky DP (1999) The endophyte *Neotyphodium coenophialum* affects root morphology of tall fescue grown under phosphorus deficiency. J Agron Crop Sci 183:53–60

Mandyam K, Jumpponen A (2005) Seeking the elusive function of the root-colonising dark septate endophytic fungi. Stud Mycol 53:173–189

Marler M, Pedersen D, Mitchell OT, Callaway RM (1999) A polymerase chain reaction method for detecting dwarf mistletoe infection in Douglas fir and western larch. Can J For Res 29:1317–1321

Millaleo R, Reyes-Díaz M, Ivanov AG, Mora ML, Alberdi M (2010) Manganese as essential and toxic element for plants: transport, accumulation and resistance mechanisms. J Soil Sci Plant Nut 10(4):470–481

Miller JD, Mackenzie S, Foto M, Adams GW, Findlay JA (2002) Needles of white spruce inoculated with rugulosin-producing endophytes contain rugulosin reducing spruce budworm growth rate. Mycol Res 106:471–479

Monnet F, Vaillant N, Hitmi A, Coudret A, Sallanon H (2001) Endophytic *Neotyphodium lolii* induced tolerance to Zn stress in *Lolium perenne*. Physiol Plant 113:557–563

Mucciarelli M, Scannerini S, Bertea C, Maffei M (2003) In vitro and in vivo peppermint (*Mentha piperita*) growth promotion by nonmycorrhizal fungal colonization. New Phytol 158:579–591

Mushin TM, Booth T, Zwain KH (1989) A fungal endophyte associated with desert parasitic plant. Kavaka Trans Mycol Soc India 17:1–5

Narisawa K, Kawamata H, Currah RS, Hashiba T (2002) Suppression of Verticillium wilt in eggplant by some fungal root endophytes. Eur J Plant Pathol 108:103–109

Omacini M, Chaneton EJ, Ghersa CM, Müller CB (2001) Symbiotic fungal endophytes control insect host-parasite interaction webs. Nature 409:78–81

Patten CL, Glick BR (2002) Role of *Pseudomonas putida* indoleacetic acid in development of the host plant root system. Appl Environ Microbiol 68:3795–3801

Patterson CG, Potter DA, Fanin FF (1991) Feeding deterrence of alkaloids from endophyte-infected grasses to Japanese beetle grubs. Entomol Exp Appl 61:285–289

Petrini O (1991) Fungal endophytes of tree leaves. In: Andrews JH, Hirano SS (eds) Microbial ecology of leaves. Springer, New York, pp 179–197

Ping L, Boland W (2004) Signals from the underground: bacterial volatiles promote growth in Arabidopsis. Trends Plant Sci 9:263–266

Pirttilä AM, Joensuu P, Pospiech H, Jalonen J, Hohtola A (2004) Bud endophytes of scots pine produce adenine derivatives and other compounds that affect morphology and mitigate browning of callus cultures. Physiol Plant 121:305–312

Ramesh R, Joshi AA, Ghanekar MP (2008) Pseudomonads: major antagonistic endophytic bacteria to suppress bacterial wilt pathogen, Ralstonia solanacearum in the eggplant (Solanum melongena L.). World J Microbiol Biotechnol 25(1):47–55

Raps A, Vidal S (1998) Indirect effects of an unspecialized endophytic fungus on specialized plant – herbivorous insect interactions. Oecologia 114:541–547

Redman RS, Sheehan KB, Stout RG, Rodrigues RJ, Henson JM (2002) Thermotolerance conferred to plant host and fungal endophyte during mutualistic symbiosis. Science 298:1581

Reid RJ, Hayes JE, Post A, Stangoulis JCR, Graham RD (2004) A critical analysis of the causes of boron toxicity in plants. Plant Cell Environ 27:1405–1414

Reis VM, Baldani JI, Baldani VLD, Dobereiner J (2000) Biological N_2 fixation in gramineae and palm trees. Crit Rev Plant Sci 19:227–247

Riedell WE, Kieckhefer RE, Petroski RJ, Powell RG (1991) Naturally occurring and synthetic loline alkaloid derivatives: insect feeding behavior modification and toxicity. J Entomol Sci 26:122–129

Rodriguez RJ, Henson J, Van Volkenburgh E, Hoy M, Wright L, Beckwith F, Kim Y, Redman RS (2008) Stress tolerance in plants via habitat-adapted symbiosis. Int Soc Microbial Ecol 2:404–416

Rodriguez RJ, White JF, Arnold AE, Redman RS (2009) Fungal endophytes: diversity and functional roles. New Phytol 182:314–330

Rouws LF, Meneses CH, Guedes HV, Vidal MS, Baldani JI, Schwab S (2010) Monitoring the colonization of sugarcane and rice plants by the endophytic diazotrophic bacterium *Gluconacetobacter diazotrophicus* marked with gfp and gusA reporter genes. Lett Appl Microbiol 51:325–330

Ryan MH, Graham JH (2002) Is there a role for arbuscular mycorrhizal fungi in production agriculture? Plant Soil 244(1):263–271

Santoyo G, Moreno-Hagelsieb G, del Carmen Orozco-Mosqueda M, Glick BR (2016) Plant growth-promoting bacterial endophytes. Microbiol Res 183:92–99

Schmid E, Oberwinkler F (1993) Mycorrhiza-like interaction between the achlorophyllous gametophyte of *Lycopodium clavatum* L. and its fungal endophyte studied by light and electron microscopy. New Phytol 124:69–81

Schulz BJE (2006) Mutualistic interactions with fungal root endophytes. In: Schulz BJE, Boyle CJC, Sieber TN (eds) Microbial root endophytes. Springer, Berlin, pp 261–280

Shah J (2009) Plants under attack: systemic signals in defence. Curr Opin Plant Biol 12:459–464

Sherwood M, Carroll G (1974) Fungal succession on needles and young twigs of old-growth Douglas fir. Mycologia 66:499–506

Siegel MR, Schardl CL (1991) Fungal endophytes of grasses: detrimental and beneficial associations. In: Andrews JH, Hirano SS (eds) Microbial ecology of leaves. Springer, New York, pp 198–221

Stone JK, Bacon CW, White JF (2000) In: Bacon CW, White JF Jr (eds) An overview of Endophytic microbes: Endophytism defined. M. Dekker, Inc, New York, pp 3–29

Stone JK, Polishook JD, White JRJ (2004) Endophytic fungi. In: Mueller G, Bills GF, Foster MS (eds) Biodiversity of fungi: inventory and monitoring methods. Elsevier, Burlington, MA, pp 241–270

Strobel GA (2003) Endophytes as sources of bioactive products. Microbes Infect 5:535–544

Sturz AV, Christie BR, Nowak J (2000) Bacterial endophytes: potential role in developing sustainable systems of crop production. CRC Crit Rev Plant Sci 19:1–30

Suryanarayanan TS, Wittlinger SK, Faeth SH (2005) Endophytic fungi associated with cacti in Arizona. Mycol Res 109:635–639

Susan DC (2004) Endophytic-host selectivity of *Disculaumbrinella* on *Quercus alba* and *Quercus rubra* characterized by infection, pathogenicity and mycelial compatibility. Eur J Plant Pathol 110:713–721

Sziderics AH, Rasche F, Trognitz F, Sessitsch A, Wilhelm E (2007) Bacterial endophytes contribute to abiotic stress adaptation in pepper plants (*Capsicum annuum*L). Can J Microbiol 53:1195–1202

Taghavi S, Garafola C, Monchy S, Newman L, Hoffman A, Weyens N, Barac T, Vangronsveld J, van der Lelie D (2009) Genome survey and characterization of endophytic bacteria exhibiting a beneficial effect on growth and development of poplar trees. Appl Environ Microbiol 75:748–757

Thakore Y (2006) The biopesticide market for global agricultural use. Ind Biotechnol 2:194–208

Tudzynski B, Sharon A (2002) Biosynthesis, biological role and application of fungal phytohormones. In: Osiewacz HD (ed) The mycota X. industrial applications. Springer, Berlin, pp 183–212

van Loon LC, Bakker PAHM, Pieterse CMJ (1998) Systemic resistance induced by rhizosphere bacteria. Annu Rev Phytopathol 36:453–483

Verma SC, Ladha JK, Tripathi AK (2001) Evaluation of plant growth promotion and colonization ability of endophytic diazotrophs from deep water rice. J Biotechnol 91:127–141

Vishwakarma K, Sharma S, Kumar V, Upadhyay N, Kumar N, Mishra R, Yadav G, Verma RK, Tripathi DK (2017a) Current scenario of root exudate–mediated plant-microbe interaction and promotion of plant growth. In: Probiotics in agroecosystem. Springer, Singapore, pp 349–369

Vishwakarma K, Mishra M, Jain S, Singh J, Upadhyay N, Verma RK, Verma P, Tripathi DK, Kumar V, Mishra R, Sharma S (2017b) Exploring the role of plant-microbe interactions in improving soil structure and function through root exudation: a key to sustainable agriculture. In: Plant-microbe interactions in agro-ecological perspectives. Springer, Singapore, pp 467–487

Vishwakarma K, Kumar V, Tripathi DK, Sharma S (2018) Characterization of rhizobacterial isolates from *Brassica juncea* for multitrait plant growth promotion and their viability studies on carriers. Env Sustain 1(3):253–265

Vu T, Hauschild R, Sikora RA (2006) *Fusarium oxysporum* endophytes induced systemic resistance against *Radopholus similis* on banana. Nematology 8:847–852

Wakelin SA, Warren RA, Harvey PR, Ryder MH (2004) Phosphate solubilization by *Penicillium spp.* closely associated with wheat roots. Biol Fertil Soil 40:36–43

Waller F, Achatz B, Baltruschat H, Fodor J, Becker K, Fischer M, Heier T, Hückelhoven R, Neumann C, Wettstein D, Franken P, Kogel KH (2005) The endophytic fungus *Piriformospora indica* reprograms barley to salt-stress tolerance, disease resistance, and higher yield. Proc Natl Acad Sci U S A 102(38):13386–13391

White PJ, Broadley MR (2001) Chloride in soils and its uptake and movement within the plant: a review. Ann Bot 88(6):967–988

White PJ, Brown PH (2010) Plant nutrition for sustainable development and global health. Annal Bot 105(7):1073–1080

Wilkinson HH, Siegel MR, Blankenship JD, Mallory AC, Bush LP, Schardl CL (2000) Contribution of fungal loline alkaloids to protection from aphids in a grass-endophyte mutualism. Mol Plant-Microbe Interact 13:1027–1033

Wimmer MA, Lochnit G, Bassil E, Mühling KH, Goldbach HE (2009) Membrane-associated, boron-interacting proteins isolated by boronate affinity chromatography. Plant Cell Physiol 50:1292–1304

Wright DP, Scholes JD, Read DJ, Rolfe SA (2005) European and African maize cultivars differ in their physiological and molecular responses to mycorrhizal infection. New Phytol 167 (3):881–896

Yazdani M, Bahmanyar M (2009) Effect of phosphate solubilization microorganisms (PSM) and plant growth promoting rhizobacteria (PGPR) on yield and yield components of corn (*Zea mays* L.). World Acad Sci Eng Technol 49:90–92

Yue C, Miller CJ, White JFJ, Richardson M (2000) Isolation and characterization of fungal inhibitors from *Epichloë festucae*. Journal Agricultural and Food Chemistry 48:4687–4692

Zakria M, Njoloma J, Saeki Y, Akao S (2007) Colonization and nitrogen-fixing ability of *Herbaspirillum* sp. Strain B501 gfp1 and assessment of its growth-promoting ability in cultivated rice. Microbiol Environ 22:197–206

Zehirov GT, Georgiev GI (2003) Effects of boron starvation on the apoplastic and total solute concentrations influencing nodule growth and acetylene reduction rate. Bulg J Plant Physiol 29:367–373

Zhang HW, Song YC, Tan RX (2006) Biology and chemistry of endophytes. Nat Prod Rep 23:753–771

Chapter 5
Dual and Tripartite Symbiosis of Invasive Woody Plants

Robin Wilgan

Abstract The invasion of alien tree species is one of the most crucial threats for biological conservation and the maintenance of biodiversity. Invasion of alien tree species can change local, continental and global biodiversity, nutrient cycles and ecosystem services and transform native habitats into novel ecosystems determined by invaders. The overwhelming majority of tree species are associated with mutualistic symbionts: arbuscular mycorrhizal fungi, ectomycorrhizal fungi, and/or N-fixing bacteria. These mutualistic symbionts play a key role in the proper development and functioning of trees, being able to profoundly influence the invasiveness of alien tree species, and even make this invasion possible. That kind of key interaction among invasive tree species and its mutualistic symbionts, what is commonly named as "co-invasion," is well known for alien ectomycorrhizal tree species and its fungal symbionts, e.g., Pinaceae species and suilloid fungi in the Southern Hemisphere.

This chapter is dedicated to the dual mycorrhizal and tripartite symbiotic relationships of alien and in particular invasive tree species. The ability to establish dual mycorrhizal symbiosis with both ectomycorrhizal (ECM) and arbuscular mycorrhizal (AM) fungi and tripartite symbiosis between N-fixing bacteria, mycorrhizal fungi (ECM and/or AM) and actinorhizal or legume trees, seems to support the global expansion of alien tree species. The main assets of dual and tripartite symbiosis for alien and invasive tree species are in seedling development, nutrient acquisition in different and unfavorable habitat conditions, so tent to naturalization outside its native range and relatively high tolerance to shift in climate conditions, such as soil temperature and large fluctuations of the water table.

The global climate changes may significantly switch the global distribution of trees, both native and invasive species, being favorable to tree species of multiple symbiotic associations. Moreover, the climate changes can increase the distribution of native tree species, thus release the ecological niches for other species, including invasive and potentially invasive tree species. Understanding the multipartite

R. Wilgan (✉)
Institute of Dendrology, Polish Academy of Sciences, Kórnik, Poland
e-mail: rwilgan@man.poznan.pl

N. Shrivastava et al. (eds.), *Symbiotic Soil Microorganisms*, Soil Biology 60,
https://doi.org/10.1007/978-3-030-51916-2_5

relationships between arbuscular fungi, ectomycorrhizal fungi, N-fixing bacteria, and invasive tree species can allow us to better predict and counteract the future invasions of alien tree species, what is necessary to maintain biodiversity and preserve native ecosystems.

Keywords Forest ecosystem · Alien tree species · Biological invasions · Arbuscular mycorrhiza · Ectomycorrhiza · Nitrogen-fixing bacteria

5.1 Introduction

The introduction of alien, i.e., non-native plant species has become a global phenomenon over the last centuries. It was resulting in the cultivation of alien tree species as an economically important branch of forestry in many countries, but at the same time, the widespread introduction of alien tree species brought the threat of its invasion worldwide. Trees, as ecosystem engineers playing the fundamental role in the functioning of the most terrestrial ecosystems and biomes, are having a severe impact on the biosphere (Barrios et al. 2017; Henry et al. 2017). Thus the expansion of invasive tree species (invaders) is important issue of the maintenance of biodiversity. The invasion of alien trees can drive the local, continental, and global-scale changes in biodiversity, nutrient cycles, and ecosystem services (Dickie et al. 2011, 2017; Nuñez et al. 2017). By modifying the biotic and abiotic conditions in local habitats, invasion of alien trees leading to habitat fragmentation and decline in the ecological niches for native, and in particular rare and endangered species and makes a "novel ecosystems", determined by invaders (Nuñez et al. 2017; Haddad et al. 2015; Morse et al. 2014). The abovementioned features make the tree invasions one of the greatest threats to global biodiversity.

The first studies of invasive plant species have been focused on the relationship between invaders and pathogenic organisms and indicates that the invasive species success is caused by the lack of their natural enemies, parasites, and pathogens in the habitat they are introduced to (enemy release hypothesis; Keane and Crawley 2002; Reinhart et al. 2003, Reinhart and Callaway 2004; Callaway et al. 2011). The role of symbiotic organisms, e.g., mycorrhizal fungi and N-fixing bacteria in the process of plant invasion has been studied since a few decades (Richardson et al. 2000; Traveset and Richardson 2014), revealed the mutualistic symbionts to be a driver of alien woody species invasion (Nuñez and Dickie 2014; Nuñez et al. 2017; Policelli et al. 2019). It follows, that plant-symbiont interaction may play a key role in the global management of invasive tree species (Dickie et al. 2017).

5.1.1 Mutualistic Symbiotic Relationship of Tree Species

The mycorrhizal symbiosis is thought to be established by almost all plant species, including numerous tree species (Brundrett, Tedersoo 2018). The most common type is arbuscular mycorrhizae, the relationship between Glomeromycota fungi and about 80% of known plant species. The ectomycorrhizal symbiosis is established by Ascomycota and Basidiomycota fungi (e.g., edible species of *Tuber*, *Boletus*, *Cantharellus*), and, in most cases, tree species such as pines, oaks, beech or eucalyptus. Dual mycorrhizal symbiosis (dual symbiosis) is considered as the plant species ability to establish both ectomycorrhizal and arbuscular mycorrhizal relationships (Fig. 5.1), what yet have been described as a characteristic feature of Myrtaceae (*Eucalyptus*) and Salicaceae (*Populus*, *Salix*). Tripartite symbiosis is an association between three types of organisms: plants, mycorrhizal fungi (arbuscular and/or ectomycorrhizal), and N-fixing bacteria (Ryc. 1). Rhizobial bacteria are associated with legume species (Fabaceae) and *Frankia* with actinorhizal plants such as *Alnus* or *Casuarina*.

This symbiotic association with mycorrhizal fungi and N-fixing bacteria benefits the plants. It is well recognized, that mycorrhizal fungus is responsible for the efficiency of nutrient acquisition by plant roots and helping plants to tolerate the environmental stress, i.e., drought, pathogens, high concentration of heavy metals in soil (Smith and Read 2008). N-fixing bacteria, by transforming atmospheric nitrogen into fixed nitrogen being usable by plants, are being provided with nitrogen availability, allows the legume and actinorhizal plants to inhabit extremely poor habitats (Benson et al. 2004; Rascio and La Rocca 2008).

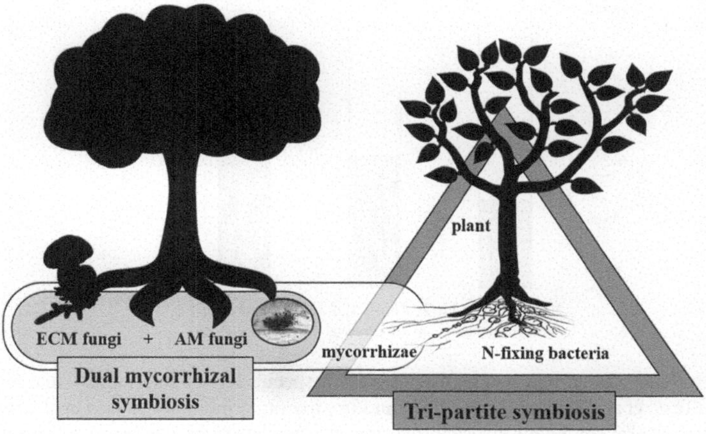

Fig. 5.1 The scheme of the multipartite symbiotic association of tree species: dual mycorrhizal symbiosis involved both ectomycorrhizal (ECM) and arbuscular mycorrhizal (AM) fungi and tripartite symbiosis between N-fixing bacteria, mycorrhizal fungi (ECM and/or AM) and actinorhizal or legume trees

5.2 Role of Mutualistic Associations in the Invasion of Alien Tree Species

The mutualistic symbiotic relationships may profoundly influence the invasiveness of alien woody species (Nunez and Dickie 2014). The lack of appropriate symbionts can be a dispersal barrier for alien species outside its native range, like the presence of appropriate symbionts can enable and support the naturalization and invasion of alien trees (Nuñez et al. 2009; Pringle et al. 2009; Dickie et al. 2010). The "co-invasion," i.e., simultaneously invasion of both alien tree species and its mutualistic symbionts, has been well-known for ectomycorrhizal fungi and trees in the southern hemisphere, e.g., *Pinus* (Nuñez et al. 2017; Policelli et al. 2019) and *Salix* (Bogar et al. 2015) and *Eucalyptus* in tropics and subtropical zone (Sulzbacher et al. 2018; Ducousso et al. 2012). These genera belong to the families with numerous invasive species, i.e., Pinaceae (33), Salicaceae (23) and Myrtaceae (35 invasive species; Rejmánek and Richardson 2013).

In contrast, the representatives of the Fagales order which involves a huge and widely distributed families of mostly ectomycorrhizal forest tree species, are seldom observed to be invaders. A few species of Fagaceae (5) and Betulaceae (3) have yet been considered as invasive, in most cases in a narrow range of expansion (one geographical region as listed in Fig. 5.2; Rejmánek and Richardson 2013). The only exceptions are *Alnus* and *Casuarina* species, which are only widespread invaders of Fagales (3–5 of 7 regions according to Fig. 5.2), and at the same time, the symbiotic partners for n-fixing bacteria and both arbuscular and ectomycorrhizal fungi. That

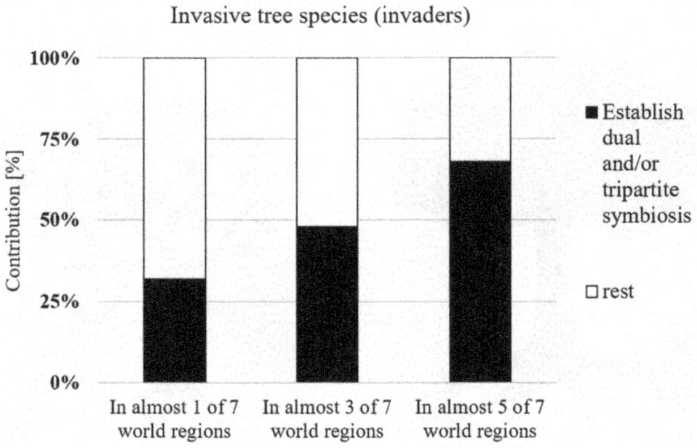

Fig. 5.2 An approximate percentage of invasive tree species known to establish dual mycorrhizal symbiosis (Teste et al. 2019; confirmed taxa) and/or tripartite symbiosis (Benson et al. 2004; Rascio and La Rocca 2008) among all invasive tree species (invaders in any of seven world regions; left bar), multi-continental (invasive in three of seven world regions; middle bar) and worldwide invaders (invasive in five of seven world regions, i.e., North America, Europe, Australasia, Asia (rest), Southern Africa, Africa (rest) with the Middle East or South America; right bar) according to Rejmánek and Richardson (2013)

relationship is rare among Fagales trees, being limited to *Alnus* and Casuarinaceae species (Benson et al. 2004).

In general, dual and/or tripartite symbiotic relationships seem to be particularly associated with widespread invasive tree species. Even up to 70% of tree species worldwide (genera, e.g., *Acacia*, *Robinia*, *Casuarina*, *Salix*, *Populus*, *Eucalyptus*, Fig. 5.2) are known to establish dual and/or tripartite symbiosis (Teste et al. 2019 (confirmed taxa only), Benson et al. 2004; Rascio and La Rocca 2008). It is around 50% among medium distributed invaders and 30% of the general number of invasive tree species (Fig. 5.2; according to Rejmánek and Richardson 2013). These data indicate that the multipartite symbiotic relations may be a determining factor for the global-scale distribution of invasive tree species.

5.3 Benefits of the Dual and Tripartite Mutualistic Relationship

The ability to establish symbiosis with arbuscular fungi may be one of the factors supporting the invasion of ectomycorrhizal tree species. Ectomycorrhizal trees that can form arbuscular symbiosis as seedlings and/or in unfavorable habitat conditions (e.g., *Salix*, *Populus*, *Eucalyptus*) are highly competitive (Brundrett and Tedersoo 2018) and more likely to become invasive. The specificity of arbuscular mycorrhizal fungi is relatively low, which is probably related to the higher naturalization success of dual and arbuscular mycorrhizal plants (Moyano et al. 2020). That appropriate arbuscular fungi are available for alien tree species (Moora et al. 2011; Majewska et al. 2015), which leads alien plants to better acclimatization outside its native range.

The legume trees are growing efficiently in symbiosis with native nitrogen-fixing bacteria and arbuscular mycorrhizal symbionts (examined for *Acacia* and *Robinia*, Rodríguez-Echeverría et al. 2009; Callaway et al. 2011). However, they are still being invasive worldwide, establishing familiar associations (sensu Dickie et al. 2017), i.e., efficient and effective relationships with local, indigenous symbionts. This ability to accept symbiotic partners from the indigenous pule of symbionts is a crucial factor of the successful acclimatization of alien tree species outside of their native range (Pringle et al. 2009; Nuñez and Dickie 2014, Dickie et al. 2017), and subsequent its naturalization (Moyano et al. 2020) and ability to being invasive.

5.3.1 Facultative Mutualistic Relationship

Facultative dual mycorrhizal symbiosis is considered to be the plant's ability to establish the different types of mycorrhizal associations in response to environmental conditions or plant age. For example, the predominance of ectomycorrhizal and presence of arbuscular structures have been examined for *Quercus rubra* seedlings inside its native range (Dickie et al. 2001) and *Carya* seedling outside its native range in Europe (Rudawska et al. 2018), where *Q. rubra* (Fagaceae, Fagales) is invasive species but *Carya* species (Juglandaceae, Fagales) are naturalized and still grown on the old European plantations abandoned for a half of a century (Paź et al. 2018; Wilgan et al. 2020). The general proportion of plant species that have naturalized outside their native range is significantly higher for plants entering the mycorrhizal symbiosis facultatively (76%) than obligatorily (52%). The obligative dual mycorrhizal plants species are in the larger part naturalized (69%) than *obligate* arbuscular (54%), *obligate* ectomycorrhizal (45%), and non-mycorrhizal (34%) plant species (Moyano et al. 2020).

Eucalyptus is one of the best-known genera of invaders among dual mycorrhizal tree species. The co-invasion of Eucalyptus and its ectomycorrhizal symbionts have been described, e.g., from Iberian Peninsula (Díez 2005), Brasil (Sulzbacher et al. 2018), China (Dell et al. 2002), Africa and Madagascar (Ducousso et al. 2012). It is considered, even more than 25% of alien ectomycorrhizal fungal species have been introduced with *Myrtaceae* (mostly Eucalyptus; Vellinga et al. 2009). However, the seedlings of *Eucalyptus* are rapidly colonized by arbuscular mycorrhizal fungi, being replaced by ectomycorrhizal fungi along with tree age (dos Santos et al. 2001; Adjoud-Sadadou and Halli-Hargas 2017), what is thought to be a major contributory factor of Eucalyptus success outside its native range (aforementioned studies; Teste et al. 2019).

The arbuscular and ectomycorrhizal fungi differ in nutrient acquisition strategy: ECM fungi are having the greater capability to release N and P from soil organic matter, but AM can transfer N after its mineralization by other organisms, e.g., N-fixing bacteria (Teste et al. 2019 and reference therein) and increase the efficiency of nitrogen uptake for plants establish tripartite symbiosis (Takács et al. 2018).

Dual mycorrhizal symbiosis (in most cases facultative dual mycorrhizal symbiosis) can be established by a significantly wider group of tree species, that it yet has been considered (Teste et al. 2019). However, it is poorly evidenced, usually once for single or a few species in genera, what raises the question, is it a typical feature of tree genera, accidental colonization or a relic of the evolutionary past.

Interestingly, the dual mycorrhizal symbiosis has been widely confirmed for 15 genera of Fabaceae tree species, e.g., 24 *Acacia* species (common tree invaders). The interactions of arbuscular and ectomycorrhizal fungi into the tripartite symbiotic relationship of invasive tree species may be an asset for worldwide invasive tree species of legume trees. Teste et al. (2019) revealed mostly neutral or positive effects of arbuscular and ectomycorrhizal fungal inoculations on the dual mycorrhizal plant species, including dual into tripartite symbiosis, e.g., *Acacia*, *Afzelia* (legume tree species) and *Alnus* (actinorhizal tree species).

Well-known dual mycorrhizal tree genera (*Populus*, *Salix*, *Alnus*, *Eucalyptus*) and tree genera of dual mycorrhizae into tripartite symbiotic relationships (numerous *Fabaceae* genera, e.g., *Acacia*) are predominant in ecosystems characterized by *la*rge fluctuations in the water level, i.e., periodic flooding and/or drought. Higher tree root colonization by arbuscular than ectomycorrhizal fungi have been observed in the soil of extreme water level, i.e., very dry and flooded. This shift may be affected by poor oxygen availability in soil, which is unfavorable for ectomycorrhizal fungi (Teste et al. 2019 and reference therein). The contribution of dead ectomycorrhizal roots (i.e., the residue of former ectomycorrhizal colonization) increases along with moisture gradient (Aučina et al. 2019) to nearly 100% in flooded hollows on pine-dominated peatland (personal observation; background described in Aučina et al. 2019). Dual mycorrhizal tree species seem to have the resilience to flooding or drought, by its ability to establish arbuscular or ectomycorrhizal symbiosis depends on the extreme (very dry, flooded) or temperate water level.

5.4 Distribution of Invasive Tree Species in the Time of Global Climate Changes

The distribution of tree species depends on the climate (Taccoen et al. 2019), thus the invasion of alien tree species can be significantly influenced by the changes in climatic conditions. The native and non-native, including invasive tree species moving into high elevation and high latitude ecosystems as a result of the climate changes (Pauchard et al. 2016; Dyderski et al. 2017). Ectomycorrhizal symbiosis is characteristic of temperature and boreal forests, while arbuscular mycorrhizal tree species are widespread in forests of the tropic and subtropic zone. Dual mycorrhizal tree species are prevalent in different climatic zones being native species (e.g., *Alnus*, *Salix*) and non-native (e.g., *Eucalyptus* in India, China, Brazil, northern Africa, south Europe). The prediction effect of climate changes on the distribution of tree species showing a high decrease of ectomycorrhizal tree species distribution in the southern range, what have not been observed for native arbuscular mycorrhizal trees (*Fraxinus excelsior*) and invasive tree species holding tripartite symbiotic relationship (*Robinia pseudoacacia*; Dyderski et al. 2017).

Teste et al. (2019) suggest, that the distribution of ectomycorrhizal tree species (mostly boreal and temperature climate) and arbuscular tree species (mostly tropics and subtropical zone; Steidinger et al. 2019) is affected by the temperature of the soil because arbuscular fungi are reduced in low soil temperatures, while ECM fungi are not (for *Alnus* species Kilpelainen et al. 2017). If it could partly explain the predominance of ECM tree species in cold climates, the same time it indicates, the expansion of alien tree species holding dual and tripartite symbiotic relationships increase in temperature and boreal zones along with a decrease of native ectomycorrhizal trees caused by climate warming.

5.5 Conclusion

The invasion of alien species is one of the crucial issues of biological conservation and the maintenance of biodiversity. In front of upcoming climate changes, the world picture of tree invasions can be significantly different, that is considered to be now. The dual and tripartite symbiotic associations seem to support the invasion of alien tree species, especially the most widespread invasive trees, which nearly all are known to establish dual and/or tripartite symbiosis.

The main assets of dual and tripartite symbiosis for alien and invasive tree species:

1. Support the development of seedlings
2. Better and usually successful naturalization outside its native range
3. Higher possibility of nutrient acquisition in different and unfavorable habitat conditions
4. Higher tolerance to shift in climate conditions (temperature of the soil, precipitation including flooding and drought)

Further research of the multipartite interactions among arbuscular fungi, ectomycorrhizal fungi, and N-fixing bacteria and alien tree species are needed for a better understanding of the relationship between invasive tree species and its symbionts. It will allow us to better predict and counteract the invasions of alien tree species invasions, what is necessary to maintain biodiversity, and preserve native ecosystems in front of upcoming climate changes.

Acknowledgment Preparation of this chapter was supported by the Institute of Dendrology, Polish Academy of Sciences.

References

Adjoud-Sadadou D, Halli-Hargas R (2017) Dual mycorrhizal symbiosis: an asset for eucalypts out of Australia? Can J For Res 47(4):500–505. https://doi.org/10.1139/cjfr-2016-0292

Aučina A, Rudawska M, Wilgan R, Janowski D, Skridaila A, Dapkūnienė S, Leski T (2019) Functional diversity of ectomycorrhizal fungal communities along a peatland–forest gradient. Pedobiologia 74:15–23. https://doi.org/10.1016/j.pedobi.2019.03.001

Barrios E, Valencia V, Jonsson M, Brauman A, Hairiah K, Mortimer P, Okubo S (2017) Contribution of trees to the conservation of biodiversity and ecosystem services in agricultural landscapes view supplementary material contribution of trees to the conservation of biodiversity and ecosystem services in agricultural landscapes. Int J Biodivers Sci Ecosyst Serv Manag 14:1–16. https://doi.org/10.1080/21513732.2017.1399

Benson DR, Vanden Heuvel BD, Potter D (2004) Actinorhizal symbioses: diversity and biogeography. In: Gillings M, Holmes A (eds) Plant microbiology. Scientific Publishers, Oxford, UK, pp 97–127

Bogar L, Dickie IA, Kennedy P (2015) Testing the co-invasion hypothesis: ectomycorrhizal fungal communities on *Alnus glutinosa* and *Salix fragilis* in New Zealand. Divers Distrib 21 (3):268–278. https://doi.org/10.1111/ddi.12304

Brundrett MC, Tedersoo L (2018) Evolutionary history of mycorrhizal symbioses and global host plant diversity. New Phytol 220:1108–1115. https://doi.org/10.1111/nph.14976

Callaway RM, Bedmar EJ, Reinhart KO, Silvan CG, Klironomos J (2011) Effects of soil biota from different ranges on Robinia invasion: acquiring mutualists and escaping pathogens. Ecology 92:1027–1103

Dell B, Malajczuk N, Dunstan WA (2002) Persistence of some Australian Pisolithus species introduced into eucalypt plantations in China. For Ecol Manag 169:271–281

Dickie IA, Koide RT, Fayish AC (2001) Vesicular-arbuscular mycorrhizal infection of *Quercus rubra* seedlings. New Phytol 151:257–264

Dickie IA, Bolstridge N, Cooper JA, Peltzer DA, Duane A (2010) Co-invasion by Pinus and its mycorrhizal fungi. New Phytol 187:475–484. https://doi.org/10.1111/j.1469-8137.2010.03277.x

Dickie IA, Yeates GW, St John MG, Stevenson BA, Scott JT, Rillig MC, Peltzer DA, Orwin KH, Kirschbaum MUF, Hunt JE et al (2011) Ecosystem service and biodiversity trade-offs in two woody successions. J Appl Ecol 48:926–934

Dickie IA, Bufford JL, Cobb RC, Desprez-Loustau ML, Grelet G, Hulme PE, Klironomos J, Makiola A, Nuñez MA, Pringle A, Thrall PH, Tourtellot SG, Waller L, Williams NM (2017) The emerging science of linked plant–fungal invasions. New Phytol 215(4):1314–1332

Díez J (2005) Invasion biology of Australian ectomycorrhizal fungi introduced with eucalypt plantations into the Iberian Peninsula. Biol Invasions 7(1):3–15

dos Santos VL, Muchovej RM, Borges AC, Neves JCL, Kasuya MCM (2001) Vesicular-arbuscular-/ecto-mycorrhiza succession in seedlings of Eucalyptus spp. Braz J Microbiol 32 (2):81–86. https://doi.org/10.1590/S1517-83822001000200002

Ducousso M, Duponnois R, Thoen D, Prin Y (2012) Diversity of ectomycorrhizal fungi associated with Eucalyptus in Africa and Madagascar. Int J For Res 2012:450715. https://doi.org/10.1155/2012/450715

Dyderski MK, Paź S, Frelich LE, Jagodziński AM (2017) How much does climate change threaten European forest tree species distributions? Glob Chang Biol 24(3):1150–1163. https://doi.org/10.1111/gcb.13925

Haddad NM, Brudvig LA, Clobert J, Davies KF, Gonzalez A, Holt RD, Lovejoy TE, Sexton JO, Austin MP, Collins CD, Cook WM, Damschen EI, Ewers RM, Foster BL, Jenkins CN, King AJ, Laurance WF, Levey DJ, Margules CR, Melbourne BA, Nicholls AO, Orrock JL, Song D-X, Townshed JR (2015) Habitat fragmentation and its lasting impact on Earth's ecosystems. Sci Adv 1(2):e1500052. https://doi.org/10.1126/sciadv.1500052

Henry RC, Palmer SCF, Watts K, Mitchell R, Atkinson N, Travis JM (2017) Tree loss impacts on ecological connectivity: developing models for assessment. Eco Inform 42:90–99

Keane RM, Crawley MJ (2002) Exotic plant invasions and the enemy release hypothesis. Trends Ecol Evol 17:164–170. https://doi.org/10.1016/S0169-5347(02)02499-0

Kilpelainen J, Barbero-Lopez A, Vestberg M, Heiskanen J, Lehto T (2017) Does severe soil drought have after-effects on arbuscular and ectomycorrhizal root colonisation and plant nutrition? Plant Soil 418:377–386

Majewska ML, Błaszkowski J, Nobis M, Rola K, Nobis A, Łakomiec D, Czachura P, Zubek S (2015) Root-inhabiting fungi in alien plant species in relation to invasion status and soil chemical properties. Symbiosis 65:101–115

Moora M, Berger S, Davison J, Öpik M, Bommarco R, Bruelheide H, Kühn I, Kunin WE, Metsis M, Rortais A, Vanatoa A, Vanatoa E, Stout JC, Truusa M, Westphal C, Zobel M, Walther G-R (2011) Alien plants associate with widespread generalist arbuscular mycorrhizal fungal taxa: evidence from a continental-scale study using massively parallel 454 sequencing. J Biogeogr 38:1305–1317

Morse N, Pellissier P, Cianciola E, Brereton R, Sullivan M, Shonka N, Wheeler T, McDowell W (2014) Novel ecosystems in the Anthropocene: a revision of the novel ecosystem concept for pragmatic applications. Ecol Soc 19(2):12. https://doi.org/10.5751/ES-06192-190212

Moyano J, Dickie I, Rodriguez-Cabal MA, Nuñez MA (2020) Patterns of plant naturalization show that facultative mycorrhizal plants are more likely to succeed outside their native Eurasian ranges. Ecography 43:648–659
Nuñez MA, Dickie IA (2014) Invasive belowground mutualists of woody plants. Biol Invasions 16:645–661
Nuñez MA, Horton TR, Simberloff D (2009) Lack of belowground mutualisms hinders Pinaceae invasions. Ecology 90:2352–2359. https://doi.org/10.1890/08-2139.1
Nuñez MA, Chiuffo MC, Dimarco RD, García RA, Moyano J, Pauchard A, Paul T, Policelli N, Raal P, Richardson DM, Torres A, van Wilgen BW (2017) Ecology and management of invasive Pinaceae around the world: progress and challenges. Biol Invasions 19(11):3099–3120
Pauchard A, Milbau A, Albihn A, Alexander J, Nun MA, Daehler C, Englund G, Essl F, Evengard B, Greenwood GBM, Haider S, Lenoir J, McDougall K, Muths E, Nuñez MA, Olofsson J, Pellissier L, Rabitsch W, Rew LJ, Robertson M, Sanders N, Kueffer C (2016) Non-native and native organisms moving into high elevation and high latitude ecosystems in an era of climate change: new challenges for ecology and conservation. Biol Invasions 18:345–353. https://doi.org/10.1007/s10530-015-1025-x
Paź S, Czapiewska N, Dyderski MK, Jagodziński AM (2018) Assessment of Carya ovata (Mill.) K. Koch introduction to the potential oak–hornbeam forest habitat in the Czerniejewo Forest District [in polish] Sylwan 162(1):41–48
Policelli N, Bruns TD, Vilgalys R, Nuñez MA (2019) Suilloid fungi as global drivers of pine invasions. New Phytol 222(2):714–725
Pringle A, Bever JD, Gardes M, Parrent JL, Rillig MC, Klironomos JN (2009) Mycorrhizal symbioses and plant invasions. Annu Rev Ecol Evol Syst 40:699–715. https://doi.org/10.1146/annurevecolsys.39.110707.173454
Rascio N, La Rocca N (2008) Biological nitrogen fixation. In: Encyclopedia of ecology, pp 412–419. https://doi.org/10.1016/B978-008045405-400273-1
Reinhart KO, Callaway RM (2004) Soil biota facilitate exotic Acer invasions in Europe and North America. Ecol Appl 14:1737–1745
Reinhart KO, Packer A, Van der Putten WH, Clay K (2003) Plant-soil biota interactions and spatial distribution of black cherry in its native and invasive ranges. Ecol Lett 6:1046–1050
Rejmánek M, Richardson DM (2013) Trees and shrubs as invasive alien species—2013 update of the global database. Divers Distrib 19(8):1093–1094
Richardson DM, Allisopp N, D'Antonio CM, Milton SJ, Rejmanek M (2000) Plant invasions-the role of mutualisms. Biol Rev 75:65–93
Rodríguez-Echeverría S, Crisóstomo JA, Nabais C, Freitas H (2009) Belowground mutualists and the invasive ability of Acacia longifolia in coastal dunes of Portugal. Biol Invasion 11 (3):651–661
Rudawska M, Leski T, Wilgan R, Karliński L, Kujawska M, Janowski D (2018) Mycorrhizal associations of the exotic hickory trees, Carya laciniosa and Carya cordiformis, grown in Kórnik Arboretum in Poland. Mycorrhiza 28:549–560
Smith SE, Read DJ (2008) Mycorrhizal symbiosis. Elsevier, Oxford
Steidinger BS, Crowther TW, Liang J, Van Nuland ME, GDA W, Reich PB, Nabuurs G, de-Miguel S, Zhou M, Picard N, Herault B, Zhao X, Zhang C, Routh D, Abegg M, Adou Yao CY, Alberti G, Almeyda Zambrano A, Alvarez-Davila E, Alvarez-Loayza P, Alves LF, Ch A, Antón-Fernández C, Araujo-Murakami A, Arroyo L, Avitabile V, Aymard G, Baker T, Bałazy R, Banki O, Barroso J, Bastian M, Bastin J-F, Birigazzi L, Birnbaum P, Bitariho R, Boeckx P, Bongers F, Bouriaud O, PHS B, Brandl S, Brearley FQ, Brienen R, Broadbent E, Bruelheide H, Bussotti F, Cazzolla Gatti R, Cesar R, Cesljar G, Chazdon R, HYH C, Chisholm C, Cienciala E, Clark CJ, Clark D, Colletta G, Condit R, Coomes D, Cornejo Valverde F, Corral-Rivas JJ, Crim P, Cumming J, Dayanandan S, de Gasper AL, Decuyper M, Derroire G, DeVries B, Djordjevic I, Iêda A, Dourdain A, Engone Obiang NL, Enquist B, Eyre T, Fandohan AB, Fayle TM, Feldpausch TR, Finér L, Fischer M, Fletcher C, Fridman J, Frizzera L, JGP G, Gianelle D, Glick HB, Harris D, Hector A, Hemp A,

Hengeveld G, Herbohn J, Herold M, Hillers A, Honorio Coronado EN, Huber M, Hui C, Cho H, Ibanez T, Jung I, Imai N, Jagodzinski AM, Jaroszewicz B, Johannsen V, Joly CA, Jucker T, Karminov V, Kartawinata K, Kearsley E, Kenfack D, Kennard D, Kepfer-Rojas S, Keppel G, Khan ML, Killeen T, Kim HS, Kitayama K, Köhl M, Korjus H, Kraxner F, Laarmann D, Lang M, Lewis S, Lu H, Lukina N, Maitner B, Malhi Y, Marcon E, Marimon BS, Marimon-Junior BH, Marshall AR, Martin E, Martynenko O, Meave JA, Melo-Cruz O, Mendoza C, Merow C, Mendoza AM, Moreno V, Mukul SA, Mundhenk P, Nava-Miranda MG, Neill D, Neldner V, Nevenic R, Ngugi M, Niklaus P, Oleksyn J, Ontikov P, Ortiz-Malavasi E, Pan Y, Paquette A, Parada-Gutierrez A, Parfenova E, Park M, Parren M, Parthasarathy N, Peri PL, Pfautsch S, Phillips O, Piedade MT, Piotto D, NCA P, Polo I, Poorter L, Poulsen AD, Poulsen JR, Pretzsch H, Ramirez Arevalo F, Restrepo-Correa Z, Rodeghiero M, Rolim S, Roopsind A, Rovero F, Rutishauser E, Saikia P, Saner P, Schall P, Schelhaas M-J, Schepaschenko D, Scherer-Lorenzen M, Schmid B, Schöngart J, Searle E, Seben V, Serra-Diaz JM, Salas-Eljatib CH, Sheil D, Shvidenko A, Silva-Espejo J, Silveira M, Singh J, Sist P, Slik F, Sonké B, Souza AF, Stereńczak K, J-CH S, Svoboda M, Targhetta N, Tchebakova N, ter Steege H, Thomas R, Tikhonova E, Umunay P, Usoltsev V, Valladares F, van der Plas F, Van Do T, Vasquez Martinez R, Verbeeck H, Viana H, Vieira S, von Gadow K, Wang H-F, Watson J, Westerlund B, Wiser S, Wittmann F, Wortel V, Zagt R, Zawila-Niedzwiecki T, Zhu Z-X, Zo-Bi IC, Peay KG (2019) Climatic controls of decomposition drive the global biogeography of forest-tree symbioses. Nature 569:404–408. https://doi.org/10.1038/s41586-019-1128-0

Sulzbacher MA, Grebenc T, Bevilacqua CB, Steffen RB, Coelho G, Silveira AO, Jacques RJ, Antoniolli ZI (2018) Co-invasion of ectomycorrhizal fungi in the Brazilian Pampa biome. Appl Soil Ecol 130:194–201. https://doi.org/10.1016/j.apsoil.2018.06.007

Taccoen A, Piedallu C, Seynave I, Perez V, Gégout-Petit A, Nageleisen L-M, Bontemps J-D, Gégout JC (2019) Background mortality drivers of European tree species: climate change matters. Proc R Soc B Biol Sci 286(1900):20190386. https://doi.org/10.1098/rspb.2019.0386

Takács T, Cseresnyés I, Kovács R, Parádi I, Kelemen B, Szili-Kovács T, Füzy A (2018) Symbiotic effectivity of dual and tripartite associations on soybean (Glycine max L. Merr.) cultivars inoculated with Bradyrhizobium japonicum and AM fungi. Front Plant Sci 9:1–14. https://doi.org/10.3389/fpls.2018.01631

Teste FP, Jones MD, Dickie IA (2019) Dual-mycorrhizal plants: their ecology and relevance. New Phytol 225(5):2020. https://doi.org/10.1111/nph.16190

Traveset A, Richardson DM (2014) Mutualistic interactions and biological invasions. Annu Rev Ecol Evol Syst 45:89–113

Vellinga EC, Wolfe BE, Pringle A (2009) Global patterns of ectomycorrhizal introductions. New Phytol 181:960–973

Wilgan R, Leski T, Kujawska M, Karliński L, Janowski D, Rudawska M (2020) Ectomycorrhizal fungi of exotic Carya ovata (Mill.) K. Koch in the context of surrounding native forests on Central European sites. Fungal Ecol 44:100908. https://doi.org/10.1016/j.funeco

Chapter 6
Eco-friendly Association of Plants and Actinomycetes

Saraswathy Nagendran, Surendra S. Agrawal, and Aryaman Girish Patwardhan

Abstract Actinomycetes are filamentous Gram-positive spore-forming largest dominant microbial population present in the soil. They are free-living rhizosphere colonizing bacteria and producers of bioactive metabolites which helps in improving the fertility of the soil, promote plant growth and development, provide biocontrol action against phytopathogens, and have the ability to withstand various environmental stress. Entophytic Actinobacteria are characterized as those that are contained within the internal structure of plants, making no obvious changes to their hosts. Entophytic actinobacteria consitute a huge part of the rhizosphere. The symbiotic association of Actinomycetes as endophytes have gained more importance because they are considered to be reservoir for potential novel bioactive compounds which finds in important applications in pharmaceutical and agricultural sectors. A notable significant feature of actinobacteria is its ability not to contaminate the environment, take active participation in pesticide degradation, phosphate solubilization, siderophores production, and nitrogen fixation. Microbial resource possesses a wide variety of plant growth potential thereby benefiting green and sustainable agriculture.

Keywords Actinomycetes · Actinobacteria · Rhizosphere · Endophytes · Bioactive metabolites · Streptomycetes · Frankia

6.1 Introduction

Actinobacteria are widely found in nature and positionedin the biggest scientific categorization inside the Bacterial domain (Kuffner et al. 2010). Actinomycetes is an important group of spore-forming rhizosphere colonizing microorganism which shows its presence in the soil, have an eco-friendly association with the

S. Nagendran (✉) · S. S. Agrawal · A. G. Patwardhan
Shobhaben Pratapbhai Patel School of Pharmacy and Technology Management, SVKM's NMIMS, Mumbai, India

N. Shrivastava et al. (eds.), *Symbiotic Soil Microorganisms*, Soil Biology 60,
https://doi.org/10.1007/978-3-030-51916-2_6

plants and acts as inhibitors against different kinds of pathogenic bacteria and fungi. These inhibitions are by producing bioactive metabolites or compounds which are harmful and toxic to disease causing pathogens, without having any harmful effects on the environment [2-3]. (Zucchi et al. 2008; Costa et al. 2013). Actinomycetes, as a biocontrol agent, play an important role in affording protection to the plants against diseases. It also aids in production plant growth promoters, participation in crop production, maintaining the integrity and protection of the ecosystem. It also plays an significant role in nitrogen fixation as it has its own commitment in cycling of nutrient, decomposition of organic matter and degradation of compounds that pollute the environment (Bhatti et al. 2017; Doumbou et al. 2001). In soil microflora association, a significant amount is engrossed by Actinomycetes. Among Actinomycetes, Streptomyces are found lavishly in the soil as an essential root colonizer when compared to genera like Nocardia, Micromonospora and Streptosporangium which are commonly present in lesser amount in comparison to Actinomycetes (Pemila Edith Chitraselvi 2018; Kamal et al. 2014). Streptomycetes as a dominant genus in the midst of Actinomycetes, has gained more importance toward their participation in production of bioactive compounds, antibiotics, and extracellular enzymes which encompasses the biocontrol and plant growth-promoting activities (Kamal et al. 2014; Olanrewaju and Babalola 2019a). Endophytic Actinobacteria are characterized by those that are contained within the internal structure of plants and making no obvious changes to their hosts. These Actinobacteria assume explicit jobs, for example, ensuring the host plants against insects and diseases. Endophytic Actinobacteria establish a huge part of the rhizosphere, that are found inside plants in which the broadly examined species are from the variety Frankia (Anandan et al. 2016). Actinorhizal symbiosis is the mutually benefiting relationship between actinobacterial species of genus Frankia and plants which can form association with these by means of nitrogen-fixing nodules present in their roots (Anandan et al. 2016; Wall 2000). The word "actinorhiza" is derived from both the filamentous bacteria Frankia which is an actinomycete and the root location of the nitrogen-fixing nodules (Wall 2000).

6.2 Actinomyces as Soil Dwellers

Actinomycetes derived from the Greek word "actis" meaning ray and "mykes" meaning fungi due to the fact that they possess the features of both fungi and bacteria (Das et al. 2008). Actinomycetes are aerobic, spore-producing bacteria filamentous Gram-positive microorganisms (Flores-Gallegos and Nava-Reyna 2019), with high G+C content Genome (Singh and Dubey 2018; Vivas et al. 2003), described by a highly intricate life cycle assuming a position within the phylum Actinobacteria, in the domain of bacteria which are included in one of the biggest ordered units among the 18 significant heredities. (Flores-Gallegos and Nava-Reyna 2019). Actinomycetes are one of the most widely observed group of soil microorganisms and surely understood for their antimicrobials generation to control the microorganisms. They are well researched for their job in natural control of plant pathogens, cooperation with plants, and their growth (Bhatti et al. 2017). Actinomycetes show their presence in various habitat in natural environments on the earth covering wide range of terrestrial and aquatic ecosystems. Strive to establish themselves under wide variety of environmental circumstances, as aerobic, anaerobic as well at temperatures ranging 5–7 °C and 45–70 °C (Bhatti et al. 2017). They are widely reported to show their establishment in soil, silt present in the water reservoirs, air as well in plant remains. As a primary inhabitant Actinomycetes are found sufficiently in soil compared to other microflora, their presence in the soil in the form of hyphal thread-like filament identical to fungi gives a peculiar "earthy" smell indicating the newly formed fresh and healthy soil (Bhatti et al. 2017; George et al. 2012; Srinivasan et al. 1991; Kuster 1968; Rowbotham and Cross 1977). Actinomycetes classification includes ten genera as follows; Actinomyces, Nocardia, Streptomyces, Thermoactinomyces, Waksmania, Thermopolyspora, Micromonospora, Thermomonospora, Actinoplanes, and Streptosporangium (Babalola et al. 2009). Among these genus, Streptomyces is considered as a significant member of Soil bacteria and gained importance due to the presence of its rich source of bioactive chemicals, antibiotics, and extracellular enzymes as well as its importance in agriculture (Olanrewaju and Babalola 2019b).

6.2.1 Rhizospheric Actinobacteria

Rhizosphere is the zone surrounding the plant root and it is an important component of the environment where the interaction between microbes and plants occur. The large area of the soil microbiome in the plant root system is occupied by rhizospheric Actinobacteria. Different Genus of the Actinomycetes present in the rhizosphere zone are reported as follows Streptomyces, Sanguibacter, Phylum Actinobacteria Rhodococcus, Pseudonocardia, Propionibacterium, Nocardia, Mycobacterium, Micrococcus, Microbacterium, Frankia, Corynebacterium, Clavibacter, Cellulomonas, Bifidobacterium, Arthrobacter, Actinomyces, and Acidimicrobium (Franco-Correa and Chavarro-Anzola 2016; Yadav and Yadav 2019). As a major resident of rhizosphere microflora, Rhizospheric Actinobacteria has received more attention in recent years due to its active participation as plant growth promoters, biocontrol agents against phytopathogens, efficient decomposers, nitrogen fixers as well as able to withstand against various environmental stress thereby promoting plant growth and productivity as well as in improving the fertility of the soil (Franco-Correa et al. 2010; Alexander and Zuberer 1991; Jeon et al. 2003; Yadav et al. 2018).

6.2.2 Endophytic Actinobacteria

Endophytes are a microbial community present inside the plant tissue without causing any adverse effects on the host plant. The presence of Endophytic microbes comprising of bacteria, fungi, and Actinomycetes are pervasive in most of the plant species grown in agricultural fields (De Bary 1866; Schulz and Boyle 2006). These endophytes proved their significant role in progressing plant growth, reduction of diseases caused by pathogens, ability to withstand unfavorable environmental stress conditions (Hasegawa et al. 2006). Endophytic Actinobacteria establish a huge part of the rhizosphere, which are likewise found inside plants in which the broadly examined species are from the variety Frankia (Wall 2000).

Recent years have seen the development of many genomes from Frankia sp. strains and the improvement of strategies for controlling plant gene expression. Understanding the bacterial interaction obligate utilization of a possibility of strategies that uncover the proteomes and transcriptomes from both refined and symbiotic Frankia. The image (Fig. 6.1) starting to rise gives some point of view on the heterogeneity of Frankial populaces in the two conditions (Schwencke and Carú 2001).

Since the acknowledgment of the name Frankia in the Approved Lists of bacterial names (1980), barely any revisions have been specified to the description of genus. Progressive releases of Bergey's Manual of Systematics of Archaea and Bacteria have comprehensively clashing suprageneric attention to the genus with no advances for subgeneric arrangement (Benson et al. 2011).

Endophytic Actinobacteria can be classified into two different subgroups: (Singh and Dubey 2018; Rosenblueth and Martínez-Romero 2006)

Fig. 6.1 Frankia; a whole Alder root nodule gall

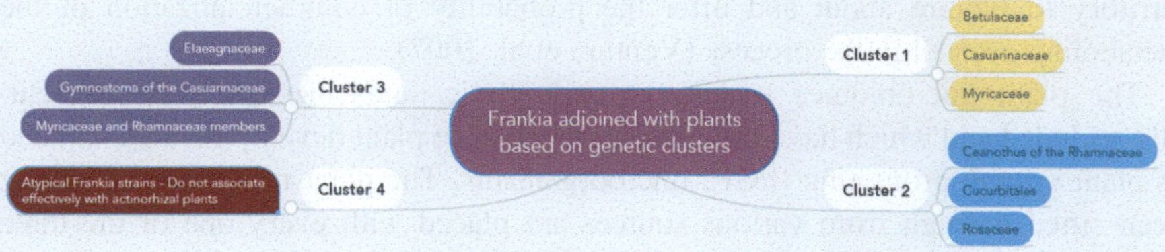

Fig. 6.2 Frankia–plant association (Sellstedt and Richau 2013)

1. Obligate—where survival of the bacteria is entirely dependent on plant metabolism and where plant-to-plant transmission occurs through various vectors or vertical transmission (Hardoim et al. 2008)
2. Facultative—where during their life cycle, in particular stages, they are not dependent on the host plant for survival and form an indirect association via surrounding soil and the environment (Abreu-Tarazi et al. 2010)

In the light of phylogenomic investigations, Frankia ought to be viewed as a single individual from the family Frankiaceae inside the monophyletic order, Frankiales. A polyphasic methodology of joining genome to genome information and omniLog® phenoarrays, together with classical methods, has permitted the assignment and an altered depiction of a type strain of the type species *Frankia alni*. This has led to the identification of 10 novel species comprising of both symbiotic and non-harmonious taxa inside the variety (Benson et al. 2011).

Frankia strains are classified into four host-specificity groups based on their ability to nodulate actinorhizal plants and into three major phylogenetic groups (Wall 2000; Baker 1987). The actinorhizal plant species can be broadly classified into four subclasses, eight dicotyledonous families (*Betulaceae*, *Casuarinaceae*, *Coriariaceae*, *Datiscaceae*, *Elaeagnaceae*, *Myricaceae*, *Rhamnaceae*, and *Rosaceae*) (Rosenblueth and Martínez-Romero 2006), and 25 genera containing more than 220 species. (Wall 2000; Schwencke and Carú 2001) A marked genetic variation in Frankia populations has been observed using molecular tools (Gtari et al. 2019a).

Frankia strains have been isolated into four hereditarily different groups dependent on 16S rRNA, nifH, and gln gene sequences (Sellstedt and Richau 2013).

The actinorhizal plants (Fig. 6.2) that are invaded by Gram-positive Frankia species are individuals from three phylogenetically related gatherings: Fagales (Betulaceae, Casuarinaceae, Myricaceae), Cucurbitales (Datiscaceae, Coriariaceae), and Rosales (Rosaceae, Elaeagnaceae, Rhamnaceae) (Sellstedt and Richau 2013; Berry et al. 2011).

The taxonomical and biological places of Actinomycetes that produce antibiotics are all around perceived for their metabolic adaptability, generally accompanied by the creation of essential and optional metabolites of financial importance. Different methodologies including classical, chemo-taxonomical, numerical taxonomy, in addition to atomic level have been routinely utilized for the recognizable proof of

Actinomycetes. The metabolic point of view of Actinomycetes gives a fascinating territory to inquire about and offer the probability of commercialization of the metabolites created in the process (Ventura et al. 2007).

The plant microbiomes (epiphytic, endophytic, and rhizospheric) are usually secluded and which have the capacity to advance plant development are alluded as plant growth-promoting (PGP) microorganisms. The plant microbiomes having been sifted through from various sources are placed with every one of the three domains archaea, bacteria, and fungi. Among these three the microbes under the bacterial domain are all around described and reported as from differing abiotic stresses, for example, basic soil, soil with high sodium chloride content, acidic soil low temperature soil, high temperature soil, and drought (Yadav and Yadav 2019).

Research that utilize exponentially developing cultures have yielded data on a complex proteolytic framework, including proteasomes, endo-and extracellular proteinases, and aminopeptidases, and furthermore on esterases, dehydrogenases, and extracellular DNAses (Abdel-Lateif et al. 2012).

Early investigations on the subatomic science of Frankia were hampered by troubles with developing enough cell material for DNA extraction and furthermore by the low effectiveness of old-style lysis strategies of Frankia cells. The nif encode the nitrogenase enzyme: nifH codes for the polypeptides of the Fe protein and nifDK codes for the alpha and beta subunits of the MoFe protein. A few hereditary and physical examinations have now evidently stated that in rapidly developing Rhizobium species the qualities associated with symbiotic relationship, especially the nif, are situated on enormous plasmids. Hypothetically DNA can be presented in beneficiary cells by four techniques: Transduction, Conjugation, Transformation, and Protoplast combination (Cervantes and Rodriguez-Barrueco 1992).

The symbiosis receptor kinase SYMRK plays a role in a signaling pathway that has been demonstrated for leguminous plants that are necessary to trigger the development of nodules and the acceptance of bacteria or fungi into their root cells as endosymbionts (Sellstedt and Richau 2013) as explained in Fig. 6.3.

Flavonoids are a gathering of secondary metabolites obtained from the phenylpropanoid pathway. They are omnipresent in the plant realm and have numerous assorted capacities including key functions at various degrees of root endosymbiosis. While there is a great deal of data on the job of specific flavonoids in the Rhizobium-legume beneficial interaction. Whereas the defined function during the foundation of arbuscular mycorrhiza and actinorhizal symbioses stays vague (Singh and Dubey 2018).

Arbuscular mycorrhiza embodies an exceptionally basic kind of root endosymbiosis, creating the groundwork of symbiosis made out of land plants and which fall under phylum Glomeromycota (Sellstedt and Richau 2013).

Frankial populaces in root knobs appear to keep up a somewhat strong metabolism that incorporates fixation of nitrogen, generous biosynthesis and pathways that create energy, alongside an altered ammonium assimilation. Until this point in time, specific bacterial qualities have not been embroiled in root knob development yet a

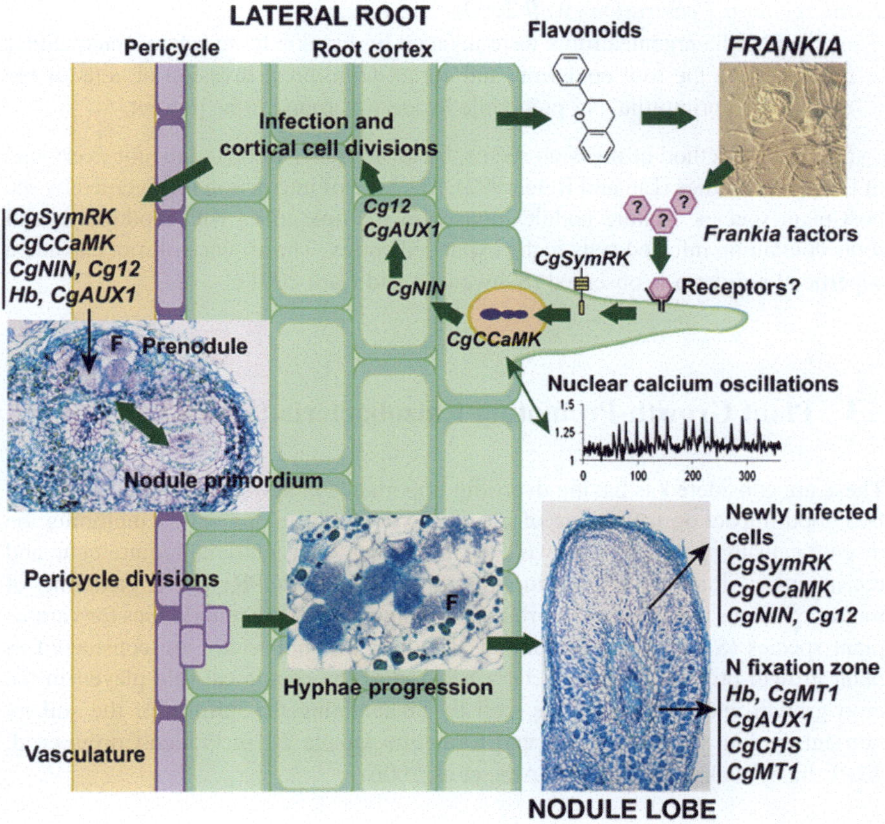

Fig. 6.3 Actinorhizal symbiosis signaling (Ostrowski and Jakubowska 2008)

few theories are rising as to how the plant and microorganism figure out how to exist together. Specifically, frankiae appear to show a nonpathogenic existence to the plant that may have the impact of limiting some plant protection reactions (Schwencke and Carú 2001).

Types of Compatible Frankia Interactions

1. *Root Hair Infection (Wall* 2000*)*

   Gymnostoma papuanum was infected by infiltration of Frankia into twisted root hairs and trailed by improvement of a prenodule region including one to many nodule lobe primordia in the root cortex. Frankia hyphae developed in a straightforward manner through cell walls from cell to cell, colonizing cells of the prenodule preceding invasion of nodule lobe cells (Racette and Torrey 1989).

2. *Intercellular Penetration (Wall* 2000*)*
   Shepherdia argentea roots were invaded by Frankia by means of intercellular entrance into the root epidermis and cortex with direct invasion of cells of the nodule lobe primordia. No prenodule locale was found to be present.

So far, the method of invasion seems, by all accounts, to be unique for every one of the plant families (Liu and Reid 1992). The type of interaction is governed by the host plant species. Mature nodules consist of multiple lobes with modified lateral roots containing infected cells in the expanded cortex. The absence of root cap and a superficial periderm is observed (Schwencke and Carú 2001).

6.3 Plant Growth-Promoting Rhizobacteria (PGPR)

These are considered as having overridig importance in agriculture due to constructive good impact on well being of plant growth and development by inhibiting the disease-causing pathogens as well as improving their nutrient requirement and consumption. Plant growth-promoting rhizobacteria (PGPR) are a gathering of microorganisms that colonize the rhizosphere and underlying foundations for various plant species (Santamarina Siurana et al. 2004). Actinobacteria are considered as plant growth-promoting rhizobacteria because of the ecological role played in the crop growth and productivity as well the maintaining the fertility of the soil by nutrient cycling (Franco-Correa and Chavarro-Anzola 2016; Franco-Correa et al. 2010; Jiang et al. 2006; Pathom-Aree et al. 2006).

6.3.1 Machinery Involved in Plant-Growth Promoting Rhizobacteria

The Plant growth-promoting rhizobacteria are influenced by several mechanisms as shown in Fig. 6.4.

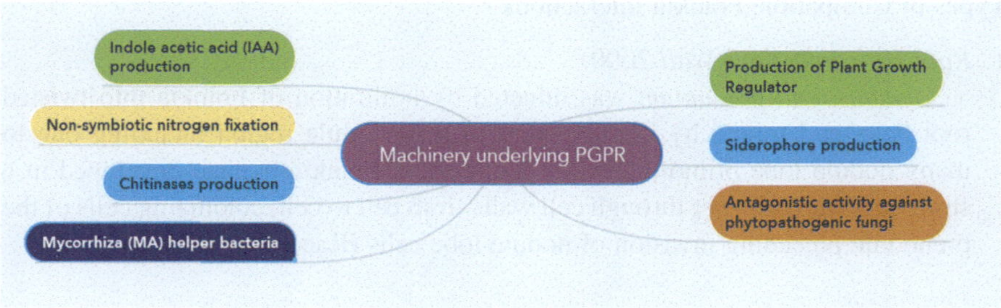

Fig. 6.4 Machinery underlying PGPR

6.3.1.1 Production of Plant Growth Regulators (PGRs)

Plant organ arrangement and their ensuing advancement are interceded by inner variables of essential significance. PGRs ("Plant Growth Regulators") are also widely recognized as plant hormones. At low concentration, PGRs are small molecules that influence plant development (Franco-Correa et al. 2010; Neilands 1995). The capability of the rhizospheric bacteria depends upon the ability to encourage the development of the root system with reference to plant growth regulators (Fett et al. 1987; Torres-Rubio et al. 2000).

Indole acetic acid (IAA) is a plant growth hormone and an active structure of auxins. It assumes a significant role in plant development through its life cycle (Vessey 2003; Ostrowski and Jakubowska 2008; Pawlowski and Demchenko 2012). IAA induces the development of the radicular framework (Liu and Reid 1992; Sylvia et al. 2005; López Nicolás et al. 2004), because of the improvement of lateral roots and divisions of the growing tip of the plant that results in roots lengthening (Rosenblueth and Martínez-Romero 2006; Abreu-Tarazi et al. 2010). The generation of IAA has been broadly contemplated in Actinobacteria (Hardoim et al. 2008). Streptomyces genus and Frankia species have been extensively investigated for their IAA producing capabilities (Péret et al. 2008; Sousa et al. 2008; Flores et al. 2003).

6.3.1.2 Production of Siderophores

Microorganisms have been constrained by natural limitations and biologic goals to deliver explicit particles that can contend positively with hydroxyl ions for ferric condition/ form of iron. Siderophores are mixes delivered by different microorganisms in soil. These living beings depend on the process of chelation to help their natural action. Siderophores are fluorescent pigments found outside the cell that bind to iron (III) in high affinity and they are soluble in water having a low molecular weight (500–1000 Da) (Prasad et al. 2011).

These compounds go about as chelate agents that act specific for ferric ion, leaving accessible the ionic structure (Fe^{+2}), which is effectively consumed by microorganisms (Vessey 2003; Pawlowski and Demchenko 2012).

Most of the nitrogen-fixing microorganisms produce siderophores to acquire iron. This is essential for the proper functioning of the enzyme nitrogenase. The compound is made out of a few protein units; a sum of 36 iron atoms are necessary for working appropriately (Froussart et al. 2016).

6.3.1.3 Non-Symbiotic Nitrogen Fixation

The Actinobacteria are heterotrophic living beings that require carbon as a source to derive the energy required for fixing N_2. Accordingly, every one of the various microbes contrasts in such a way that carbon is processed with characteristic capacity of nitrogen fixation, demonstrating various rates of acetylene reduction assay (ARA). This test depends on identifying in an indirect fashion, the presence of nitrogenase protein, which is responsible for the reduction of nitrogen (N_2) to ammonium ion. ARA evaluates the enzymatic reduction of acetylene to ethylene (NH_4^+) (Aranibar et al. 2003). Some non-leguminous, nitrogen-fixing plants convey dual symbiotic infection including both mycorrhizal parasites and actinorhizal actinomycetes. Examples include Arthrobacter, Agromyces, Corynebacterium, Mycobacterium, Micromonospora, Propionibacteria, and Streptomyces (Sellstedt and Richau 2013).

6.3.1.4 Chitinase Production

The chitin is a homopolymer containing N-acetyl-D-Glucosamine residues with α-1, 4 bonds. It is extensively found in the environment as a basic component of fungi (22–44%), insects and crustaceans (25–58%), and protozoa (Azcón-Aguilar and Barea 1997; Hoster et al. 2005; Nehl and Knox 2006; Bhattacharyya 2012). The chitin is hydrolyzed by a complex chitinase that contains three proteins which are exochitinase, endochitinase, and N-acetyl-D-Glucosamine (Franco-Correa and Chavarro-Anzola 2016).

Actinobacteria are considered as the predominant creatures associated with the deterioration of chitin in soil (Ames 1989) and furthermore encouraging antagonistic agents for biocontrol because of the hydrolysis response over the mycelium fungi (Gomes et al. 2001). The species that come under Streptomyces class are considered as the primary chitinolytic microbe in soil, because of its ability to breakdown this polymer (Metcalfe et al. 2003).

6.3.1.5 Antagonistic Activity Against Phytopathogenic Fungi (Franco-Correa and Chavarro-Anzola 2016)

In phytopathology, antagonism refers to the action of any organism that suppresses or interfere with the normal growth and activity of a plant pathogen, such as the main parts of bacteria or fungi.

Different Actinobacteria may behave as hostile microorganisms toward *F. oxysporum* by antimicrobial synthesis (antibiosis). These synthesized chemicals diffuse through the medium hindering the development of phytopathogenic growth.

Molano et al. (2000) decided in vitro inhibition of *Fusarium oxysporum* development by actinomycin, an anti-infection created by Nocardia sp., strain confined from rhizosphere soil sample lichen (Mosquera, Colombia). The generation of such optional metabolites was lethal to the phytopathogenic organism.

6.3.1.6 Mycorrhiza (MA) Helper Bacteria (Franco-Correa and Chavarro-Anzola 2016)

By and large, the capacity of specific microorganisms to impact the arrangement and working of the symbiotic MA through different types of activities, for example, initiation of parasitic propagules in infective pre-symbiotic stages (Azcón-Aguilar and Barea 1997; Giovannetti et al. 2002), and encourage the development of sources of input points into the root and thereby augument the development rate. (Giovannetti et al. 2002; Carpenter-Boggs et al. 1995). The portrayal of the variety Frankia, the microsymbiont in the actinorhizal advantageous interaction, is moving toward maturity however the ideal use of techniques previously created requires the thought of the genuine commitment of mycorrhiza related with the actinorhizal symbioses. Mycorrhizal species that are found in most actinorhizal plant genera so far inspected in both ectomycorrhizal and vesicular-arbuscular mycorrhizal symbioses. The nearness of a mycorrhizal beneficial interaction in a plant species relies generally upon many soil attributes. In this way, contamination of Myrica gale by both ecto- and vesicular-arbuscular mycorrhiza is limited to well-drained soils, being missing in wet soils (Gtari et al. 2019b).

6.4 Ecological Contributions of Actinomycetes

For quite a long time plants have been broadly utilized as the reservoir of bioactive chemicals for their therapeutic effect. In the present day scenario, plant-related microorganisms have been observed to create substances of high therapeutic potential (Vivas et al. 2003; Hoster et al. 2005). The microorganisms living inside plant tissues, generally in symbiosis, may incorporate different communities, for instance, fungi and Actinobacteria (Singh and Dubey 2018; Pimentel et al. 2011).

Customarily, serious research has been conducted on free dwelling soil actinobacteria. Generally, advantageous Actinobacteria dwelling as endophytes inside the plant tissues have produced enormous enthusiasm as a potential wellspring of novel mixes, which may find applications in medication, horticulture, and environment as given in Fig. 6.5 (Kuffner et al. 2010).

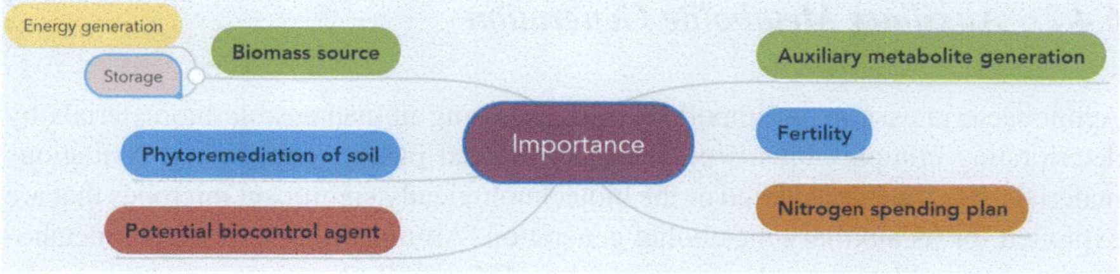

Fig. 6.5 Ecological importance of Actinomycetes

6.4.1 *Handling of Abiotic Stresses*

There are advancements in research on transgenic actinorhizal plants on actinorhizal-explicit qualities and proteins (actinorhizins) associated with symbiotic communications, infectivity, and host particularity. Actinorhizal plants are quickly developing species, ready to grow in N-poor soils, and for specific species, in unforgiving natural pressure conditions (Vivas et al. 2003).

6.4.1.1 Salinity

Antioxidative defence system and entire cell proteome description is dependent on spectrophotometric investigation. SDS-PAGE and 2-dimensional gel electrophoresis have been investigated among salt-tolerant and salt-sensitive Frankia strains. Proteomic premise giving basis for salinity tolerance in the recently identified Frankia strains from Hippophae salicifolia. Salt-tolerant strain HsIi10 shows greater quantity of superoxide dismutase, catalase, and ascorbate peroxidase when compared with salt-sensitive strain HsIi8. Differential 2-DGE profile has uncovered differential profiles for salt-tolerant and salt-sensitive strains. Proteomic affirmation of salt resilience in the strains with inbuilt proficiency of flourishing in nitrogen-insufficient regions is a distinctive benefit for these organisms. This would be similarly valuable for the development of soil nitrogen status. Proficient protein modulation in HsIi10 recommends further investigation for its latent capacity to be used as biofertilizer in saline soils, strains (Srivastava et al. 2017).

6.4.1.2 pH

Neutrophiles and acidophiles are two characteristically different groups observed in Streptomyces species of two pine forest soils and it is recognized that the acidophilic group is more widely found and of significance to the soil more than what was known previously (Williams et al. 1971).

6.4.2 *Auxiliary Metabolite Generation*

Actinobacteria assume an important role in reusing unmanageable biomaterials by deteriorating complex blends of polymers in dead plants, animals and contagious materials. They are considered as the biotechnologically significant microbes that are exploited for its auxiliary metabolite generation. Around, 10,000 bioactive metabolites are created by Actinobacteria, which is 45% of all bioactive microbial metabolites found (Anandan et al. 2016).

6.4.3 Fertility

Actinorhizal plants play a very important role in improving fertility of agroforestry ecosystems, as sources of biomass for generating energy or for carbon storage, for remediation of soils in harsh environmental stress conditions, reclamation of land, amenity planning for preventing erosions and as coastal windbreaks (Pemila Edith Chitraselvi 2018; Kamal et al. 2014). The actinorhizal advantageous interaction is a significant supporter of the worldwide nitrogen spending plan, assuming a prevailing job in environmental progressions following unsettling influences. The components included are still inadequately known; however aided in understanding of kinase that transmit the symbiotic signal and thereby transmission of the Rhizobium Nod signal in legumes. On the microbial side complementation with Frankia DNA of Rhizobium nod mutants prevented identification of genes thought to be symbiotic (Simonet et al. 2018).

Of late land degradation has expanded extensively because of climatic reasons and human mediation bringing about a decrease in fertility, biodiversity, and profitability. Since tropical nations are portrayed by increasing population density, the populace reliance on biological systems is bringing about environment degradation. Consequently, the restoration of these lands is essential. To beat the issue of the absence of fertility in soil, rapidly developing nitrogen-fixing trees, for example, actinorhizal plants in blend with biofertilizers are utilized (Diagne et al. 2013; Mahdi et al. 2010; Sayed 2011). PGPRs have high economical benefit and ecological significance as biofertilizers. Example: Azospirillum (Fuentes-Ramirez and Caballero-Mellado 2005).

6.4.4 Phytoremediation

Phytoremediation of Cd mediated in sunflower plant by *S.tendae* and mobilization of Zinc(Zn) and Cadmium(Cd) by secondary metal-binding metabolites of *Salix caprea* clearly define the scope of endophytic Actinobacteria in heavy metal removal from the soil and increased mobilization in soils contaminated by heavy metals (Das et al. 2008; Dimkpa et al. 2009; Baoune et al. 2018). Endophytic Streptomyces sp. have been accounted for improving phytoremediation efficiency of oil-polluted soil by their petroleum degradation property (Karthikeyan et al. 2018).

6.4.5 Miscellaneous Contributions

Auxin, siderophore production, solubilization of phosphates, and complementing VA Mycorrhizal fungi are also some of the major functions of actinomycetes and their isolates. They are also potential biocontrol agents and all of these properties play an essential role in improving agricultural productivity and quality and they should therefore be taken into consideration as an alternative tool for promotion of eco-friendly as well as sustainable farming practices unlike harmful chemicals which are detrimental to the environment (Pemila 2018; Sharma et al. 2014).

6.5 Conclusion

Plant and actinomycetes and their eco-friendly association form an integral and essential component of the ecosystem as a whole. Actinomycetes are classified using classical and taxonomic classification and can be obligative or facultative. The symbiotic relationship is supported due to various complex signaling pathways between the actinorhizal plant and the particular strain of actinomycetes, with different species, each acting in a unique and characteristic way. The interaction between the plant and actinomycetes occurs either by infection into root hair or penetration between the cells of host plant. Actinomycetes, once they form the symbiotic relationship are a major influencer of the plant growth and regulates various synthesis pathways and systems such as auxin and chitinase production and nitrogen fixation. With the increasing concern for the environment, actinomycetes has been considered as possible alternative for a number of objectives such as Biofertilizers or in land reclamation or in phytoremediation. Therefore, the existence of Actinomycetes and their eco-friendly association with plants forms the essential basis and foundation for future research including their benefit to the ecosystem.

References

Abdel-Lateif K, Bogusz D, Hocher V (2012) The role of flavonoids in the establishment of plant roots endosymbioses with arbuscular mycorrhiza fungi, rhizobia and Frankia bacteria. Plant Signal Behav 7(6):636–641

Abreu-Tarazi MF, Navarrete AA, Andreote FD, Almeida CV, Tsai SM, Almeida M (2010) Endophytic bacteria in long-term in vitro cultivated "axenic" pineapple microplants revealed by PCR–DGGE. World J Microbiol Biotechnol 26(3):555–560

Alexander DB, Zuberer DA (1991) Use of chrome azurol S reagents to evaluate siderophore production by rhizosphere bacteria. Biol Fertil Soils 12(1):39–45

Ames RN (1989) Mycorrhiza development in onion in response to inoculation with chitin decomposing actinomycetes. New Phytol 112(3):423–427

Anandan R, Dharumadurai D, Manogaran GP (2016) An introduction to actinobacteria. In: Actinobacteria-basics and biotechnological applications. IntechOpen, London

Aranibar JN, Anderson IC, Ringrose S, Macko SA (2003) Importance of nitrogen fixation in soil crusts of southern African arid ecosystems: acetylene reduction and stable isotope studies. J Arid Environ 54(2):345–358

Azcón-Aguilar C, Barea JM (1997) Arbuscular mycorrhizas and biological control of soil-borne plant pathogens–an overview of the mechanisms involved. Mycorrhiza 6(6):457–464

Babalola OO, Kirby BM, Roes-Hill L, Cook AE, Cary SC, Burton SG, Cowan DA (2009) Phylogenetic analysis of actinobacterial populations associated with Antarctic Dry Valley mineral soils. Environ Microbiol 11(3):566–576

Baker DD (1987) Relationships among pure cultured strains of Frankia based on host specificity. Physiol Plant 70(2):245–248

Baoune H, El Hadj-Khelil AO, Pucci G, Sineli P, Loucif L, Polti MA (2018) Petroleum degradation by endophytic Streptomyces spp. isolated from plants grown in contaminated soil of southern Algeria. Ecotoxicol Environ Saf 147:602–609

Benson DR, Brooks JM, Huang Y, Bickhart DM, Mastronunzio JE (2011) The biology of Frankia sp. strains in the post-genome era. Mol Plant-Microbe Interact 24(11):1310–1316

Berry AM, Mendoza-Herrera A, Guo YY, Hayashi J, Persson T, Barabote R, Demchenko K, Zhang S, Pawlowski K (2011) New perspectives on nodule nitrogen assimilation in actinorhizal symbioses. Funct Plant Biol 38(9):645–652

Bhattacharyya PN (2012) Plant growth-promoting rhizobacteria (PGPR): emergence in agriculture. World J Microbiol Biotechnol 28:1327–1350

Bhatti AA, Haq S, Bhat RA (2017) Actinomycetes benefaction role in soil and plant health. Microb Pathog 111:458–467

Carpenter-Boggs L, Loynachan TE, Stahl PD (1995) Spore germination of Gigaspora margarita stimulated by volatiles of soil-isolated actinomycetes. Soil Biol Biochem 27(11):1445–1451

Cervantes E, Rodriguez-Barrueco C (1992) 22 relationships between the mycorrhizal and actinorhizal symbioses in non-legumes. In: Methods in microbiology, vol 24. Academic, Amsterdam, pp 417–432

Costa FG, Zucchi TD, Melo ISD (2013) Biological control of phytopathogenic fungi by endophytic actinomycetes isolated from maize (Zea mays L.). Braz Arch Biol Technol 56(6):948–955

Das S, Lyla PS, Khan SA (2008) Distribution and generic composition of culturable marine actinomycetes from the sediments of Indian continental slope of Bay of Bengal. Chin J Oceanol Limnol 26(2):166–177

De Bary A (1866) Morphologie und physiologie der pilze, flechten und myxomyceten. Engelmann, Leipzig

Diagne N, Arumugam K, Ngom M, Nambiar-Veetil M, Franche C, Narayanan KK, Laplaze L (2013) Use of Frankia and actinorhizal plants for degraded lands reclamation. Biomed Res Int 2013:948258

Dimkpa CO, Merten D, Svatoš A, Büchel G, Kothe E (2009) Siderophores mediate reduced and increased uptake of cadmium by Streptomyces tendae F4 and sunflower (Helianthus annuus), respectively. J Appl Microbiol 107(5):1687–1696

Doumbou C, Hamby Salove M, Crawford D, Beaulieu C (2001) Actinomycetes, promising tools to control plant diseases and to promote plant growth. Phytoprotection 82(3):85–102. https://doi.org/10.7202/706219ar.

Fett WA, Osman SF, Dunn MF (1987) Auxin production by plant-pathogenic Pseudomonads and Xanthomonads. Appl Environ Microbiol 53(8):1839–1845. https://doi.org/10.1371/journal.pone.0119867

Flores FJ, Rincón J, Martín JF (2003) Characterization of the iron-regulated desA promoter of Streptomyces pilosus as a system for controlled gene expression in actinomycetes. Microb Cell Factories 2(1):5

Flores-Gallegos AC, Nava-Reyna E (2019) Plant growth-promoting microbial enzymes. In: Enzymes in food biotechnology. Academic, Amsterdam, pp 521–534

Franco-Correa M, Chavarro-Anzola V (2016) Actinobacteria as plant growth promoting rhizobacteria. In: Actinobacteria-basis and biotechnological application. InTech, Croatia, pp 249–270

Franco-Correa M, Quintana A, Duque C, Suarez C, Rodríguez MX, Barea JM (2010) Evaluation of actinomycete strains for key traits related with plant growth promotion and mycorrhiza helping activities. Appl Soil Ecol 45(3):209–217

Froussart E, Bonneau J, Franche C, Bogusz D (2016) Recent advances in actinorhizal symbiosis signaling. Plant Mol Biol 90(6):613–622
Fuentes-Ramirez LE, Caballero-Mellado J (2005) Bacterial biofertilizers. In: PGPR: biocontrol and biofertilization. Springer, Dordrecht, pp 143–172
George M, Anjumol A, George G, Hatha AM (2012) Distribution and bioactive potential of soil actinomycetes from different ecological habitats. Afr J Microbiol Res 6(10):2265–2271
Giovannetti M, Sbrana C, Avio L (2002) Arbuscular mycorrhizal fungal mycelium: from germlings to hyphal networks. In: In Mycorrhizal technology in agriculture. Birkhäuser, Basel, pp 49–58
Gomes RC, Semedo LTAS, Soares RMA, Linhares LF, Ulhoa CJ, Alviano CS, Coelho RRR (2001) Purification of a thermostable endochitinase from Streptomyces RC1071 isolated from a cerrado soil and its antagonism against phytopathogenic fungi. J Appl Microbiol 90(4):653–661
Gtari M, Benson DR, Nouioui I, Dawson JO, Ghodhbane-Gtari F (2019a) 19th International Meeting on Frankia and Actinorhizal Plants.
Gtari M, Nouioui I, Sarkar I, Ghodhbane-Gtari F, Tisa LS, Sen A, Klenk HP (2019b) An update on the taxonomy of the genus Frankia Brunchorst, 1886, 174 AL. Antonie Van Leeuwenhoek 112 (1):5–21
Hardoim PR, van Overbeek LS, van Elsas JD (2008) Properties of bacterial endophytes and their proposed role in plant growth. Trends Microbiol 16(10):463–471
Hasegawa S, Meguro A, Shimizu M, Nishimura T, Kunoh H (2006) Endophytic actinomycetes and their interactions with host plants. Actinomycetologica 20(2):72–81
Hoster F, Schmitz JE, Daniel R (2005) Enrichment of chitinolytic microorganisms: isolation and characterization of a chitinase exhibiting antifungal activity against phytopathogenic fungi from a novel Streptomyces strain. Appl Microbiol Biotechnol 66(4):434–442
Jeon JS, Lee SS, Kim HY, Ahn TS, Song HG (2003) Plant growth promotion in soil by some inoculated microorganisms. J Microbiol 41(4):271–276
Jiang Y, Li WJ, Xu P, Tang SK, Xu LH (2006) Study on Actinomycete diversity under salt and alkaline environments. Wei Sheng Wu Xue Bao 46:191–195. https://doi.org/10.1038/sj/jim/7000176
Kamal R, Gusain YS, Kumar V (2014) Interaction and symbiosis of AM fungi, actinomycetes and plant growth promoting rhizobacteria with plants: strategies for the improvement of plants health and defense system. Int J Curr Microbial Appl Sci 3(7):564–585
Karthikeyan N, Pandiyan K, Sahu PK, Srinivasan R, Singh UB (2018) Actinomycetes: a promising tool for plant growth promotion and disease control. Int J Curr Microbiol App Sci 7 (7):2418–2429
Kuffner M, De Maria S, Puschenreiter M, Fallmann K, Wieshammer G, Gorfer M, Strauss J, Rivelli AR, Sessitsch A (2010) Culturable bacteria from Zn-and Cd-accumulating Salix caprea with differential effects on plant growth and heavy metal availability. J Appl Microbiol 108 (4):1471–1484
Kuster E (1968) Taxonomy of soil actinomycetes and related organisms. In: Ecology of soil bacteria. Liverpool University press, Liverpool, pp 322–336
Liu JH, Reid DM (1992) Adventitious rooting in hypocotyls of sunflower (Helianthus annuus) seedlings. IV. The role of changes in endogenous free and conjugated indole-3-acetic acid. Physiol Plant 86(2):285–292
López Nicolás JI, Acosta M, Sánchez-Bravo J (2004) Role of basipetal auxin transport and lateral auxin movement in rooting and growth of etiolated lupin hypocotyls. Physiol Plant 121 (2):294–304
Mahdi SS, Hassan GI, Samoon SA, Rather HA, Dar SA, Zehra B (2010) Bio-fertilizers in organic agriculture. J Phytology 2
Metcalfe AC, Williamson N, Krsek M, Wellington EM (2003) Molecular diversity within chitinolytic actinomycetes determined by in situ analysis. Actinomycetologica 17(1):18–22
Molano A, Algecira N, Bernal J, Franco-Correa M (2000) Evaluación y Selección de un Medio de Cultivo a partir de Actinomycetes. In: Memorias del II Congreso Internacional de Microbiología Industrial. Pontificia Universidad Javeriana, Mayo, pp 10–12

Nehl DB, Knox OG (2006) Significance of bacteria in the rhizosphere. In: Microbial activity in the Rhizoshere. Springer, Berlin, Heidelberg, pp 89–119

Neilands JB (1995) Siderophores: structure and function of microbial iron transport compounds. J Biol Chem 270(45):26723–26726

Olanrewaju OS, Babalola OO (2019a) Streptomyces: implications and interactions in plant growth promotion. Appl Microbiol Biotechnol 103(3):1179–1188

Olanrewaju OS, Babalola OO (2019b) Streptomyces: implications and interactions in plant growth promotion. Appl Microbiol Biotechnol 103(3):1179–1188

Ostrowski M, Jakubowska A (2008) Identification of enzyme activity that conjugates indole-3-acetic acid to aspartate in immature seeds of pea (Pisum sativum). J Plant Physiol 165 (5):564–569

Pathom-Aree W, Stach JE, Ward AC, Horikoshi K, Bull AT, Goodfellow M (2006) Diversity of actinomycetes isolated from challenger deep sediment (10,898 m) from the Mariana Trench. Extremophiles 10(3):181–189. https://doi.org/10.1007/s00792-005-0482-z

Pawlowski K, Demchenko KN (2012) The diversity of actinorhizal symbiosis. Protoplasma 249 (4):967–979

Pemila ECR (2018) Actinomycetes: dependable tool for sustainable agriculture. Curr Invest Agri Curr Res 1(5):128–130. https://doi.org/10.32474/CIACR.2018.01.000122.

Pemila Edith Chitraselvi R (2018) Actinomycetes: dependable tool for sustainable agriculture. Curr Invest Agri Curr Res 1(5):128–130. https://doi.org/10.32474/CIACR.2018.01.000122.

Péret B, Svistoonoff S, Lahouze B, Auguy F, Santi C, Doumas P, Laplaze L (2008) A role for auxin during actinorhizal symbioses formation? Plant Signal Behav 3(1):34–35

Pimentel MR, Molina G, Dionísio AP, Maróstica Junior MR, Pastore GM (2011) The use of endophytes to obtain bioactive compounds and their application in biotransformation process. Biotechnol Res Int 2011:576286

Prasad S, Manasa P, Buddhi S, Singh SM, Shivaji S (2011) Antagonistic interaction networks among bacteria from a cold soil environment. FEMS Microbiol Ecol 78(2):376–385

Racette S, Torrey JG (1989) Root nodule initiation in Gymnostoma (Casuarinaceae) and Shepherdia (Elaeagnaceae) induced by Frankia strain HFPGpI1. Can J Bot 67(10):2873–2879

Rosenblueth M, Martínez-Romero E (2006) Bacterial endophytes and their interactions with hosts. Mol Plant-Microbe Interact 19(8):827–837

Rowbotham TJ, Cross T (1977) Rhodococcus coprophilus sp. nov.: an aerobic nocardioform actinomycete belonging to the 'rhodochrous' complex. Microbiology 100(1):123–138

Santamarina Siurana MP, Garcia Breijo FJ, Rosello Caselles JL. Biologia y botanica. UPV. 2004.

Sayed WF (2011) Improving Casuarina growth and symbiosis with Frankia under different soil and environmental conditions. Folia Microbiol 56(1):1–9

Schulz B, Boyle C (2006) What are endophytes. In: Microbial root endophytes. Springer, Berlin, Heidelberg, pp 1–13

Schwencke J, Carú M (2001) Advances in actinorhizal symbiosis: host plant-Frankia interactions, biology, and applications in arid land reclamation. A review. Arid Land Res Manag 15 (4):285–327

Sellstedt A, Richau KH (2013) Aspects of nitrogen-fixing Actinobacteria, in particular free-living and symbiotic Frankia. FEMS Microbiol Lett 342(2):179–186

Sharma M, Dangi P, Choudhary M (2014) Actinomycetes: source, identification, and their applications. Int J Curr Microbiol App Sci 3(2):801–832

Simonet P, Normand P, Hirsch AM, Akkermans AD (2018) The genetics of the Frankia-actinorhizal symbiosis. In: Molecular biology of symbiotic nitrogen fixation. CRC, Boca Raton, FL, pp 77–109

Singh R, Dubey AK (2018) Diversity and applications of endophytic actinobacteria of plants in special and other ecological niches. Front Microbiol 9:1767

Sousa CDS, Soares ACF, Garrido MDS (2008) Caracterização de estreptomicetos com potencial para promoção de crescimento de plantas e biocontrole. Sci Agric 65(1):50–55

Srinivasan MC, Laxman RS, Deshpande MV (1991) Physiology and nutritional aspects of actinomycetes: an overview. World J Microbiol Biotechnol 7(2):171–184

Srivastava A, Singh A, Singh SS, Mishra AK (2017) Salt stress–induced changes in antioxidative defense system and proteome profiles of salt-tolerant and sensitive Frankia strains. J Environ Sci Health A 52(5):420–428

Sylvia DM, Fuhrmann JJ, Hartel PG, Zuberer DA (2005) Principles and applications of soil microbiology (No. QR111 S674 2005). Pearson, Upper Saddle River, NJ

Torres-Rubio MG, Valencia SA, Bernal J, Martínez P (2000) Isolation of Enterobacteria, Azotobacter sp. and Pseudomonas sp., producers of Indole-3-acetic acid and siderophores, from Colombian rice rhizosphere. Revist Latin Am Microbiol 42:171–176

Ventura M, Canchaya C, Tauch A, Chandra G, Fitzgerald GF, Chater KF, van Sinderen D (2007) Genomics of Actinobacteria: tracing the evolutionary history of an ancient phylum. Microbiol Mol Biol Rev 71(3):495–548

Vessey JK (2003) Plant growth promoting rhizobacteria as biofertilizers. Plant Soil 255(2):571–586

Vivas A, Azcón R, Biró B, Barea JM, Ruiz-Lozano JM (2003) Lead-polluted soil and their interactions with arbuscular mycorrhizae on the growth of Trifolium pratense L. under lead toxicity. Can J Microbiol 49:10

Wall LG (2000) The actinorhizal symbiosis. J Plant Growth Regul 19(2):167–182

Williams ST, Davies FL, Mayfield CI, Khan MR (1971) Studies on the ecology of actinomycetes in soil II. The pH requirements of streptomycetes from two acid soils. Soil Biol Biochem 3 (3):187–195

Yadav N, Yadav AN (2019) Actinobacteria for sustainable agriculture. J Appl Biotechnol Bioeng 6 (1):38–41

Yadav AN, Verma P, Kumar S, Kumar V, Kumar M, Sugitha TCK, Singh BP, Saxena AK, Dhaliwal HS (2018) Actinobacteria from rhizosphere: molecular diversity, distributions, and potential biotechnological applications. In: New and future developments in microbial biotechnology and bioengineering. Elsevier, Amsterdam, pp 13–41

Zucchi TD, De Moraes LAB, De Melo IS (2008) Streptomyces sp. ASBV-1 reduces aflatoxin accumulation by Aspergillus parasiticus in peanut grains. J Appl Microbiol 105(6):2153–2160

Chapter 7
The Arbuscular Mycorrhizal Symbiosis of Trees: Structure, Function, and Regulating Factors

Leszek Karliński

Abstract The majority of terrestrial plants establish symbiotic relationships with fungi, which mostly co-create various types of mycorrhizae, differing in phylogenetic history, anatomy, and functional features. Arbuscular mycorrhiza is evolutionary the oldest and most common form of the symbiosis of plants and fungi belonging to phylum Glomeromycota. Arbuscular mycorrhiza plays a significant role in mineral nutrition of plants, protection against stress factors and pathogens, and supports the soil structure and fertility. Arbuscular fungi dominate in relatively nutrient-rich and phosphorus-limited soils. The highest diversity of arbuscular fungi was observed in the tropics, where fungal communities revealed significant spatial heterogeneity and non-random associations with the different hosts. Arid areas and wetlands are also often inhabited by shrubs or trees associated with arbuscular mycorrhizal fungi, presenting high tolerance to environmental conditions. A few tree species form tripartite symbiotic associations with arbuscular and ectomycorrhizal fungi, whose contribution in the microbiome of trees is mainly shaped by soil conditions and tree genotype.

This review presents a brief summary of the main types and evolutional history of mycorrhizal associations with trees with special emphasis on arbuscular fungi, their characteristic structures, taxonomical classification, distribution, and factors impacting them.

Keywords Symbiosis · Microorganisms · Host genotype · Glomeromycota

---

L. Karliński (✉)
Institute of Dendrology, Polish Academy of Sciences, Kórnik, Poland
e-mail: leszekk@man.poznan.pl

N. Shrivastava et al. (eds.), *Symbiotic Soil Microorganisms*, Soil Biology 60,
https://doi.org/10.1007/978-3-030-51916-2_7

7.1 Types of Mycorrhiza

The majority of terrestrial plants establish symbiotic relationships with fungi, which mostly co-create four main types of mycorrhizae (Smith and Read 2008), differing in phylogenetic history, and anatomical and functional features (Brundrett and Tedersoo 2018). Arbuscular mycorrhiza is the most common type of plant–fungi symbiosis, occurring in more than 71–90% of colonized vascular plant species. The presence of structures of arbuscular mycorrhizae has been demonstrated in angiosperms, gymnosperms, and pteridophytes having roots as well as gametophytes of some mosses, lycopods, Psilotalus, which do not have true roots (Smith and Read 2008; Błaszkowski 2012). Next to arbuscular mycorrhiza is Orchid mycorrhiza, which is found in 10% of plant species (Orchidaceae) and characterized by specific for Orchidaceae, hyphal peletons within root cells, colonization from the root surface mycelia or neighboring cells. Two percentage of plant species (2 Gymnosperm and 28 Angiospermae lineages) establish symbiotic relationships with ectomycorrhizal fungi, characterized by external differentiated hyphal mantle, current Hartig network and lack of intracellular colonization of the plant host. An even smaller group of species are plants belonging to Ericaceae and Diapensiaceae and forming together with fungi ericoid mycorrhizas (1.5%) characterized by structures such as hyphal coils in cells, cells colonized separately from the root surface, presence of mantle (Brundrett and Tedersoo 2018). Sometimes two other types of mycorrhizae are also distinguished, i.e., mycoheterotrophic monotropoid mycorrhiza, specific for the Monotropa genus and arbutoid mycorrhiza found in genera Arctostaphylos and Arbutus (Garg and Chandel 2010). However, both types of mycorrhiza are included in ectomycorrhizas (Smith and Read 2008).

7.2 Evolution of Mycorrhizal Symbiosis

Arbuscular mycorrhiza, apart from the fact that it is most common in the world of plants, is also evolutionarily the oldest form of the symbiosis of fungi with plants, probably accompanying them from the moment the plants enter the land. Arbuscular mycorrhiza could appear in plant tissue structures over 450 million years ago in Ordovician (Brundrett 2002). However, problems with the interpretation of these findings mean that the first well-preserved fossils of mycorrhizal rhizomes are structures similar to modern counterparts from early Devon, whose age is estimated at 407 million years old. The Devonian land plants revealed the presence of mycorrhiza-like intracellular structures similar to arbuscules of symbiotic Glomeromycotina and hyphae reminiscent Mucoromycotina, having the same ancestor somewhere between 358 and 508 million years ago (Martin et al. 2017). However, the first AM found fungus-like spores are 50 Million years older and come from Ordovician (Brundrett and Tedersoo 2018). In contrast, molecular data of plant symbiosis genes indicated that the relationships between fungi and plants are much older than the presence of plants on land and already occurred in their ancestors in the aquatic environment (Martin et al. 2017). Even if arbuscular fungi did not directly participate in leaving the aquatic environment by plants, they significantly supported the plants to colonize the terrestrial environment in a relatively short period (Chen et al. 2018; Delaux 2017). Along with the changes in environmental conditions on Earth and the ongoing process of plant species radiation, the development or decline of symbiotic associations with fungi has also taken place. The presence of arbuscular mycorrhizal structures has been found, among others, in late Carboniferous (over 300 Million years) in woody plants Lepidodendron and early relatives of conifers—Cordaites (Martin et al. 2017). Then the symbiotic relationships of plants with arbuscular fungi develop in gymnosperms cycads and conifers, evolve, and constitute the main form of mycorrhizal symbiosis from Triassic to Cretaceous. Next, in the Early Cretaceous (145 Million years) the arbuscular mycorrhiza also occurs in basal angiosperm trees, and monocots, and eudicots. During the Cretaceous, most of the non-tree plant species established symbiotic associations with arbuscular fungi and this relationship continues to this day (Martin et al. 2017). Earlier, in the Jurassic, for the first time, ectomycorrhiza had evolved from arbuscular mycorrhiza, establishing relationships with the early Pinaceae. The next evolutionary wave takes place in Crataceus with the origins of the Orchidaceae, Ericaceae, and multiple families with ECM, parasitic plants, carnivorous plants, and nitrogen-fixing symbioses (Brundrett 2002; Brundrett and Tedersoo 2018; Martin et al. 2017). Over the last 65 Million years (Paleogene), another wave of radiation has been underway mainly associated with ectomycorrhizal and non-mycorrhizal species, leading to complete transformations of mycorrhizal types from arbuscular mycorrhiza to ectomycorrhiza, but also to the joint coexistence of both mycorrhizal types on the roots of the same tree species (e.g., the genus Salix, Casuarinaceae, Alnus, Eucalyptus), whose contribution is conditioned by many environmental and tree-host-genetic factors (Brundrett and Tedersoo 2018).

7.3 Classification of Arbuscular Fungi

Arbuscular fungi as one of the most abundant groups of soil organisms, establish symbiotic relationships with more than 200 thousand of plants (Lee et al. 2013). Despite their common prevalence, arbuscular fungi are known to show low species diversity. About 240 species of arbuscular fungi are described mostly based on their morphology and grouped in the phylum Glomeromycota (Błaszkowski 2012; Lee et al. 2013). According to classification presented by Błaszkowski (2012), phylum Glomeromycota consists of four orders: Archaeosporales, Diversisporales, Glomerales, and Paraglomerales, comprising nine families and thirteen genera, belonging to the class Glomeromycetes of the phylum Glomeromycota. The only Geosiphon pyriformis of the family Geosiphonaceae (Archaeosporales) forms endocytosymbiosis with cyanobacteria (Nostoc sp.) but does not form arbuscular mycorrhizae (Błaszkowski 2012). Members of Glomeromycota phylum turned out to be closer to fungi belonging to Ascomycota and Basidiomycota than to those from Zygomycota as previously thought (Błaszkowski 2012).

The factors limiting the estimation of species diversity of arbuscular fungi and their taxonomical position were the difficulty in extracting their spores from the soil and maintaining the fungi in cultures (Błaszkowski 2012). However, molecular methods used in recent years, have suggested that diversity of arbuscular fungi may be underestimated (Husband et al. 2002; Wubet et al. 2003; Baltruschat et al. 2019). Also, it has been shown that individual species and single spores present high genetic diversity, resulting in differentiation of important functions such as colonization rates, growth of extraradical hyphae, or phosphorus uptake by arbuscular fungi (Lee et al. 2013). The highest diversity of arbuscular fungi was observed in the tropics, where fungal communities revealed significant spatial heterogeneity and non-random associations with the different hosts. Fungal communities presented also a high variation of species composition in time (Husband et al. 2002).

7.4 Structure of Arbuscular Mycorrhizae

The term arbuscular mycorrhiza has evolved over many years with the development of science. Originally, arbuscular mycorrhiza was called endomycorrhiza. However, after finding its distinctness in terms of functionality and evolution in comparison to the other types of endomycorrhiza, the name vesicular-arbuscular mycorrhiza (VAM) was adopted and used for several years. Because only 80% of VAM forms vesicles, this name was finally abandoned and the new name arbuscular mycorrhiza (AM) has been started to use without indicating the existing structural traits.

The arbuscular mycorrhiza is formed of its characteristic extra- and intraradical structures. The extraradical structures are spores, extraradical hyphae penetrating the soil and connecting spores with roots, and auxiliary cells (formed instead of the vesicles—only by species belonging to Gigasporaceae and Pacisporaceae) (Błaszkowski 2012). Extraradical hyphae penetrate to the roots through the appressorium. Colonization of plant roots by arbuscular fungi is linked with the presence of intraradical hyphae located between the cells of the plant root (intercellular) or penetrating inside the cells (intracellular). Other structures whose arbuscular fungi owe their name are arbuscules—intracellular tree-like transformations of mycelium, responsible for the exchange of nutrients between the host and the fungus. Some arbuscular fungi also create intraradical oval structures—vesicles. Vesicles containing abundant lipids and numerous nuclei are thought to be important storage organs and as propagules play a role in recolonization of roots in the soil. Fungal species in the genera Gigaspora and Scutellospora produce auxiliary vesicles (sometimes called auxiliary bodies or cells) in the extraradical mycelium (Peterson et al. 2004). Among arbuscular fungi, there are two main anatomical types—Arum-type and Paris-type, taking their names from the names of plants in which they were first observed by Gallaud at the beginning of the twentieth century (Gallaud 1904, 1905). In the case of the Arum-type, hyphae penetrate the epidermis generally forms a coil either in the epidermal cell or first cortical cell layer before it enters the intercellular spaces of the cortex. Hyphae also branch many times. Inki Arum-type, roots can be rapidly colonized along the root axis due to the free movement of hyphae in the intercellular spaces (Peterson et al. 2004). Arbuscules arise depending on the species of fungus inside cortical cells. Vesicles may also be present. In the case of woody species, the presence of Arum-type was found, among others, in the roots of *Sorbus torminalis* (Rosaceae) (Bzdyk et al. 2016).

Paris-type is characterized by the presence of intracellular coiled hyphae spreading from cell to cell. Plant species containing this type of mycorrhiza show a lack of conspicuous intercellular spaces and only intracellular hyphae is present. Coils develop from that intracellular hyphae and next, the coils give rise to fine lateral branches. These coils are often accompanied by arbuscules or arbusculate coils. The hyphal coils and the fine branches have a perifungal membrane and an interfacial matrix of host-derived cell wall components (Peterson et al. 2004). The Paris-type is the most common type of arbuscular mycorrhiza in the plant world (Smith and Read 2008). It occurs in several herbaceous plants present in forests as well as in woody plants such as *Acer saccharum, Ginkgo biloba,* or *Taxus baccata* (Smith and Read 2008). In addition to both types, there are also many intermediate types containing characteristic traits of both Arum- and Paris-type. One should also take into account the seasonal variability of mycorrhizal structures, their durability, and the impact of soil conditions. An example would be arbuscules, whose life span may be limited to 1–15 days (Harley and Harley 1987). For example, in the case of horse chestnut depending on the period of observation of colonization of roots by arbuscular fungi, both arbuscules and coils or only coils were observed (Karliński et al. 2014; Tyburska-Woś et al. 2019). Also, the common presence of vesicles in fine roots of *Aesculus hippocastanum* in late autumn is characteristic for the end of the growing season (Peterson et al. 2004; Karliński et al. 2014). Hence, to fully assess the structure of fungal symbionts in tree root tissues it is important to repeat observations at different time points.

7.5 Global Distribution of Arbuscular Mycorrhizal Trees

Arbuscular mycorrhiza is the most common form of symbiosis in the plant community and thus also among woody species (Table 7.1). The main factors determining global distribution and diversity of trees colonized by arbuscular fungi seem to be the soil nutrients abundance, climatic conditions, and the evolutionary heritage of arbuscular symbiosis (Chen et al. 2018; Tedersoo and Bahram 2019; Soudzilovskaia et al. 2019). The long history of the evolution of arbuscular symbiosis and their prevalence in different (often extreme) habitats underline high adaptation and good metabolic fit of arbuscular fungi and trees (Chen et al. 2018). Also, the low host specificity supports the potential success of arbuscular fungi in the establishment of symbiotic relations with different tree partners (Smith and Read 2008). On the other hand, arbuscular fungi are known for their intraspecific diversity of spores (Lee et al.

Table 7.1 Selected shrubs and trees, and their mycorrhizal symbionts

Type of mycorrhiza	Shrubs and trees
AM	*Acer* sp.; *Aesculus hippocastanum* L.; *Araucaria* sp.; *Arecaceae*; *Artemisia* sp.; *Asimina* Adans.; Buxus sp.; *Camellia sinensis* (L.) Kuntze; *Carica papaya* L.; *Carya ovata* (Mill.) K. Koch; *Cedrus* Trew; *Citrus* sp.; *Cocos nucifera*; *Coffea* sp.; *Cornus* florida L.; *Cupressus* sp.; *Elaeagnus angustifolia* L.; *Erythrina caffra* Thunb.; *Euonymus*; *Euphorbia pulcherrima* Willd. ex Klotzsch; *Fagus* sp.; *Ficus* sp.; *Fraxinus* sp.; *Ginkgo biloba* L.; *Hevea brasiliensis*; *Juglans nigra* L.; *Juniperus* sp.; *Ligustrum* sp.; *Liquidambar styraciflua* L.; *Magnolia* L.; *Malus* sp.; *Mangifera indica* L.; *Musa* sp.; *Myrica* L.; *Olea europaea* L.; *Oxydendrum arboreum*; *Parthenium argentatum*; *Persea americana*; *Pistacia vera* L.; *Pittosporum tobira* (Thunb.) W.T.Aiton; *Platanus* L.; *Podocarpus* Persoon; *Prosopis* L.; *Prunus* sp.; *Rhaphiolepis* Lindl.; Rhus L.; *Robinia pseudoacacia*; *Robus* sp.; *Rosa* sp.; *Sequoia sempervirens* (lamb. Ex D. Don) Endl.; *Sequoiadendron giganteum*; *Shibataea kumasasa*; *Sorbus torminalis* (L.) Crantz; *Taxus baccata* L.; *Theobroma cacao*; *Triadica sebifera*
AM and ECM	*Acacia* sp.; *Alnus* sp.; *Carya laciniosa* (F.Michx.) G.Don; *Carya cordiformis* (Wangenh.) K.Koch; *Casuarina* sp.; *Eucalyptus sp.; Myrtaceae;* Populus sp.; *Quercus rubra* L.; *Salix* sp.

2013), which may selectively favor different plants (Chen et al. 2018). A more diverse arbuscular fungal community can maintain plant diversity (van der Heijden et al. 1998; Husband et al. 2002; Lovelock et al. 2003).

Arbuscular fungi dominate in relatively nutrient-rich and phosphorus-limited soils of tropics (Tedersoo and Bahram 2019; Soudzilovskaia et al. 2019), in contrast to temperate and northern nitrogen-limited habitats dominated by ectomycorrhizal forests. The picture of temperate areas seems to be the most complex because of human-driven land transformations (Soudzilovskaia et al. 2019). It can be assumed that due to the rapid global climate changes the contribution of trees establishing symbiotic relationships with arbuscular fungi will increase in this area. Arid areas and wetlands are also often inhabited by arbuscular mycorrhizal shrubs or trees because of high tolerance of arbuscular fungi to unfavorable conditions.

7.6 Factors Impacting Trees and Their Fungal Symbionts

Soil conditions, their diversity, and frequent changes (much higher than in the aquatic environment) have been the main factor that arbuscular fungi have faced since the beginning of their colonization of the lands. The impact of environmental factors on the communities of fungi associated with roots of plants is well reflected by the multi-seasonal woody plants. In this respect particularly interesting is the small group of trees that establish symbiotic associations with arbuscular as well as ectomycorrhizal fungi. The presence of both groups of fungal symbionts on the roots of the one host tree allows to analyzing not only the mutual relationship of trees and fungi but also to compare the reactions between these two different fungal worlds and host trees.

Environmental conditions determine the global distribution of mycorrhizal fungi from the tropical forests being the habitat of mainly arbuscular mycorrhiza and rich plant communities, to the less differentiated northern forests, dominated by ectomycorrhiza and much slower nutrient cycling than in the tropics. Soil conditions are thought to be an important factor in mycorrhizal colonization (Gonçalves and Martins-Loução 1996; Smith and Read 2008; Karliński et al. 2010, 2013, 2020). Arbuscular fungi are characterized by a greater amplitude of tolerance to adverse soil conditions than ectomycorrhizal fungi (Smith and Read 2008). In case of poplars as well as the majority of other members of Salicaceae, they primarily establish symbiotic associations with arbuscular fungi. Later, arbuscular fungi are partly replaced by ectomycorrhizas (Dominik 1958; Smith and Read 2008). Arbuscular fungi giving way to ectomycorrhizal fungi, tend to partially move into deeper soil layers with limited oxygen and nutrients concentration (Neville et al. 2002; Karliński et al. 2010). Also, the floods are important disturbances impacting dual mycorrhizal colonization symbiosis of trees, e.g., *Populus, Salix, Alnus,* or some species of *Quercus* (Watson et al. 1990; Teste et al. 2019). Wet and poorly aerated conditions favor arbuscular fungi and negatively impact ectomycorrhizas of *Alnus* (Truszkowska 1953), *Populus*, *Salix* (Lodge 1989) or *Quercus rubra* (Watson et al. 1990). On the other hand, also drought stress negatively impacts the ectomycorrhizas in contrast to dominating arbuscular fungi (Gehring et al. 2006; Quereyeta et al. 2009; Kilpeläinen et al. 2017). Also, forest fires to a greater extent reduce ectomycorrhizal inoculum than arbuscular fungi, which are able more quickly to support the regeneration of disturbed areas (Lapeyrie and Chilvers 1985; Horton et al. 1998; Teste et al. 2019). The neighborhood of other trees may also affect the root colonization of young seedlings. The proximity of ectomycorrhizal trees supports the higher ectomycorrhizal colonization of seedlings roots and vice versa, the trees colonized by arbuscular fungi may increase arbuscular mycorrhizal colonization of seedlings (Dickie et al. 2001).

A significant impact of soil temperature on the range of root colonization by both fungal groups was also found. In high soil temperatures, ectomycorrhizal colonization was found to be decreased (McGee 1988). On the other hand, the low temperature was the limiting factor for arbuscular fungi, but ectomycorrhizal fungi were not affected (Kilpeläinen et al. 2016). Different factors have been suggested to influence the arbuscular and ectomycorrhizal colonization such as litter accumulation negatively impacted arbuscular fungi (Conn and Dighton 2000; Piotrowski et al. 2008) or N and P fertilization (Baum and Makeschin 2000).

The negative impact of heavy metal pollutants on the biomass of arbuscular fungi and other groups of fungi has been reported by several authors (e.g., Frostegård et al. 1993; Hagerberg et al. 2011; Pennanen et al. 1996). The analyses of microbial communities in the rhizosphere of mature poplars growing in the vicinity of the copper smelter revealed that the negative effect of heavy metal pollution was more pronounced for the biomass of fungal part of soil microbiome than for bacterial biomass (Karliński et al. 2020). This result is in line with the other observations obtained near copper or zinc smelters, which showed a decrease in phospholipid fatty acids (PLFAs) as indicators of the fungal biomass (e.g., Pennanen et al. 1996). Lower tolerance of fungi than bacteria to heavy metal pollution (Cu, Zn) was also confirmed in laboratory experiments (Rajapaksha et al. 2004).

The negative impact of heavy metal pollution was observed especially in the case of the biomass of arbuscular fungi in the rhizosphere of poplars (Karliński et al. 2020). To a lesser extent, soil pollution decreased the biomass of ectomycorrhizal fungi. On the other hand, the colonization of poplar roots by fungal endophytes was significantly higher at the polluted site than at the unpolluted sites (Karliński et al. 2010). This group of fungi seems to be more tolerant of high concentrations of heavy metals and to compete with arbuscular fungi in the root colonization of poplars. The increase of dark septate endophytes in the roots at polluted sites was also observed by Routsalainen et al. (2007) and Jumpponen and Trappe (1998).

Plants, as primary suppliers of carbon exudates and other plant-derived materials, are important players in shaping the microbiome of the soil environment (e.g., Bulgarelli et al. 2015; Lottmann et al. 2010; Schweitzer et al. 2008). The role of plant genotypes in shaping fungal communities associated with plants has mainly focused on short-lived crop plants, e.g., barley, tomato, cucumber, sweet pepper, and chickpea (Bulgarelli et al. 2015; Ellouze et al. 2013; Ravnskov and Larsen 2016). To a lesser extent, the impact of the genotype has been analyzed in case of trees, especially those growing in different environmental conditions and impacted by the long-lasting natural or anthropogenic factors (Karliński et al. 2020). The observations of different poplar clones and hybrids revealed their different response to environmental conditions, which was particularly evident in the polluted area. Poplars clearly reflected different rooting strategies expressed in differentiated fine roots production and their different distribution in the soil profile (Karliński et al. 2010), perhaps due to the increased environmental pressure caused by heavy metal contamination of soil. At the poplar rhizosphere, the biomass of arbuscular fungi and other soil microorganisms was mainly determined by site conditions and to a lesser extent by genotype. But the contribution of microorganisms in poplars microbiome was significantly impacted by tree genotype (Karliński et al. 2020). The poplar genotype determined the contribution of the biomass ratio of arbuscular fungi and other soil fungi despite their different character of interactions with host trees. Also, the biomass ratio of fungi (including arbuscular and ectomycorrhizal symbionts of poplars) and bacteria (F:B ratio) in poplar rhizosphere appeared to be a good indicator of the poplar genotype (Karliński et al. 2020). However the F:B ratio was also shown as a suitable indicator of changes in the microbial community depending on the soil moisture gradient and changes in soil management or soil pollution (Bailey et al. 2002; Zhang et al. 2016). Generally, fungi (including arbuscular fungi) seem to be more related to poplar genotype, while bacteria to a greater extent depend on the site (Karliński et al. 2020). On the other hand, the soil microbiomes of the poplars showed relatively stable proportions of groups of microorganisms different sites, which confirm high adaptability of poplars to different soil conditions and their ability to shape communities of microorganisms associated with them (Karliński et al. 2020).

Both arbuscular and ectomycorrhizal fungal species may belong to commonly occurring generalists/cosmopolitans and to the specialists, whose occurrence is limited by several environmental, geographical, or physiological factors. Up to now, more studies describing fungal communities have been concentrated on ectomycorrhizal fungi associated with woody species in natural conditions, nurseries, or at disturbed areas (e.g., Karliński et al. 2013; Leski et al. 2019; Rudawska et al. 2018, 2019). Communities of arbuscular fungi, despite their widespread occurrence, were thought to be less diverse (Smith and Read 2008). However, the use of new molecular techniques causes this image to change in recent years (e.g. Öpik et al. 2008, Helgason and Fitter 2009; Baltruschat et al. 2019). However, as in case of ectomycorrhizal fungi, also in the case of arbuscular fungi, the main factors shaping their communities are soil conditions. The tree genotype only plays a limited role here (Karliński et al. 2013). The main role in the colonization of young plants by mycorrhizal fungi plays the local network of mycorrhizal fungi, shaped by geographical and ecological factors. Most of the fungi as a generalist establish symbiotic relationships with many plant species. Thus, poplars growing in different conditions show a common set of root bacteria as well as endophytic and ectomycorrhizal fungi associated with them, and no single taxon or consortium of microbes is indicative of a particular Populus genotype (Karliński et al. 2013; Bonito et al. 2019). This means that possible differences between communities of mycorrhizal fungi associated to poplar genotypes are marked rather locally, in specific soil conditions and general trends can be expected at the level of groups of organisms rather than individual species

References

Bailey VL, Smith JL, Bolton H Jr (2002) Fungal-to-bacterial biomass ratios in soils investigated for enhanced carbon sequestration. Soil Biol Biochem 34:997–1007

Baltruschat H, Santos VM, Alves da Silva DK, Schellenberga I, Deubel A, Sieverdinge E, Oehl F (2019) Unexpectedly high diversity of arbuscular mycorrhizal fungi in fertile Chernozem croplands in Central Europe. Catena 182:104135

Baum C, Makeschin F (2000) Effect of nitrogen and phosphorus fertilization on mycorrhizal formation of two poplar clones (*Populous trichocarpa* and *P. tremula* × *tremuloides*). Plant Nutr Soil Sci 163:491–497

Błaszkowski J (2012) Glomeromycota W. Szafer Institute of Botany, Polish Academy of Sciences, Kraków, Poland, pp 303

Bonito G, Benucci GMN, Hameed K, Weighill D, Jones P, Chen K-H, Jacobson D, Schadt C, Vilgalys R (2019) Fungal-bacterial networks in the Populus rhizobiome are impacted by soil properties and host genotype. Front Microbiol 10:481

Brundrett MC (2002) Coevolution of roots and mycorrhizas of land plants. New Phytol 154:275–304

Brundrett MC, Tedersoo L (2018) Evolutionary history of mycorrhizal symbioses and global host plant diversity. New Phytol 220:1108–1115

Bulgarelli D, Garrido-Oter R, Münch PC, Weiman A, Droge J, Pan Y, McHardy AC, Schulze-Lefert P (2015) Structure and function of the bacterial root microbiota in wild and domesticated barley. Cell Host Microbe 17:392–403

Bzdyk RM, Kohler J, Olchowik J, Aleksandrowicz-Trzcińska M, Kirisits T (2016) Arum-type of arbuscular mycorrhizae, dark septate endophytes and olpidium spp. in fine roots of container-grown seedlings of *Sorbus torminalis* (Rosaceae). Acta Soc Bot Pol 85:3495
Chen M, Arato M, Borghi L, Nouri E, Reinhardt D (2018) Beneficial services of arbuscular mycorrhizal fungi – from ecology to application. Front Plant Sci 9:1270
Conn C, Dighton J (2000) Litter quality influences on decomposition, ectomycorrhizal community structure and mycorrhizal root surface acid phosphatase activity. Soil Biol Biochem 32:489–496
Delaux PM (2017) Comparative phylogenomics of symbiotic associations. New Phytol 213:89–94
Dickie IA, Koide RT, Fayish AC (2001) Vesicular-arbuscular mycorrhizal infection of *Quercus rubra* seedlings. New Phytol 151:257–264
Dominik T (1958) Study of mycotrophism of the genus Populus. Pr Inst Bad Leśn 181:117–172
Ellouze W, Hamelb C, Vujanovic V, Ganb Y, Bouzid S, St-Arnaud M (2013) Chickpea genotypes shape the soil microbiome and affect the establishment of the subsequent durum wheat crop in the semiarid North American Great Plains. Soil Biol Biochem 63:129–141
Frostegård Å, Tunlid A, Bååth E (1993) Phospholipid fatty acid composition, biomass, and activity of microbial communities from two soil types experimentally exposed to different heavy metals. Appl Environ Microbiol 59:3606–3617
Gallaud I (1904) Études sur les Mycorrhizes Endotrophes. le Bigot Frères, Lille
Gallaud I (1905) Études sur les mycorrhizen endotrophes. Révue Generale de Botanique 17:5–48, 66–83, 123–136, 223–239, 313–325, 425–433, 479–500
Garg N, Chandel S (2010) Arbuscular mycorrhizal networks: process and functions. A review. Agron Sustain Dev 30:10.1051
Gehring CA, Mueller RC, Witham TG (2006) Environmental and genetic effects on the formation of ectomycorrhizal and arbuscular mycorrhizal associations in cottonwoods. Oecologia 149:158–164
Gonçalves MT, Martins-Loução MA (1996) Mycotrophic conditions of populus nigra and P. × euramericana. Evolution of endo vs ectomycorrhiza. Evaluation of endo vs ectomycorrhizal. In: Azcon-Aguilar C, Barea JM (eds) Mycorrhizas in integrated systems from genes to development. Proceedings of fourth European symposium on mycorrhizas. ECSC-EC-EAEC, Brussels, pp 118–120
Hagerberg D, Manique N, Brandt KK, Larsen J, Nybroe O, Olsson S (2011) Low concentration of copper inhibits colonization of soil by the arbuscular mycorrhizal fungus *Glomus intraradices* and changes the microbial community structure. Microb Ecol 61:844–852
Harley JL, Harley EL (1987) A check-list of mycorrhiza in the British flora-addenda, errata and index. New Phytol 107:741–749
Helgason T, Fitter AH (2009) Natural selection and the evolutionary ecology of the arbuscular mycorrhizal fungi (Phylum Glomeromycota). J Exp Bot 60:24652480
Horton TR, Cázares E, Bruns TD (1998) Ectomycorrhizal, vesicular-arbuscular and dark septate fungal colonization of bishop pine (*Pinus muricata*) seedlings in the first 5 months of growth after wildfire. Mycorrhiza 8:11–18
Husband R, Herre EA, Turner SL, Gallery R, Young JPW (2002) Molecular diversity of arbuscular mycorrhizal fungi and patterns of host association over time and space in a tropical forest. Mol Ecol 11:2669–2678
Jumpponen A, Trappe JM (1998) Dark septate endophytes: a review of facultative biotropic root colonizing fungi. New Phytol 140:295–310
Karliński L, Rudawska M, Kieliszewska-Rokicka B, Leski T (2010) Relationship between genotype and soil environment during colonization of poplar roots by mycorrhizal and endophytic fungi. Mycorrhiza 20:315–324
Karliński L, Rudawska M, Leski T (2013) The influence of host genotype and soil conditions on ectomycorrhizal community of poplar clones. Eur J Soil Biol 58:51–58
Karliński L, Jagodziński AM, Leski T, Butkiewicz P, Brosz M, Rudawska M (2014) Fine root parameters and mycorrhizal colonization of horse chestnut trees (*Aesculus hippocastanum* L.) in urban and rural environments. Landsc Urban Plan 127:154–163

Karliński L, Ravnskov S, Rudawska M (2020) Soil microbial biomass and community composition relates to poplar genotypes and environmental conditions. Forests 11:262

Kilpeläinen J, Vestberg M, Repo T, Lehto T (2016) Arbuscular and ectomycorrhizal root colonisation and plant nutrition in soils exposed to freezing temperatures. Soil Biol Biochem 99:85–93

Kilpeläinen J, Barbero-Lopez A, Vestberg M, Heiskanen J, Letho T (2017) Does severe soil drought have after-effects on arbuscular and ectomycorrhizal root colonisation and plant nutrition? Plant Soil 418:377–386

Lapeyrie FF, Chilvers GA (1985) An endomycorrhiza-ectomycorrhiza succession associated with enhanced growth of *Eucalyptus dumosa* seedlings planted in a calcareous soil. New Phytol 100:93–104

Lee EH, Eo JK, Ka KH, Eom AH (2013) Diversity of arbuscular mycorrhizal fungi and their roles in ecosystems. Mycobiology 41:121–125

Leski T, Rudawska M, Kujawska M, Stasińska M, Janowski D, Karliński L, Wilgan R (2019) Both forest reserves and managed forests help maintain ectomycorrhizal fungal diversity. Biol Conserv 238:108206

Lodge DJ (1989) The influence of soil moisture and flooding on formation of VA-endo- and ectomycorrhizae in Populus and Salix. Plant Soil 117:243–253

Lottmann J, O'Callaghan M, Baird D, Walter C (2010) Bacterial and fungal communities in the rhizosphere of field-grown genetically modified pine trees (*Pinus radiata* D.). Environ Biosaf Res 9:25–40

Lovelock CE, Andersen K, Morton JB (2003) Arbuscular mycorrhizal communities in tropical forests are affected by host tree species and environment. Oecologia 135:268–279

Martin FM, Uroz S, Baker DG (2017) Ancestral alliances: plant mutualistic symbioses with fungi and bacteria. Science 356:eaad4501

McGee PA (1988) Vesicular-arbuscular and ectomycorrhizas on the annual composite. *Podotheca angustifolia*. Symbiosis 6:271–280

Neville J, Tessier JL, Morrison I, Scarratt J, Canning B, Klironomos JN (2002) Soil depth distribution of ecto- and arbuscular mycorrhizal fungi associated with Populus tremuloides within a 3-year-old boreal forest clear-cut. Appl Soil Ecol 19:209–216

Öpik M, Moora M, Zobel Z, Saks U, Wheatley R, Wright F, Daniell T (2008) High diversity of arbuscular mycorrhizal fungi in a boreal herb-rich coniferous forest. New Phytol 179:867–876

Pennanen T, Frostegård Å, Fritze H, Bååth E (1996) Phospholipid fatty acid composition and heavy metal tolerance of soil microbial communities along two heavy metal-polluted gradients in coniferous forests. Appl Environ Microbiol 62:420–428

Peterson RL, Massicotte HB, Melville LH (2004) Mycorrhizas: anatomy and cell biology. NRC Research Press National Research Council of Canada, Ottawa

Piotrowski J, Morford S, Rillig M (2008) Inhibition of colonization by a native arbuscular mycorrhizal fungal community via *Populus trichocarpa* litter, litter extract, and soluble phenolic compounds. Soil Biol Biochem 40:709–717

Quereyeta J, Egerton-Warburton LM, Allen MF (2009) Topographic position modulates the mycorrhizal response of oak trees to interannual rainfall variability. Ecology 90:649–662

Rajapaksha RMCP, Tobor-Kapłon MA, Bååth E (2004) Metal toxicity affects fungal and bacterial activities in soil differently. Appl Environ Microbiol 70:2966–2973

Ravnskov S, Larsen J (2016) Functional compatibility in cucumber mycorrhizas in terms of plant growth performance and foliar nutrient composition. Plant Biol 18:816–823

Routsalainen AL, Markkola A, Kozlov MV (2007) Root fungal colonization in Deschampsia flexuosa: effect of pollution and neigbouring trees. Environ Pollut 147:723–728

Rudawska M, Leski T, Wilgan R, Karliński L, Kujawska M, Janowski D (2018) Mycorrhizal associations of the exotic hickory trees, *Carya laciniosa* and *Carya cordiformis*, grown in Kórnik Arboretum in Poland. Mycorrhiza 28:549–560

Rudawska M, Kujawska M, Leski T, Janowski D, Karliński L, Wilgan R (2019) Ectomycorrhizal community structure of the admixture tree species *Betula pendula*, *Carpinus betulus*, and *Tilia cordata* grown in bare-root forest nurseries. For Ecol Manag 43:113–125

Schweitzer JA, Bailey JK, Fischer DG, LeRoy CJ, Lonsdorf EV, Whitham TG, Hart SC (2008) Plant-soil-microorganism interactions: heritable relationship between plant genotype and associated soil microorganisms. Ecology 89:773–781

Smith SE, Read D (2008) Mycorrhizal symbiosis. Academic, London

Soudzilovskaia NA, van Bodegom PM, Terrer C, van't Zelfde M, McCallum I, McCormack ML, Fisher JB, Brundrett MC, César de Sá N, Tedersoo L (2019) Global mycorrhizal plant distribution linked to terrestrial carbon stocks. Nat Commun 10:5077

Tedersoo L, Bahram M (2019) Mycorrhizal types differ in ecophysiology and alter plant nutrition and soil processes. Biol Rev 94:1857–1880

Teste FP, Jones MD, Dickie IA (2019) Dual-mycorrhizal plants: their ecology and relevance. New Phytol 225:1835–1851. https://doi.org/10.1111/nph.16190

Truszkowska W (1953) Mycotrophy of Alneta in the Bualowieża National Park and in Domaszyn near Wroclaw. Acta Soc Bot Pol 22:737–752

Tyburska-Woś J, Nowak K, Kieliszewska-Rokicka B (2019) Influence of leaf damage by the horse chestnut leafminer (*Cameraria ohridella* Deschka & Dimić) on mycorrhiza of *Aesculus hippocastanum* L. Mycorrhiza 29:61–67

van der Heijden MGA, Klironomos JN, Ursic M, Moutoglis P, Streitwolf-Engel R, Boller T, Wiemken A, Sanders IR (1998) Mycorrhizal fungal diversity determines plant biodiversity, ecosystem variability and productivity. Nature 396:69–72

Watson G, von der Heide-Spravka K, Howe V (1990) Ecological significance of endo-ectomycorrhizae in the oak subgenus Erythrobalanus. Arboricult J 14:107–116

Wubet T, Weiss M, Kottke I, Teketay D, Oberwinkler F (2003) Molecular diversity of arbuscular mycorrhizal fungi in Prunus africana, an endangered medicinal tree species in dry Afromontane forests of Ethiopia. New Phytol 161:517–528

Zhang Q, Wu J, Yang F, Lei Y, Zhang Q, Cheng X (2016) Alternations in soil microbial community composition and biomass following agricultural land use change. Sci Rep 6:36587

Chapter 8
Effectiveness of Arbuscular Mycorrhizas in Improving Carob Culture in the Mediterranean Regions

Abdellatif Essahibi, Laila Benhiba, Cherki Ghoulam, and Ahmed Qaddoury

Abstract Carob (*Ceratonia siliqua* L.) tree is considered among the most important forest-fruit species native to the Mediterranean region. It has various uses and great valorization potential, all parts of this plant could be exploited as a source of income, as human food or livestock fodder as well as source of raw materials for pharmaceutical, cosmetic, or food industries. Moreover, due to its particular agroecological features, carob tree offers the advantage of growing in poor and unfertile soils in the Mediterranean and Mediterranean-like regions of the world. Thus, carob trees are suitable for the rehabilitation of marginal and sub-marginal areas, helping to compensate for the expanding land desertification in these regions where it can play the role of pioneer and productive species. Carob has been intermittently explored over the last 20 years as a potential tree crop industry in low rainfall areas. The importance of developing the industrial agroforestry potential of the carob tree is hurdled by the lack of options for agroforestry, especially in Mediterranean regions with low rainfall (below 500 mm), and by the need to develop suitable practices for the sustainable management of natural resources. Viable commercial carob cultivation will require mastering efficient farming practices with detailed attention to water requirements and soil fertility. It would improve agricultural productivity in low rainfall areas, help manage water and land degradation, diversify farmers' incomes, and contribute to the development of export industries contributing to balance the economy of the country. This chapter will provide current knowledge regarding the use of mycorrhizal symbiosis for the improvement of carob culture and productivity in the context of Mediterranean ecosystems. An overview on the multipurpose potential of the carob tree and how spreading its cultivation will benefit people and the environment in marginal areas is highlighted.

Keywords Carob tree · Multiplication · Mycorrhiza · Biofertilizer · Sustainability · Abiotic stresses

---

A. Essahibi · L. Benhiba · C. Ghoulam · A. Qaddoury (✉)
Plant Biotechnology and Bioengineering Laboratory, Department of Biology, Faculty of Sciences & Technology, University Cadi Ayyad, Marrakech, Morocco
e-mail: a.qaddoury@uca.ma

N. Shrivastava et al. (eds.), *Symbiotic Soil Microorganisms*, Soil Biology 60,
https://doi.org/10.1007/978-3-030-51916-2_8

8.1 Introduction

The carob tree (*Ceratonia siliqua* L.) is a flowering woody species belonging to the legume family, *Fabaceae*, native to the Mediterranean region. This evergreen forest-fruit tree may reach, under favorable conditions, 20 m in height, up to 3.5 m in trunk girth, and up to 200 years of longevity (Ait Chitt et al. 2007). Carob trees have been widely exploited for food and forage since antiquity. It is an important component of the arboreal flora of the Mediterranean region which is widely cultivated for industrial, ecological, ornamental, environmental, agricultural, and land restoration purposes (Batlle and Tous 1997). It has various uses and great valorization potential: human food, livestock fodder as well as ornamental, carpentry, beekeeping, and traditional medicines purposes (Morton 1987; Rejeb et al. 1991; Ait Chitt et al. 2007). The nutritional value of carob fruits (seeds and pulp) is indisputable; it is in high demand for pharmaceutical, chemical, food, and cosmetic industries (Rejeb et al. 1991; Makris and Kefalas 2004; Konate 2007; Custodio et al. 2011a, b). On the other hand, carob tree develops strong and deep roots, which probably account for its ability to grow under harsh conditions including drought, salinity, and nutrient-poor soils (Vertovec et al. 2001; Sakcali and Ozturk 2004; Correia et al. 2010; Ozturk et al. 2010). In addition, due to its ability to preserve and enrich the soil, carob cultivation facilitates the establishment of other plant species, being particularly useful for the rehabilitation of difficult areas where it can simultaneously play the role of pioneer and productive species. In fact, this tree has been used to restore numerous marginal and semiarid and arid lands around the Mediterranean basin (Ozturk et al. 2010; Ouahmane et al. 2012; Bakry et al. 2013). Nowadays, it has caused to lost due to problems as urban sprawl, illegal cut of trees and overgrazing, and global climate changes, leading to damage of natural flora and ecosystems. In addition, carob cultivation in the Mediterranean region is limited to marginal and/or semiarid areas characterized by a harsh climate and poor soil conditions that negatively affected plants establishment, thus limiting the success of reforestation programs. Moreover, the large scale cultivation of carob is hampered by the traditional methods of its propagation that fail to meet the growing demand for high value plants. The detrimental effects of these constraints exacerbated by the global climate changes caused not only reduction in the carob production, a considerable source of incomes for millions of humans in the marginal and semiarid areas, but also an imbalance of the natural ecosystems causing a serious threat to plant resources and sustainable development in these difficult environments.

While carob is reported to be tolerant of water shortage and saline soil, a commercial carob cultivation will require reliable yields that can only be achieved with rainfall of 500 mm or greater. In areas of low or unpredictable rainfall, irrigation will be necessary to achieve required yields. In addition, correct agroforestry practices including irrigation and fertilizer applications will also be required to achieve optimum yields. One of the main challenges for the sustainable development of carob cultivation is the combination of the proper agroforestry practices based on plant technology innovations and the management of the agroecosystem services that has been at the forefront of generating and promoting sustainable agricultural production. In this context, arbuscular mycorrhizal fungi (AMF), often referred to as "ecosystem engineers," "biocontrol agents," "biofertilizers," or "bioenhancers," are the most important providers of these ecological services. Arbuscular mycorrhizal fungi can form symbiotic associations (mycorrhiza), to the mutual benefit of the host plant and fungus, with roots of 80% of terrestrial plant species including trees, shrubs, herbs, and crop plants (Harley and Smith 1983; Smith and Read 2008). Mycorrhizal plants show not only high nutrient and water acquisition efficiency, but also less susceptibility to disease and high tolerance to adverse conditions (Evelin et al. 2012; Beltrano et al. 2013; Ruiz-Lozano et al. 2016; Wu 2017). The benefits from AMF are thought to be highest when the colonization occurs as early as possible during the vegetative growth of the plant (Scagel 2001; Essahibi et al. 2017). Carob tree is among the vascular plants that can establish mycorrhizal associations and strongly depends on its functioning to grow under severe environmental constraints (Manaut et al. 2015; Essahibi et al. 2018, 2019). This chapter will provide an overview regarding the multipurpose potential of the carob tree and how spreading its cultivation will benefit people and environment in marginal areas. It discusses also the current knowledge concerning the use of mycorrhizal symbiosis for the improvement of carob culture and productivity in the context of Mediterranean ecosystems. The most effective applications of AMF for improving carob performance in terms of multiplication, growth, nutrition, and protection against adverse effects of environmental stresses are highlighted.

8.2 Carob Origins and Cultivation

Carob tree has been grown since antiquity in most countries of the Mediterranean basin. The origin of carob is not clear, it was considered by De Candolle (1883) and Vavilov (1951) as native to the eastern Mediterranean region, mainly Turkey and Syria. However, according to Schweinfurth (1894) carob is native to the southern Arabia (Yemen). Recently, Zohary (1973) considered carob as originating from Indo-Malesian flora. The only known carob-related species is *Ceratonia oreothauma*, which is considered to be native to the southeast Arabia (Oman) and the African horn (north of Somalia) (Hillcoat et al. 1980). The origin of carob populations throughout the Mediterranean basin has been associated with an historical process of dissemination by humans since its domestication in the Middle-East around 4000 BC. Its value was recognized by the ancient Greeks, who brought it to Greece and Italy and by Arabs who disseminated it along the North African coast and into Spain and Portugal. Lately, it has also been successfully introduced in Australia, South Africa, the USA, Chile, Argentina, Philippines, and Iran (Fig. 8.1) (Morton 1987; Batlle and Tous 1997; Sbay and Abourouh 2006). This indicates that the general environmental conditions (e.g., rainfall, soils, temperature, etc.) of these regions do not impose significant limitations on carob growth. However, commercial cultivation will require detailed attention to carob needs. For example, while the

Fig. 8.1 World carob distribution (Batlle and Tous 1997)

literature often reports that carobs are drought tolerant carob trees have little yield under low rainfall conditions but increase with increasing rainfall; 500 mm/year considered necessary for reliable yields.

8.3 Economic Potential of Carob Tree

Carob is predominantly grown in the Mediterranean and Mediterranean-like climate regions of the world. The World production of carob is estimated at about 420,000 t/year, and the main producers of pulp and seeds, respectively, are Spain (36%, 28%), Morocco (24%, 38%), Italy (10%, 8%), Portugal (10%, 8%), Greece (8%, 6%), Turkey (4%, 6%), and Cyprus (3%, 2%) (Bouhadi et al. 2017). The economic importance of carob tree comes from the multiple uses of its products in pharmaceutical, chemical, food, and cosmetic industries (Makris and Kefalas 2004; Custodio et al. 2011a, b; El Kahkahi et al. 2016). Indeed, Carob tree has various uses and great valorization potential; all parts of the tree are valued as food or as a source of income. Carob fruits or pods consist of a pulp wrapping seeds, presenting respectively 90 and 10% of the total fruit weight. The pulp is currently used in the food industry for the production of juices, biscuits, and cocoa, and as a substituent for the manufacture of chocolate, since it contains no alkaloids, caffeine, or theobromine (Craig and Nguyen 1984; Sbay and Abourouh 2006). It is also used for the production of alcohol (ethanol) and citric acid (Rejeb et al. 1991). Moreover, the gum extracted from the endosperm is used as stabilizer, fixer, and gelling agent in many industrial sectors including food (cheese, mayonnaise), cosmetics (creams, toothpaste), and pharmaceuticals (medicines, syrups) (Batlle and Tous 1997). Carob pod is rich in carbohydrates, sucrose (437.3 mg/g dry weight), glucose (395.3 mg/g dry weight) and fructose (42.3 mg/g dry weight), and proteins (5–8 g protein per 100 g dry weight). Carob pod contains also vitamins A and B, and minerals (K, P,

Ca, and Mg, Fe, Mn, Zn, and Cu), and is fat-free (Khatib and Vaya 2010). Carob flowers are very useful for beekeepers; leaves serve for animal feed, while the wood of carob tree is valued in carpentry and charcoal manufacturing (Rejeb et al. 1991; Gharnit et al. 2006). Carob processing could contribute to regional employment and enhance the viability of rural communities, and would help increase the potential of export industries to compete in the global carob bean gum market valued at $100 million/year (Hogan 1995), contributing to balance the economy of the country. Viable commercial carob cultivation will require mastering efficient farming practices with meticulous attention to water requirements and soils fertility. It would improve agricultural productivity in low rainfall areas, help manage water and land degradation, and diversify and increase farmers' incomes. Moreover, economic returns need maintaining access to higher value markets. The economic returns will fall in direct relationship with the decline in tree yields because of the major fall in profitability due to the low rainfall dryland scenario.

8.4 Agroecological Potential of Carob Tree

The Mediterranean basin is a global hotspot of biological diversity. However, the various environmental biotic and abiotic constraints prevailing in this region negatively influence plant growth and development. Drought it is a normal recurrent feature of the climate that occurs in both developing and developed countries, especially in arid and semiarid regions. This multidimensional stress is the consequence of several factors including low rainfall, salinity, high temperatures, and high intensity of light. Carob is reported to be highly tolerant of water scarcity and saline soil. The optimum annual rainfall for the carob tree is approximately five hundred millimeters. It is generally assumed that poor quality pod production is predictable under conditions of less than 300 mm per annum; however, carob trees are able to survive in areas with a rainfall of 250 mm/year (Esbenshade and Wilson 1986). Carobs have rarely been intensively cropped and there is little information available on optimum water requirements. According to Curtis and Race (1998), an average yield of 6-7 t/ha was obtained by mature trees receiving 300 L of water per tree per year in addition to the natural rainfall (500 mm/year) in the Catalina region of Spain. For low rainfall areas, irrigation (or access to groundwater) will be a critical factor in obtaining consistent yields and establishing desirable growth characteristics during the early years of development (Esbenshade 1994). Thus, the carob's ability to survive under marginal conditions should not be the basis for commercial production; long-term monitoring will be required to determine optimum irrigation practices. In any case, due to its particular agroecological features, this robust and rustic tree has a high capacity to withstand the harsh climatic and edaphic conditions of the Mediterranean regions (Correia et al. 2001) and is considered suitable for the rehabilitation of marginal and sub-marginal lands. According to Nunes et al. (1989), *Ceratonia* leaves can, to some extent, maintain turgor under situation of soil drought, using different strategies according to the season. The most common strategy adopted by carob to withstand drought is through water stress avoidance which is mostly related to its ability to maintain cell water homeostasis mainly by prohibiting water loss and increasing the water inlet to the cells, which eventually leads to normal cell functions (Correia et al. 2001; Vertovec et al. 2001; Essahibi et al. 2018). A variety of other particular morphological and physiological characters appeared to be involved in carob tolerance to drought: (1) the efficient regulation of the transpiration rate and gas exchanges depending on the general plant water status, (2) the occurrence (only on the epidermis of the lower (abaxial) leaf surface), the size (very small), and the frequency (about 217 stomata mm^{-2}) of stomata (Nunes and Linskens (1980), (3) the importance of wax deposition on leaf cuticle, which contributes in improving stomatal resistance (El Kahkahi et al. 2016), and (4) the extensive root system with quite long tap root which penetrates deeply into the sol (down to twenty meters) maintaining a continuous flow of water to the leaves even during an extended period of the topsoil drying (Rhizopoulou and Davies 1991), and the high efficiency of its xylem in term of water transport (Kikuta et al. 1997). In addition, carob tree appears to be very resistant to cavitation, tolerating leaf water potential values below the turgor loss point, with minor losses in hydraulic conductivity. It can, therefore, quickly restore the continuity of the water film within the xylem after re-watering (Salleo and Gullo 1989).

On the other hand, salinization affects about 16 million hectare of soils in the Mediterranean basin, more than 20% of arable lands in arid and semiarid areas are affected by increased salinization, of which 350,000 ha in Morocco (Hamdy and Lacirignola 1999). The presence of salt in the soil results in reduction of the water availability. Thus crops reaction to salinization is very similar to their reaction to drought stress; despite the presence of water in the soil, the plants react as if the soil was dry or almost dry. In fact, in addition to striving to alleviate the toxic effect due to Na^+ and Cl^- ions, plants seek to reduce water loss and maintain normal vital functions (Zraibi et al. 2012). Carob tree is characterized by its high tolerance to salinity; it appears to maintain almost all its physiological processes and grow successfully under NaCl concentrations up to 3% (Winer 1980; Batlle and Tous 1997; Correia et al. 2010); it may have considerable merit as a cash crop for saline lands. According to El Kahkahi et al. (2015), Carob conserves relatively stable contents of proline and soluble sugar under salt stress.

Carob tree is indifferent to the nature of the substrate, having the ability to adapt to several types of soils and wide range of pH (from 6.2 to 8.6) (Eshghi et al. 2018). Esbenshade and Wilson (1986) reported that carob prefer a calcareous, well-drained, low-clay soil, but can grow on almost any type of soil which is well drained and aerated, including sands, clay loams, limestone, and alkaline or moderately acid soils. However, carob cannot tolerate acidic soil or overly wet conditions (Esbenshade and Wilson 1986; Sbay and Abourouh 2006).

Thus carob, as pioneer and productive forest-fruit tree has great potential for reestablishing vegetation in degraded areas around the Mediterranean basin and in other Mediterranean-like dry areas of the world and has considerable merit for improving productivity in lands which would be marginal for other crops.

8.5 Importance of Mycorrhiza in Improving Carob Cultivation

Along with root nodules, mycorrhiza is considered to be the most important symbiosis that helps feed the world's population, conserve ecosystems, and preserve biodiversity. Arbuscular mycorrhiza is a highly evolved mutualistic symbiotic relationship that forms between soil-borne fungi and plant roots. The fungus colonizes the root system of a host plant, providing increased water and nutrient absorption capabilities and increased protection against biotic and abiotic stresses, while the plant provides the mycorrhizal fungus with carbohydrates formed from photosynthesis. Nowadays, the fundamental role of arbuscular mycorrhizal fungi (AMF) at the interface soil–roots is well documented. The potential of arbuscular mycorrhiza to enhance the multitrophic and protective interactions that affect plants productivity, competitiveness, and survival both in natural ecosystems and in managed fields has been widely demonstrated (Baslam et al. 2014; Fouad et al. 2014; Benhiba et al. 2015; Essahibi et al. 2018, 2019). Arbuscular mycorrhizal fungi (AMF) can form symbiotic associations with roots of over 80% of terrestrial plant species including most agricultural and horticultural crop species (Harley and Smith 1983; Smith and Read 2008). Fossil evidence (Remy et al. 1994) and DNA sequence analyzes (Simon et al. 1993) suggest that arbuscular mycorrhizal symbiosis appeared 400–460 million years ago when the first terrestrial plants started colonizing dry land, suggesting that AMF helped plants to colonize and adapt to diverse terrestrial habitats (Morton 1990). Arbuscular mycorrhizal fungi are multifunctional and provide a variety of ecosystem benefits. The extraradical mycelium of the fungus grows into the surrounding soil (reaching up to 10–14 m cm^{-1}), developing an extensive mycelium network exploring the environment for water and mineral nutrients (Smith and Read 2008), thus increasing the root absorbing surface area 100 or even 1000 fold (Barea et al. 2011). Moreover, due to their narrow diameter, the thin hyphae are able to extend into the microsites of rocks and soil pores that are inaccessible to roots or even root hairs (Barea et al. 2011). Nowadays, it is well established that AM symbiosis provides complementary characteristics that increase host plant's tolerance to water-related stresses including drought and salinity (Evelin et al. 2012; Ruiz-Lozano et al. 2016; Wu 2017). It is also well known that AMF play an important role in phosphate (P) acquisition mainly in soils with low availability and/or mobility of P (Bagayoko et al. 2000; Beltrano et al. 2013). In addition, benefits from AMF are not limited only to the host plant, but concern the entire ecosystem; strong potential of AMF for ecosystems engineering leading to sustainable improvement in soil quality and plant growth in semiarid and/or degraded areas have been demonstrated (Rillig 2004; Manaut et al. 2015; Chitarra et al. 2016; Tyagi et al. 2017; Chen et al. 2018).

8.5.1 Occurrence of Mycorrhizal Symbiosis in Carob Tree

The majority of wild plant species in natural ecosystems forms mycorrhizal symbiosis. Difficult habitats with extreme environmental conditions are generally lacking in water and nutrient and most plants growing there are endowed with adaptive strategies allowing acquisition and conservation of water and nutrient; mycorrhizal symbiosis being one of them. Regarding the adverse conditions prevailing in the area of carob distribution, it can be assumed that carob has evolved highly mycotrophic as a strategy to overcome the environmental constraints. Currently, the mycotrophic status of carob tree is well established (Ouahmane et al. 2012; El Asri et al. 2014; Essahibi 2018). In fact, Carob tree always forms mycorrhizal associations regardless of its age, the area of growth or the type of soil (Mohammad et al. 2003; El Asri et al. 2014; Essahibi 2018). Moreover, the roots of carob tree have been shown receptive not only to the arbuscular mycorrhizal fungi but also to other endophytes (El Asri et al. 2014; Mohammad et al. 2003). In fact, roots of adult trees of *Ceratonia siliqua* sampled from different Mediterranean regions showed very high mycorrhizal frequency (> 90%) confirming its high mycotrophy. Moreover, carob roots naturally colonized by AMF showed the presence of all arbuscular mycorrhizal structures (vesicles, arbuscules, and hyphae), proving the presence of functioning mycorrhiza (Mohammad et al. 2003; El Asri et al. 2014; Essahibi 2018). The high mycorrhizal status of the adult carob tree is related to the abundance and diversity of the community of mycorrhizal fungi in the rhizospheric soil of the carob groves. In fact, high spore density (2100 spores/100 g of soil) was reported in Ourika valley, regions of Marrakech, Morocco (Ouahmane et al. 2012), while less than 160 spores/ 100 g of soil) were reported in Zaio and Afourar, in the center of Morocco (El Asri et al. 2014). The identification of the native AMF species based on spore morphology showed between 12 and 18 AMF morphotypes belonging to six genera: *Glomus*, *Acaulospora*, *Scutellospora*, *Gigaspora*, *Entrophospora,* and *Pacispora*. The *Glomus* genera is the most abundant (Ouahmane et al. 2012; El Asri et al. 2014; Essahibi 2018), confirming its occurrence in a wide range of environmental and soil conditions, particularly in arid and semiarid areas (Stutz et al. 2000). The high mycotrophy of the carob tree along with the abundance and diversity of AMF community in the rhizospheric soil of the carob groves prove the strong mycorrhizal-dependency of this tree. Thus, the integration of AMF in the commercial propagation of carob (nurseries activities), the economic carob cultivation, and the restoration and/or creation of new carob plantations would have considerable merit.

8.5.2 Importance of AMF in Improving Carob Propagation

In general, Carob plants can be propagated by seeds germination or asexual propagation techniques such as grafting and cutting. Carob seeds are difficult to germinate and require scarification; its coat is extremely hard and does not absorb water. In addition, seedlings show high heterozygoty and 50% of them are potentially non-productive males. Moreover, seedlings are not true to type and normally result in fruits of varying size and quality. Propagation by cuttings is problematic because carob has been described as one of the most difficult to root species (Hartmann et al. 1997); this method is not commercially used Batlle and Tous (1997). Grafting is, generally, used to maximize the number of female trees, by grafting explants from a female or hermaphrodite tree onto a male one (Curtis and Race 1998). Carob grafting takes more than two years to obtain grafted nursery material Batlle and Tous (1997). The first attempts of in vitro propagation of carob tree were performed by Martins-Louçao and Rodriguez-Barrueco (1981). Then, many efforts have been made to elaborate efficient protocol for carob micropropagation using explants from different origin, diversified culture media, and hormonal combinations (Romano et al. 2002; Radi et al. 2013; Aguinaz et al. 2017). More recently, the plant production potential of the cuttings method has been substantially improved through controlling cuttings water status (using fog system), increasing rooting ability (using auxin treatment and arbuscular mycorrhizal fungi), and determining the best sampling period (El-Soda et al. 2016; Essahibi et al. 2017; Essahibi 2018).

Currently, it is well known that the integration of AMF into the plants' propagation cycle is one of the main practical and effective applications of these microorganisms (Scagel 2001, 2004; Davies 2008; Fouad 2015). The utilization of AMF in the different phases of carob propagation by cutting strongly improved the potential of this method. Indeed, the rooting capacity of carob cuttings was strongly enhanced by the inoculation of the rooting substrate with a mixture of AMF strains (*Funneliformis mosseae, Rhizophagus fasciculatus, and Rhizophagus intraradices*) (Essahibi et al. 2017). In fact, rooting performance in terms of rooting percentage and number of roots per cutting were higher in the presence of AMF compared to the control. Several studies have reported the positive effect of AMF on the rooting ability of cuttings in other plants species (Scagel 2004; Davies 2008; Fouad 2015). Scagel (2004) explained the positive effects of AMF on adventitious roots formation before colonization by the existence of a pre-colonization signal between cuttings and the propagules of AMF. This signal, relatively similar to that existing in the presence of host plant roots is triggered in the basal ends of the cuttings by the liberation of CO_2 or other metabolites able to activate AMF propagules (Gadkar et al. 2001; Tamasloukht et al. 2003). AMF exudates released in response to these metabolites may induce changes in cuttings' metabolism, thereby enhancing adventitious roots initiation (Larose et al. 2002); after the colonization of the first roots AMF induces the formation of new roots (Larose et al. 2002). Inoculation with AMF has also positive effects on the acclimatization of carob rooted-cuttings. According to Essahibi et al. (2017), AMF increased rooted-cuttings survival to the transplantation and hardening shocks. Moreover, mycorrhizal rooted-cuttings showed higher performance in terms of growth, physiology and biomass production and high tolerance to water and nutrient deficiency during the post-acclimation development compared to non-mycorrhizal plantlets (Essahibi et al. 2017; Essahibi 2018). The positive effects of AMF on cuttings performance have been reported in other plants species (Binet et al. 2007; Fouad 2015). According to Ouahmane et al. (2012), mycorrhizal seedlings of carob exhibited high growth and biomass production and high open field establishment capacity, being more suitable for the restoration of degraded areas.

8.5.3 Importance of AMF in Enhancing Carob Tolerance to Drought

In the Mediterranean basin, especially the southern regions, drought is one of the most serious abiotic stress limiting plants growth and development. In fact, due to the global climate changes effects, which are more exacerbated in these areas, drought is becoming longer, more extreme, and more frequent negatively impacting all plants functions. Due to its particular agroecological features, carob trees can stand severe conditions (drought, salinity, and nutrient-poor and degraded soils), being suitable for the rehabilitation of marginal and sub-marginal areas of the Mediterranean basin (Correia et al. 2001). Carob can survive dry climates without irrigation and it is well adapted to dry environments with annual average rainfall between 250 and 500 mm year^{-1}. Although carob tree can withstand long periods of drought, large amount of water is required for vigorous growth and high yields; in these regions, the groundwater is the main source to meet carob water needs. Therefore, limited precipitation along with the overexploitation of groundwater can significantly threatened the cultivation and the long-term productivity of carob trees. Bibliographic data agree on the importance of the mycorrhizal inoculation in strengthening carob tolerance to drought and other abiotic constraints (El Asri et al. 2014; Essahibi 2018; Essahibi et al. 2018, 2019; Boutasknit et al. 2020). In fact, mycorrhizal carob plants subjected to severe water deficit have maintained almost normal growth and biomass production (Essahibi et al. 2018; Boutasknit et al. 2020). The performance of mycorrhizal plants under water stress implies several tolerance strategies including (1) improved water balance and nutrient status, (2) maintained stomatal conductance at high levels (3) preserved cellular turgor by increasing cell wall rigidity and osmolytes accumulation, and (4) enhanced defense systems involved in oxidative stress alleviation, including enzymatic and nonenzymatic compounds (Essahibi et al. 2018). Indeed, due to their highly developed extrametrical hyphae network, AMF showed high efficiency in taking and transporting water improving thereby the host plant water status even in soils with low water availability. According to Smith et al. (2010), the length of the AMF hyphae associated with roots is estimated at more than 100 m per gram of soil. In addition, the average diameter of these hyphae varies between 2 and 20 μm which is one or two degrees lesser than the diameter of the absorbing hairs (Ruiz-Lozano et al. 2012). This very large and very fine extraradical hyphae network can get water from pores inaccessible to the non-mycorrhizal roots, allowing better water supply to the host plant. Thus mycorrhizal plants could maintain stomatal conductance at higher level compared to non-mycorrhizal plants (Chitarra et al. 2016). According to these authors, the high stomatal conductance in mycorrhizal plants is associated with low ABA concentration in the xylem sap. Indeed, under the same conditions of water stress, the roots of mycorrhizal plants perceive less the effect of water stress and therefore produce less ABA than non-mycorrhizal plants (Chitarra et al. 2016). The high stomata conductance and increased water and mineral nutrient acquisition in mycorrhizal plants allow maintaining high photosynthetic activity. The photosynthesized carbon products are then used both for growth and osmotic adjustment as water potential of the soil becomes low. In addition, AMF provide host plant with complementary characteristics to avoid the oxidative stress induced by water stress through their action on the antioxidant system involved in the elimination of reactive oxygen species (ROS). AMF increase both the activity of antioxidant enzymes, superoxide dismutase, ascorbate peroxidase, guaiacol peroxidase, and catalase (Baslam et al. 2014; Fouad et al. 2014; Benhiba et al. 2015), as well as the content of abscisic acid, glutathione, carotenoids, and anthocyanins (Baslam and Goicoechea 2012). At the molecular level, AMF regulate the expression of *p5cs* genes encoding a rate-limiting enzyme in the biosynthesis of proline (Porcel and Ruiz-Lozano 2004), genes encoding aquaporins (Porcel et al. 2006), and *nced* genes encoding a key enzyme in the biosynthesis of ABA (Aroca et al. 2008), allowing mycorrhizal plants to maintain a better water status (Porcel et al. 2004).

8.5.4 Importance of AMF in Improving Carob Tolerance to P Deficiency

In addition to drought, another serious limitation of plants development and crop production in the Mediterranean regions is the deficiency of available phosphorus (P). In fact, the high affinity of P for soil particles and the facility of its precipitation by free Al^{3+} and Fe^{3+} in acidic soils and by Ca^{2+} in alkaline soil (Havlin et al. 2005) make it inaccessible to plants. The importance of AMF in improving P acquisition by the host plant, particularly in soils with low P availability and/or mobility has been demonstrated by several investigations (Cardoso et al. 2006). This is due to the narrow extra-radical mycelial network, which penetrates wider and deeper into the soil (Turk et al. 2006), increasing thereby P absorption of the host plant. In addition, AMF have the ability to mobilize other forms of P unavailable to plants through the solubilization of inorganic phosphorus (Pi) (Duponnois et al. 2005; Gholamhoseini et al. 2013). In fact, AMF can interact with phosphate-solubilizing bacteria to improve P availability and can produce with the associated roots organic acids and hydrolytic enzymes involved in the decomposition of organic matter and dissolution of Pi (Gholamhoseini et al. 2013; Goussous and Mohammad 2009). It is also well established that mycorrhizal symbiosis induces the expression of Pi transporters in plants (Xie et al. 2013; Walder et al. 2015). In tomato, three Pi transporter (*PT*) genes (*LePT3*, *LePT4*, and *LePT5*) are upregulated in colonized roots (Nagy et al. 2005). *MtPT4* genes are regulated in root cells of *Medicago truncatula* to absorb P supplied by mycorrhizae (Javot et al. 2007). According to Essahibi et al. (2019), under P deficiency, arbuscular mycorrhizal symbiosis significantly improved growth and biomass production of carob plants. Moreover, mineral nutrient acquisition, photosynthetic activity, stomatal conductance, total chlorophyll content, and soluble sugar accumulation were also strongly improved in mycorrhizal plants in comparison with non-mycorrhizal ones. In addition, under P deficiency, mycorrhizal plants showed strongly increased acid phosphatase activity in the roots and the rhizospheric soil than Non-mycorrhizal plants. Furthermore, mycorrhizal plants maintained high membrane integrity (over 80%) and low hydrogen peroxide (H_2O_2) and malondialdehyde (MDA) contents, associated with increased activities of superoxide dismutase (SOD), ascorbate peroxidase (APX), guaiacol peroxidase (G-POD), and catalase (CAT) compared to non-mycorrhizal plants.

8.5.5 Importance of AMF in Improving Field Establishment of Carob Tree

Ecosystem restoration using pioneer plants is considered to be necessary and useful and has achieved great progress in recent years. Because of its hardiness and adaptation to environmental constraints, carob tree is included in the list of priority as forest resources for conservation in many countries in the Mediterranean basin. It is often used for afforestation and reforestation of areas affected by erosion and desertification (Rejeb et al. 1991). However, successful restoration is very difficult, because of the severity of abiotic stress, affecting negatively plants installation and making them growing hardly. The potentialities of ecological engineering strategy based on the use of AMF for improving afforestation programs with carob trees in degraded environments have been demonstrated (Manaut et al. 2015). These authors revealed the high potential of this approach in improving sustainably the growth and nutrient status of carob trees plantation. Indeed, in addition to benefit host plants' growth via alleviating various environmental stresses, AMF improve aggregate stability, through their fine hyphal mesh and glomalin production thereby enhancing soil quality and stability (Rillig 2004; Chen et al. 2018).

8.6 Conclusion

Carob (*Ceratonia siliqua* L.) tree is one of the most important species existing in the Mediterranean regions due to its exceptional socioeconomic and environmental benefits. It is evident from the above discussion that arbuscular mycorrhizal fungi are very effective for promoting carob culture in these regions. Indeed, this mycorrhizal technology has been proved useful for promoting the successful mass propagation of carob (a rooting-recalcitrant species) using herbaceous semi-herbaceous cuttings under mist system. Moreover, AM symbiosis has been shown to provide complementary characteristics that increase carob plant's tolerance to drought and P deficiency, the most serious abiotic stresses limiting plants growth and development in the Mediterranean regions. The positive effects of AMF on carob growth and physiology were not only observed in greenhouse but also in open field conditions, showing their high potential for reinforcing the reforestation programs of degraded areas. While the mycorrhizal fungus is effective when used singly or in combination, the specificity of strain in this regard is also important so that maximum benefits can be obtained from AM symbiosis.

References

Aguinaz H, Qaddoury A, Anjarne M (2017) Shoot formation efficiency of mature carob trees (*Ceratonia siliqua* L.) based on plant growth regulators pre-culture. Res J Biotechnol 12 (8):56–62

Ait Chitt M, Belmir H, Lazrak A (2007) Production de plants sélectionnés et greffés de caroubier. Bulletin mensuel d'information et de liaison du PNTTA MAPM/DERD 153:1–4

Aroca R, Vernieri P, Ruiz-Lozano JM (2008) Mycorrhizal and non-mycorrhizal Lactuca sativa plants exhibit contrasting responses to exogenous ABA during drought stress and recovery. J Exp Bot 59(8):2029–2041

Bagayoko M, George E, Römheld V, Buerkert A (2000) Effects of mycorrhizae and phosphorus on growth and nutrient uptake of millet, cowpea and sorghum on a West African soil. J Agric Sci 135:399–407

Bakry M, Lamhamedi MS, Caron J (2013) Changes in the physical properties of two Acacia compost-based growing media and their effects on carob (*Ceratonia siliqua* L.) seedling development. New For 44:827–847

Barea JM, Palenzuela J, Cornejo P, Sánchez-Castro I, Navarro-Fernández C, Lopéz-García A (2011) Ecological and functional roles of mycorrhizas in semi-arid ecosystems of Southeast Spain. J Arid Environ 75:1292–1301

Baslam M, Goicoechea N (2012) Water deficit improved the capacity of arbuscular mycorrhizal fungi (AMF) for inducing the accumulation of antioxidant compounds in lettuce leaves. Mycorrhiza 22:347–359

Baslam M, Qaddoury A, Goicoechea N (2014) Role of native and exotic mycorrhizal symbiosis to develop morphological, physiological and biochemical responses coping with water drought of date palm, *Phoenix dactylifera* L. Trees 28:161–172

Batlle I, Tous J (1997) Carob tree *Ceratonia siliqua* L. International Plant Genetic Resources Institute, Rome, Italy, pp 92

Beltrano J, Ruscitti M, Arango MC, Ronco M (2013) Effects of arbuscular mycorrhiza inoculation on plant growth, biological and physiological parameters and mineral nutrition in pepper grown under different salinity and p levels. J Soil Sci Plant Nutr 13:123–141

Benhiba L, Fouad MO, Essahibi A, Ghoulam C, Qaddoury A (2015) Arbuscular mycorrhizal symbiosis enhanced growth and antioxidant metabolism in date palm subjected to long-term drought. Trees 29(6):1725–1733

Binet MN, Lemoine MC, Martin C, Chambon C, Gianinazzi S (2007) Micropropagation of Olive (*Olea europaea* L.) and application of mycorrhiza to improve plantlet establishment. In Vitro Cell Dev Biol Plant 43(5):473–478

Bouhadi D, Hariri A, Ould Yerou K, Benattouche Z, Sahnouni F, Hadjari M (2017) Kinetics of batch production of lactic acid from carob pods syrup. Bull Pharm Res 7(1):140

Boutasknit A, Baslam M, Ait-El-Mokhtar M, Anli M, Ben-Laouane R, Douira A, El Modafar C, Mitsui T, Wahbi S, Meddich A (2020) Arbuscular mycorrhizal fungi mediate drought tolerance and recovery in two contrasting Carob (*Ceratonia siliqua* L.) ecotypes by regulating stomatal, water relations, and (in) organic adjustments. Plan Theory 9(1):80

Cardoso IM, Boddington CL, Janssen BH, Oenema O, Kuyper TW (2006) Differential access to phosphorus pools of an oxisol by mycorrhizal and non mycorrhizal maize. Commun Soil Sci Plant Anal 37:1537–1551

Chen M, Arato M, Borghi L, Nouri E, Reinhardt D (2018) Beneficial services of arbuscular mycorrhizal fungi – from ecology to application. Front Plant Sci 9:1270

Chitarra W, Pagliarani C, Maserti B, Lumini E, Siciliano I, Cascone P, Schubert A, Gambino G, Balestrini R, Guerrieri E (2016) Insights on the impact of arbuscular mycorrhizal symbiosis on tomato tolerance to water stress. Plant Physiol 171(2):1009–1023

Correia MJ, Coelho D, David MM (2001) Response to seasonal drought in three cultivars of *Ceratonia siliqua*: leaf growth and water relations. Tree Physiol 21:645–653

Correia PJ, Gama F, Pestana M, Martins-Loução MA (2010) Tolerance of young (*Ceratonia siliqua* L.) carob rootstock to NaCl. Agr Water Manage 97:910–916

Craig WJ, Nguyen TT (1984) Caffeine and theobromine level in cocoa and carob products. J Food Sci 49:302–305

Curtis A, Race D (1998) Carob agroforestry in the low rainfall Murray valley: a market and economic assessment. Publication No. 98/8. Rural Industry Research and Development Corporation (RIRDC), ISBN 0642540306. Australia

Custodio L, Escapa AL, Fernandes E, Fajardo A, Aligué R (2011a) Phytochemical profile, antioxidant and cytotoxic activities of the carob tree (*Ceratonia siliqua* L.) germ flour extracts. Plant Food Hum Nutr 66:78–84

Custodio L, Fernandes E, Escapa AL, Fajardo A, Aligue R (2011b) Antioxidant and cytotoxic activities of carob tree fruit pulps are strongly influenced by gender and cultivar. J Agric Food Chem 59:7005–7012

Davies FT (2008) Opportunities from down under: How mycorrhizal fungi can benefit nursery propagation and production systems. In: The International Plant Propagators Society Combined Proceedings (Seattle WA, ed.). International Plant Propagators Society, Combined Proceedings 58:538–548.

De Candolle A (1883) L'origine des plantes cultivées. Balière, Paris, France

Duponnois R, Colombet A, Hien V, Thioulouse J (2005) The mycorrhizal fungus *Glomus intraradices* and rock phosphate amendment influence plant growth and microbial activity in the rhizosphere of *Acacia holosericea*. Soil Biol Biochem 37:1460–1468

El Asri A, Talbi Z, Ait Aguil F, Chliyeh M, Sghir F, Touati J, Ouazzani Touhami A, Benkirane R, Douira A (2014) Arbuscular mycorrhizal fungi associated with rhizosphere of carob tree (Ceratonia siliqua L.) in Morocco. Int J Pure App Biosci 2:286–297

El Kahkahi M, Mouhajir S, Bachir A, Lemrhari R, Zouhair M, Ait Chitt M, Errakhi R (2015) Morphological and physiological analysis of salinity stress response of carob (*Ceratonia siliqua* L.) in Morocco. Sci Int 3:73–81

El Kahkahi R, Moustaine M, Mouhajir A, Bachir S, Lemrhari A, Zouhair R, Errakhi R (2016) Technical sheet on the culture carob tree (*Ceratonia Siliqua* L.) in Morocco

El-Soda AS, Elhusseiny AM, Hammad AA, El-Arabi NI (2016) Using different methods to produce vegetative carob seedlings. Egypt J Hort 43(2):241–257

Esbenshade H (1994) Olives and carobs for landcare and profit in South Australia. In: Proceedings of national symposium on olive

Esbenshade HW, Wilson G (1986) Growing Carobs in Australia. Goddard and Dobson Publishers, Box Hill, VIC

Eshghi S, Rostami AA, Jamali B (2018) Carob tree: a suitable species for the future. Acta Hortic 1190:67–70

Essahibi A (2018) Amélioration des performances des boutures du caroubier en termes d'enracinement, d'acclimatation, et de tolérance aux contraintes du milieu en post-acclimatation: importance des champignons mycorhiziens à arbuscules. Doctorate dissertation, FST-Marrakech, Morocco

Essahibi A, Benhiba L, Fouad MO, Babram M, Ghoulam C, Qaddoury A (2017) Improved rooting capacity and hardening efficiency of carob (*ceratonia siliqua* L.) cuttings using arbuscular mycorrhizal fungi. Arch Biol Sci 69(2):291–298

Essahibi A, Benhiba L, Babram M, Ghoulam C, Qaddoury A (2018) Influence of arbuscular mycorrhizal fungi on the functional mechanisms associated with drought tolerance in carob (*Ceratonia siliqua* L.). Trees 32:87–97

Essahibi A, Benhiba L, Fouad MO, Babram M, Ghoulam C, Qaddoury A (2019) Responsiveness of carob (*Ceratonia siliqua* L.) plants to arbuscular mycorrhizal symbiosis under different phosphate fertilization levels. J Plant Growth Regul 38:1243–1254

Evelin H, Giri B, Kapoor R (2012) Contribution of *Glomus intraradices* inoculation to nutrient acquisition and mitigation of ionic imbalance in NaCl-stressed *Trigonella foenum-graecum*. Mycorrhiza 22:203–217

Fouad MO (2015) Déterminant de la capacité des boutures d'olivier à surmonter les opérations d'acclimatation et de transplantation: importance des champignons mycorhiziens à arbuscules. Thèse Doctorat, FST-Marrakech, 157 p

Fouad MO, Essahibi A, Benhiba A, Qaddoury A (2014) Effectiveness of arbuscular mycorrhizal fungi in the protection of olive plants against oxidative stress induced by drought. Span J Agric Res 12(3):763–771

Gadkar V, David-Schwartz R, Kunik T, Kapulnik Y (2001) Arbuscular mycorrhizal fungal colonization. Factors involved in host recognition. Plant Physiol 127:1493–1499

Gharnit N, El Mtili N, Ennabili A, Sayah F (2006) Importance socio-éconmique du caroubier (*Ceratonia siliqua* L.) dans la province de Chefchaouen (Nord-Ouest du Maroc). J Bot Soc Bot France 33:43–48

Gholamhoseini M, Ghalavand A, Dolatabadian A, Jamshidi E, Khodaei-Joghan A (2013) Effects of arbuscular mycorrhizal inoculation on growth, yield, nutrient uptake and irrigation water productivity of sunflowers grown under drought stress. Agric Water Manag 117:106–114

Goussous SJ, Mohammad MJ (2009) Comparative effect of two arbuscular mycorrhizae and N and P fertilizers on growth and nutrient uptake of onions. Int J Agric Biol 11:463–467

Hamdy A, and Lacirignola C (1999) Mediterranean water resources: major challenges towards the 21st century. Proceedings of the international seminar on Mediterranean water resources: major challenges towards the 21st century, March, Cairo, Egypt, pp 562–562

Harley JL, Smith SE (1983) Mycorrhizal symbiosis. Academic, New York, p 483

Hartmann HT, Kester DE, Davies FT, Geneve RL (1997) Plant propagation: principles and practices, 6th edn. Prentice-Hall, New Jersey, p 770

Havlin JL, Beaton JD, Tisdale SL, Nelson WL (2005) Soil fertility and fertilizers: an introduction to nutrient management. Pearson Prentice Hall, Upper Saddle River, pp 97–141

Hillcoat D, Lewis G, Verdcourt B (1980) A new species of *Ceratonia* (*Leguminosea-Caesalpinoideae*) from Arabia and the Somali Republic. Kew Bull 35:261–271

Hogan M (1995) Carob. Unpublished notes, Uncle Bens of Australia, Wodonga, VIC

Javot H, Penmetsa RV, Terzaghi N (2007) A *Medicago truncatula* phosphate transporter indispensable for the arbuscular mycorrhizal symbiosis. Proc Natl Acad Sci 104:1720–1725

Khatib S, Vaya J (2010) Fig, Carob, Pistachio, and health in bioactive foods in promoting health. In: Watson RR (ed) Fruits and vegetables. Elsevier, Amsterdam, pp 245–263

Kikuta SB, Gullo MA, Nardini A, Richter H, Salleo S (1997) Ultrasound acoustic emissions from dehydrating leaves of deciduous and evergreen trees. Plant Cell Environ 20:1381–1390

Konate I (2007) Diversité phénotypique et moléculaire du Caroubier (*Ceratonia siliqua* L.) et des bactéries endophytes qui lui sont associées. Dissertation, University of Mohammed V

Larose G, Chênevert R, Moutoglis P, Gagné S, Piché Y, Vierheilig H (2002) Flavonoid levels in roots of *Medicago sativa* are modulated by the developmental stage of the symbiosis and the root colonizing arbuscular mycorrhizal fungus. J Plant Physiol 159:1329–1339

Makris DP, Kefalas P (2004) Carob pods (*Ceratonia siliqua* L.) as a source of polyphenolic antioxidants. Food Technol Biotechnol 42:105–108

Manaut N, Sanguin H, Ouahmane L, Bressan M, Thioulouse J, Baudoin E, Galiana A, Hafidi M, Prin Y, Duponnois R (2015) Potentialities of ecological engineering strategy based on native arbuscular mycorrhizal community for improving afforestation programs with carob trees in degraded environments. Ecol Eng 79:113–119

Martins-Louçao MA, Rodriguez-Barrueco C (1981) Establishment of proliferation callus from roots, cotyledons and hypocotyls of carob seedlings. Z Pflanzenphysiol 103:297–303

Mohammad MJ, Hamad SR, Malkawi HI (2003) Population of arbuscular mycorrhizal fungi in semiarid environment of Jordan as influenced by biotic and abiotic factors. J Arid Environ 53:409–417

Morton JF (1987) Carob. In: Morton JF (ed) Fruits of warm climates. Julia F. Morton, Miami, FL, pp 65–69

Morton JB (1990) Species and clones of arbuscular mycorrhizal fungi (Glomales, Zygomycetes): their role in macro- and microevolutionary processes. Mycotaxon 37:493–515

Nagy R, Karandashov V, Chague V (2005) The characterization of novel mycorrhiza-specific phosphate transporters from *Lycopersicon esculentum* and *Solanum tuberosum* uncovers functional redundancy in symbiotic phosphate transport in solanaceous species. Plant J 42:236–250

Nunes MA, Linskens HF (1980) Some aspects of the structure and regulation of Ceratonia siliqua L. stomata. Portug Acta Biol 16:165–174

Nunes MA, Catarino FM, Pinto E (1989) Strategies for acclimation to seasonal drought in *Ceratonia siliqua* leaves. Physiol Plant 77(1):150–156

Ouahmane L, Ndoye I, Morino A, Ferradous A, Sfairi Y, Al Faddy MN, Abourouh M (2012) Inoculation of *Ceratonia siliqua* L. with native arbuscular mycorrhizal fungi mixture improves seedling establishment under greenhouse conditions. Afr J Biotechnol 11:16422–16426

Ozturk M, Dogan Y, Sakcali MS, Doulis A, Karam F (2010) Ecophysiological responses of some maquis (*Ceratonia siliqua* L., *Olea oleaster Hoffm. & Link, Pistacia lentiscus* and *Quercus coccifera* L.) plant species to drought in the East Mediterranean ecosystem. J Environ Biol 31:233–245

Porcel R, Ruiz-Lozano JM (2004) Arbuscular mycorrhizal influence on leaf water potential, solute accumulation and oxidative stress in soybean plants subjected to drought stress. J Exp Bot 55:1743–1750

Porcel R, Aroca R, Azcon R (2006) PIP aquaporin gene expression in arbuscular mycorrhizal *Glycine max* and *Lactuca sativa* plants in relation to drought stress tolerance. Plant Mol Biol 60 (3):389–404

Radi A, Echchgadda G, Ibijbijen I, Rochd M (2013) In vitro propagation of Moroccan carob (*Ceratonia siliqua* L.). J Food Agri Environ 11(1):1103–1107

Rejeb MN, Laffray D, Louguet P (1991) Physiologie du caroubier (*Ceratonia siliqua* L.) en Tunisie. Physiologie des arbres et arbustes en zones arides et semi-arides. Group d'Etude de l'Arbre, Paris, France, pp 417–426

Remy W, Taylor TM, Hass H et al (1994) Proc Natl Acad Sci USA 91:11841–11843

Rhizopoulou S, Davies WJ (1991) Influence of soil drying on root development, water relations and leaf growth of *Ceratonia siliqua* L. Oecologia 88:41–47

Rillig MC (2004) Arbuscular mycorrhizae, glomalin, and soil aggregation. Can J Soil Sci 84:355–363

Romano A, Barros S, Martins-Loucao MA (2002) Micropropagation of the Mediterranean tree *Ceratonia siliqua*. Plant Cell Tissue Organ Cult 68:35–41

Ruiz-Lozano JM, Porcel R, Azcón C, Aroca R (2012) Regulation by arbuscular mycorrhizae of the integrated physiological response to salinity in plants: new challenges in physiological and molecular studies. J Exp Bot 63:4033–4044

Ruiz-Lozano JM, Aroca R, Zamarreño ÁM, Molina S, Andreo-Jiménez B, Porcel R (2016) Arbuscular mycorrhizal symbiosis induces strigolactone biosynthesis under drought and improves drought tolerance in lettuce and tomato. Plant Cell Environ 39:441–452

Sakcali MS, Ozturk M (2004) Eco-physiological behaviour of some Mediterranean plants as suitable candidates for reclamation of degraded areas. J Arid Environ 57:141–153

Salleo S, Gullo ML (1989) Different aspects of cavitation resistance in *Ceratonia siliqua*, a drought-avoiding Mediterranean tree. Ann Bot 64:325–336

Sbay H, Abourouh M (2006) Apport des espèces à usages multiples pour le développement durable: cas du pin pignon et du caroubier. Centre de Recherche Forestière, Haut-Commissariat aux Eaux et Forêts et à la Lutte Contre la Désertification, Rabat 9 p.

Scagel CF (2001) Cultivar specific effects of mycorrhizal fungi on the rooting of miniature rose cuttings. J Environ Hortic 19(1):15–20

Scagel CF (2004) Enhanced rooting of kinnikinnick cuttings using mycorrhizal fungi in rooting substrate. Hort Technol 14(3):355–363

Schweinfurth G (1894) Sammlung arabischaethiopischer Pflanzen, Ergebnisse von Reisen in dem Jahren 1881, 1888-89, 1891-92. Bull Herb Boissier 2:1–114

Simon L, Levesque RC, Lalonde M (1993) Identification of endomycorrhizal fungi colonizing roots by fluorescent single-strand conformation polymorphism polymerase chain reaction. Appl Environ Microbiol 59:4211–4215

Smith SE, Read DJ (2008) Mycorrhizal symbiosis, 3rd edn. Academic, Waltham

Smith SE, Facelli E, Pope S, Smith FA (2010) Plant performance in stressful environments: interpreting new and established knowledge of the roles of arbuscular mycorrhizas. Plant Soil 326:3–20

Stutz JC, Copeman R, Martin CA, Morton JB (2000) Patterns of species composition and distribution of arbuscular mycorrhizal fungi in arid regions of southwestern North America and Namibia, Africa. Can J Bot 78:237–245

Tamasloukht MB, Séjalon-Delmas N, Kluever A, Jauneau A, Roux C, Bécard G, Franken P (2003) Root factors induce mitochondrial-related gene expression and fungal respiration during the developmental switch from asymbiosis to presymbiosis in the arbuscular mycorrhizal fungus *Gigaspora rosea*. Plant Physiol 131:1468–1478

Turk MA, Assaf TA, Hameed KM, Al-Tawaha AM (2006) Significance of mycorrhizae. World J Agric Sci 2:16–20

Tyagi J, Varma A, Pudake RN (2017) Evaluation of comparative effects of arbuscular mycorrhiza (*Rhizophagus intraradices*) and endophyte (*Piriformospora indica*) association with finger millet (*Eleusine coracana*) under drought stress. Eur J Soil Biol 81:1–10

Vavilov NI (1951) The origin, variation, immunity and breeding of cultivated plants [translated from the Russian by K.S. Chester]. The Ronald Press, New York

Vertovec M, Sakçali S, Ozturk M, Salleo S, Giacomich P, Feoli E, Nardini A (2001) Diagnosing plant water status as a tool for quantifying water stress on a regional basis in Mediterranean drylands. Ann For Sci 58:113–125

Walder F, Brule D, Koegel S et al (2015) Plant phosphorus acquisition in a common mycorrhizal network: regulation of phosphate transporter genes of the Pht1 family in sorghum and flax. New Phytol 205:1632–1645

Winer N (1980) The potential of the carob tree (*Ceratonia siliqua*). Int Tree Crops J 1(1):15–26

Wu QS (2017) Arbuscular mycorrhizas and stress tolerance of plants. In: Wu QS (ed) Arbuscular mycorrhizas and stress tolerance of plants. Springer, Singapore

Xie X, Huang W, Liu F, Tang N, Liu Y, Lin H, Zhao B (2013) Functional analysis of the novel mycorrhiza-specific phosphate transporter AsPT1 and PHT1 family from *Astragalus sinicus* during the arbuscular mycorrhizal symbiosis. New Phytol 198:836–852

Zohary M (1973) Geobotanical foundations of the Middle East, 2 vols. Gustav Fisher Verlag, Stuttgart

Zraibi L, Nabloussi A, Merimi J, El Amrani A, Kajeiou M, Khalid A, Caid HS (2012) Effet du stress salin sur des paramétres physiologiques et agronomiques de différentes variétes de carthame (*Carthamus tinctorius* L.). Al Awamia 125–126:15–40

Chapter 9
Leaf Endophytes and Their Bioactive Compounds

Parikshana Mathur, Payal Mehtani, and Charu Sharma

Abstract Endophytes can reside in every part of a plant and show mutualistic relationship but leaf tissue is considered as an ideal habitat for such microorganisms. Endophytic microflora is a great source of secondary metabolites that can have various biological activities like antibacterial, antifungal, anticancerous, etc. This chapter focuses on such endophytic microbes that inhabit leaf tissues and produce numerous bioactive compounds. Further various classes of secondary metabolites obtained from leaf endophytes are discussed along with some applications of the bioactive compounds.

Keywords Anticancer · Antimicrobial · Endophytic fungi · Mutualism · Plant · Secondary Metabolites

9.1 Introduction

Endophytes are microorganisms that colonize internal parts of plant at least once in their lifetime and cause asymptomatic and imperceptible infections in the plant. Endophytes have proved to play an important role in the adaptation and evolution of plants to their environment. With due course of time plants have developed coping mechanisms to deal with external stress but still most plants fail to survive in hostile environment. However, due to the presence of microorganisms in association with plants, they have developed various adaptations to deal with stressful environmental conditions. In order to help plants in surviving in an unfriendly environment, endophytes are hold responsible in host secondary metabolite synthesis either partially or completely (Chutulo and Chalannavar 2018). Every plant is found to cohabit endogenous microbes like bacteria, fungi, algae, actinomycetes, etc. that are

P. Mathur · P. Mehtani (✉) · C. Sharma
Department of Biotechnology, IIS (Deemed to be University), Gurukul Marg, SFS, Mansarovar, Jaipur, Rajasthan, India
e-mail: payal.mehtani@iisuniv.ac.in

N. Shrivastava et al. (eds.), *Symbiotic Soil Microorganisms*, Soil Biology 60,
https://doi.org/10.1007/978-3-030-51916-2_9

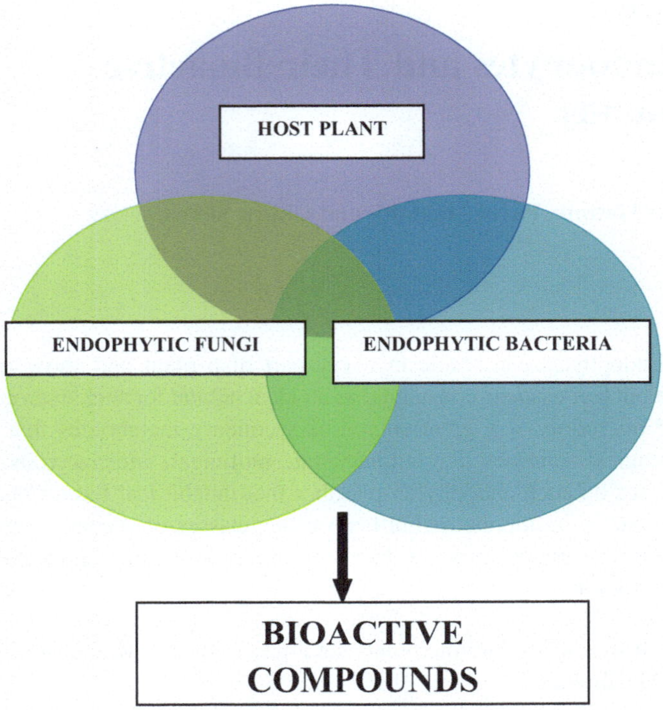

Fig. 9.1 Association between host plant and endophytic microflora

located in all plant parts including stem, leaves, root, flowers, etc. It has also been said that the phenotypic expression of plants is also a result of their interactions with the microorganisms inhabiting them. In vitro and in vivo studies have confirmed that removal of endophytes from the plant has led to loss of their adaptative and resistant properties toward an unfavorable environment where they survive with presence of the endophyte (Li et al. 2019). Variation is found in the population of endophyte in plants depending on host species, developmental stage of the host, and the environmental conditions (Gouda et al. 2016).

The bioactive compounds synthesized by endophytes as a part of their defense and survival mechanisms have shown potential applications in medicine, pharmaceutical industries, agriculture, etc. (Fig. 9.1). With the extraction of Taxol from endophytic fungi *Taxomyces andreanae* residing in the plant *Taxus brevifolia* and its application as an anticancer compound, the interest of researchers shifted toward finding novel bioactive products from endophytic flora and exploring their applications (Stierle et al. 1993; Strobel et al. 2004). Research in the last few decades has led to the exploration of many valuable applications of these bioactive compounds as anticancer agents, insecticides, and antimicrobial agents (Zhang et al. 2006). This chapter primarily focuses on the endophytes dwelling in the leaves of the plants and the bioactive compounds they synthesize. We have also discussed some applications of the bioactive products produced by the leaf endophytes.

9.2 Leaf Endophytes

Taylor et al. (1999), have discovered endophytic fungi from fossils of stems and leaves. Leaf tissue is reported to be ideal for exploring endophytic microflora and their secondary metabolites as it has shown the highest species richness and endophyte colonization. Colonization of endophytic microbes is majorly found in xylem tissues, leaf veins, mesophyll cells trichomes, cut sections of leaf pieces, and substomatal areas (Fig. 9.2). In different studies on leaf endophytes of *Azadirachta indica* basal leaflets showed higher colonizing frequency than apical or middle leaflets with significant increase during the rainy season (Verma et al. 2011). Similar results were obtained by Chareprasert et al. in 2006 on endophytes of *Tectona grandis* L and *Samanea saman* Merr. with more colonization frequency in mature leaves compared to young leaves. Eighty-five endophytic fungal isolates were isolated from the leaves of *Azadirachta indica* (Tenguria and Khan 2011). Higher isolation frequency from petiole than blade segments from leaves of *Neolitsea sericea* were reported in a study conducted by Hata and Sone in 2008 though blade segments showed more variation in endophyte pattern.

The presence of *Burkholderia* sp. strain PsJN has been reported in xylem and substomatal chambers of inoculated leaves of grapevine plants by FISH and other microscopy techniques. *Klebsiella variicola* was found in the mesophyll cells of sugarcane leaves; *Herbaspirillum* sp. in young leaves and shoots of wild rice; *Herbaspirillum seropedicae* Z67 in leaf vein, mesophyll cells, and substomatal cavities of rice leaves and *Serratia marcescens* in the leaf sheaths and leaf aerenchyma of rice plants (Kandel et al. 2017). Lepanthes orchids of tropical region have

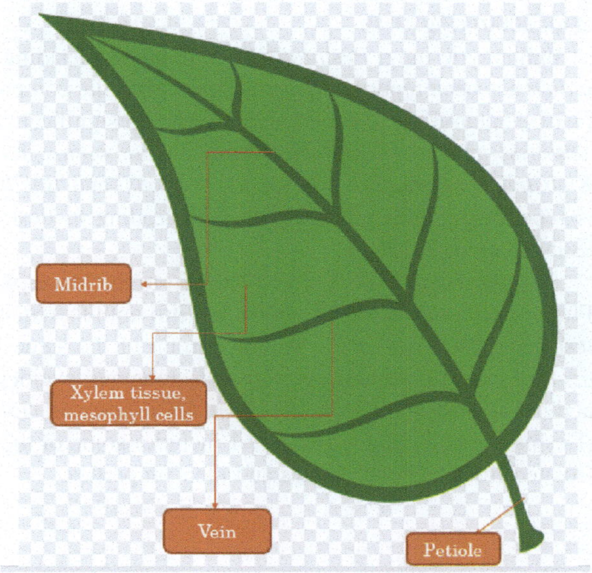

Fig. 9.2 Diagrammatic representation of parts of leaf which may have colonization of endophytes

reported the presence of *Xylaria* and *Rhizoctonia* fungi in the leaves of the plant more than the roots (Bayman et al. 1997).

9.3 Endophytic Secondary Metabolites and Their Biological Activities

Secondary metabolites are compounds biosynthetically derived from primary metabolites and are classified on the basis of their biosynthetic origin (Fig. 9.3). The following section throws light on some of the bioactive compounds obtained from endophytes isolated from leaves and their applications.

9.3.1 Phenolic Compounds

Phenolic compounds comprise the largest class of secondary metabolites. The term phenolic compounds include a wide range of plant substances which are water-soluble and have a common aromatic ring having hydroxyl substituents. Almost all plant phenolics are produced by phenylpropanoid metabolism from products of the shikimic acid pathway. This class includes important subclasses like lignans and flavonoids. Flavonoids further include subclasses like anthocyanin pigments, tannins, isoflavonoids, flavanones, flavones, leucoanthocyanidins, catechins, chalcones, and aurones. Tannins are feeding deterrents and wood protectants. Isoflavonoids are used as signaling molecules and also used in defense mechanism of plants (Croteau et al. 2000) (Fig. 9.4).

Endophytic fungi *Pestalotiopsis mangiferae* residing in the leaves of *Mangifera indica* has been reported to synthesize a phenolic compound 4-(2,4,7-trioxa-bicyclo [4.1.0]heptan-3-yl) phenol which has shown significant antibacterial and antifungal activities against *Candida albicans*, *Pseudomonas aeruginosa, Escherichia coli, Klebsiella pneumoniae*, *Micrococcus luteus*, and *Bacillus subtilis*. The compound is reported to form pores in cell wall leading to destruction of the bacterial cell (Subban et al. 2013).

Phomopsis sp. BCC 1323 isolated from the leaves of *Tectona grandis* produces Phomoxanthones A and B shows in vitro antitubercular activity against *Mycobacterium tuberculosis* strain. Another strain of the same fungi isolated from *Laurus azorica* leaves produces cycloepoxylactone and cycloepoxytriol B has also shown activity against *Bacillus megaterium, Microbotryum violaceum*, and *Chlorella fusca* (Deshmukh et al. 2015). Table 9.1 includes some of the other phenolic compounds isolated from leaf endophytes and their properties.

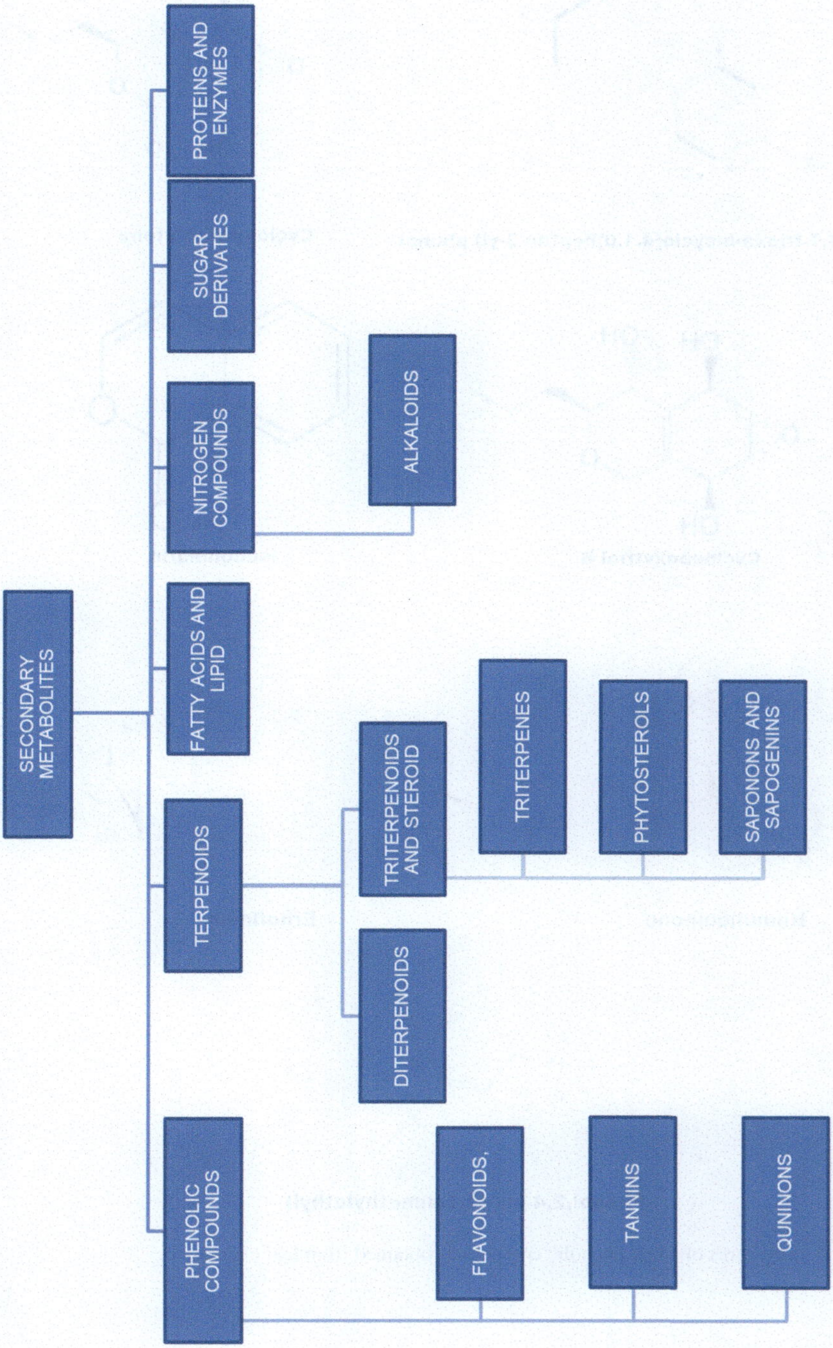

Fig. 9.3 Major classes of secondary metabolites

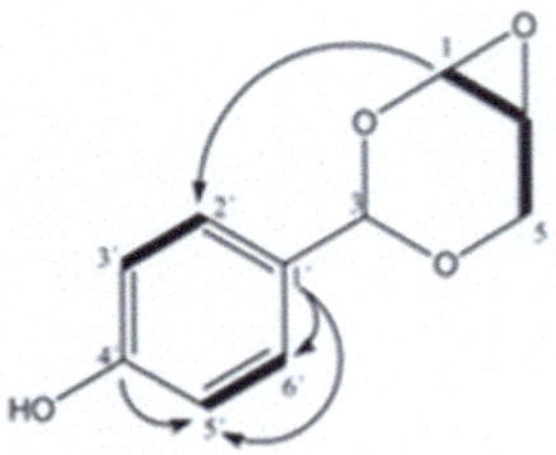

4-(2,4,7-trioxa-bicyclo[4.1.0]heptan-3-yl) phenol

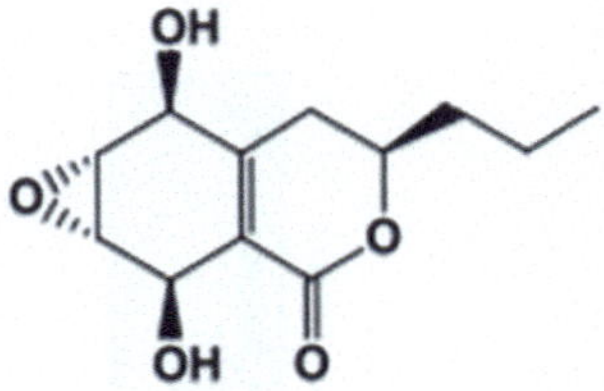

Cycloepoxylactone

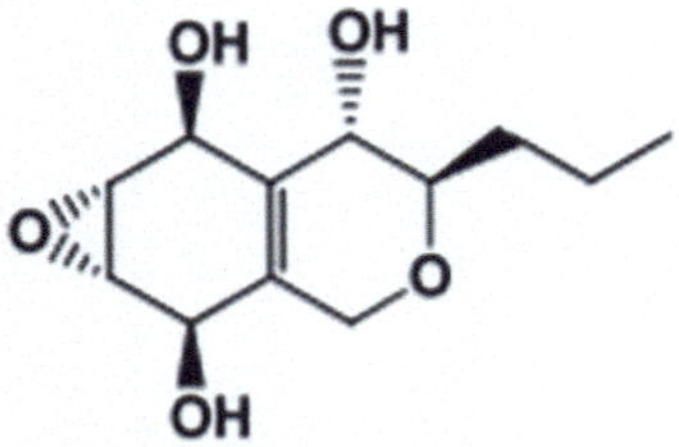

Cycloepoxytriol B

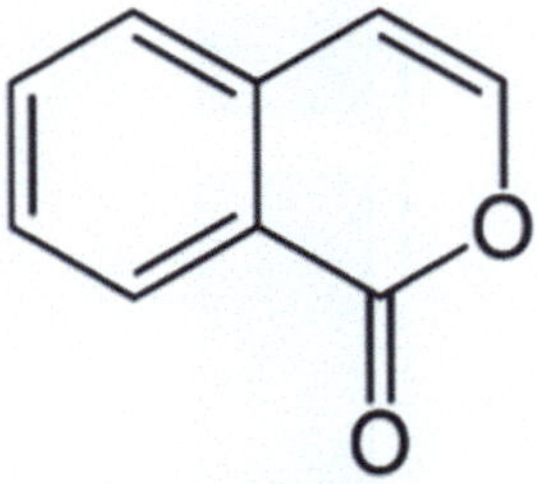

Isocoumarin

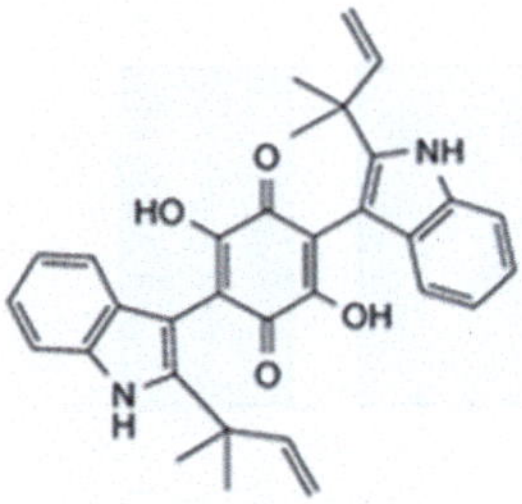

Hinnuliquinone

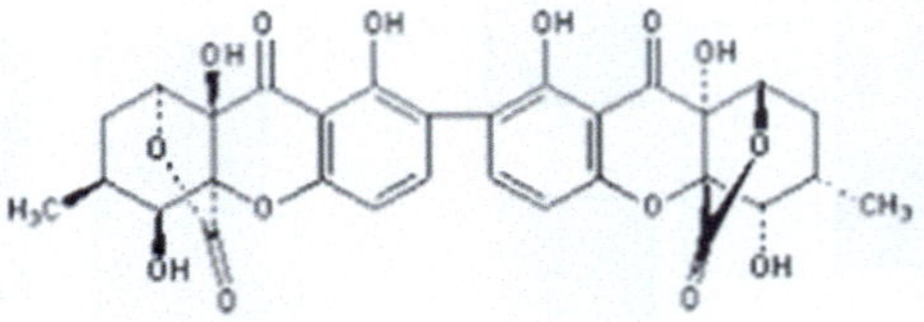

Ergoflavin

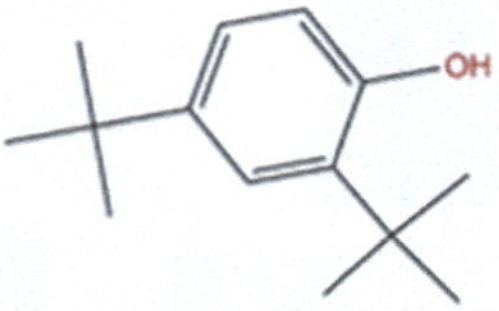

Phenol,2,4-bis (1,1-dimethylethyl)

Fig. 9.4 Structures of some phenolic compounds obtained from leaf endophytes

Table 9.1 Some phenolic compounds produced by leaf endophytes

Endophyte	Source plant	Compound	Property	References
Trichoderma harzianum	*Ficus elastica*	Isocoumarin	Antimicrobial	Ding et al. (2019)
Unknown fungi	*Quercus coccifera*	Hinnuliquinone	Antiviral	Geetanjali (2017)
Fungi PM0651480	*Mimusops elengi*	Ergoflavin	Anticancerous	Deshmukh et al. (2009)
Colletotrichum gloeosporioides	*Phlogacanthus thyrsiflorus*	Phenol,2,4-bis (1,1-dimethylethyl)	–	Devi and Singh (2013)

9.3.2 Terpenoids and Steroids

Terpenoids and steroids are major class of substance biosynthetically derived from isopentenyl diphosphate synthesized by way of the acetate/mevalonate pathway or the glyceraldehyde 3-phosphate/pyruvate pathway. Terpenoids are commercially important in the fragrance industry and food industry as flavoring agents (Ohloff 1990). Terpenoids are classified by the number of five-carbon units(isoprene) they contain. Diterpenes commonly include phytol, gibberellin hormones, and resins. Triterpenoids include phytoalexins and toxins whereas sesquiterpenes majorly include essential oils (Croteau et al. 2000). Diterpenoids, triterpenoids, and sesquiterpenes produced by endophytic fungi are mainly responsible for their antimicrobial activity (Geetanjali 2017). More than ten varieties of terpenes synthesized by endophytic fungi *Alternaria alternata* isolated from the leaves *Azadirachta indica* are reported to exhibit antimicrobial activity against *Bacillus subtilis, Listeria monocytogenes, Staphylococcus aureus, Salmonella typhimurium,* and *Escherichia coli.* The same compounds also showed strong antioxidant activity (Chatterjee et al. 2019) (Fig. 9.5). Antimicrobial activity is reported against *Enterococcus faecalis* by compound Pestalotiopens A produced by endophytic fungi *Pestalotiopsis* sp. isolated from the leaves of *Rhizophora mucronate* (Deshmukh et al. 2015) (Fig. 9.6).

9.3.3 Alkaloids

Alkaloids is one of the most important fungal metabolites from pharmaceutical and industrial point of view. They are a class of nitrogenous organic compounds biosynthesized from amino acids. They are known for their diverse biological properties, such as antifungal, anticancer and antiviral activities and are an important source for drugs (Wang et al. 2011). Production of spiroquinazoline alkaloids alanditrypinone, alantryphenone, alantrypinene, and alantryleunone by endophytic fungi *Eupenicillium* sp. isolated from leaves of *Murraya paniculata* is reported by Barros and Rodrigues-Filho in 2005 (Fig. 9.7). Fungal endophyte *Neotyphodium* sp. are known for the production of ergot alkaloids (Panaccione et al. 2006). Ergot

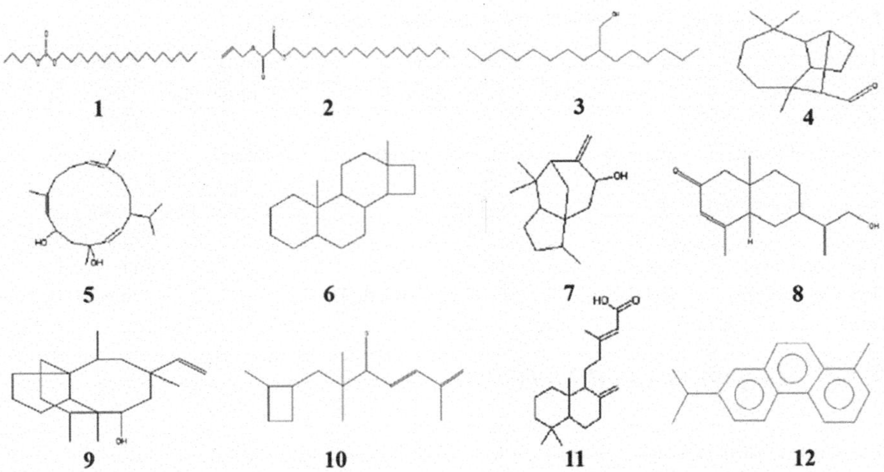

Fig. 9.5 Structures of terpenes obtained from leaf endophytes of *Azadirachta indica* (Chatterjee et al. 2019)

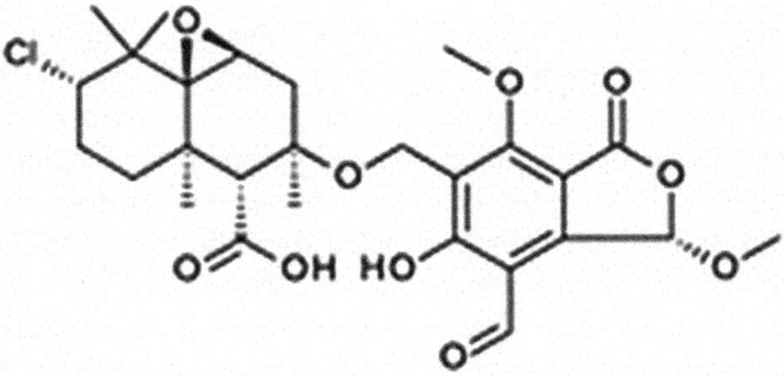

Fig. 9.6 Structure of Pestalotiopens A

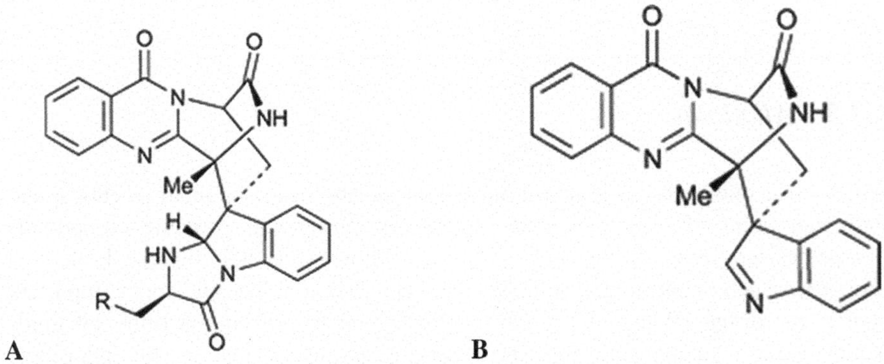

Fig. 9.7 Structures of some spiroquinazoline alkaloids obtained from endophytic fungi *Eupenicillium* sp. isolated from leaves of *Murraya paniculate.* (**a**) Alanditrypinone (R= 3-indolyl); Alantryphenone (R= Ph); Alantryleunone (R= $CHMe_2$), (**b**) Alantrypinene

alkaloid synthesis is also reported by an endophytic fungus of tall fescue *Acremonium coenophialum* (Bacon 1988).

9.3.4 Lipids

Lipids consist of group of natural compounds like waxes, essential oils, fixed oils, sterols, fat-soluble vitamins (such as vitamins A, D, E, and K), phospholipids and others. Waxes are long aliphatic chains having one or functional groups and are esters of fatty acids. Very popular Jojoba wax is reported to have wound healing and antiaging properties. They are also used in the cosmetic industry. Essential oils and fatty acids are derived from propionyl CoA, acetyl CoA, and methyl malonyl CoA precursors. Fixed oils are high molecular aliphatic long-chain fatty acids, e.g., oleic acids. They have antioxidant and anti-inflammatory properties. Essential oils are low molecular weight volatile compounds like linalool, menthol, menthone, camphor, etc. They possess sedative, analgesic, antiseptic, anesthetic, and spasmolytic activities (Hussein and El-Anssary 2018). Linoleic acid and cyclodecasiloxane produced by endophytic fungi *Alternaria* sp. isolated from *Pelargonium sidoides* showed antibacterial activity against *Bacillus cereus, Enterococcus faecium,* and *E. gallinarum* (Manganyi et al. 2019) (Fig. 9.8). 1-Hexadecene, 1-Hexadecanol, Hexadecanoic acid, octadecanoic acid methyl ester, and 1-Nonadecene synthesized by endophytic fungi *Colletotrichum gloeosporioides* residing in leaves of *Phlogacanthus thyrsiflorus* is reported by Devi and Singh (2013).

Apart from the abovementioned reports, there are many other studies and applications of leaf endophytes. However, these studies were conducted on crude extracts and not on purified compounds. Majority of these applications include antimicrobial activity of the crude extract as summarized in Table 9.2. One hundred and sixty-one fungal endophytes were isolated from leaves of *Calotropis procera* with 35.1% colonization rate. All these isolates were tested for antifungal activity against

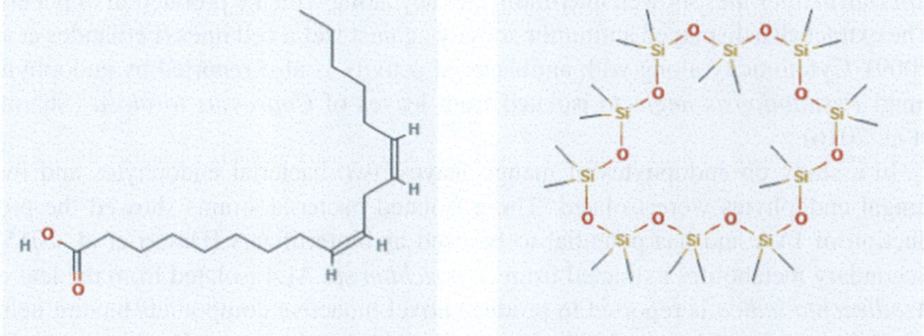

A B

Fig. 9.8 Structures of (**a**) Linoleic acid and (**b**) Cyclodecasiloxane obtained from leaf endophytes of *Pelargonium sidoides*

Table 9.2 Antimicrobial activity of certain leaf endophytes

Endophyte	Plant source	Property	References
Penicillium sp	*Azadirachta indica*	Antibacterial and antifungal	Abubakar et al. (2017)
Trichoderma longibrachiatum and *Syncephalastrum racemosum*	*Markhamia tomentosa*	Antifungal and antiproliferative	Ibrahim et al. (2017)
Unidentified fungal isolate	*Mangifera indica*	Antibacterial	Dasari et al. (2015)
Cochliobolus intermedius	*Sapindus saponaria*	Antibacterial	Garcia et al. (2012)
Pestalotiopsis neglecta	*Cupressus torulosa*	Antibacterial	Sharma et al. (2016)
Cladosporium sp. and *Curvularia* sp.	*Cupressus torulosa*	Antibacterial and antifungal	Bisht et al. (2016)
Pestalotiopsis sp.	*Pinus caneriensis*	Antibacterial and antifungal	Bagyalakshmi et al. (2012)
Alternaria alternata	*Coffea arabica*	Antibacterial and antifungal	Fernandes et al. (2009)
Chaetomium globosum and *Myrothecium verrucaria*	*Calotropis procera*	Antifungal	Gherbawy and Gashgari (2014)
Nigrospora sphaerica and *Pestalotiopsis maculans*	*Indigofera suffruticosa*	Antibacterial	Santos et al. (2015)

Alternaria alternata, *Fusarium oxysporum*, *Botrytis cinerea*, and *Pythium ultimum.* Out of all the isolates, four strains of *Chaetomium globosum* and three strains of *Myrothecium verrucaria* showed maximum efficacy against the selected pathogenic fungi (Gherbawy and Gashgari 2014). Leaves of *Indigofera suffruticosa* are inhabited by sixty-five fungal endophytes out which *Nigrospora sphaerica* and *Pestalotiopsis maculans* presented antimicrobial activity against *Staphylococcus aureus* (Santos et al. 2015). Twenty-two endophytic fungal isolates were isolated from *Coffea arabica.* Their crude extracts showed antimicrobial activity against *Staphylococcus aureus*, *Escherichia coli,* and *Candida albicans.* The isolate *Alternaria alternata* showed maximum efficacy along with the production of phenol. The extract also displayed antitumor activity against HeLa cell lines (Fernandes et al. 2009). Cytotoxicity along with antibacterial activity is also reported by endophytic fungi *Pestalotiopsis neglecta* isolated from leaves of *Cupressus torulosa* (Sharma et al. 2016).

In a study on endophytes of mango leaves, two bacterial endophytes and five fungal endophytes were isolated. These isolated bacterial strains showed the production of IAA and has potential to be used as biofertilizers (Dasari et al. 2015). Secondary metabolites extracted from *Geotrichum* sp. AL4 isolated from the leaf of *Azadirachta indica* is reported to produce novel bioactive compounds having nematicidal activities (Li et al. 2007). Lipophilic antimicrobial peptides from an endophytic bacterial strain EML-CAP3 isolated from *Capsicum annuum* is reported to show antiangiogenic activity (Jung et al. 2015).

9.4 Conclusion

Endophytes are still considered a lesser exploited source of secondary metabolite extraction. Leaf endophytes are easier to isolate and produce all types of secondary metabolites. As per the abovementioned reports, there are several classes of secondary metabolites and all of them have various biological activities. Therefore, the endophytic microflora of leaves can be explored further to obtain novel bioactive compounds. These can also be studied for gene expression of specific genes that contribute to such type of plant–microbe interaction.

References

Abubakar S, Ndana RW, Afolabi AS (2017) Bioprospective potentials of endophytic fungi *Penicillium* spp. isolated from leaves of *Azadirachta indica* (A. Juss). Int J Biol Res 5:15–21

Bacon CW (1988) Procedure for isolating the endophyte from tall fescue and screening isolates for ergot alkaloids. Appl Environ Microbiol 54(11):2615–2618

Bagyalakshmi TA, Ramesh V, Arivudainambi USE, Rajendran A (2012) A novel endophytic fungus *Pestalotiopsis* sp. inhibiting *Pinus canariensis* with antibacterial and antifungal potential. Int J Adv Life Sci 1:1–7

Barros FAP, Rodrigues-Filho E (2005) Four spiroquinazoline alkaloids from *Eupenicillium* sp. isolated as an endophytic fungus from leaves of *Murraya paniculata* (Rutaceae). Biochem Syst Ecol 33(3):257–268

Bayman P, Lebron L, Tremblay RL, Lodge DJ (1997) Variation in endophytic fungi from roots and leaves of Lepanthes (Orchidaceae). New Phytol 135:143–149

Bisht R, Sharma D, Agarwal PK (2016) Antagonistic and antibacterial activity of endophytic fungi isolated from needle of *Cupressus torulosa* D. Don. Asian J Pharm Clin Res 9(3):282–288

Chareprasert S, Piapukiew J, Thienhirun S, Whalley AJS, Sihanonth P (2006) Endophytic fungi of teak leaves *Tectona grandis* L. and rain tree leaves *Samanea saman* Merr. World J Microbiol Biotechnol 22(5):481–486

Chatterjee S, Ghosh R, Mandal NC (2019) Production of bioactive compounds with bactericidal and antioxidant potential by endophytic fungus *Alternaria alternata* AE1 isolated from *Azadirachta indica* A. Juss. PLoS One 14(4):e0214744. https://doi.org/10.1371/journal.pone.0214744

Chutulo EC, Chalannavar RK (2018) Endophytic mycoflora and their bioactive compounds from *Azadirachta indica*: a comprehensive review. J Fungi 4:42

Croteau R, Kutchan TM, Lewis NG (2000) Natural products (secondary metabolites). In: Buchanan B, Gruissem W, Jones R (eds) Biochemistry & molecular biology of plants. American Society of Plant Physiologists, pp 1250–1318

Dasari TR, Inamdar SM, Pawar KV (2015) Study on production of bioactive compounds and plant promoting ability of endophytes isolated from *Rosa* sp. and *Mangifera indica*. Int J Curr Microbiol App Sci 2:136–143

Deshmukh SK, Mishra PD, Kulkarni A (2009) Anti-inflammatory and anticancer activity of ergoflavin isolated from an endophytic fungus. Chem Biodivers 6(5):784–789

Deshmukh SK, Verekar SA, Bhave SV (2015) Endophytic fungi: a reservoir of antibacterials. Front Microbiol 5(715):1–43

Devi NN, Singh MS (2013) GC-MS analysis of metabolites from endophytic fungus *Colletotrichum gloeosporioides* isolated from *Phlogacanthus thyrsiflorus* Nees. Int J Pharm Sci 23(2):392–395

Ding Z, Tao T, Wang L, Zhao Y, Huang H, Zhang D, Liu M, Wang Z, Han J (2019) Bioprospecting of novel and bioactive metabolites from endophytic fungi isolated from rubber tree *ficus elastica* leaves. J Microbiol Biotechnol 29(5):731–738

Fernandes MDRV, Silva TAC, Pfenning LH, Costa-Neto CMD, Heinrich TA et al (2009) Biological activities of the fermentation extract of the endophytic fungus *Alternaria alternata* isolated from *Coffea arabica* L. Braz J Pharm Sci 45(4):677–685

Garcia A, Rhoden SA, Bernardi-Wenzel J, Orlandelli RC, Azevedo JL, Pamphile JA (2012) Antimicrobial activity of crude extracts of endophytic fungi isolated from medicinal plant *Sapindus saponaria* L. J Appl Pharm Sci 2(10):35–40

Geetanjali (2017) Exploring the endophytic fungi for bioactive metabolites: an emerging paradigm. IJARSE 6(12):1815–1822

Gherbawy Y, Gashgari R (2014) Molecular characterization of endophytic fungi from *Calotropis procera* plants in Taif region (Saudi Arabia) and their antifungal activities. Plant Biosyst 148 (6):1085–1092

Gouda S, Das G, Sen SK, Shin HS, Patra JK (2016) Endophytes: a treasure house of bioactive compounds of medicinal importance. Front Microbiol 7:1538

Hata K, Sone K (2008) Isolation of endophytes from leaves of *Neolitsea sericea* in broadleaf and conifer stands. Mycoscience 49(4):229–232

Hussein RA, El-Anssary AA (2018) Plants secondary metabolites: the key drivers of the pharmacological actions of medicinal plants. Intech, London. https://doi.org/10.5772/intechopen.76139

Ibrahim M, Kaushik N, Sowemimo A, Chhipa H, Koekemoer T, Venter M, Odukoya OA (2017) Antifungal and antiproliferative activities of endophytic fungi isolated from the leaves of *Markhamia tomentosa*. Pharm Biol 55(1):1590–1595

Jung HJ, Kim Y, Lee HB, Kwon HJ (2015) Antiangiogenic activity of the lipophilic antimicrobial peptides from an endophytic bacterial strain isolated from red pepper leaf. Mol Cells 38 (3):273–278

Kandel SL, Joubert PM, Doty SL (2017) Bacterial endophyte colonization and distribution within plants. Microorganisms 5:77

Li GH, Yu ZF, Li X, Wang XB, Zheng LJ, Zhang KQ (2007) Nematicidal metabolites produced by the endophytic fungus *Geotrichum* sp. AL4. Chem Biodivers 4:1520–1524

Li F, He X, Sun Y, Zhang X, Tang X, Li Y, Yi Y (2019) Distinct endophytes are used by diverse plants for adaptation to karst regions. Sci Rep 9:5246. https://doi.org/10.1038/s41598-019-41802-0

Manganyi MC, Tchatchouang CDK, Regnier T, Bezuidenhout CC, Ateba CN (2019) Bioactive compound produced by endophytic fungi isolated from *Pelargonium sidoides* against selected bacteria of clinical importance. Mycobiology 47(3):335–339

Ohloff G (1990) Riechstoffe und Geruchssinn. Die molekulare Welt der Düfte, Springer, Berlin, ISBN 3-540-52560-2. English translation: Scent and Fragrances: The Fascination of Odors and Their Chemical Perspectives. New York: Springer; 1994

Pannacione DG, Cipoletti JR, Sedlock AB, Blemings KP, Schardi CL, Machado C, Seidel GE (2006) Effects of ergot alkaloids on food preference and satiety in rabbits, as assessed with gene-knockout endophytes in perennial ryegrass (*Lolium perenne*). J Agric Food Chem 54 (13):4582–4587

Santos IP, Silva NL, Silva MV, Araujo JM, Cavalcant MSI, Lima VM (2015) Antibacterial activity of endophytic fungi from leaves of *Indigofera suffruticosa* Miller (Fabaceae). Front Microbiol 6 (350):1–7

Sharma D, Pramanik A, Agarwal PK (2016) Evaluation of bioactive secondary metabolites from endophytic fungus *Pestalotiopsis neglecta* BAB-5510 isolated from leaves of *Cupressus torulosa* D.Don. 3 Biotech 6(2):210

Stierle A, Strobel G, Stierle D (1993) Taxol and taxane production by Taxomyces andreanae, an endophytic fungus of Pacific yew. Science 260:214–216

Strobel G, Daisy B, Castillo U, Harper J (2004) Natural products from endophytic microorganisms. J Nat Prod 67:257–268

Subban K, Subramani R, Johnpaul M (2013) A novel antibacterial and antifungal phenolic compound from the endophytic fungus *Pestalotiopsis mangiferae*. Nat Prod Res 27 (16):1445–1449. https://doi.org/10.1080/14786419.2012.722091

Taylor JE, Hyde KD, Jones EBG (1999) Endophytic fungi associated with the temperate palm *Trachycarpus fortunei* within and outside its natural geographical range. New Phytol 142 (2):335–346

Tenguria RK, Khan FN (2011) Distribution of endophytic fungi in leaves of *Azadirachta indica* A. Juss. (Neem) of Panchmarhi biosphere reserve. Curr Bot 2:27–29

Verma VC, Gond SK, Kumar A, Kharwar RN, Boulanger LA, Strobel GA (2011) Endophytic fungal flora from roots and fruits of an Indian Neem plant *Azadirachta indica* A. Juss, and impact of culture media on their isolation. Indian J Microbiol 51:469–476

Wang LW, Zhang YL, Lin FC, Hu YZ, Zhang CL (2011) Natural products with antitumor activity from endophytic fungi. Mini Rev Med Chem 11(12):1056–1074. https://doi.org/10.2174/138955711797247716

Zhang HW, Song YC, Tan RX (2006) Biology and chemistry of endophytes. Nat Prod Rep 23:753–771

Chapter 10
Role of Endophytic Fungus *Piriformospora indica* in Nutrient Acquisition and Plant Health

Neha Sharma and Ajit Varma

Abstract *Piriformospora indica* is a unique endophytic fungus belongs to sebacinales order of Basidiomycetes. This endophytic fungus colonized roots of many plant species including model plants like *Arabidopsis thaliana*, *Hordeum vulgare* (barley), and various monocots and dicots. *P. indica* has a beneficial impact on plant growth promotion. Interestingly, symbiotic relationship of *P. indica* does not depend on host specificity and thus it can form symbiotic associations with wide variety of terrestrial plants. Colonization of *P. indica* improves crop productivity and enhances the tolerance of host plants against different biotic (root pathogens) and abiotic (drought, salinity, cold, high temperature) stresses. The positive interactions of *P. indica* with different model plants are used to explore the molecular mechanism of plant–microbe interactions. Furthermore, *P. indica* also has a significant role in nutrient uptake and transport. In this chapter, we discussed the possible beneficial role of *P. indica* in difference aspects of plant growth.

Keywords *Piriformospora indica* · Root endophyte · Crop health · Biotic and abiotic stress · Phytohormones · Plant growth promotion

10.1 Introduction

Piriformospora indica, the entophytic fungus that colonized roots of monocots as well as dicots plants (Verma et al. 1998). This symbiotic fungus has growth-promoting activity in wide variety of plants (Verma et al. 1998). This entophytic fungus can also function as a supporting factor in the enhancement of the host tolerance in plants against abiotic and biotic stresses (Waller et al. 2005). In barley, *P. indica* inoculated plants become more resistance to stress and various root pathogens. Plant with *P. indica* inoculation also induces disease resistance in plants

N. Sharma (✉) · A. Varma
Amity Institute of Microbial Technology, Amity University, Noida, Uttar Pradesh, India
e-mail: nsharma27@amity.edu

N. Shrivastava et al. (eds.), *Symbiotic Soil Microorganisms*, Soil Biology 60,
https://doi.org/10.1007/978-3-030-51916-2_10

(Serfling et al. 2007; Waller et al. 2005; Waller et al. 2008). Different from other mycorrhizal fungi *P. indica* is easy growing on different substrates.

P. indica, belongs to the Basidiomycota division behaves similar in many ways like Arbuscular Mycorrhizal Fungi (AMF) that also belong to Sebacinaceae family (Qiang et al. 2012). Similar to AMF, *P. Indica* also promotes plants growth, increases the resistance in colonized plants against fungal pathogens, and to various abiotic stress and overall beneficial for plant growth (Harman 2011). It can modify secondary metabolites of many economical relevant plants and also can promote seed production in many plants. Similar to AMF, *P. indica* did not invade two strains of Brassica and myc mutants of glycine max. However, different from AMF, *P. indica* can cultivate axenically (Varma et al. 2012a) and it also has the potential to colonize the model plant *Arabidopsis thaliana*.

Plant colonized with *P. indica* exhibits enhancement of plant growth. This entophytic fungus increases nutrient uptake, makes plant susceptible to survive under salt stress, water stress and temperature stress and also provides resistance to heavy metals, toxic elements, pathogens. Many researchers have reported that *P. indica* treatment can improve plant health and biomass production of diverse host (Varma et al. 1999, 2001; Achatz et al. 2010; Sun et al. 2010). More than hundred important medicinal, agricultural, horticultural plants are reported to interact with *P. indica* (Sahay and Varma 1999, 2000; Varma et al. 2012b; Sun et al. 2010).

Genome of *P. indica* has been sequenced and publically available since 2011 (Zuccaro et al. 2011). It possesses 6 chromosomes with genome size of 15.4 to 24 Mb. Genome of *P. indica* contains 50.7% of GC content and 11,768 putative protein coding genes which are responsible for mutualistic interactions, early biotrophic and late necrotrophic phases of the fungus. Moreover, *P. indica* can be transformed stably by random genomic integration of foreign DNA (Zuccaro et al. 2009). *P. indica* acquires a relatively small genome as compared to other members of the Basidiomycota that makes it good model system to study symbiotic relationship in plants (Zuccaro et al. 2009).

10.2 Role of *P. indica* in Plant Growth Promotion

Symbiotic fungus *P. indica* interacts beneficially with large number of crops to promote growth by enhancing root and shoot biomass, secondary root formation, increase in vegetative growth and early and increase flower and seed production (Oelmüller et al. 2009; Qiang et al. 2012; Varma et al. 2012c, 2014; Johnson et al. 2014; Shrivastava et al. 2018). Once endosymbiosis is initiated in the roots, fungus gets access to various plant nutrients, which ultimately promotes colonization and proliferation of symbiotic fungus and thus significantly increases plant growth (Oelmüller et al. 2009; Johnson et al. 2014; Su et al. 2017).

P. indica interaction promotes growth in almost all economically important crops like wheat (*Triticum aestivum*), maize (*Zea mays*), rice (*Oryza sativa*), sugarcane (*Saccharum officinarum*), tomato (*Solanum lycopersicum*), potato (*Solanum

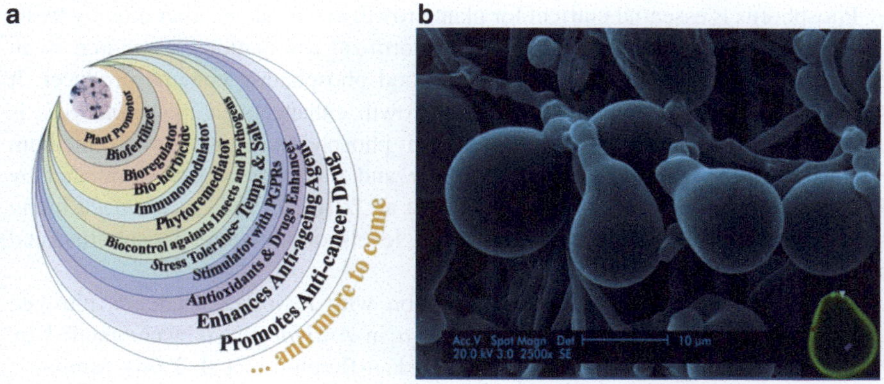

Fig. 10.1 (**a**) Functional annotation of *P. indica*. (**b**) Ultra structure of *P. indica* spores (Shrivastava et al. 2019)

tuberosum), and gram (*Cajanus cajan*). Also in model plant *Arabidopsis thaliana*, in medicinal plants *Abrus precatorius, Bacopa monnieri, Chlorophytum tuberosum, Curcuma longa, Stevia rebaudiana* growth enhancement have been observed (Franken 2012; Varma et al. 2012c, 2013). *P. indica* also efficiently colonized roots of winter wheat varieties and further result in enhanced root and shoot biomass specifically when wheat crops were grown under nutrient limiting conditions (Serfling et al. 2007). Interaction of *P. indica* with barley also results in more grain yield (Waller et al. 2005). Similarly in maize colonization with *P. indica* exhibits enhanced biomass production, root length and root number (Kumar et al. 2009, 2012). *P. indica* also promotes growth of tropical leguminous plants like chickpea, mung bean, peas, and soybean (Varma et al. 2012c). *P. indica* is considered as potential candidate to enhance the biomass production along with various value additions in the form of active ingredients. It significantly increases the vegetative growth of plant and boost immunity to sustain in diverse environment as compared to control. The important pharmaceutically important metabolites are also found increased many folds in fungal treated *A. vera* host. Various functional properties imparted from *P. indica* during interaction with host is depicted in Fig. 10.1.

10.3 Role of *P. indica* in Nutrient Transport

Symbiotic plant–microbe interaction is mainly helpful in assimilation of nitrogen, phosphorus, and other vital micronutrient that help in plant growth (Varma et al. 2013). Symbiotic fungus *P. indica* also promotes plant growth by transportation and absorption of nutrients from soil. This fungus is able to extract and transport macronutrients like nitrogen, phosphorus, potassium, sulfur and magnesium and micronutrients like iron, zinc, manganese, and copper (Shahollari et al. 2007).

Phosphorus is essential nutrient for plant growth and is taken either directly by its own transporters or indirectly through mycorrhizal associations. *P. indica* is an excellent phosphate mobilizer and also a good phosphatase enzyme producer. It has been shown that *P. indica* mediated growth enhancement in *Arabidopsis* is associated with an uptake of radiolabeled phosphorus from growth medium (Shahollari et al. 2007). Phosphorus uptake and transport is also stimulated by *P. indica* colonized roots of maize (Yadav et al. 2010). *P. Indica* colonized mung beans plants also exhibit significantly high levels of nitrogen, phosphorus, and potassium (Kumar et al. 2012).

It has been reported that root colonization with *P. indica* increases nitrogen uptake. Colonization of roots with *P. indica* in *Arabidopsis* is accompanied by requisition of nitrogen from environment (Peskan-Berghofer et al. 2004). Similarly, *P. indica* colonized tomato plants also exhibit better nitrogen acquisition (Cruz et al. 2013). Moreover, *P. indica* stimulated various plastid localized genes involved in sulfur metabolism in *A. thaliana* (Oelmüller et al. 2009). *P. indica* can efficiently mobilize micronutrient from soil and increase their availability to plants through its colonization with roots (Achatz et al. 2010). Growth enhancement in mung bean through *P. indica* colonization is correlated with the significant uptake of nitrogen, phosphorus, and potassium from the soil (Kumar et al. 2012). Overall, *P. indica* plays a significant role in delivering various nutrients for plants growth and development.

10.4 Role of *P. indica* in Phytohormones Regulation

Phytohormones like auxin, cytokinin, gibberellin, and ethylene play significant role in *P. indica* induced growth enhancement in *Arabidopsis* and *Hordeum vulgare* (Sirrenberg et al. 2007; Schäfer et al. 2009). It has been found that *P. indica* inoculated plants exhibit significant promotion in growth that is associated with the higher levels of auxin and gibberellin in colonized roots. Furthermore, cytokinins also play an important role in growth promotion in *P. indica* colonized *Arabidopsis* (Vadassery et al. 2008).

Interestingly, it has been found that in that many genes involved in auxin signaling and metabolism are upregulated in the *P. indica* colonized roots of Chinese cabbage (Lee et al. 2011; Johnson et al. 2013). Significant role of gibberellins has been reported in *P. indica* colonized roots of barley plants (Schäfer et al. 2009). Although, ethylene biosynthesis and signaling that are necessary for plant growth are inhibited by *P. indica* (Schäfer et al. 2009; Khatabi et al. 2012). Abscisic acid and brassinosteroids are also synthesized or modulated by *P. indica* (Schäfer et al. 2009). Thus, *P. indica* colonized plants have better compatibility for endosymbiosis between this endophyte and host plant. Apart from that *P. indica* promotes robust root architecture by producing IAA.

10.5 *P. indica* as a Biocontrol Agent in Disease Resistance

Symbiotic fungus like AMF is reported to induce systemic resistance in hosts (Pham et al. 2004). Similarly, *P. indica* is also beneficial in protecting host from pathogenic fungus (Waller et al. 2005). *P. indica* provides bioprotection against root parasite *Fusarium verticillioides* in maize plants (Kumar et al. 2009). This fungus inhibits the colonization by *F. verticillioides*. Antioxidant enzyme Catalase activity was also found to be significantly high in *F. verticillioides-P. indica* colonized roots. This minimized the probabilities of oxidative burst and ultimately *F. verticillioides* could be protected from the oxidative defense system during colonization.

P. indica can act as a bioprotector and biofertilizer in barley plants (Waller et al. 2005). Similarly, in wheat, *P. indica* is beneficial as a biocontrol agent against pathogenic fungus *Pseudocercosporella herpotrichoides*. Overall, *P. indica* has the potential as a beneficial biocontrol against major cereal pathogens those are harmful for economical important crops. *P. indica* also biosynthesizes important secondary metabolites like hydroxamic acids which can function as natural pesticides (Varma et al. 2001).

10.6 Role of *P. indica* in Stress Response

P. indica can help in stress tolerance like salt and nutritional stress in plants. Tomato plants co-cultivated with *P. indica* promotes stress resistance through antioxidant metabolism activation. Furthermore, *P. indica* inoculated fruit plants can sustain lycopene content better independent of growth conditions. *P. indica* also found beneficial in survival of plantlet, phosphorus content, and nutrient acquisition in *Chlorophytum* sp. (Gosal et al. 2010). Similar results were found in sugarcane plantlets in which *P. indica* inoculation enhanced the survival rate up to 12% after transfer to soil. There was a significant effect on cane yield, cane height, and tillering in *P. indica* inoculated sugarcane plants. Also in ratoon crop Fe and Cu uptake were promoted in P indica colonized plants. Moreover, the cell wall extract of *P. indica* was found significant in increasing cadmium tolerance in rice plants (Varma et al. 2012c). Thus *P. indica* can make plants resistance to various stresses.

10.7 Interaction of *P. indica* with Model Plants *Arabidopsis thaliana* and *Nicotiana attenuate*

Different from AMF, *P. indica* also interacts with the non-mycorrhizal host like *Arabidopsis thaliana* and promotes plant growth in this model plant (Peskan-Berghofer et al. 2004). This growth enhancement effect was explained during the entire life cycle of plant. First, *A. thaliana* seedlings were co-inoculated with the

P. indica under in vitro conditions, there was enhancement found in root and shoot biomass production of seedlings. When these seedlings were transferred to soil leaves were larger in size. Later on, the *A. thaliana* plants grew faster with more number of leaves and early flowering occurs. Also seed yield per plant was higher in *P. indica* inoculated plants (Peskan-Berghofer et al. 2004; Johnson et al. 2011).

Similar observations were also seen in another model plant *Nicotiana attenuata*. Cocultivation of *N. attenuata* results in increased seed germination and enhanced stalk elongation and plant growth (Barazani et al. 2005). Seedlings with *P. indica* colonization exhibit significant increase in root and shoot biomass in soil experiments (Johnson et al. 2011).

The ability of this entophytic fungus to maintain endosymbiosis with well-established model plants like *A. thaliana*, *N. attenuata*, *N. tabacum*, *H. vulgare* makes these plants a powerful model system to study plant–microbe interactions.

10.8 Effect of *P. indica* on Transgenic Plants

P. indica inoculation found to be significant in wide variety of plants. *P. indica* found to be associated with an increase in fresh and dry matter of shoot in rice plants. *P. indica* has a stimulatory effect on eco-physiological parameters like carboxylation efficiency, photosynthetic rate, stomatal conductance, intrinsic water use ability. However, research done on transgenic rice, overexpressing the vacuolar H^+-PPase shows that *P. indica* inoculation was more pronounced in wild type plants in compare to transgenic rice plants (Bertolazi et al. 2019). In wild type rice plants, *P. indica* inoculation was promoting stimulation of Hc pumps while in transgenic plants ATPase were found to be inhibited. Interestingly, nutrient uptake in roots and shoots of wild type and transgenic plants were increased. But nutrient uptake was less in transgenic rice plants compare to transgenic plants (Bertolazi et al. 2019). This shows that in transgenic rice plants *P. indica* colonization is inefficient. Reasons possible for this insufficient symbiosis need to study more.

10.9 Conclusion and Discussion

In brief, *P. indica* can be utilized as a plant promoter, biofertilizer, bioprotector, bioregulator, and biotization agent. *P. indica* can significantly affect plants growth and total biomass. Easy cultivation of *P. indica* with variety of synthetic media makes it an easy to use as biofertilizer. The root endophyte *P. indica* colonization has potential to enhance plant growth, better nutrient uptake, allows plants to survive under various stress like salt stress, temperature stress, and protection from pathogens. *P. indica* is also helpful in stimulating growth and seed production. *P. indica* very well interact with model plants like *Arabidopsis thaliana* and *Nicotiana attenuata* and makes it an excellent model system to study plant–microbe interactions. *P. indica* also plays an important role in phytohormones regulation.

More knowledge of *P. indica* properties will create new horizons for various biotechnological application of this multifunctional fungus especially for agriculture to enhance the crop resistance against different abiotic stresses like temperature and drought. Deeper understanding of molecular mechanism and biomolecules involved in the symbiotic plant–fungus association will give provide significant input for future biotechnological applications of *P. indica*. Overall, *P. indica* is an excellent eco-friendly biofertilizer that can improve crop health, soil health, and can help in fighting hunger problems across the globe.

References

Achatz B, Ruden SV, Andrade D, Neumann E, Kuhnemann JP, Kogel KH, Philipp F, Waller F (2010) Root colonization by Piriformospora indica enhances grain yield in barley under diverse nutrient regimes by accelerating plant development. Plant Soil 22:9–70

Barazani O, Benderoth M, Groten K, Kuhlemeier C, Baldwin IT (2005) *Piriformospora indica* and *Sebacina vermifera* increase growth performance at the expense of herbivore resistance in *Nicotiana attenuata*. Oecologia 146:234–243

Bertolazi AA, Bastos de Souza S, Ruas KF, Campostrini E, Rezende C, Cruz C, Melo J, Colodete CM, Varma A, Ramos AC (2019) Inoculation with Piriformospora indica is more efficient in wild-type rice than in transgenic rice over-expressing the vacuolar H+-PPase. Front Microbiol 10:1087

Cruz C, Fegghi Z, Amelia M, Martins-Lousao MA, Varma A (2013) Plant nitrogen use efficiency may be improved through symbiosis with *Piriformospora indica*. In: Varma A et al (eds) *Piriformospora indica*: Sebacinales and their biotechnological applications, vol 33. Springer, Berlin, pp 285–293

Franken P (2012) The plant strengthening root endophyte *Piriformospora indica*: potential application and the biology behind. Appl Microbiol Biotechnol 96:1455–1464

Gosal SK, Karlupia A, Gosal SS, Chhibba IM, Varma A (2010) Biotization with *Piriformospora indica* and *Pseudomonas fluorescens* improves survival rate, nutrient acquisition, field performance and saponin content of micropropagated *Chlorophytum* sp. Indian J Biotechnol 9:289–297

Harman GE (2011) Multifunctional fungal plant symbiont: new tools to enhance plant growth and productivity. New Phytol 189:647–649

Johnson JM, Sherameti I, Ludwig A, Nongbri PL, Sun C, Varma A, Oelmüller R (2011) Protocols for *Arabidopsis thaliana* and *Piriformospora indica* cocultivation - a model system to study plant beneficial traits. Endocyt Cell Res 21:101–113

Johnson JM, Sherameti I, Nongbri PL, Oelmüller R (2013) Standardized conditions to study beneficial and nonbeneficial traits in the *Piriformospora indica/Arabidopsis thaliana* interaction. In: Varma A et al (eds) *Piriformospora indica*: Sebacinales and their biotechnological applications, vol 33. Springer, Berlin, pp 325–343

Johnson JM, Alex T, Oelmuller R (2014) Piriformospora indica: the versatile and multifunctional root endophytic fungus for enhanced yield and tolerance to biotic and abiotic stress in crop plants. J Trop Agric 52:103–122

Khatabi B, Molitor A, Lindermayr C, Pfiffi S, Durner J, von Wettstein D, Kogel KH, Schäfer P (2012) Ethylene supports colonization of plant roots by the mutualistic fungus *Piriformospora indica*. PLoS One 7(4):e35502

Kumar M, Yadav V, Tuteja N, Johri AK (2009) Antioxidant enzyme activities in maize plants colonized with *Piriformospora indica*. Microbiology 155:780–790

Kumar V, Sarma MVRK, Saharan K, Srivastava R, Kumar L, Sahai V, Bisaria VS, Sharma AK (2012) Effect of formulated root endophytic fungus *Piriformospora indica* and plant growth promoting rhizobacteria fluorescent pseudomonads R62 and R81 on *Vigna mungo*. World J Microbiol Biotechnol 28:595–603

Lee Y-C, Michal Johnson J, Chien C-T, Sun C, Cai D, Lou B, Oelmüller R, Yeh K-W (2011) Growth promotion of Chinese cabbage and arabidopsis by Piriformospora indica is not stimulated by mycelium-synthesized auxin. Mol Plant Microbe Interact 24:421–421

Oelmüller R, Sherameti I, Tripathi S, Varma A (2009) *Piriformospora indica*, a cultivable root endophyte with multiple biotechnological applications. Symbiosis 49:1–17

Peskan-Berghofer T, Shahollari B, Giang PH, Hehl S, Markert C, Blanke V, Varma AK, Oelmüller R (2004) Association of Piriformospora indica with *Arabidopsis thaliana* roots represents a novel system to study beneficial plant-microbe interactions and involves early plant protein modifications in the endoplasmatic reticulum and at the plasma membrane. Physiol Plant 122:465–477

Pham GH, Singh AN, Malla R, Kumari R, Prasad R, Sachdev M, Rexer K-H, Kost G, Luis P, Kaldorf M, Buscot F, Herrmann S, Pesˇkan T, Oelmuller R, Saxena AK, Declerck S, Mittag M, Stabentheiner E, Hehl S, Varma A (2004) Interaction of Piriformospora indica with diverse microorganisms and plants. In: Varma A, Abbott L, Werner D, Hampp R (eds) Plant surface microbiology. Springer, Berlin, pp 227–265

Qiang X, Weiss M, Kogel KH, Schafer P (2012) Piriformospora indica - a mutualistic basidiomycete with an exceptionally large plant host range. Mol Plant Pathol 13(5):508–518

Sahay NS, Varma A (1999) Piriformospora indica: a new biological hardening tool for micropropagated plants. FEMS Microbiol Lett 181:297–302

Sahay NS, Varma A (2000) Biological approach towards increasing the survival rates of micropropogated plants. Curr Sci 78:126–129

Schäfer P, Pfiffi S, Voll LM, Zajic D, Chandler PM, Waller F, Scholz U, Pons-Kühnemann J, Sonnewald S, Sonnewald U, Kogel KH (2009) Manipulation of plant innate immunity and gibberellin as factor of compatibility in the mutualistic association of barley roots with *Piriformospora indica*. Plant J 59:461–474

Serfling A, Wirsel SGR, Lind V, Deising HB (2007) Performance of the biocontrol fungus Piriformospora indica on wheat under greenhouse and field conditions. Phytopathology 97:523–531

Shahollari B, Vadassery J, Varma A, Oelmüller R (2007) A leucine-rich repeat protein is required for growth promotion and enhanced seed production mediated by the endophytic fungus *Piriformospora indica* in *Arabidopsis thaliana*. Plant J 50:1–13

Shrivastava N, Li J, Pan L, Sharma AK, Sanling W, Pandey R, Gao QK, Lou BG (2018) Proteomic approach to understand molecular physiology of symbiotic interaction between *Piriformospora indica* and *Brassica napus*. Sci Rep 8(1):5773

Shrivastava N, Mahajan S, Jain A, Sharma P, Kharakwal AC, Varma A (2019) Mutualistic Interaction of *Piriformospora indica* (*Serendipita indica*) with Aloe vera, the wonder plant for modern living. Am J Plant Sci 10(11):2002–2011

Sirrenberg A, Göbel C, Grond S, Czempinski N, Ratzinger A, Karlovsky P, Santos P, Feussner I, Pawlowski K (2007) *Piriformospora indica* affects plant growth by auxin production. Physiol Plant 131:581–589

Sun C, Johnson JM, Cai D, Sherameti I, Oelmuller R, Lou B (2010) Piriformospora indica confers drought tolerance in Chinese cabbage leaves by stimulating antioxidant enzymes, the expression of drought-related genes and the plastid localized CAS protein. J Plant Physiol 167:1009–1017

Su ZZ, Wang T, Shrivastava N, Chen YY, Liu X, Sun C, Yin Y, Gao QK, Lou BG (2017) *Piriformospora indica* promotes growth, seed yield and quality of *Brassica napus* L. Microbiol Res 199(1):29–39

Vadassery J, Ritter C, Venus Y, Camehl I, Varma A, Shahollari B, Novk O, Strnad M, Ludwig-Muller J, Oelmuller R (2008) Piriformospora indica mediated growth promotion in Arabidopsis is sensitive to high auxin levels, requires trans-cytokinin biosynthesis and the cytokinin receptor combination CRE1/AHK2. Mol Plant Microbe Interact 21:1271–1282

Varma A, Verma S, Sudha SN, Britta B, Franken P (1999) Piriformospora indica – a cultivable plant growth promoting root endophyte with similarities to arbuscular mycorrhizal fungi. Appl Environ Microbiol 65:2741–2744

Varma A, Singh A, Sudha SN, Sharma J, Roy A, Kumari M, Rana D, Thakran S, Deka D, Bharti K, Franken P, Hurek T, Blechert O, Rexer K-H, Kost G, Hahn A, Hock B, Maier W, Walter M, Strack D, Kranner I (2001) Piriformospora indica: a cultivable mycorrhiza-like endosymbiotic fungus. In: Hock B (ed) Mycota IX. Springer, Berlin, pp 125–150

Varma A, Bakshi M, Lou B, Hartmann A, Oelmueller R (2012a) Piriformospora indica: a novel plant growth-promoting mycorrhizal fungus. Agric Res 1(2):117–131

Varma A, Kharkwal A, Agarwal A, Bajaj R, Prasad R (2012b) Piriformospora indica: the model microbe for organic green revolution. Biofertiliser Newsl 20(1):3–8

Varma A, Sherameti I, Tripathi S, Prasad R, Das A, Sharma M, Bakshi M, Johnson JM, Bhardwaj S, Arora M et al (2012c) The symbiotic fungus Piriformospora indica: review. In: Hock B (ed) Fungal associations. Springer, Berlin, pp 231–254

Varma A, Kost G, Oelmüller R (2013) *Piriformospora indica*: Sebacinales and their biotechnological applications. Life Sci 33:1–397

Varma A, Sowjanya Sree K, Arora M, Bajaj R, Prasad RC, Kharkwal A (2014) Functions of novel symbiotic fungus - Piriformospora indica. Proc Indian Natl Sci Acad 80:429–441

Verma S, Varma A, Rexer KH, Hassel A, Kost G, Sarbhoy A, Bisen P, Butehorn B, Franken P (1998) Piriformospora indica, gen. et sp. nov., a new root-colonizing fungus. Mycologia 90:896–902

Waller F, Achatz B, Baltruschat H, Fodor J, Becker K, Fischer M, Heier T, Huckelhoven R, Neumann C, von Wettstein D, Franken P, Kogel KH (2005) The endophytic fungus Piriformospora indica reprograms barley to salt-stress tolerance, disease resistance, and higher yield. Proc Natl Acad Sci USA 102:12286–12291

Waller F, Mukherjee K, Deshmukh SD, Achatz B, Sharma M, Schafer P, Kogel KH (2008) Systemic and local modulation of plant responses by Piriformospora indica and related Sebacinales species. J Plant Physiol 165:60–70

Yadav V, Kumar M, Deep DK, Kumar H, Sharma R, Tripathy T, Tuteja N, Saxena AK, Johri AK (2010) A phosphate transporter from a root endophytic fungus *Piriformospora indica* plays a role in the phosphate transfer to the plants. J Biol Chem 285:26532–26544

Zuccaro A, Basiewicz M, Zurawska M, Biedenkopf D, Kogel K-H (2009) Karyotype analysis, genome organization, and stable genetic transformation of the root colonizing fungus Piriformospora indica. Fungal Genet Biol 46:542–550

Zuccaro A, Lahrmann U, Guldener U, Langen G, Pfiffi S, Biedenkopf D, Wong P, Samans B, Grimm C, Basiewicz M, Murat C, Martin F, Kogel KH (2011) Endophytic life strategies decoded by genome and transcriptome analyses of the mutualistic root symbiont Piriformospora indica. PLoS Pathog 7:e1002290

Chapter 11
The Role of Symbiotic Fungi in Nutri-Farms

Saumya Singh and Ajit Varma

Abstract Small holder farms comprise about 78% of the entire country's agricultural system. These farms are very important for Indian economy, as it produces around 41% of the country's total food grain production. The Indian government and different agricultural institutes have started taking initiatives to convert these farms into nutrition farms, by growing biofortified varieties. Nutri-farms are supposed to be an important link in order to bridge the wide gap between agriculture and nutrition. However, the use of biofortified varieties can be further assisted with various other biofortification techniques, such as use of arbuscular mycorrhizal fungi (AMF) to enhance agricultural output on a broader spectrum. AMF form symbiotic association with the majority of the cultivated crops and play an important role in improving plant nutrition, abiotic and biotic stress tolerance and soil fertility. Incorporation of AMF as bio-fertilizers can help farmers to support a cost-effective and sustainable agriculture.

Keywords Arbuscular mycorrhizal fungi · Nutri-farms · Biotic and abiotic stresses · Bio-fertilizer · Sustainable agriculture

11.1 Introduction

Traditionally, economists believed that health of people in the developing countries was directly related to increased calorie intake. By increased calorie, they meant increased energy to work. In order to attain this objective, the main concern was to increase agricultural productivity. Higher energy intake was thought to enable labours work more efficiently, which would ultimately increase their wages and higher wages would empower them to buy enough food. This was the economic cycle, thought to be governing the society (Meenakshi 2016). This was the main

S. Singh (✉) · A. Varma
Amity Institute of Microbial Technology, Amity University, Noida, Uttar Pradesh, India
e-mail: ajitvarma@amity.edu

N. Shrivastava et al. (eds.), *Symbiotic Soil Microorganisms*, Soil Biology 60,
https://doi.org/10.1007/978-3-030-51916-2_11

underlying idea of the "Green Revolution" in the late 1960s. Dr. M. S. Swaminathan started Green Revolution in India, which helped farmers to increase their agricultural output, mainly cereals such as wheat, rice and maize, thereby, freeing India from the threat of prolonged hunger. Very less importance was given to the nutritional impact of agricultural output on human health.

It is only in recent times, that agricultural researchers have started paying greater attention to the impact of agricultural performance on nutritional outcomes. It has been found that vitamins and micronutrients in non-staple foods and animal products were more strongly correlated to human health benefits. Whereas in developing countries, the poorer and rural communities are majorly dependant on staple foods. In order to overcome malnutrition, frequently called as 'hidden hunger', major challenge is improving the nutritional quality of these staple food crops.

There can be several way-outs to fight this hidden hunger. First, several fortified products, supplemented with essential minerals and vitamins are available off-the-shelf, worldwide. A very common example in the category is, common salt enriched with Fe and I, developed by the National Institute of Nutrition, Hyderabad, India (Sesikaran and Ranganathan 2009). Second, dietary patterns can change in favour of increased consumption of red meat, fruits and vegetables (Prasad et al. 2014). The major drawback for the two approaches is, that these are restricted to urban and semi-urban masses in the developing countries. It is in this context, that the biofortification of staple crops has the potential to avoid calorie-micronutrient trade-off in the underprivileged section of the society. Considering the adversity of the problem, different research universities and government programmes have come up with genetic biofortification. Genetic biofortification involves both traditional breeding and biotechnological tools. Traditional breeding is labour and time intensive. However, the biofortified cultivars thus produced must be essentially high yielding, to be accepted by farmers, particularly in developing countries. Further, the genetically modified crops developed through genetic engineering face problems in the acceptance in several countries. On the contrary, the use of ferti-fortification, that is use of chemical fertilizers to increase yield as well as micronutrients in grains have gained popularity among agronomists. However, only 3–5% of the applied fertilizer can be utilized by plants (Singh et al. 2017). Therefore, repetitive use of chemical fertilizers leads to environmental pollution and disturbance in the native ecological niche. Another less intervened approach to increase micronutrients and vitamins content in grains could be using symbiotic fungi or other biological agents (Rana et al. 2012).

There are two important aspects of agriculture in India, i.e. family farming and smallholder farming. FAO 2014 has declared that family farming plays an important role in world agriculture (Haque 2016). It exists in both developed as well as developing countries. Though, the relative size of family farms is small in developing countries in the Asia-Pacific region. On the other hand, smallholder farms, in India particularly, are less than 2.0 ha (Singh et al. 2002). They are alone responsible for country's 41% of the food grain production. In both farming styles, farmers have adapted the local techniques to support their agriculture. Major advantages of these farming systems are agroforestry, crop rotation and intercropping (Oruru and Njeru 2016). These practices help to improve soil texture, health and microflora. Therefore,

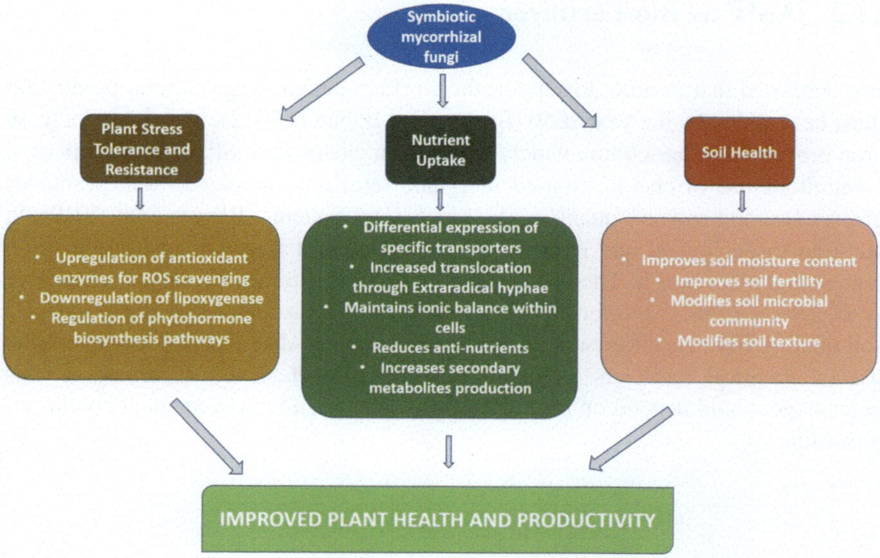

Fig. 11.1 A diagrammatic representation of symbiotic fungal functions to improve plant health, nutrition and productivity (Begum et al. 2019)

there has been surging interest to incorporate sustainable agriculture to these farms and convert these farms to nutritional farms or Nutri-farms.

Nutri-farms of fruits and vegetables were supposed to provide the household with direct access to important micronutrients that may otherwise not be accessible within their economic reach. Therefore, Nutri-farms were thought to be one of the alternatives to enhance access and generate conditions for better consumption of nutritive food. Influenced by the idea, in the budget 2013–2014, United Progressive Alliance (UPA) government, elicited a proposal to establish nutrition farms, also known as Nutri-farms in India (Singla and Grover 2017). This initiative was supported by Harvest Plus, an international organization that works on Nutri-farms. Biofortified staple crops, such as maize (rich in lysine and tryptophan), wheat and rice (rich in zinc) and pearl millet (rich in iron) were grown (Ravi and Usha 2017). Nutri-farms were thought to improve nutritional availability for poorer people especially women and children. Therefore, Nutri-farms create a cost-effective way to challenge the escalating problem of malnutrition in India.

Nutri-farms are supposed to involve low input cropping systems. Therefore, the use of bio-fertilizers can contribute to control plant pathogens and crop productivity for a wider range of crops. Symbiotic mycorrhizal fungi form an association with about 80% of the terrestrial plants, thus can prove to be useful in improving nutritional values of crops and protection from different abiotic and biotic stresses (Fig. 11.1).

11.2 AMF as Bio-Fertilizer

It is supposed that in order to support the world population, agricultural production must be doubled by the year 2050 (Igiehon and Babalola 2017). In order to increase crop productivity, agriculture widely depends on chemical fertilizers and pesticides. Continuous use of chemical-based inorganic fertilizers, herbicides and pesticides deteriorate food and soil quality and air and water systems (Begum et al. 2019). In addition to ecological and environmental problems, it causes harmful effects on human and plant health. Therefore, using mixture of naturally occurring substances or microorganisms as bio-fertilizers can prove to be a sustainable source to improve soil quality. Further, different studies have shown that AMF can possibly lower the use of chemical fertilizers by about 50% (Begum et al. 2019). However, this percentage would depend on host and fungal species and environmental conditions prevailing.

11.3 AMF and Mineral Nutrition

Rigorous studies have shown that AMF symbiosis can enhance the accumulation of macro and micronutrients in plants. These in turn increase photosynthate production in host plants which in turn increases biomass production. Sometimes nutrients are insufficiently phytoavailable in soil. This may be due to either low absolute nutrient amounts or formation of metal complexes which makes them less mobile. Solubility of these metal complexes increases with increasing soil pH and this is a major problem as 30% of the agricultural soil worldwide is alkaline (Lehmann and Rillig 2015). AMF develop symbiosis with host plants to obtain carbon resource from them and in return provide them with essential nutrients such as N, P, K, Zn, Fe, Cu, Mg, Mn, Ca and S.

P is an important mineral that limits plant growth. AMF have been found to beneficially affect plants, in the conditions where soil P levels are low. This may be associated with extra radicle hyphae (ERH) formed by AMF. ERH helps plants to explore greater soil volume, particularly beyond the depletion zone, as compared to root hairs of non-AMF plants. This has been experimentally proved by many researchers. Rhodes and Gerdemann (1978) and Marschner and Dell (1994) performed pot experiments with onions and clover, respectively, and observed that AMF colonized roots were able to explore greater distance in soil, under P deficiency. Since pot conditions are different from field conditions, Mai et al. (2019) planned experiments with AMF colonized cotton under field conditions. They also found that AMF colonization in roots increased their P acquisition range by 15 folds, under P deficiency.

Furthermore, not all P present in the soil is available for plants. In particular, 20–80% of the total P in soil consists of organic phosphates and insoluble inorganic salts (Sato et al. 2019). Thus, the hydrolysis of these phosphates into plant-available form is an important aspect. It has been reported that some plants secrete alkaline phosphatases in their root exudates, to hydrolyse unavailable P and increase plant-available inorganic P (Pi) pool. Zhang et al. (2011) found that *Glomus mosseae* increased maize growth even in the presence of organic fertilizers. Increased ability of plants to absorb P, infected by AMF, is predicted to be caused by the production of certain organic acids and phosphatase enzymes (Aswitha et al. 2019). Similar study was carried out by Pel et al. (2018) to establish if extraradical hyphae of different *Glomus* strains were capable of releasing Pi from resilient apatite matrix.

It has been found that mobility of inorganic P is much lower than inorganic N. Therefore, the role of AMF to increase acquisition of N is less significant than for the acquisition of P. However, there may be situations when available nitrogen in soil limits plant growth. For instance, AMF can be useful to increase N uptake in semiarid and arid conditions, when the mobility of nitrate is reduced substantially or under water stress conditions (Miransari 2011; Shrivastava et al. 2018).

Moreover, apart from the direct use of nitrogen by plants, AMF have been observed to enhance the activities of certain enzymes like, nitrate reductase, glutamine synthetase and glutamine synthase (Miransari 2011). These enzymes help plants to sustain through drought conditions. AM fungi release certain hydrolytic enzymes such as pectinases, xyloglucanase, in the mycorrhizosphere, to organic N patches present in soil. In addition to this, AMF sometimes affect the activities of other soil microbes, which may in turn enhance mineralization of organic matter in soil.

Moreover, AMF balance the dynamic ratio of Ca^{+2} and Na^{+}, important to improve plants' overall performance (Begum et al. 2019). Tran et al. (2019) found that inoculation of AM fungi *Rhizopus irregularis* with durum wheat, improved the plant's availability of zinc and iron. Farzaneh et al. (2011) found that AMF inoculation had positive impact on acquisition of P, K, Fe, Mn and Cu in chickpea. However, they observed that amount of N decreased under the effect of AMF. This was supposed to be due to the dilution effect due to increased biomass. Liu et al. (2000) also found that colonization of *Glomus intraradices* had a positive influence on uptake of nutrients such as Zn, Cu, Mn and Fe in *Zea mays*. Subramanian et al. (2013) have observed that AMF colonization help plants to increase Fe and Zn uptake in both calcareous and non-calcareous soils. In order to hydrolyse the tightly bound nutrients, AMF acidifies surrounding soil.

11.4 AMF and Anti-Nutrients

Phytates (inositol hexakisphosphate) is the most abundant source of organic P in soil. This P cannot be utilized by plants directly. In order to hydrolyse this phytate organisms use various phosphatases, mainly phytase and phosphomonoesterase e.g. acid phosphatases. Early researchers believed that fungal hyphae only produced acid phosphatases in order to hydrolyse lower order inositol phosphates. However, later it was found that fungal hyphae possess phytase activity along with acid phosphatases. Though, the phytase activity was less than 5% of the total acid phosphatase activity (Wang et al. 2017). Subramanian et al. (2013) demonstrated that use of AMF not only increased the accumulation of zinc and iron in maize grains but also helped in circumventing the impact of anti-nutritional compounds like phytate. Zhang et al. (2018) carried out tripartite experiments using AMF strain *Rhizopus irregularis*, a phosphate solubilizing bacteria *Rahnella aquatilis* and two plants root organ cultures of carrot and whole plant of *Medicago truncatula*. They demonstrated that few rhizospheric bacteria are associated with AMF. These rhizobacteria provide P to AMF in return to C. These sugars not only work as C source but they also play an important role as signal molecules for rhizobacteria. These signal molecules stimulate expression of phosphatase genes in bacterium.

11.5 AMF and Abiotic Stresses

11.5.1 Heavy Metals

Researchers have been suggesting, that strigolactones are the molecules responsible for the germination of fungal spores and formation of extra radical hyphae. Further, it was found that optimal or increased concentration of various nutrients in rhizosphere, restrain the secretion of strigolactones from the plant roots (Konieczny and Kowalska 2017). These in turn reduce the mycorrhizal colonization and formation of hyphae. Thus, enabling AM fungi to act as a buffer and inhibit accumulation of any nutrient beyond a required amount. This suggests that AMF increases micronutrient content under nutrient deficiency but prevents over accumulation of minerals when their concentration in soil is close to being toxic.

Joner et al. (2000) demonstrated through experiments, that fungal hyphae are extremely capable of binding heavy metals. Kaldorf et al. (1999) performed experiments with different strains of Glomus and maize. They found that AM fungi colonized plants when grown in heavy metal contaminated soils, increased the concentration of essential minerals such as K, P and Mg but acquisition of heavy metals (HM) such as Ni, Fe, Zn or Cu were decreased in aerial parts of plants, as compared to non-AMF colonized controls. Microbeam analysis to study element localization have shown that heavy metals which entered roots, were deposited in the root parenchyma cells. As a result of this study, it was difficult to establish, whether the HM were deposited in plant roots or fungal cells as, most of the fungal structures such as intraradical hyphae, arbuscules and vesicles are present in root parenchyma cells. In later studies, electron dispersive X-ray spectrometry confirmed that the HM were deposited in electron-dense granules present in fungal cell cytoplasm or their vesicles (Hildebrandt et al. 2007). AM fungal hyphae were found to produce some insoluble glycoproteins called as glomalin, which was capable of binding HM and toxic elements.

11.5.2 Drought

Drought stress adversely affects plant growth by affecting its enzyme activity, ion uptake and nutrient absorption (Begum et al. 2019). Various studies have established strong evidence that AMF helps to lessen drought stress in different crops such as wheat, barley, maize, soybean, strawberry and onion (Begum et al. 2019). Wu and Zou (2017) have outlined potential mechanisms regarding AMF induced drought tolerance in plants under different headings such as morphology, physiology, biochemistry and molecular.

AMF colonization reduced epicuticular wax and lower cuticle weight in rose leaves during drought stress. This led to the abscission of leaves (Henderson and Davies 1990). Further, increased deposition of starch in palisade mesophyll tissues enables mycorrhized plants to recover more quickly from wilting after drought recovery (Henderson and Davies 1990). Recently, Liu et al. (2016) found that mycorrhization influence root morphology of host plants by increasing its total root length, projected area, surface area, root diameter and volume under both well-watered and drought circumstances (Fig. 11.2). These changes in the root morphology may be due to regulation of endogenous polyamine metabolism and phytohormone equilibrium such as root putrescine synthetases and ornithine decarboxylase, IAA (Wu and Zou 2017).

Further, mycorrhizal hyphae have been well known to increase transport and uptake of micronutrients from soil. However, researches have now shown that fungal hyphae mediate to transport water to host plants, along with different nutrients. Fungal hyphae are thinner than root hairs and hence can penetrate soil pores,

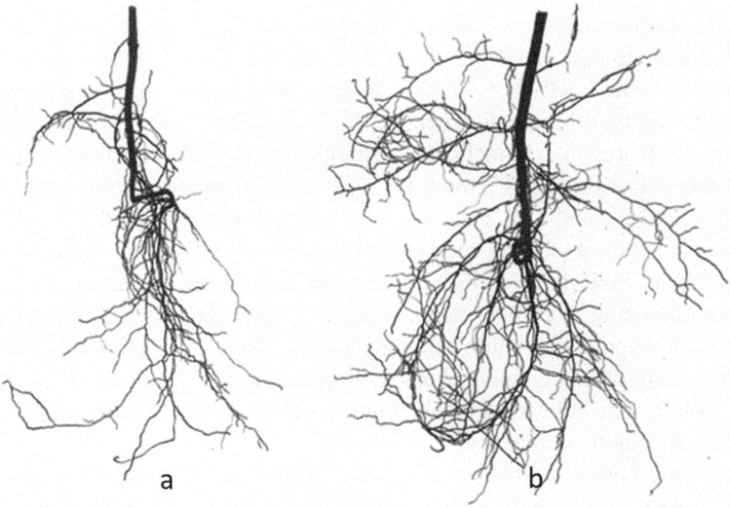

Fig. 11.2 Difference in root morphology of plants under drought stress (**a**) without AMF, (**b**) with AMF (Wu and Zou 2017)

inaccessible to root hairs. Furthermore, due to less or no septa in hyphae, it transports water to plant cells from distant places.

Liu et al. (2016) found that mycorrhization increased levels of following phytohormones in trifoliate orange seedlings: IAA, ABA, methyl jasmonate (MeJA) and zeatin riboside (ZR) under well-watered conditions and IAA, ABA, MeJA, ZR and brassinosteroids under drought conditions. Variations in these phytohormones could be responsible for drought tolerance in plants. Further, mycorrhization causes increased osmotic adjustments in plants. This leads to increased accumulation of soluble starch, glucose, sucrose and K^+ and Ca^{+2} in leaves and roots (Wu and Zou 2017). These osmolytes tend to protect and stabilise macromolecules and cellular structures from osmotic damage and maintain a water potential gradient for water absorption from soil to roots.

AMF not only imparts resistance against drought to plants but also helps plants to be drought tolerant. AMF colonization enhances antioxidant enzymatic and non-enzymatic activities in plants, thereby protecting plant organs against oxidative damage of reactive oxygen species, thus enhancing drought resistance (Wu and Zou 2017).

Nichols (2008) demonstrated an indirect influence of AMF to prevent plants against drought stress. He found that AMF release Glomalin-related soil protein (GRSP) which helps to maintain a good structures soil in comparison to non-mycorrhizal soils. GRSP forms a hydrophobic layer on fungal hyphae which helps to lessen water loss in soil aggregates.

11.5.3 Salinity

Increased concentration of Na^+ and Cl^- in plants adversely affects its normal growth by damaging its foliage, causing nutrient deficiencies, inhibiting carbonic anhydrase and nitrate reductase activities, damage to cellular organelles, decreased photosynthesis by damaging PS II reaction centre (Borde et al. 2017). $[Na^+] > 40$ mM within the root zone creates an extremely hyperosmotic condition for the plants which leads to damaging effect on plants. This leads to an ionic imbalance within plant system, by increasing Na^+ and Cl^- concentration and decreasing K^+, Ca^{+2}, NO_3^- and inorganic phosphorous. Under such stress conditions AMF colonization help plants to maintain the ionic balance by reducing Na^+ uptake and increasing K^+ uptake (Porras-Soriano et al. 2009). K^+ helps to preserve pH balance within plant cells and increase osmotic potential within cellular vacuoles. Porras-Soriano et al. (2009) have shown that AMF colonized olive plants have the highest K^+ in shoots. Thus, K^+ as an osmotic solute was able to maintain high tissue water level even under adverse conditions of osmotic deficiency.

In addition to osmotic imbalance, salinity stress leads to increased production of reactive oxygen species (ROS) like superoxide radicals (O_2^-), hydroxyl radical (OH^-) and singlet oxygen (O_1^-) within plant cells. High levels of these species negatively affect plant cellular metabolism through oxidative damage to lipids, proteins and nucleic acids (He et al. 2007). In response to these destructive elements, plants produce ROS-scavenging enzymes such as superoxide dismutase (SOD), peroxidase (POD), APX and catalase (CAT) in order to neutralize ROS, produced as a result of stress. SOD converts superoxide to oxygen and H_2O_2, and CAT converts this H_2O_2 to oxygen and water. He et al. (2007), reported that AMF colonization in tomato plants helps to enhance host's antioxidant defense mechanism.

11.5.4 Temperature

Temperature stress (high or low) inhibits plants growth and biomass production. Zhu et al. (2011) reported that mycorrhized maize had higher water conservation potential and relative water content. However, it was observed that low or high temperature stress reduces plant's ability to take up water. AMF colonization improves this water status by enhancing water uptake through extraradical hyphae and improving hydraulic conductivity of roots (Zhu et al. 2017).

High temperatures also affect plant's photosynthetic ability. Mathur et al. (2018) investigated effect of AMF colonization in maize under high temperature stress. They reported that mycorrhization increased number of active PS II reaction centres, improved quantum efficiency of PS II, linear e^- transport, excitation energy trapping, thus enhancing the overall photosynthesis rate even at 44 °C. AMF colonization increased total chlorophyll content in leaves.

Furthermore, temperature stress affects integrity of plant plasma membrane by altering its composition and structure (Zhu et al. 2017). Such damages increase membrane permeability for electrolytes. However, Evelin et al. (2009) have reported that mycorrhization helps plants to maintain ionic balance by improving stability of plasma membrane. In addition to these effects, temperature stress also leads to production of ROS. Mycorrhized plants' response to these ROS have already been explained previously.

11.5.5 Biotic Stress

AMF symbiosis have been widely accepted to reduce the damage caused by soil-borne pathogens such as *Fusarium*, *Rhizoctonia*, *Erwinia carotovora*, *Phytophthora*, *Pythium* (Whipps 2004), certain root parasitic plants such as *Striga* and *Orobanche* (Bouwmeester et al. 2003) and to some extent above ground diseases (Whipps 2004).

Several mechanisms have been proposed that may be responsible for the enhanced resistance offered to mycorrhized plants. It may be possible that AMF offers a strong competition for the photosynthates to the soil-borne pathogens (Cordier et al. 1998). In addition to this, AMF colonization results in modification of root architecture, morphology and composition of root exudates. These changes may be responsible for alterations in the dynamics of infection. Moreover, AMF colonization reduces production of strigolactones from plant roots. This reduced production of strigolactones can be positively correlated to the reduced susceptibility of hosts to parasitic plants (Pozo et al. 2010).

11.6 AMF and Secondary Metabolites

These endosymbionts only improve plants' nutritional status and their tolerance for different abiotic and biotic stresses, as described above, but also enhance quality of crops. For example, Bajaj et al. 2014 and Su et al. 2017 observed that colonization of *Curcuma longa* and *Brassica napus* with *Piriformospora indica* increased productivity along with volatile oil and curcumin content. Curcumin has several medicinal and ROS scavenging properties. Dry rhizomes have importance as flavouring and colouring agent in Asian diets. Similarly, Liu et al. (2019) reported that symbiotic association of *Funneliformis mosseae* with *Astragalus membranaceus* promotes its medicinal properties by increasing active ingredients such as astragaloside IV, calycosin-7-glucoside, astragalus polysaccharide and Se. Similarly, CRIŞAN et al. (2018) described several AMF and host combinations, that have been used to demonstrate usefulness of symbiotic fungi to increase active secondary metabolites in hosts.

11.7 Conclusion and Future Prospects

Over the years several researchers have published work related to beneficial work of AMF to increase plant productivity and tolerance for different abiotic and biotic stress. Therefore, this chapter aims to combine the existing information regarding AMF and their potential use in Nutri-farms. By combining the concepts of Nutri-farms and use of AMF could possibly provide nutritional security among populations of developing nations. Green Revolution particularly focused to increase crop yield by employing improved quality seeds to farmers, increased use of chemical fertilizers and pesticides and improved irrigation facilities. But even after Green Revolution, there is a missing link between increased productivity and nutritional security. Therefore, there is a need of Nutritional Revolution, specially in developing countries.

In order to meet this need, converting small holder farms to nutritional farms can prove to be useful. In order to achieve this, the Indian Government has already taken initiatives to promote more establishment of Nutri-farms. However, only biofortified varieties are being employed in these farms, as of now. The major drawback of this could be, wide acceptability of the limited biofortified varieties thus available. Thus, measures could be taken to increase the effectiveness of this concept, by involving other methods of biofortification like the use of AMF.

However, there are some points of concern that should be taken care of, in order to make this technique more acceptable. Most of the smallholding farmers of a particular geographic area, use old landraces having similar properties and genetic composition. Therefore, these genotypes could be tested for their response efficiency to AMF colonization. Moreover, AMF diversity patterns need to be mapped and their efficiency determined in order to encourage different agroecosystems. Farmers should be educated for the beneficial and correct use of AMF so that they are more likely to adopt this sustainable technique. Apart from training the farmers, it is important to mass cultivate AMF for its cost-effective availability to farmers. Cheap fungal inocula need to be produced through different techniques such as mass in vitro production, on-farm multiplication and use of nursery inoculated crops. Large scale production of AMF is a cost-intensive technique, which is borne by small farmers and nursery owners. An alternative for this technique could be on-farm method to develop mycorrhizal inoculum of either indigenous or introduced AMF strains. Thus, all these methods would collectively help to increase number of Nutri-farms in developing countries, thereby bridging the gap between agriculture and nutrition.

References

Aswitha K, Malarvizhi P, Meena S, Kalaiselvi T (2019) Interaction effect of vesicular arbuscular mycorrhiza on increasing phosphorus availability in alkaline soil. J Pharmacogn Phytochem 8:2408–2412

Bajaj R, Agarwal A, Rajpal K, Asthana S, Kumar R, Prasad R, Kharkwal AC, Sherameti I, Oelmüller R, Varma A (2014) Co-cultivation of *Curcuma longa* with Piriformospora indica enhances the yield and active ingredients. Am J Curr Microbiol 2:6–17

Begum N, Qin C, Ahanger MA, Raza S, Khan MI, Ahmed N, Ashraf M, Zhang L (2019) Role of arbuscular mycorrhizal fungi in plant growth regulation: implications in abiotic stress tolerance. Front Plant Sci 10:1068

Borde M, Dudhane M, Kulkarni M (2017) Role of arbuscular mycorrhizal fungi (AMF) in salinity tolerance and growth response in plants under salt stress conditions. In: Varma A, Prasad R, Tuteja N (eds) Mycorrhiza - eco-physiology, secondary metabolites, nanomaterials. Springer, Cham, pp 71–86

Bouwmeester HJ, Matusova R, Zhongkui S, Beale MH (2003) Secondary metabolite signalling in host-parasitic plant interactions. Curr Opin Plant Biol 6:358–364

Cordier C, Pozo MJ, Barea JM, Gianinazzi S, Gianinazzi-Pearson V (1998) Cell defense responses associated with localized and systemic resistance to *Phytophthora* induced in tomato by an arbuscular mycorrhizal fungus. Mol Plant Microbes Interact 11:1017–1028

Crişan I, Vidican R, Stoian V (2018) Induced modifications on secondary metabolism of aromatic and medicinal plants: an endomycorrhizal approach. Hop Med Plants 26:15–29

Evelin H, Kapoor R, Giri B (2009) Arbuscular mycorrhizal fungi in alleviation of salt stress: a review. Ann Bot 104:1263–1280

Farzaneh M, Vierheilig H, Lössl A, Kaul HP (2011) Arbuscular mycorrhiza enhances nutrient uptake in chickpea. Plant Soil Environ 57:465–470

Haque T (2016) Sustainability of small family farms in Asia-Pacific countries. Family Farming 2016:17–40

He Z, He C, Zhang Z, Zou Z, Wang H (2007) Changes of antioxidative enzymes and cell membrane osmosis in tomato colonized by arbuscular mycorrhizae under NaCl stress. Colloids Surf B Biointerfaces 59:128–133

Henderson JC, Davies FT (1990) Drought acclimation and the morphology of mycorrhizal Rosa hybrida L cv Ferdy is independent of leaf elemental content. New Phytol 115:503–510

Hildebrandt U, Regvar M, Bothe H (2007) Arbuscular mycorrhiza and heavy metal tolerance. Phytochemistry 68:139–146

Igiehon NO, Babalola OO (2017) Biofertilizers and sustainable agriculture: exploring arbuscular mycorrhizal fungi. Appl Microbiol Biotechnol 101:4871–4881

Joner EJ, Briones R, Leyval C (2000) Metal binding capacity of arbuscular mycorrhizal mycelium. Plant Soil 226:227–234. https://doi.org/10.1023/A:1026565701391

Kaldorf M, Kuhn AJ, Schroder WH, Hildebrandt U, Bothe H (1999) Selective element deposits in maize colonized by a heavy metal tolerance conferring arbuscular mycorrhizal fungus. J Plant Physiol 154:718–728

Konieczny A, Kowalska I (2017) Effect of arbuscular mycorrhizal fungi on the content of zinc in lettuce grown at two phosphorus levels and an elevated zinc level in a nutrient solution. J Elem 22:761–772

Lehmann A, Rillig MC (2015) Arbuscular mycorrhizal contribution to copper, manganese and iron nutrient concentrations in crops – a meta-analysis. Soil Biol Biochem 81:147–158

Liu A, Hamel C, Hamilton RI, Ma BL, Smith DL (2000) Acquisition of Cu, Zn, Mn and Fe by mycorrhizal maize (*Zea mays* L.) grown in soil at different P and micronutrient levels. Mycorrhiza 9:331–336

Liu J, Guo C, Chen ZL, He JD, Zou YN (2016) Mycorrhizal inoculation modulates root morphology and root phytohormone responses in trifoliate orange under drought stress. Emirates J Food Agric 28:251–256

Liu J, Cai J, He R, Zhang X (2019) Influences of Funneliformis mosseae on the photosynthetic parameters and active secondary metabolites contents of Astragalus membranaceus and Astragalus membranaceus var. mongholicus. ScienceAsia 45:324–331

Mai W, Xue X, Feng G, Yang R, Tian C (2019) Arbuscular mycorrhizal fungi–15-fold enlargement of the soil volume of cotton roots for phosphorus uptake in intensive planting conditions. Eur J Soil Biol 90:31–35

Marschner H, Dell B (1994) Nutrient uptake in mycorrhizal symbiosis. Plant Soil 159:89–102

Mathur S, Sharma MP, Jajoo A (2018) *Improved photosynthetic efficacy of maize (Zea mays)* plants with arbuscular mycorrhizal fungi (AMF) under high temperature stress. J Photochem Photobiol B 180:149–154

Meenakshi JV (2016) Nutri-farms: a review of the evidence on nutritional. Family Farming 2016:195–212

Miransari M (2011) Arbuscular mycorrhizal fungi and nitrogen uptake. Arch Microbiol 193:77–81

Nichols KA (2008) Indirect contributions of AM fungi and soil aggregation to plant growth and protection. In: Siddiqui ZA, Akhtar MS, Futai K (eds) Mycorrhizae: sustainable agriculture and forestry. Springer, Berlin, pp 177–194

Oruru MB, Njeru EM (2016) Upscaling arbuscular mycorrhizal symbiosis and related agroecosystems services in smallholder farming systems. Biomed Res Int. https://doi.org/10.1155/2016/4376240

Pel R, Dupin S, Schat H, Ellers J, Kiers ET, van Straalen NM (2018) Growth benefits provided by different arbuscular mycorrhizal fungi to *Plantago lanceolata* depend on the form of available phosphorus. Eur J Soil Biol 88:89–96

Porras-Soriano A, Soriano-Martin ML, Porras-Piedra A, Azcon R (2009) Arbuscular mycorrhizal fungi increased growth, nutrient uptake and tolerance to salinity in olive trees under nursery conditions. J Plant Physiol 166:1350–1359

Pozo MJ, Jung SC, López-Ráez JA, Azcón-Aguilar C (2010) Impact of arbuscular mycorrhizal symbiosis on plant response to biotic stress: the role of plant defence mechanisms. In: Koltai H, Kapulnik Y (eds) Arbuscular mycorrhizas: physiology and function. Springer, Dordrecht, pp 193–207

Prasad R, Shivay YS, Kumar D (2014) Agronomic biofortification of cereal grains with iron and zinc. In: Sparks DL (ed) Advances in agronomy, vol 125. Academic Press, New York, pp 55–91

Rana A, Joshi M, Prasanna R, Shivay YS, Nain L (2012) Biofortification of wheat through inoculation of plant growth promoting rhizobacteria and cyanobacteria. Eur J Soil Biol 50:118–126

Ravi Y, Usha R (2017) Nutri-farms for nutritional security of farm women: a study in Chamarajanagara district of Karnataka. Mysore J Agric Sci 51:516–521

Rhodes LH, Gerdemann JW (1978) Translocation of calcium and phosphate by external hyphae of vesicular-arbuscular mycorrhizae. Soil Sci 126:125–126

Sato T, Hachiya S, Inamura N, Ezawa T, Cheng W, Tawaraya K (2019) Secretion of acid phosphatase from extraradical hyphae of the arbuscular mycorrhizal fungus Rhizophagus clarus is regulated in response to phosphate availability. Mycorrhiza 29:599–605

Sesikaran B, Ranganathan S (2009) Salt fortification for micronutrient security. In: Symposium on nutrition security for India—issues and way forward

Shrivastava N, Li J, Pan L, Sharma AK, Sanling W, Pandey R, Gao QK, Lou BG (2018) Proteomic approach to understand molecular physiology of symbiotic interaction between *Piriformospora indica* and *Brassica napus*. Sci Rep 8(1):5773

Singh RB, Kumar P, Woodhead T (2002) Smallholder farmers in India: food security and agricultural policy. FAO, Bangkok, Thailand

Singh D, Rajawat MVS, Kaushik R, Prasanna R, Saxena AK (2017) Beneficial role of endophytes in biofortification of Zn in wheat genotypes varying in nutrient use efficiency grown in soils sufficient and deficient in Zn. Plant Soil 416:107–116

Singla P, Grover K (2017) Biofortified cereal crops: a sustainable approach for food and nutritional security. Curr J Appl Sci Technol 24:1–13

Subramanian KS, Balakrishnan N, Senthil N (2013) Mycorrhizal symbiosis to increase the grain micronutrient content in maize. Aust J Crop Sci 7:900–910

Su ZZ, Wang T, Shrivastava N, Chen YY, Liu X, Sun C, Yin Y, Gao QK, Lou BG (2017) *Piriformospora indica* promotes growth, seed yield and quality of *Brassica napus* L. Microbiol Res 199(1):29–39

Tran BT, Cavagnaro TR, Watts-Williams SJ (2019) Arbuscular mycorrhizal fungal inoculation and soil zinc fertilisation affect the productivity and the bioavailability of zinc and iron in durum wheat. Mycorrhiza 29(5):445–457

Wang XX, Hoffland E, Feng G, Kuyper TW (2017) Phosphate uptake from phytate due to hyphae-mediated phytase activity by arbuscular mycorrhizal maize. Front Plant Sci 8:684

Whipps JM (2004) Prospects and limitations for mycorrhizas in biocontrol of root pathogens. Can J Bot 82:1198–1227

Wu QS, Zou YN (2017) Arbuscular mycorrhizal fungi and tolerance of drought stress in plants. In: Wu QS (ed) Arbuscular mycorrhizas and stress tolerance of plants. Springer, Singapore, pp 25–41

Zhang GY, Zhang LP, Wei MF, Liu Z, Fan QL, Shen QR, Xu GH (2011) Effect of arbuscular mycorrhizal fungi, organic fertilizer and soil sterilization on maize growth. Acta Ecol Sin 31:192–196

Zhang L, Feng G, Declerck S (2018) Signal beyond nutrient, fructose, exuded by an arbuscular mycorrhizal fungus triggers phytate mineralization by a phosphate solubilizing bacterium. ISME J 12(10):2339–2351

Zhu XC, Song FB, Liu SQ, Liu TD (2011) Effects of arbuscular mycorrhizal fungus on photosynthesis and water status of maize under high temperature stress. Plant Soil 346:189–199

Zhu X, Song F, Liu F (2017) Arbuscular mycorrhizal fungi and tolerance of temperature stress in plants. In: Wu QS (ed) Arbuscular mycorrhizas and stress tolerance of plants. Springer, Singapore, pp 163–194

Part II
Bacterial Symbiosis

Chapter 12
Understanding the Evolution of Plant Growth-Promoting Rhizobacteria

Pratyusha Sambangi, Vadlamudi Srinivas, and Subramaniam Gopalakrishnan

Abstract Soil is an integral part of the complicated natural environment which is very much alive with complex ecosystem of microbes. Among them, the symbiotic association of rhizobacteria with plants especially on agriculturally important crops is very much advantageous in improving the soil and plant health. These plant growth-promoting rhizobacteria (PGPR) have evolved over the years and involved in many plant functions such as growth promotion, root development, colonization, production of metabolites and in eliciting plant defence mechanism against abiotic and biotic agents. The PGPR's ability to fix the atmospheric nitrogen, solubilize phosphate, potassium and zinc, produce siderophore along with wide variety of phytohormones and secondary metabolites such as antibiotics have attributed to their significance as biocontrol agents. These functions lead to their application as biofertilizers, biopesticides, bioprotectants and phytostimulators. The employment of these PGPR is very much important in agricultural fields as they reduce the burden of chemical fertilizers and pesticides to the farmers and in turn promises an increased crop yield. This chapter discusses the symbiotic association of PGPR with plants in detail including their direct and indirect mechanisms and basis of their induced systemic defence mechanism. It also highlights the use of bioinoculants and nano-formulations of PGPR as an effective tool towards enhanced agricultural production and to combat the plant diseases in an eco-friendly manner.

Keywords Rhizobacteria · PGPR · Symbiosis · Biocontrol · Antifungal · Agriculture

---

P. Sambangi · V. Srinivas · S. Gopalakrishnan (✉)
International Crops Research Institute for the Semi-Arid Tropics (ICRISAT), Patancheru, Telangana, India
e-mail: s.pratyusha@cgiar.org; s.vadlamudi@cgiar.org; s.gopalakrishnan@cgiar.org

N. Shrivastava et al. (eds.), *Symbiotic Soil Microorganisms*, Soil Biology 60,
https://doi.org/10.1007/978-3-030-51916-2_12

12.1 Introduction

The plant root system interacts with a large mixture of microorganisms and these interactions define the extent of association between the plant and microbe. This relationship between the soil bacteria i.e. rhizobacteria and plant is very precise and often influences a lot of factors such as plant growth, soil health, microbiome and the environment (Muller et al. 2016). Rhizobacteria such as rhizobia, root nodule bacteria of the many leguminous plants, undergo symbiotic association and facilitate in biological nitrogen fixation (Peix et al. 2015). In the arid and semi-arid regions, this rhizobia-legume symbiosis is extensively investigated and many studies have reported for their significant source of Nitrogen (N) input in the agricultural fields (del Pozo et al. 2000; Buhian and Bensmihen 2018). The bioavailability of nutrients and minerals at a given soil location is highly dependent on the type of residing rhizobacteria. These nutrient transformations occur depending on the variety of plant–microbial symbiosis. The symbiotic association between the plant and microbe is the key driving factor for the plant growth and even affects the local soil ecosystem (van der Heijden et al. 2008; Verbon and Liberman 2016).

With greater demand for sustainable agriculture, the application of PGPR to crops is beneficial and essential. These root-associated bacteria are diverse in nature and colonize a wide variety of agricultural crops. The genera of rhizobacteria that exhibits the plant growth promotion include *Azospirillum*, *Azotobacter*, *Bacillus*, *Burkholderia*, *Enterobacter*, *Erwinia*, *Flavobacterium*, *Mycobacterium*, *Mesorhizobium*, *Pseudomonas*, *Rhizobium* and *Streptomyces*. These ecological engineers have an association with many agriculturally important crops namely barley, corn, canola, chickpea, groundnut, oats, maize, wheat, rice, lentils, peas, rye and radicchio (Podile and Kishore 2006). These PGPR directly synthesize compounds and provide them to the plant to assist in their well-being. Sometimes they indirectly also facilitates the plant root system to absorb certain nutrients from the soil environment. In this manner, either directly or indirectly the rhizobacteria symbiotically benefit the plant and also fight-off against disease-causing pathogens (Maksimov et al. 2011).

During recent years, several research studies have reported the significance of rhizosphere microbes in playing an important role in the plant growth promotion, in formation of important microbial consortia and disease resistance in host plants (Bhattacharyya and Jha 2012; Alekhya and Gopalakrishnan 2017; Vijayabharathi et al. 2018; Anusha et al. 2019; Gopalakrishnan and Vadlamudi 2019; Kim et al. 2019). Also, they emerged as a potential alternative for chemical fertilizers and have shown promising crop yield outputs in agricultural fields (Laslo and Mara 2019). These PGPR, with high abundance and low cost could be further exploited for their applicative advantages in sustainable agriculture over conventional practices.

PGPR are mainly known to associate with the agriculturally important cereals and leguminous crops. These cereals and legumes are the vital food source for humans and they were also widely used for the livestock (FAO 2018). Hence, the PGPR role as cereals/leguminous plant growth and yield promoters is very much in need of the hour with ever-increasing global population. In the present chapter, the mechanisms and applications of the rhizobacteria were comprehensively analysed to further exploit them and to understand the pre- and post-colonization strategies with changing times. The underlying cellular and molecular mechanisms of rhizosphere microbiota were also studied for their efficacy as bioinoculants.

12.2 Biology of PGPR

Rhizobacteria are a soil bacterium that infects the host plant root system and help plants in many ways. They form a symbiotic association with the host plants and facilitate its growth and development by the exchange of nutrients and metabolites. The main mechanisms that aid in these host-microbe interactions are symbiosis, nitrogen fixation and growth promotion.

12.2.1 Symbiosis

The symbiosis between the rhizobia and legumes triggers the nodulation process and they are fully functional in 3–4 week old plants. The specificity of these nodules usually depends on the type of microbe associated to it. The plant flavonoids are the important metabolites that are utilized by these soil rhizobia to recognize the host system and initiate the symbiotic nodule association. Especially, the aglycones play a key role in the activation of rhizobial *Nod* genes namely *NodA*, *NodB* and *NodC* (Perret et al. 2000). Apart from the flavonoids, the levels of calcium also alter the plant roots hair structure to develop the nodules (Ehrhardt et al. 1996; Downie and Walker 1999). It is reported that *NIN*, a transposon, is involved in the nodule formation and it is the first cloned gene to successfully develop nodules in the host plant (Schauser et al. 1999). During the formation of these nodules an infection thread is formed between the symbiotic bacteria cell surface and host plant wall. From many studies, it is evident that the root lectins facilitated this attachment of rhizobial infection thread and the nodulation process in the plants (Kijne et al. 1997; Van Rhijn et al. 2001). Galibert et al. (2001) reported higher content of G + C in the nodulating and nitrogen fixation genes.

Rhizobia present in soil usually reside in large colonies and it is very much essential to communicate among them. It is evident from various research studies that rhizobacteria have a complex quorum sensing system to form a symbiotic relationship with the host plant. Some of the rhizobia namely, *Rhizobium fredii*, *Rhizobium leguminosarum* and *Sinorhizobium meliloti* are known to have a well-established quorum sensing signalling for nodulation and nitrogen fixation. Mainly this quorum sensing is mediated by the production of *N*-acyl homoserine lactones (AHLs) by rhizobial strains, which involves the chemical crosstalk (González and Marketon 2003). The purpose of understanding this diverse legume-rhizobial chemical and molecular symbiosis is important because of their ability to fix atmospheric nitrogen and pathogen suppression. By recognizing the evolution of host-rhizobial symbioses in agriculturally important crops, we may have better applicative value towards sustainable agriculture.

The biological nitrogen fixation is a major contribution of soil rhizobia to the plant kingdom. Nitrogen is one of the major atmospheric gases and very much essential for the plant growth and photosynthesis (Wagner 2012). But the available form of nitrogen (NH_3) is only made available by these rhizobacteria through the process of biological nitrogen fixation in the root nodules of the plant. Rhizobia are the best-known group of symbiotic soil bacteria that fix nitrogen in relation with a wide variety of agriculturally important crops (Peoples et al. 1995; Dawson 2008; Lindstrom and Mousavi 2010). Applications of chemical fertilizers have significantly reduced in agriculturally important crops due to rhizobia being an efficient source of nitrogen and nutrients (David and Ian 2000). The process of biological nitrogen fixation is regulated by a group of bacterial *nif* (*nifH*, *nifD* and *nifK*) genes. The structure and function of this *nif* gene are similar in many diazotrophs such as *Azotobacter vinelandii*, *Bradyrhizobium japonicum*, *Herbaspirillum seropedicae* and *Pseudomonas stutzeri* (Fischer 1994). Due to this property of biological nitrogen fixation, many rhizobacterial strains were inoculated in legume plants and a significant increase in nodulation and nitrogen fixation was observed. The size, weight and number of nodules and fixed nitrogen were found to significantly enhance in the PGPR inoculated plants compared to un-inoculated plants (Islam et al. 2013; Kuan et al. 2016; Gopalakrishnan et al. 2017, 2018). Hence, these bacteria are regarded as renewable source of nitrogen in the fields and environment that majorly contribute to the conservation of the soil health.

12.2.2 Growth Promotion

Many research studies have reported that the treatment of PGPR has enhanced the plant growth and nutrition status. This proves the fact that these PGPR have the ability to increase the soil fertility and microbial diversity through the production of various root exudates namely extracellular metabolites, hormones, signal compounds and antibiotics (Van Loon 2007; Wani and Gopalakrishnan 2019). PGPR have the capacity to synthesize the phytohormones that directly aid in the plant development. Many reports have identified the synthesis of indole acetic acid (IAA) by PGPR which is mainly responsible for the maturation of plant root system (Patten and Glick 2002; Remans et al. 2008). Cytokinins were also observed and regulated the cell division and root–shoot development of the host plants when inoculated with PGPR (Hussain and Hasnain 2009). In addition, ethylene, gibberellic acid and abscisic acid were also emitted by PGPR that aids in the plant development (Dodd et al. 2010).

The importance of these soil bacteria is mainly attributed to their ability to produce siderophores, which greatly assists in the iron uptake of the host plants. In the presence of metal competition these siderophores aid in the solubilization and diffusion of iron into the plant cell walls (Crowley 2006). This siderophore production also assists in the PGPR colonization by evading other microbial and fungal pathogens. In iron-deficient agricultural fields, the siderophore expressing PGPR are promising alternative for the bioavailability of iron to soil and plants (Sayyed et al. 2013). Fertilizer applications will secure enough phosphorous (P) in the agricultural fields but, these PGPR are the main solubilizers that provide soluble form of phosphate to the crops. PGPR such as *Bacillus*, *Pseudomonas*, *Rhizobium* and *Streptomyces* species are known to enhance the phosphate uptake in the inoculated plants (Ramaekers et al. 2010). Apart from the plant growth and development, these PGPR also influence the yield of the crops, by increasing the mineral density of the seeds (Sathya et al. 2016). Enhanced content of Fe, Zn, Mg, Mn, Ca and Cu were reported in the seeds of PGPR inoculated crops such as wheat, rice chickpea and pigeon pea (Rana et al. 2012; Sharma et al. 2013; Gopalakrishnan et al. 2016a). This ability of biofortification by the PGPR in the agricultural inoculated crops will provide greener choices for better nutrition intake in humans.

In evidence, various greenhouse and field research studies conducted at the International Crops Research Institute for the Semi-Arid Tropics (ICRISAT), based at Patancheru, Hyderabad, India, with different strains of PGPR have shown multiple growth potentials by enhancing the plant growth-promoting and biocontrol traits in the PGPR inoculated host crops (Table 12.1). This emphasizes the significance of PGPR utilization in the future for sustainable agricultural practices in legume crops.

12.3 Role of PGPR as Biocontrol Agents

It is reported that PGPR have antagonistic activities against wide array of bacterial, viral and fungal pathogens. These soil rhizobacteria, within its habitation, exhibit a variety of defence mechanisms to control and fight against the invaders. They trigger the host plant induced systemic resistance i.e. alters the plant cell wall, pathways and metabolites in response to the pathogen infection. Hence, the utilization of PGPR for the management of soil-borne pathogens is highly beneficial in the agriculturally important crops, as it reduces the use of chemical fungicides and eco-friendly in nature (Gopalakrishnan and Vadlamudi 2019). The anti-oxidant enzymes namely, peroxidase, phenylalanine ammonia-lyase, superoxide dismutase and polyphenol oxidase are elicited in the infected host plants by the PGPR to trigger the defence pathways (Gopalakrishnan et al. 2019). This in turn initiates the production of plant defence metabolites such as phenolic compounds, phytoalexins, lytic enzymes and antibiotics (Conrath et al. 2001; Walters et al. 2005). The antibiotics produced by these soil bacteria, especially by *Bacillus* spp. were known to suppress many pathogenic bacteria (Maksimov et al. 2011). Metabolic compounds such as lipids produced by *Bacillus* and *Pseudomonas* are effective biocontrol agents against many bacteria, fungi and protozoans (Raaijmakers et al. 2010). PGPR especially, the

Table 12.1 In vitro evaluation of PGPR strains for PGP and biocontrol traits

PGPR	NCBI no.	PGP properties			Biocontrol properties						Crops evaluated (greenhouse/field)	References
		IAA	*Sid.*	*HCN*	*Cel.*	*Lip.*	*Pro.*	*Chi.*	$\beta - 1,3$	P Sol		
Acinetobacter tandoii (SRI-305)	JQ247013	–	+	–	+	+	+	+	–	+	Chickpea; Rice	Gopalakrishnan et al. (2012), Sreevidya and Gopalakrishnan (2015), Anusha et al. (2019)
Bacillus sp. (VBI-4)	KM250376	+	+	+	+	+	+	+	+	–		
Bacillus sp. (VBI-19)	KM250377	+	+	+	+	+	+	+	+	–		
Bacillus sp. (VBI-23)	KM250378	+	+	+	+	+	+	+	+	–		
Bacillus sp. (SBI-23)	KM250375	+	+	+	+	+	+	+	+	–		
Bacillus altitudinis (SRI-178)	JQ247010	+	+	+	+	+	+	+	–	+		
Bacillus xiamenensis (BS-10)	MF359733	+	–	–	+	+	–	–	+	–		
Bacillus safensis (BS-15)	MF359733	+	+	+	+	+	+	+	+	–		
Bacillus subtilis (BS-17)	MF359737	+	–	+	+	+	+	+	+	–		
Bacillus altitudinis (BS-19)	MF370070	+	–	+	+	+	+	+	+	–		
Bacillus altitudinis (BS-20)	MF370069	+	+	+	+	–	+	+	+	–		
Brevibacterium antiquum (SRI-158)	JQ247009	+	+	–	+	–	+	+	–	+		

Chryseobacterium indologenes (ICKM-4)	KX583496	+	+	+	+	+	+	–	+	–	Chickpea	Gopalakrishnan et al. (2017)
Chryseobacterium sp. (ICKM-17)	KX611375	+	+	–	+	+	+	+	+	–		
Chryseobacterium indologenes (ICS-31)	KY800376	+	–	–	+	+	+	+	–	+		
Enterobacter ludwigii (SRI-211)	JQ247011	+	+	+	+	–	+	+	–	+	Rice	Gopalakrishnan et al. (2012)
E. ludwigii (SRI-229)	JQ247012	+	+	+	+	+	+	+	–	+		
Pseudomonas plecoglossicida (SRI-156)	JQ247008	+	+	+	+	+	+	+	–	+	Rice; chickpea	Gopalakrishnan et al. (2017, 2018)
Pseudomonas monteilii (SRI-360)	JQ247014	+	+	+	+	+	+	+	–	–		
Paraburkholderia kururiensis (IC-76A)	MF373465	+	+	+	+	–	+	–	+	–		
Pantoea dispersa (ICKM-1)	KX583493	+	+	+	+	+	+	–	+	–		
Pseudomonas geniculata (ICKM-7)	KX583495	+	+	+	+	+	+	+	+	+		
Pseudomonas geniculata (ICKM-12)	KX583492	+	+	+	+	+	+	+	+	+		

(continued)

Table 12.1 (continued)

PGPR	NCBI no.	PGP properties			Biocontrol properties						Crops evaluated (greenhouse/ field)	References
		IAA	*Sid.*	*HCN*	*Cel.*	*Lip.*	*Pro.*	*Chi.*	$\beta - 1{,}3$	P Sol		
P. geniculata (ICKM-14)	KX611373	+	+	+	+	+	+	+	+	–		
Pseudomonas geniculata (ICS-30)	KX611376	+	+	+	+	+	+	+	–	+		
Rhizobium pusense (IC-59)	MF372582	+	+	+	+	+	+	–	+	–		
Stenotrophomonas maltophilia (IC-2002)	MF372584	+	+	+	+	–	+	–	+	–		
Stenotrophomonas pavanii (ICKM-9)	KX583494	+	+	+	+	+	+	+	+	+		
Stenotrophomonas maltophilia (ICKM-15)	KX611374	+	+	+	+	+	+	+	–	+		
Stenotrophomonas acidaminiphila (ICS-32)	KX611377	+	+	+	+	+	+	+	+	–		

IAA Indole Acetic Acid (μg/ml); *β, 1–3* β-1,3-glucanase (mg/ml); *Sid* siderophore; *HCN* hydrocyanic acid; *Cel* cellulase; *Lip* lipase; *Pro* protease; *Chi* chitinase. For HCN production, the following rating scale was used: 0 = no colour change; 1 = light reddish brown; 2 = medium reddish brown; and 3 = dark reddish brown

Streptomyces spp. produce various hydrolytic enzymes and acids that show antifungal ability against different agriculturally important fungal pathogens (Alekhya and Gopalakrishnan 2017; Vijayabharathi et al. 2018; Kim et al. 2019; Gopalakrishnan et al. 2019). This PGPR-plant interaction also enhances the jasmonic acid, salicylic acid and ethylene production, which in turn activates the induced and systemic acquired resistance to subdue the disease (Vleesschauwer and Höfte 2009). As the agriculturally important crops namely chickpea, pigeon pea, groundnut and soybean are more prone to the soil-borne pathogens, it is essential to utilize these rhizobacteria for their broad spectrum of biocontrol and plant growth-promoting activities.

12.4 Application of PGPR in Agriculture

As these soil rhizobacteria are advantageous in many ways and confer multiple benefits to the agriculture, application of these beneficial microbes led to their exploitation as biofertilizers and biopesticides. Nowadays many PGPR based bioproducts with high competence are prevailing in the agri-market. *Actinorhizobium* spp., *Azotobacter* spp., *Azospirillum* spp. and *Rhizobium* spp. based biofertilizers are the promising nitrogen suppliers in the agriculture fields (Marketsandmarkets 2014). The *Bacillus* spp. and *Pseudomonas* spp. were also widely used as biopesticides to increase the plant growth and suppress the pathogen (Sallam et al. 2013). Due to their unique specificity and less toxicity, PGPR are further formulated with inoculants to enhance their shelf life. Inoculants such as peat, compost, talc, alginate and chitosan are widely used to entrap these beneficial microbes (Vijayabharathi et al. 2016). PGPR bioinoculants, especially *Rhizobium* spp. could increase the bioavailability and shelf life of the bacteria in the field conditions and also protects against the adverse climate conditions and native microbial flora (Gopalakrishnan et al. 2016b).

In recent years, with the development of nanotechnology, techniques like micro and nano-encapsulation are also being utilized for the efficient delivery of these rhizobacteria. Nanoparticles such as silver, silica and chitosan, which are known to enhance the plant growth are used to encapsulate these rhizobacteria and their metabolites. This will aid in the improved efficacy and better management of plant growth and yield (Nayana et al. 2020). Nowadays, nanofibers are also being used to immobilize these microbial cells for targeted delivery (John et al. 2011). In order to maintain the viability of these beneficial PGPR, the inoculum will be coated over the seeds using spun nanofibers (De Gregorio et al. 2017). Apart from these, new applicative approaches must be identified in the future to explore these PGPR for more innovative bioproducts.

12.5 Commercialization

Various research studies have confirmed the potential of PGPR such as *Azospirillum*, *Bacillus*, *Pseudomonas*, *Serratia* and *Streptomyces* as growth-promoting and biocontrol agents in many horticultural and agricultural crops (Reddy 2014; Wani and Gopalakrishnan 2019). These strategies have led to their successful commercialization as bioproducts in the agri-market. From strain discovery, lab to field, formulations and mass production many steps are to be undertaken to successfully commercialize a PGPR product. Procedures such as documentation, regulations and registrations of the PGPR bioproduct are the main challenges during commercialization. Over the years, the manufacturing industry of these rhizobial inoculants has increased steadily in many countries. They are successfully applied either as single inoculants or co-inoculants to various crops namely legumes, maize, wheat, rice and sugarcane (Santos et al. 2019). But, factors such as long term efficacy, viability and expenditure act as the limiting factors in their commercialization (Hungria et al. 2005). Apart from that, the global market of PGPR products is also influenced by the different patenting policies and legislations in each continent of EU, US and Asia (Backer et al. 2018). But, measures are being undertaken to establish new legislations, alternative technologies, educating farmers and to provide financial support for successful product commercialization, as these PGPR reduce the cost of synthetic agrochemicals for the farmers and lead towards low-cost agricultural practices.

12.6 Conclusion

It is a well-established fact that these soil rhizobacteria are evolving day-by-day in their plant–microbe association. Hence, for sustainable agricultural practices, the researchers and entrepreneurs are attracted to these symbiotic microbes for their plant growth and development, biofertilization, rhizo-remediation, biofortification and disease resistance properties. The bioproducts of these PGPR are successfully enhancing the agricultural yields but are often discouraged by their inefficacy and less viability over a long time. Environmental factors also affect the growth and proliferation of these rhizobacteria in the field conditions. Hence, these limitations should be addressed by multidisciplinary research team for better PGPR formulations and crop protection.

Acknowledgement We thank Mr. PVS Prasad for his significant contribution in collecting the literatures.

Conflict of Interest None.

References

Alekhya G, Gopalakrishnan S (2017) Biological control and plant growth-promotion traits of *Streptomyces* species under greenhouse and field conditions in chickpea. Agric Res 6:410–420
Anusha B, Gopalakrishnan S, Naik MK, Sharma M (2019) Evaluation of *Streptomyces* spp. and *Bacillus* spp. for biocontrol of Fusarium wilt in chickpea (*Cicer arietinum* L.). Arch Phytopathol Plant Prot 52:1–26
Backer R, Rokem JS, Ilangumaran G, Lamont J, Praslickova D, Ricci E, Subramaniam S, Smith DL (2018) Plant growth-promoting rhizobacteria: context, mechanisms of action, and roadmap to commercialization of biostimulants for sustainable agriculture. Front Plant Sci 9:1473
Bhattacharyya P, Jha D (2012) Plant growth-promoting rhizobacteria (PGPR): emergence in agriculture. World J Microbiol Biotechnol 28:1327–1350
Buhian WP, Bensmihen S (2018) Mini-review: nod factor regulation of phytohormone signaling and homeostasis during rhizobia-legume symbiosis. Front Plant Sci 9:1247
Conrath U, Thulke O, Katz V, Schwindling S, Kohler A (2001) Priming as a mechanism in induced systemic resistance of plants. Eur J Plant Pathol 107:113–119
Crowley DE (2006) Microbial siderophores in the plant rhizospheric. In: Barton LL, Abadía J (eds) Iron nutrition in plants and rhizospheric microorganisms. Springer, Dordrecht, pp 169–198
David H, Ian R (2000) Breeding for enhanced nitrogen fixation in crop legumes. Field Crops Res 65:229–248
Dawson JO (2008) Ecology of actinorhizal plants. In: Pawlowski K, Newton W (eds) Nitrogen-fixing Actinorhizal symbioses. Nitrogen fixation: origins, applications, and research Progress. Springer, Dordrecht, the Netherlands, pp 199–234
De Gregorio PR, Michavila G, Ricciardi Muller L, de Souza BC, Pomares MF, Saccol de Sá EL, Pereira C, Vincent PA (2017) Beneficial rhizobacteria immobilized in nanofibers for potential application as soybean seed bioinoculants. PLoS One 4(12):e0176930
del Pozo A, Garnier E, Aronson J (2000) Contrasted nitrogen utilization in annual C3 grass and legume crops: physiological explorations and ecological considerations. Acta Oecol 21:79–89
Dodd IC, Zinovkina NY, Safronova VI, Belimov AA (2010) Rhizobacterial mediation of plant hormone status. Ann Appl Biol 157:361–379
Downie JA, Walker SA (1999) Plant responses to nodulation factors. Curr Opin Plant Biol 2:483–489
Ehrhardt DW, Wais R, Long SR (1996) Calcium spiking in plant root hairs responding to rhizobium nodulation signals. Cell 85:673–681
FAO (2018) World Food and Agriculture—Statistical Pocketbook 254
Fischer HM (1994) Genetic regulation of nitrogen fixation in rhizobia. Microbiol Rev 58:352–386
Galibert F, Finan TM, Long SR, Pühler A, Abola P, Ampe F, Barloy-Hubler F, Barnett MJ et al (2001) The composite genome of the legume symbiont *Sinorhizobium meliloti*. Science 293:668–672
González JE, Marketon MM (2003) Quorum sensing in nitrogen-fixing rhizobia. Microbiol Mol Biol Rev 67:574–592
Gopalakrishnan S, Vadlamudi S (2019) Management of soil-borne diseases of grain legumes through broad-spectrum actinomycetes having plant growth-promoting and biocontrol traits. In: Plant microbe interface. Springer, New York, pp 129–144. https://doi.org/10.1007/978-3-030-19831-2_5
Gopalakrishnan S, Pagidi H, Vadlamudi S, Vijayabharathi R, Bhimineni R, Rupela O (2012) Plant growth-promoting traits of Streptomyces with biocontrol potential isolated from herbal vermicompost. Biocont Sci Technol 22. https://doi.org/10.1080/09583157.2012.719151
Gopalakrishnan S, Vadlamudi S, Samineni S, Sameer Kumar CV (2016a) Plant growth-promotion and biofortification of chickpea and pigeonpea through inoculation of biocontrol potential bacteria, isolated from organic soils. SpringerPlus 5:1882
Gopalakrishnan S, Sathya A, Vijayabharathi R, Vadlamudi S (2016b) Formulations of plant growth-promoting microbes for field applications. In: Microbial inoculants in sustainable

agricultural productivity: functional applications. Springer, Berlin, Germany, pp 239–251. https://doi.org/10.1007/978-81-322-2644-4_15
Gopalakrishnan S, Vadlamudi S, Samineni S (2017) Nitrogen fixation, plant growth and yield enhancements by diazotrophic growth-promoting bacteria in two cultivars of chickpea (*Cicer arietinum* L.). Biocatal Agric Biotechnol 11:116–123
Gopalakrishnan S, Vadlamudi S, Samineni S, Rathore A (2018) Influence of diazotrophic bacteria on nodulation, nitrogen fixation, growth promotion and yield traits in five cultivars of chickpea. Biocatal Agric Biotechnol 15:35–42
Gopalakrishnan S, Srinivas V, Naresh N, Alekhya G, Sharma R (2019) Exploiting plant growth-promoting *Amycolatopsis* sp. for bio-control of charcoal rot of sorghum (*Sorghum bicolor* L.) caused by *Macrophomina phaseolina* (Tassi) Goid. Arch Phytopathol Plant Prot 52:543–559
Hungria M, Loureiro FM, Mendes IC, Campo RJ, Graham PH (2005) Inoculant preparation, production and application. In: Werner D, Newton WE (eds) Nitrogen fixation agriculture, forestry, ecology and the environment. Springer, Dordrecht, pp 224–253
Hussain A, Hasnain S (2009) Cytokinin production by some bacteria: its impact on cell division in cucumber cotyledons. Afr J Microbiol Res 3:704–712
Islam MR, Sultana T, Joe M, Yim W, Cho J-C, Sa T (2013) Nitrogen-fixing bacteria with multiple plant growth-promoting activities enhance growth of tomato and red pepper. J Basic Microbiol 53:1004–1015
John RP, Tyagi RD, Brar SK, Surampalli RY, Prévost D (2011) Bio-encapsulation of microbial cells for targeted agricultural delivery. Crit Rev Biotechnol 31:211–226
Kijne JW, Bauchrowitz MA, Díaz CL (1997) Root lectins and rhizobia. Plant Physiol 115:869–873
Kim YJ, Kim JH, Rho JY (2019) Antifungal activities of *Streptomyces blastmyceticus* strain 12-6 against plant pathogenic fungi. Mycobiology 49:329–334
Kuan KB, Othman R, Abdul Rahim K, Shamsuddin ZH (2016) Plant growth-promoting rhizobacteria inoculation to enhance vegetative growth, nitrogen fixation and nitrogen remobilisation of maize under greenhouse conditions. PLoS One 11:e0152478
Laslo É, Mara G (2019) Is PGPR an alternative for NPK fertilizers in sustainable agriculture? In: Singh D, Gupta V, Prabha R (eds) Microbial interventions in agriculture and environment. Springer, Singapore, pp 51–62
Lindstrom K, Mousavi SA (2010) Rhizobium and other N-fixing symbioses. Encyclopaedia of Life Science (ELS). Wiley, Chichester, UK
Maksimov IV, Abizgil'dina RR, Pusenkova LI (2011) Plant growth promoting rhizobacteria as alternative to chemical crop protectors from pathogens. Appl Biochem Microbiol 47:333–345
Marketsandmarkets (2014) Biopesticides market by active ingredient, by types, by application, by formulation, by crop type and by geography. Marketsandmarkets, Pune
Muller DB, Vogel C, Bai Y, Vorholt JA (2016) The plant microbiota: systems-level insights and perspectives. Annu Rev Genet 50:211–234
Nayana AR, Joseph BJ, Jose A, Radhakrishnan EK (2020) Nanotechnological advances with PGPR applications. In: Hayat S, Pichtel J, Faizan M, Fariduddin Q (eds) Sustainable agriculture reviews, vol 41. Springer, Cham, pp 163–180
Patten CL, Glick BR (2002) Role of *Pseudomonas putida* indoleacetic acid in development of the host plant root system. Appl Environ Microbiol 68:3795–3801
Peix A, Ramírez-Bahena MH, Velázquez E, Bedmar EJ (2015) Bacterial associations with legumes. Crit Rev Plant Sci 34:17–42
Peoples MB, Ladha JK, Herridge DF (1995) Enhancing legume N2 fixation through plant and soil management. Plant Soil 174:83–101
Perret X, Stahelin C, Broughton WJ (2000) Molecular basis of symbiotic promiscuity. Microbiol Mol Biol Rev 64:180–201
Podile AR, Kishore GK (2006) Plant growth-promoting rhizobacteria. In: Gnanamanickam SS (ed) Plant-associated bacteria. Springer, Dordrecht, pp 195–230

Raaijmakers JM, De Bruijn I, Nybroe O, Ongena M (2010) Natural functions of lipopeptides from bacillus and pseudomonas: more than surfactants and antibiotics. FEMS Microbiol Rev 34:1037–1062

Ramaekers L, Remans R, Rao IM, Blair MW, Vanderleyden J (2010) Strategies for improving phosphorus acquisition efficiency of crop plants. Field Crops Res 117:169–176

Rana A, Joshi M, Prasanna R, Shivay YS, Nain L (2012) Biofortification of wheat through inoculation of plant growth-promoting rhizobacteria and cyanobacteria. Eur J Soil Biol 50:118–126

Reddy P (2014) Plant growth promoting rhizobacteria for horticultural crop protection. Springer, New York 10, pp 978–81

Remans R, Beebe S, Blair M, Manrique G, Tovar E, Rao IM, Croonenborghs A, Torres Gutiérrez R, El-Howeity M, Michiels J, Vanderleyden J (2008) Physiological and genetic analysis of root responsiveness to auxin-producing plant growth-promoting bacteria in common bean (*Phaseolus vulgaris* L.). Plant Soil 302:149–161

Sallam NA, Riad SN, Mohamed MS, El-Eslam AS (2013) Formulations of *Bacillus* spp. and *Pseudomonas fluorescens* for biocontrol of cantaloupe root rot caused by *Fusarium solani.* J Plant Prot Res 53:295–300

Santos MS, Nogueira MA, Hungria M (2019) Microbial inoculants: reviewing the past, discussing the present and previewing an outstanding future for the use of beneficial bacteria in agriculture. AMB Exp 9:205

Sathya A, Vijayabharathi R, Gopalakrishnan S (2016) Exploration of plant growth-promoting actinomycetes for biofortification of mineral nutrients. A new avenue for enhancing the productivity and soil fertility of grain legumes. Springer, New York. https://doi.org/10.1007/978-981-10-0707-1_17

Sayyed RZ, Chincholkar SB, Reddy MS, Gangurde NS, Patel PR (2013) Siderophore producing PGPR for crop nutrition and phytopathogen suppression. In: Maheshwari D (ed) Bacteria in agrobiology: disease management. Springer, Berlin, pp 449–471

Schauser L, Roussis A, Stiller J, Stougaard J (1999) A plant regulator controlling development of symbiotic root nodules. Nature 402:191–195

Sharma A, Shankhdhar D, Shankhdhar SC (2013) Enhancing grain iron content of rice by the application of plant growth-promoting rhizobacteria. Plant Soil Environ 59:89–94

Sreevidya M, Gopalakrishnan S (2015) Direct and indirect plant growth-promoting abilities of bacillus species on chickpea, isolated from compost and rhizosphere soils. Org Agric 7. https://doi.org/10.1007/s13165-015-0141-3

Van der Heijden MGA, Bardgett RD, Van Straalen NM (2008) The unseen majority: soil microbes as drivers of plant diversity and productivity in terrestrial ecosystems. Ecol Lett 11:296–310

Van Loon LC (2007) Plant responses to plant growth-promoting rhizobacteria. Eur J Plant Pathol 119:243–254

Van Rhijn P, Fujishige NA, Lim PO, Hirsch AM (2001) Sugar-binding activity of pea lectin enhances heterologous infection of transgenic alfalfa plants by *Rhizobium leguminosarum* biovar viciae. Plant Physiol 126:133–144

Verbon EH, Liberman LM (2016) Beneficial microbes affect endogenous mechanisms controlling root development. Trends Plant Sci 21:218–229

Vijayabharathi R, Sathya A, Gopalakrishnan S (2016) Formulation and commercialization of rhizobia: Asian scenario. In: Agriculturally important microorganisms. Springer, New York, pp 47–67

Vijayabharathi R, Gopalakrishnan S, Arumugam S, Vadlamudi S, Sharma M (2018) Deciphering the tri-dimensional effect of endophytic *Streptomyces* sp. on chickpea for plant growth promotion, helper effect with *Mesorhizobium ciceri* and host-plant resistance induction against *Botrytis cinerea*. Microbiol Pathog 122:98–107. https://doi.org/10.1016/j.micpath.2018.06.019

Vleesschauwer D, Höfte M (2009) Rhizobacteria-induced systemic resistance. Adv Bot Res 51:223–281

Wagner SC (2012) Biological nitrogen fixation. Nat Educ Knowl 3:15

Walters D, Newton A, Lyon G (2005) Induced resistance: helping plants to help themselves. Biologist 52:28–33

Wani S, Gopalakrishnan S (2019) Plant growth-promoting microbes for sustainable agriculture. In: Plant growth promoting rhizobacteria (PGPR): prospects for sustainable agriculture. Springer, Singapore, pp 449–471. https://doi.org/10.1007/978-981-13-6790-8_2

Chapter 13
Rhizobia–Legume Symbiosis During Environmental Stress

Sriram Shankar, Ekramul Haque, Tanveer Ahmed, George Seghal Kiran, Saqib Hassan, and Joseph Selvin

Abstract The rhizobia are microorganisms present in the soil that interact with leguminous plants. Legumes form a unique symbiotic relationship with bacteria known as rhizobia, which they allow to infect their roots. This in turn leads to root nodule formation where bacteria are accommodated for carrying out the process of nitrogen fixation. This symbiotic nitrogen fixation allows legumes to thrive in habitats with limited nitrogen availability. In the present time various environmental stresses such as desiccation, alkalinity, acidity, toxic doses of fertilizer, global rise in temperature, salinity, etc. not only suppress the growth and symbiotic characteristics of rhizobia but also change the effect the nodulation processes. Extreme environmental conditions have been found to adversely affect rhizobia–legume interactions wherein rhizobial partner utilizes endogenous or exogenous osmolytes and secretes specific proteins to alleviate the problem of aridity, salinity and toxicity. In this chapter, we discuss about the diversity of rhizobia in soil, environmental stresses affecting the rhizobia–legume symbiosis and their regulation.

Keywords Rhizobia · Nodulation · Salinity · Legume · Symbiosis

S. Shankar · E. Haque · S. Hassan · J. Selvin (✉)
Department of Microbiology, School of Life Sciences, Pondicherry University, Puducherry, India

T. Ahmed
Department of Biotechnology, School of Life Sciences, Pondicherry University, Puducherry, India

G. S. Kiran
Department of Food Science and Technology, School of Life Sciences, Pondicherry University, Puducherry, India

N. Shrivastava et al. (eds.), *Symbiotic Soil Microorganisms*, Soil Biology 60,
https://doi.org/10.1007/978-3-030-51916-2_13

13.1 Introduction

The chief concern for the global agricultural market lies in offering sustainable food supplies to approximately 8 billion people, wherein 40% of the global agricultural lands are affected by the environmental stress related to drought, salinity and soil toxicity. The cultivation of short duration crops, use of excessive fertilizers and the extensive usage of herbicides and pesticides which played an important part in the green revolution are the primary reasons for the degradation of soil quality. The adverse soil conditions negatively affect the rhizobia–legume symbiosis. The agricultural significance of rhizobia–legume symbiotic interactions cannot be taken lightly. This symbiotic relationship fixes 45×10^6 metric tons of N_2 per year and gives almost half the amount of nitrogen utilized in agriculture (Brockwell et al. 1995). For maintaining a sustainable supply of food, the amount of N_2 required to meet the global requirements is expected to double by 2030 (Tilman 1999). Therefore, the rhizobia–legume symbiosis will have positive impact on both environment and agriculture. Moreover, generation of reactive oxygen species (ROS) such as superoxide ion, hydrogen peroxide and hydroxyl radical can negatively affect the survival of the cells of rhizobia. Drought area is mostly characterized by relatively low humidity, high temperature, high evaporation and scanty rainfall, all of which lead to soil salinity. It also results in the accumulation of salts and solutes, hyperosmotic stress, deterioration of metabolism due to reduction in the water activity and generation of ROS. In the symbiotic relationship, the legumes maintain the production of protein-rich seeds while enhancing the efficiency of various cereals and different crops on the basis of crop rotation (Graham and Vance 2003). These exclusive characteristic features make legume plant farming very important for sustainable agriculture which improves the quality of soil. However, the symbiotic relationship is often adversely affected by many natural conditions, like drought and soil salinity, soil pH, temperature, etc. (Zahran 1999). It is recorded that about 60% production of legume in the developing countries happens in extreme drought conditions (Graham and Vance 2003).

13.2 Diversity of Rhizobia in Soil

Rhizobia refers to polyphyletic group of genera that produces root nodules and fix atmospheric nitrogen in a symbiotic relationship with legumes. They belong to the largest and most remarkable metabolically distinct order *Rhizobiales* of the phylum Alphaproteobacteria. The Genera includes *Ochrobactrum, Bradyrhizobium, Azorhizobium, Methylobacterium, Mesorhizobium, Devosia, Sinorhizobium, Phyllobacterium and Rhizobium.* Generally, rhizobia are isolated by smashing nodules, spreading their contents on solid media rich in nutrients and examining the colonies for the observance of abundant amount of exopolysaccharide slime. There is a noticeable specificity among the species of rhizobia and legume. A specific rhizobial species can infect only a particular species of a legume. A group of similar legumes which can be infected by a specific rhizobial species is known as cross-inoculation group. N_2 fixing nodules get established on roots only when inoculated with an appropriate rhizobial strain.

Learning about rhizobial diversity is an indispensable biological resource and endeavours to discover bacterial strains with fascinating features to boost agricultural productivity (Dai et al. 2012). Moreover, rhizobial residents in surface soil are diverse because of environmental factors. Also, the distribution and genotype of legume plants influence the existence and prevalence of rhizobia in soil. For instance, *Mimosa affinis*, *Phaseolus vulgaris* vary in its nodulation selectivity for legumes. While, *Phaseolus vulgaris* is quite nonselective for its nodulating companion, it gets nodulated by different rhizobia that include *Rhizobium giardinii, R. gallicum, R.etli, R. tropici Bradyrhizobium spp.* and *R. leguminosarum bv. phaseoli.* In contrast, *M. affinis* is quite selective and becomes nodulated by *R. etli* only (Wang et al. 1999). *Bradyrhizobium* strains nodulate Genistoid host plant (brooms) of Morocco and Spain which is a part of four different evolutionary ancestries that contain *B. canariense*, *B. japonicum* and two anonymous genospecies (Vineusa et al. 2005). Further, the diversity of rhizobial players in soil get affected by abiotic factors which include pH, soil type, rainfall and temperature, and the type of soil can further disturb the constitution of rhizobial community. This can be typically demonstrated by a host plant where a legume growing in distinct terrestrial sites gets nodulated with diverse rhizobial genera. For instance, Glycine max (soybean) is frequently nodulated by *B. japonicum* worldwide; Nevertheless, at Xinjiang region in China, *Sinorhizobium fredii* and *Mesorhizobium tianshanense have* been retrieved from this legume. Furthermore, *Rhizobium leguminosarum bv. trifolii* and *bv. viciae* were noted in bean nodules inhabiting Leon, France, while, additionally *S. fredii, R. etli*, and *R. gallicum* also have been documented in bean plants thriving in Andalusia region (Velázquez et al. 2001). Conventionally *Mesorhizobium mediterranean* and *M. ciceri* were documented from nodules of *Cicer arietinum*, on the other hand, *Ensifer meliloti* (previously *Sinorhizobium meliloti*) has been documented from nodules of this host plant growing in water-deficient conditions in Tunisia (Romdhane et al. 2009). Likewise, *E. meliloti* was also reported in *C. arietinum* host plant inhabiting Terai and Almora area of Uttarakhand Himalayas, India (Rajwar et al. 2013). The belief that diversity of rhizobia is determined by the type of soil is ascertained by recognition of distinct rhizobia from the host Caragana plant inhabiting multiple ecological regions in China contrasting in type of soil. *Mesorhizobium genospecies* I, II, IV, VI and VII were identified from host *Caragana* inhabiting sandy soils of Mongolia. *R. yanglingense, M. tianshanense, M. temperatum, M. septentrionale, Rhizobium sp. IV* and *M. genospecies III* were identified in saline/alkaline soils and *Rhizobium sp. IV, M. genospecies VII* and *V* and *M. plurifarium* growing in nutrient-rich forest soils of Northwestern Yunnan locality (Lu et al. 2009). Furthermore, by studying rhizobial diversity stress-tolerant natural isolates can be identified very easily (Zahran 2001). Moreover, when non-hostile environments are tested, inhabitants regularly hold strains tolerant to non-acting stresses, possibly as flexibility to acclimatize to novel challenges (Giller et al. 1997).

13.3 Interaction Between Legume and Rhizobia

One of the significant mutualistic interactions of great importance to humans is that of legume nitrogen-fixing bacteria, a plant bacterial symbiosis. Angiosperms that bear their seeds in pods which include plants of agricultural significance such as peas, clover, soybeans, alfalfa and beans are termed as legumes. Food and agricultural sectors greatly depend on legumes and the ability of the legumes to grow without the addition of nitrogen encourages agriculturalists economically as there is no need to spend huge amount of money towards fertilizers and also minimizes the land pollution which is caused due to the addition of chemical fertilizers. The symbiosis partners are called symbionts and nitrogen-fixing symbionts are called rhizobia. Formation of root nodules takes place when rhizobia infect the legume roots where the bacteria help in nitrogen fixation. The steps in formation of root nodules are as follows:

1. Both plants and the bacterium recognize the correct partner followed by the adherence of bacteria to the root hairs.
2. Bacterium secrets signalling molecules known as Nod factors.
3. Invasion into the root hairs by the bacterium.
4. Formation of infection thread and migration to the main root.
5. Formation of bacteroids (modified bacterial cells) within plant cells followed by the cell division to form a mature root nodule where the nitrogen fixation takes place under anaerobic conditions involving the scavenging of oxygen molecules by leghaemoglobin.

13.4 Role of Exopolysaccharides in Legume–Rhizobia Interaction

The development of an efficient symbiosis does not rely only on the genes that are required for symbiosis but also on genes required for the synthesis of different kinds of cell surface polysaccharides. The rhizobia cell surface comprises of many different polysaccharides, which include), cyclic β-(1,2) glucans, lipopolysaccharide (LPS), outer membrane-confined capsule polysaccharide (CPS exopolysaccharides (EPS), gel-forming polysaccharide (GPS), neutral polysaccharide (NP or glucomannan) and K-antigen polysaccharide (KPS) (Skorupska et al. 2006; Janczarek and Skorupska 2011; Laus et al. 2006). LPS is present in the outer membrane and is composed of lipid A, O-antigen polysaccharide and a core oligosaccharide, for symbiosis to occur lipopolysaccharide is very important (Fraysse et al. 2003). Neutral CPS is present as an insoluble polysaccharide matrix and is synthesized in the idiophase stage of *Rhizobium trifolii* and *R.leguminosarum* growth (Zevenhuizen 1986). Whereas, periplasmic space contains cyclic neutral β-(1, 2) glucans and play a prominent role in plant infection and hypo-osmotic adaptation (Breedveld and Miller 1994). Rhizobial KPS structurally is similar to K-antigens of *E. coli* (Becker et al. 2005). Whereas, glucomannan, that is entirely localized towards a particular pole of the bacterial cell, provides high-affinity adherence to lectin and nodulation of rhizobia (Laus et al. 2006; Williams et al. 2008). Rhizobia synthesizes heteropolymeric exopolysaccharides that are chemically different, species specific and made of linear or branched monomers comprising monosaccharides, like D-galactose, D-glucose, L-rhamnose, D-mannose acid and D-galacturonic acid typically substituted with non-carbohydrate residues (e.g. pyruvyl, acetyl, 3-hydroxybutanoyl groups and succinyl,) and D-glucuronic (Cremers et al. 1991; Laus et al. 2005; O'Neill et al. 1991). Moreover, during stress, EPS supports the survival of microorganisms by excluding toxic compounds such as chloride ions and by forming a barrier to oxygen that encourages the higher persistence of rhizobia (Lloret et al. 1998).

The secreted polysaccharides are given extraordinary importance due to their functions (biofilm formation, safeguarding environmental stresses, adherence to abiotic surfaces and roots and nutrient acquirement) which confers an adaptive environmental advantage to the bacteria (Downie 2010; Skorupska et al. 2006). Although the exact role played by EPS in symbiosis has been studied in detail, the mechanism of action is still under investigation. Very few studies have been reported about their mechanism in symbiosis. A possible mechanism is that EPS plays a significant role in the advanced stages of infection thread initiation and in bacterial release during symbiosis with *Lotus japonicas* (Kelly et al. 2013). The role played by EPS in initiating the infection process has been intensively explored using *Rhizobium leguminosarum* and *S meliloti* symbioses. Mutants incapable of producing EPS give rise to root hair curlings however, they lack the capacity to form infection thread and nitrogen-fixing nodules (Cheng and Walker 1998; Leigh et al. 1985).

13.5 Role of ROS in Nodulation

The plant–microbe interaction consists of both the symbiotic relationship and plant-phytopathogen association. The plant's response to both the interactions are quite similar in the initial stages. Microbes adhere to a compatible host plant tissue and initiate the invasion process wherein the plant protects by the generation of reactive oxygen species (ROS). ROS include singlet oxygen (1O2), superoxide radical (·O2−), (H2O2), hydroxyl radical (•OH) and hydrogen peroxide. ROS are highly known to have a negative impact on biomolecules like protein and nucleic acids by oxidizing them. The plant's defence mechanism reacts strongly to the high built up of ROS that includes the restricting the spread of microorganisms through the formation of strong cell wall barriers and induction of hypersensitive response which activates the secondary cell signalling pathways that is responsible for additional defensive mechanisms. For instance, hydrogen peroxide is reported to be the most important signal molecule for the induction of plant defensive responses to both biotic and abiotic stresses (Bartoli et al. 2013; Djébali et al. 2011). Even in the legume–rhizobia symbiosis the conventional plant defence response is carried out. When rhizobia infects the legume host, the generation of hydrogen peroxide and superoxide radical has been reported at the site of infection (Santos et al. 2001). This resembles the hypersensitive response of plants in incompatible plant–pathogen interactions. Even though this resembles the regular hypersensitive responses exhibited by the plant, legume–rhizobia the host plant implements a distinct defence mechanism (Mithöfer 2002). The nodule, an organ formed to confine the invading bacteria has been widely reported as a defence mechanism in the legume–rhizobia interaction. This is well understood when there is a defect in the surface polysaccharides of rhizobia it in turn affects nodulation leading to susceptibility towards pathogenic responses (Oldroyd and Downie 2008). So a controlled ROS level must be maintained by both the invading bacteria as well as by the legume host to weaken the oxidative stress response, thus, the legume host and the invading rhizobia inherently possess a collection of antioxidant molecules in the form of enzymes that include superoxide dismutase, catalase and different peroxidases (Becana et al. 2000; Matamoros et al. 1999).

A few studies have reported that *Sinorhizobium meliloti* contains an array of antioxidant enzymes which includes three catalases and two superoxide dismutases (Minchin et al. 2008; Becana et al. 2010; Jamet et al. 2007). So the absence of any of the three catalases or two superoxide dismutases in the mutants leads to amplified sensitivity of *S.meliloti* to ROS. However, the prevention of nodule formation was not observed (Jamet et al. 2007). Also, nod factors (NFs, signalling molecule) released by rhizobia inhibits the efflux of ROS from the roots of the legume host (Shaw and Long 2003). Nod factors play a significant role in suppressing the generation of ROS by the plant which is highly essential for the compatible association among the host plant and rhizobia (Chang et al. 2009). Thus, for a successful nodulation, the host defence system against the oxidative stress exhibited by ROS is highly essential, as catalase deficit due to RNA interference in *Medicago truncatula* roots, has been found to reduce nodulation, particularly during osmotic stress situations. Even though the suppressing ROS is essential for nodule formation and the compatible symbiosis, ROS like hydrogen peroxide is vital for the transpeptidation and strengthening of plant cell wall formation. Previously, Chang et al. (2009) has reported that the root hair curling and infection thread formation is prevented when ROS production is inhibited. Finally, the difference in the ROS accumulation differentiated at distinct time period during the symbiosis describes the indispensable role of ROS at multiple steps in the nodulation.

13.6 Environmental Stresses and Their Regulation

13.6.1 Temperature Stress

The temperatures close to the soil surface may be too high in arid and semiarid areas. Egyptian sandy soils can reach 59 °C when the atmospheric temperature is 39 °C. The temperature of soil declines quickly through penetration, reaching a modest temperature of 35 °C, at 0.15 m. It looks as if rhizobial strains are unaffected by high temperatures in soil than in laboratory conditions. The temperature plays a vital role in the transfer of environmental signals among rhizobia and their host plant, consequently decreasing nodule formation. Low temperature prevents inter-organismal signalling among the symbiotic associates. It has been revealed that low temperature obstructs the synthesis and secretion of ligands (signal molecules) required in initiating the nodulation process (Abd-Alla 2001).

The ideal temperature for rhizobial growth is 24–30 °C (Zhang et al. 1995). Most of the research on rhizobial temperature stress tolerance concentrates on common bean and soybean microsymbionts. Soybean isolates develop feebly at 40 °C and none of the isolates was found to be capable to grow at 45 °C (Chen et al. 2002). Temperature stress can be of two types, cold and heat shock. The heat shock response is analogous to the acid stress response considering the synthesized proteins. The heat shock proteins (HSPs) assist in tolerating heat by providing protection against heat to the bacteria thus maintaining internal homeostasis (Yura 2000). Cold shock is just the reverse of heat shock. Instead of proteins denaturing and misfolding, cells experiencing cold shock have to struggle with disorientation of membrane integrity, cytosol fluidity and with the maintenance of secondary structures of DNA/RNA (Phadtare et al. 2000). The impact of temperature stress affecting nodulation and nitrogen fixation has long been recognized, since the first studies on this topic can be traced back to 1960s. Even prior to the nodule formation, the temperature of root zone affects the rhizobia prevalence in soil, along with the mutual transfer of molecular signals among the symbiotic associates (Sadowsky 2005). High temperature has been found to have an inhibitory influence on adherence of rhizobia to root hairs, root hair development and infection thread initiation (Hungria and Vargas 2000). Nodule functioning is also affected by high temperatures. Specifically, affecting the rate of reactions involved in the production of leghemoglobin or in nitrogenase activity. Low temperatures also influence nodulation, as cold temperatures interrupt nodulation initiation or even totally inhibit the process and appear to affect nodule occupancy (Graham 1992). Also, prior studies with *B. japonicum* highlight the fact that Nod factor synthesis markedly drops at 17 °C or 15 °C, in spite of the fact that its biological action remains unchanged (Duzan et al. 2006). Soybean and bean have related threshold, whereas lentil is extra tolerant and nodulation is considerably delayed only at subordinate temperatures (10 °C) (Junior et al. 2005)

13.6.2 Salt Stress

Salinity is among the chief ecological threats towards agriculture and affects about 7% of the global total land space (Türkan and Demiral 2009). Biomass and defence mechanisms are restricted due to salt stress (Zheng et al. 2009), and it also obstructs the mechanism of symbiotic biological nitrogen fixation of the legume plants. The negative impact of salt on biological nitrogen fixation in legumes has been widely described, for instance, the effects include reduced photosynthate substrate supply to the nodule, minimum supply of respiratory requirements to the bacteroids, and variations in the oxygen diffusion barrier (Soussi et al. 1998).

The salt stress has been found to affect the activity of nodules in a few ways. The higher concentration of solutes can hinder nodulation and affect nitrogenase activity. There is a threshold level of few ions in the soil and if it increases beyond that level, it acts as a factor that can contribute to salt stress. Previous studies using soybean demonstrated that the contact between *Bradyrhizobium japonicum* and root hairs in increasing salt concentrations (0% to 1.8%) affected the properties of the nodule. Inhibition was apparent in the flexibility of the hairs at a concentration of 1%, a phenomenon noticeably emphasized at increasing concentrations. Growth was gradual at 0.2% salinity and the nodulation was affected at 1.2% salinity.

A few studies using *Vigna unguiculata* (cowpea) and *Vicia faba* (pea) demonstrated that the administration of various concentrations of salt at various stages of development of root hairs decreases the colonization of roots, coiled hairs, organisation of the hypodermic hair and hair cells, the quantity and weight of nodules and activity of nitrogenase (Zahran and Sprent 1986; Georgiev and Atkins 1993). Another study documented the impact of various NaCl concentrations on growth and accumulation of polyhydroxybutyrate (PHB) in four distinct *Sinorhizobium* strains (Arora et al. 2006). Oxygen flux into the nodule gets affected due to the salt stress that alters the oxygen diffusion barrier (Serraj and Drevon 1998). The drop in N_2 fixation is due to the minimum supply of carbon to bacteroids, principally because of malate limitation and salinity-induced inhibition of nodule through the inhibition of the enzyme sucrose synthase (Salah et al. 2009). The accumulation of compatible solutes like sucrose, D-pinitol and proline, in the nodules of some legumes such as cowpea and alfalfa has been described and they employ an osmoregulatory activity during salt stress (Irigoyen et al. 1992).

In the legume- rhizobia symbiosis, the bacterial partner can tolerate salt stress efficiently than their host which displays disparity in the tolerance level (Zurayk et al. 1998). 100 mM NaCl can inhibit the growth of several rhizobia (Yelton et al. 1983), but the previous studies have described that several species can withstand salt concentration ranging from 300 to 700 mM (Mohammad et al. 1991). Another study reported that *R. leguminosarum* can survive even upto 350 mM NaCl concentration (Breedveld et al. 1991). The rapidly growing strains of chickpea and soybean rhizobia can also survive upto a concentration of 340 mM of NaCl related to slow-growing bacterial strains (Elsheikh and Wood 1995). Mashhady et al. (1998) have documented that *S. meliloti* developed productive mutualism with *M. sativa* in salinity conditions, i.e. 100 mM NaCl (Mashhady et al. 1998). Among the tree species, *Prosopis articulata*, *P. pallida* and *P. tamarugo* can grow and fix nitrogen in less than 300 mM NaCl concentration (Felker et al. 1981). Australian Salt-tolerant Acacia (A. *stenophylla* and A. *auriculiformis*) can grow in the range from 1.7 to 1.8 M NaCl (Aswathappa et al. 1987).

In *Phaseolus vulgaris* (common bean), a negative impact of NaCl on the expression of nod genes and nodulation factors by *Rhizobium etli*, *Rhizobium tropici* was observed (Dardanelli et al. 2008). The preincubation of *B. japonicum* using the signal molecule genistein, in saline conditions, was designated as a method to improve the stressful impact of salt on soybean–*B. japonicum* symbiosis (Miransari and Smith 2009). High salt (NaCl) concentrations modify the metabolism of rhizobia and also influence nitrogen fixation (Dowling and Broughton 1986). Nevertheless, rhizobia strains, that are salt-tolerant and effective nitrogen fixers, have been isolated (Rai 1983). The hunt for strains of rhizobia that can tolerate salinity can perhaps be an alternative for successful symbiosis in conditions wherein crop growth is not altered (Douka et al. 1984; Rosas et al. 1996).

13.6.3 Drought Stress

Crop production and symbiotic N_2 fixation are affected due to drought. The biological N_2 fixation is also extremely sensitive to water deficit. When exposed to soil water deficit, several rhizobia exhibit a drop in nitrogen fixation (Zahran 1999; Pimratch et al. 2008). Regulation of water loss is the primary mechanism by which plants escape drought stress and is controlled either by morphological changes or by physiological adaptations. Structural changes include stiffening of waxy cuticle and expansion of root systems. Moreover, responsive stomata, leaf rolling and plant hormone secretions are the physiological adaptations that plants carry out to combat drought (Turner et al. 2001). Therefore, plants with water potential lower than that of soil persist (Tiaz and Zeiger 2002). Both root development and root hairs have been reported to be greatly affected by drought stress that results in the inhibition of nodule formation, it has also been observed that watering the soil helped in the renewal of growth. Reduction in the soil water potential has been observed to affect the formation of infection threads and has been reported to cause complete inhibition of nodulation (Worrall and Roughley 1976). After the start of infection, a restricted supply of water may slow the nodulation and quicken its senescence. There is also a reduction in nitrogenase activity complemented by the reduction in the respiration of common bean and nodules of soybean (Weisz et al. 1985; Ramos et al. 2003) A curb in the metabolism of bacteroids and oxidative damage of cellular constituents are causative factors for the inhibition of nitrogenase enzyme activity in nodules of alfalfa (Naya et al. 2007). Furthermore, the transfer of fixed nitrogen out of the nodule is lessened, because of an inadequate supply of photosynthates in leaves and stems that are under stress (Huang et al. 1975). A subordinate rate of water flow out of the nodule throughout drought stress can control the export of N_2 fixation metabolites and restrict nitrogenase activity through a feedback mechanism (Serraj and Drevon 1998). Oxidative stress in nodules is stimulated due to drought. This results in an overall reduction in the antioxidant performances that are associated with nodule senescence (Porcel et al. 2003). It has been reported that water stress prevalent during vegetative growth is unfavourable for nodule formation and biological nitrogen fixation than the water stress that occurs during the reproductive phase (Pena-Cabriales and Castellanos 1993). The exact factors regulating desiccation are unknown. A few mentioned mechanisms are the capacity to limit cellular metabolism, the better catalase activity and also the existence of particular plasmids for drought tolerance. A substantial known evidence has demonstrated the changes in desiccation tolerance of distinct rhizobial species (Al-Rashidi et al. 1982; Vriezen et al. 2007). Free-living rhizobia have the potential to survive during lower water potential or drought stress (Fuhrmann et al. 1986). Although, population densities tend to decrease during the maximum desiccated conditions and increase as the conditions improve. The survival and activity of microbes rely on their distribution within microhabitats and variations in soil moisture (Orchard and Cook 1983).

A few studies have presented that a mutualistic association can happen while working with legumes and rhizobia strains selected for desiccation tolerance (Soria et al. 1996), while, most legumes are sensitive to surplus water (flooding). The capability of aerobes to use nitrogenous oxides, as terminal electron acceptors, allows them to persist and grow during stages of anoxia. This may be beneficial to the survival of rhizobial species in soils (Zablotowicz et al. 1978). Jenkins et al. (1987) found a rhizobial species that nodulate arid legumes (Jenkins et al. 1987). Another study demonstrated that osmotolerant rhizobial species can migrate even during scarce moisture conditions (Wadisirisuk et al. 1989). Athar and Johnson (1996) identified that osmotolerant strains of *S. meliloti* performed well than that of the non- tolerant alfalfa rhizobial species and developed effective symbiotic association during drought stress (Athar and Johnson 1996).

13.6.4 Agrochemical Based Stress

The expanding global population causes insufficiency in natural resources, especially in the requirement of food. Therefore, to improve crop protection and increase production, numerous agrochemicals are being utilized. Agrochemicals, such as pesticides are highly essential in crop protection against several pests. The extreme utilization of pesticides over many years has triggered environmental complications such as aggregation of agrochemicals in groundwater and soil. Moreover, merely 15% reach the target when agrochemicals are applied to a plant. Moreover, factors like leaching, soil sorption, volatilization, surface runoff and their acquisition by plants determine their destiny in the soil. Agrochemicals in soil influence both micro and macroorganisms comprising bacteria, fungi, actinomycetes, earth-worms,

Table 13.1 Frequently used agrochemicals and their impact on rhizobia

S. no.	Pesticides	Effects
1.	Thiamethoxan (I) Glyphosate (H) Metribuzin (H)	Reduction in the synthesis of Auxin (IAA) and siderophore
2.	Mancozeb (F)	Reduction in the growth of rhizobia by altering membrane composition
3.	Captan (F)	Inhibit the growth of *Sinorhizobium*, *Rhizobium*, *Bradyrhizobium*, *Mesorhizobium*
4.	Thiaram (F)	Reduction in rhizobium and Mesorhizobium population
5.	2,4-D(H)	Alteration of membrane fluidity
6.	Lindane (I)	Inhibit the growth of *Rhizobium japonicum*
7.	Atrazine (H)	Reduction in nodule numbers
8.	Hexaconazole (F)	Inhibition of dehydrogenase system in *Rhizobium*

I insecticides; *F* fungicides; *H* herbicides; *IAA* indole acetic acid

crustaceans, nematodes, arthropods and legumes. Soil microbes have a crucial role in the detritus organic matter degradation along with nitrification, nitrogen fixation and discharge of numerous nutrients from the soil (Pandey and Singh 2004; Seghers et al. 2003). Therefore, the applicability of agrochemicals poses a main risk to beneficial soil microorganisms and affects the sustainability of agronomy. Likewise, they unite with rhizobia and decrease the number of existing root sites that are required for initiating infection (Anderson et al. 2004). Additionally, agrochemicals block the transfer of signals among rhizobia and legumes and hinder the communication between them (Fox et al. 2007).

Pesticides are agrochemicals that safeguard agricultural plants from insects, weeds or pathogens. Although pesticides are utilized for protecting crops, they eventually reach soil system and affect microbes and plants. Pesticides can have an impact on the diversity of rhizobia in soil and rhizobia–legume symbiosis (Ahemad and Khan 2013). Pesticides affect rhizobia–legume symbiosis in various means. Their impacts are summarized in Table 13.1. Firstly, they result in decrease of root biomass as there is reduction in the sites available for initiating infection which ultimately affects the transport of sugars to the prevailing nodules with subsequent reduction in the capability of rhizobia to initiate infection. Accordingly, the activity of meristem falls which is crucial for nodule development (Anderson et al. 2004). Secondly, they inhibit nodulation by blocking the signalling molecules (Flavanoids and nod factors). Finally, the competence of pesticides and rhizobia for common adhering site in the roots.

For instance, glyphosate [*N*-(phosphonomethyl) glycine] is a nonspecific herbicide that competes for adherence sites on plant roots and gets transported via phloem. Glyphosate blocks synthesis of aromatic amino acids and other phenolic compounds like flavones (Ishikura et al. 1986) cinnamic acid (Cañal et al. 1987) by targeting shikimate acid pathway that ultimately reduces plant growth. The activity of this herbicide reduces starch synthesis (Geiger and Bestman 1990) causes inequity in phytohormones (Lee 1984), decreases synthesis of chlorophyll (Kitchen et al. 1981) and its precursor δ-aminolevulinic acid (ALA), and affects the performance of nitrogenase enzyme complex (Mallik and Tesfai 1984). The quantity of total nitrogen fixed by leguminous plants is indirectly decreased and photosynthesis is prevented by the action of herbicide. Fluchloralin and Metribuzin prevent the movement of electrons from compound Q to plasto-quinone in the electron transport chain of photosystem II, inhibiting reduction of NADP+ which is required in Calvin cycle (Fedtke 1982). Although, in the presence of herbicide the activity of nitrogenase remains unaffected.

A huge number of studies are carried out using soybeans which is the first legume cultivated on a large-scale globally; and in various instances the results related to agrochemicals were contrary. For instance, trifluralin, a herbicide has been reported to have an inhibitory effect on nodulation in soybean (Parker and Dowler 1976). Furthermore, it has been found that trifluralin influences the number and weight of nodules formed and activity of nitrogenase enzyme (Bollich et al. 1985), whereas Eldin et al., previously proposed that greater dosage of trifluralin than generally administered promoted soybean nodulation (Eldin et al. 1981). Likewise, a study suggested that pentachlorobenzene (PCNB), a fungicide decreases nodulation in soybean (Curley and Burton 1975). Whereas another study indicated that it is harmless even when ten times higher dosages were administered than normal (Mallik and Tesfai 1984). Laboratory scale studies also exhibited the negative impact of pesticides and fungicides on the persistence of microorganisms.

13.6.5 pH Stress

Rhizobia are generally known to grow in optimum pH between 6 and 7 (Hungria and Vargas 2000). They greatly vary in tolerating acidic conditions. Although, a few mutants of *Rhizobium leguminosorum* can grow at a pH near to 4.5 (Chen et al. 1993). *Sinorhizobium meliloti* has been found to grow only when pH is near to 5.5 (Foster 2000). S. *fredii* has a higher range (4 to 9.5) of pH for survival, whereas *Bradyrhizobium japonicum* lack the extreme extend of pH for their survival (Fujihara and Yoneyama 1993). The efficacy of symbiosis between legume–rhizobia is significantly determined by the pH of the soil (Glenn and Dilworth 1994). The limiting factor for a successful symbiotic association is the pH of the soil as rhizobial strains can grow only in a neutral or acidic soil. *Rhizobium* Alkaline or acidic soil has a negative impact on the survival or multiplication of rhizobia and can affect both symbiotic companions (Zahran 1991). Most of the agricultural lands are alkaline having a pH ranging from 7.0 to 8.5. Thus, stress conferred by the alkalinity can interfere with Rhizobium growth and successive formation of nitrogen-fixing symbiosis with legumes. Hence, it is logical to select rhizobia isolates that can withstand the alkaline pH which results in the capability of the rhizobia to nodulate (Farissi et al. 2014). Well characterized rhizobial defence mechanism has been recorded at acidic pH and can also be applicable to basic pH (Fujihara and Yoneyama 1993). A common mechanism employed by the bacterium such pH stress is by raising the internal pH of the cell with the help of ABC transport system and other transport mechanisms either expelling H^+ ions out of the cell or by influx of basic ions into the cell (Priefer et al. 2001). Alternative mechanism against this acid shock is by synthesizing acid shock proteins (ASPs) by the bacterium. However, they do not modify the internal pH of the cell but provide acid protection (Foster 1993). Kurchak et al. (2001) reported almost 20 genes that play a key role in protecting *Rhizobium leguminosarum* from acid stress and are termed as act genes (Kurchak et al. 2001). A form of cellular signalling must be carried out by the bacteria to sense the acid stress. The key players involved in this type of signalling consist of a sensor and a regulator and a typical acid stress sensing signal coded by actR gene (Regulator) and actS (Sensor) in *S.meliloti* (Tiwari et al. 1996).

13.6.6 Waterlogging Stress

A few research studies have highlighted the effect of waterlogging on nodulation capacity. Legumes like alfafa, soybean and pea display reduction in nodule weight when they are treated under hypoxic conditions (Minchin and Pate 1975; Arrese-Igor et al. 1993; Sung 1993).When 0.1% of oxygen concentration was maintained *Medicago truncatula* exhibited 45% reduction in nodulation capacity and was not having any impact when treated under 4.5% of oxygen concentration (Pucciariello et al. 2019). Effective nodulation was demonstrated after few weeks by 21 species of legumes that included waterlogging sensitive species like *Medicago sativa* and *Melilotus albus* (Nichols et al. 2008). It is uncertain whether the type, nature of nodule could demonstrate distinct mechanisms of combating waterlogging stress as *Medicago* spp., *Melilotus* spp. and *Pisum* spp. possessing indeterminate nodules show persistent meristem and continuous growth characteristics compared with *Glycine* spp., *Lotus* spp. and *Vigna* spp exhibiting non-persistent and limited growth characteristics. A few phenotypical and physiological adaptations can be seen in nodulation process of flood-tolerating legumes. Nodules formed on *Melilotus siculus*, in the course of waterlogging stress has been observed on adventitious roots (Rhizopoulou and Psaras 2003).

13.7 Conclusion

The effect of environmental factors on legume–rhizobia interactions is of great importance as they control and regulate the symbiotic process. Survival and proliferation of rhizobia in rhizosphere and surface soil are greatly affected by abiotic factors as they can restrict the infection thread formation, metabolism of nodule and alteration of legume growth. The rhizobial population is distinctive in their efficiency to tolerate major environmental stresses like pH, temperature, salinity, drought and agrochemicals, thus selecting resistant strains is considerably preffered option. Many strains of rhizobia that can tolerate various environmental stresses have been isolated for different crops and wild legumes and this seems to be a favourable strategy that may promote the efficient interaction among rhizobia and legume resulting in increased agricultural productivity. Moreover, innovative and efficient mixtures of formulation using tolerant strains and factors (Flavanoids, Nod factors) involved in the nodulation process appears to be a promising technological tool for the betterment of agronomy. These products can be used to extend the cultivation of native or naturalized legumes and increase the efficiency and potential of rhizobia–legume interaction which can increase agricultural productivity.

References

Abd-Alla MH (2001) Regulation of nodule formation in soybean-*Bradyrhizobium* symbiosis is controlled by shoot or/and root signals. Plant Growth Regul 34(2):241–250

Ahemad M, Khan MS (2013) Pesticides as antagonists of rhizobia and the legume-*Rhizobium* symbiosis: a paradigmatic and mechanistic outlook. Biochem Mol Biol 1:63–75

Al-Rashidi RK, Loynachan TE, Frederick LR (1982) Desiccation tolerance of four strains of *Rhizobium japonicum*. Soil Biol Biochem 14:489–493

Anderson A, Baldock JA, Rogers SL, Bellotti W, Gill G (2004) Influence of chlorsulfuron on rhizobial growth, nodule formation, and nitrogen fixation with chickpea. Aust J Agric Res 55 (10):1059–1070

Arora NK, Singhal V, Maheshwari DK (2006) Salinity-induced accumulation of poly-β-hydroxybutyrate in rhizobia indicating its role in cell protection. World J Microbiol Biotechnol 22(6):603–606

Arrese-Igor C, Royuela M, De Lorenzo C, De Felipe MR, Aparicio-Tejo PM (1993) Effect of low rhizosphere oxygen on growth, nitrogen fixation and nodule morphology in lucerne. Physiol Plant 89(1):55–63

Aswathappa N, Marcar NE, Thomson LA (1987) Salt tolerance of Australian tropical and subtropical acacias in Australian acacias in developing countries. ACIAR Proc 16:70

Athar M, Johnson DA (1996) Influence of drought on competition between selected *Rhizobium meliloti* strains and naturalized soil rhizobia in alfalfa. Plant Soil 184:231–241

Bartoli CG, Casalongué CA, Simontacchi M, Marquez-Garcia B, Foyer CH (2013) Interactions between hormone and redox signalling pathways in the control of growth and cross tolerance to stress. Environ Exp Bot 94:73–88

Becana M, Dalton DA, Moran JF, Iturbe-Ormaetxe I, Matamoros MA, Rubio MC (2000) Reactive oxygen species and antioxidants in legume nodules. Physiol Plant 109(4):372–381

Becana M, Matamoros MA, Udvardi M, Dalton DA (2010) Recent insights into antioxidant defenses of legume root nodules. New Phytol 188(4):960–976

Becker A, Fraysse N, Sharypova L (2005) Recent advances in studies on structure and symbiosis-related function of rhizobial K-antigens and lipopolysaccharides. Mol Plant Microbe Interact 18 (9):899–905

Bollich PK, Dunigan EP, Jadi AW (1985) Effects of seven herbicides on N 2 (C 2 H 2) fixation by soybeans. Weed Sci 33(4):427–430

Breedveld MW, Zevenhuizen LP, Zehnder AJ (1991) Osmotically-regulated trehalose accumulation and cyclic β-(1, 2)-glucan excretion by *Rhizobium leguminosarum biovar trifolii* TA-1. Arch Microbiol 156(6):501–506

Brockwell J, Bottomley PJ, Thies JE (1995) Manipulation of rhizobia microflora for improving legume productivity and soil fertility: a critical assessment. Plant Soil 174:143–180

Breedveld MW, Miller KJ (1994) Cyclic beta-glucans of members of the family Rhizobiaceae. Microbiol Rev 58(2):145–161

Cañal MJ, Tamés RS, Fernández B (1987) Glyphosate-increased levels of indole-3-acetic acid in yellow nutsedge leaves correlate with gentisic acid levels. Physiol Plant 71(3):384–388

Chang C, Damiani I, Puppo A, Frendo P (2009) Redox changes during the legume–*Rhizobium* symbiosis. Mol Plant 2(3):370–377

Chen H, Richardson AE, Rolfe BG (1993) Studies of the physiological and genetic basis of acid tolerance in *Rhizobium leguminosarum biovar trifolii*. Appl Environ Microbiol 59 (6):1798–1804

Chen L, Figueredo A, Villani H, Michajluk J, Hungria M (2002) Diversity and symbiotic effectiveness of rhizobia isolated from field-grown soybean nodules in Paraguay. Biol Fertil Soils 35 (6):448–457

Curley RL, Burton JC (1975) Compatibility of *Rhizobium japonicum* with chemical seed protectants 1. Agron J 67(6):807–808

Cheng HP, Walker GC (1998) Succinoglycan is required for initiation and elongation of infection threads during nodulation of alfalfa by Rhizobium meliloti. J Bacteriol 180:5183–5191

Cremers HC, Stevens K, Lugtenberg BJ, Wijffelman CA, Batley M, Redmond JW, Breedveld MW, Zevenhuizen LP (1991) Unusual structure of the exopolysaccharide of Rhizobium leguminosarum bv. viciae strain 248. Carbohydrate research 218:185–200

Dardanelli MS, de Cordoba FJ, Espuny MR, Carvajal MA, Díaz ME, Serrano AM, Okon Y, Megías M (2008) Effect of *Azospirillum brasilense* coinoculated with *Rhizobium* on *Phaseolus vulgaris* flavonoids and Nod factor production under salt stress. Soil Biol Biochem 40(11):2713–2721

Djébali N, Mhadhbi H, Lafitte C, Dumas B, Esquerré-Tugayé MT, Aouani ME, Jacquet C (2011) Hydrogen peroxide scavenging mechanisms are components of *Medicago truncatula* partial resistance to *Aphanomyces euteiches*. Eur J Plant Pathol 131(4):559

Douka CE, Xenoulis AC, Paradellis T (1984) Salinity tolerance of a *rhizobium meliloti* strain isolated from salt affected soils. Folia Microbiol 29(4):316–324

Dowling DN, Broughton WJ (1986) Competition for nodulation of legumes. Annu Rev Microbiol 40(1):131–157

Duzan HM, Mabood F, Souleimanov A, Smith DL (2006) Nod Bj-V (C18: 1, MeFuc) production by *Bradyrhizobium japonicum* (USDA110, 532C) at suboptimal growth temperatures. J Plant Physiol 163(1):107–111

Dai J, Liu X, Wang Y (2012) Genetic diversity and phylogeny of rhizobia isolated from Caragana microphylla growing in desert soil in Ningxia, China. Genet Mol Res 11(3):2683–2693

Downie JA (2010) The roles of extracellular proteins, polysaccharides and signals in the interactions of rhizobia with legume roots. FEMS Microbiol Rev 34:150–170

Eldin A, Mahmoud SA, Makawi A, Herzallah NA (1981) Effect of pre-emergence application of some herbicides on nodulation, nitrogen fixation and growth of soybean [*Glycine max*]. Pesquisa Agropecuaria Brasileira

Elsheikh EA, Wood M (1995) Nodulation and N2 fixation by soybean inoculated with salt-tolerant rhizobia or salt-sensitive *bradyrhizobia* in saline soil. Soil Biol Biochem 27(4-5):657–661

Farissi M, Bouizgaren A, Aziz F, Faghire M, Ghoulam C (2014) Isolation and screening of rhizobial strains nodulating alfalfa for their tolerance to some environmental stresses. Pacesetter J Agric Sci Res 2:9–19

Fedtke C (1982) Photosynthesis in biochemistry and physiology of herbicide action. Springer, Berlin, pp 19–113

Felker P, Clark PR, Laag AE, Pratt PF (1981) Salinity tolerance of the tree legumes: Mesquite (*Prosopis glandulosa var. torreyana, P. velutina and P. articulata*) Algarrobo (*P. chilensis*), Kiawe (*P. pallida*) and Tamarugo (*P. tamarugo*) grown in sand culture on nitrogen-free media. Plant Soil 61(3):311–317

Foster JW (1993) The acid tolerance response of *Salmonella typhimurium* involves transient synthesis of key acid shock proteins. J Bacteriol 175(7):1981–1987

Foster JW (2000) Microbial responses to acid stress. In: Bacterial stress responses. ASM Press, Washington, DC, pp 99–115

Fox JE, Gulledge J, Engelhaupt E, Burow ME, McLachlan JA (2007) Pesticides reduce symbiotic efficiency of nitrogen-fixing rhizobia and host plants. Proc Natl Acad Sci 104(24):10282–10287

Fraysse N, Couderc F, Poinsot V (2003) Surface polysaccharide involvement in establishing the rhizobium–legume symbiosis. Eur J Biochem 270(7):1365–1380

Fuhrmann J, Davey CB, Wollum AG (1986) Desiccation tolerance of clover rhizobia in sterile soils. Soil Sci Soc Am J 50(3):639–644

Fujihara S, Yoneyama T (1993) Effects of pH and osmotic stress on cellular polyamine contents in the soybean rhizobia *Rhizobium fredii* P220 and *Bradyrhizobium japonicum* A1017. Appl Environ Microbiol 59(4):1104–1109

Geiger DR, Bestman HD (1990) Self-limitation of herbicide mobility by phytotoxic action. Weed Sci 38(3):324–329

Georgiev GI, Atkins CA (1993) Effects of salinity on N2 fixation, nitrogen metabolism and export and diffusive conductance of cowpea root nodules. Symbiosis 15:239–255

Glenn AR, Dilworth MJ (1994) The life of root nodule bacteria in the acidic underground. FEMS Microbiol Lett 123(1–2):1–9

Graham PH (1992) Stress tolerance in *Rhizobium* and *Bradyrhizobium*, and nodulation under adverse soil conditions. Can J Microbiol 38(6):475–484

Graham PH, Vance CP (2003) Legumes: importance and constraints to greater use. Plant Physiol 131:872–877

Giller KE, Beare MH, Lavelle P, Izac AMN, Swift MJ (1997) Agricultural intensification, soil biodiversity and agroecosystem function. Appl Soil Ecol 6:3–16

Huang CY, Boyer JS, Vanderhoef LN (1975) Limitation of acetylene reduction (nitrogen fixation) by photosynthesis in soybean having low water potentials. Plant Physiol 56(2):228–232

Hungria M, Vargas MA (2000) Environmental factors affecting N2 fixation in grain legumes in the tropics, with an emphasis on Brazil. Field Crops Res 65(2-3):151–164

Irigoyen JJ, Einerich DW, Sánchez-Díaz M (1992) Water stress induced changes in concentrations of proline and total soluble sugars in nodulated alfalfa (*Medicago sativa*) plants. Physiol Plant 84(1):55–60

Ishikura N, Teramoto S, Takeshima Y, Mitsui S (1986) Effects of glyphosate on the shikimate pathway and regulation of phenylalanine ammonia-lyase in Cryptomeria and Perilla cell suspension cultures. Plant Cell Physiol 27(4):677–684

Jamet A, Mandon K, Puppo A, Hérouart D (2007) H_2O_2 is required for optimal establishment of the *Medicago sativa/Sinorhizobium meliloti* symbiosis. J Bacteriol 189(23):8741–8745

Jenkins MB, Virginia RA, Jarrell WM (1987) Rhizobial ecology of the woody legume mesquite (*Prosopis glandulosa*) in the Sonoran desert. Appl Environ Microbiol 53(1):36–40

Junior MD, Lima AS, Arruda JR, Smith DL (2005) Effect of root temperature on nodule development of bean, lentil and pea. Soil Biol Biochem 37(2):235–239

Janczarek M, Skorupska A (2011) Modulation of rosR expression and exopolysaccharide production in Rhizobium leguminosarum bv. trifolii by phosphate and clover root exudates. Int J Mol Sci 12:4132–4155

Kitchen LM, Witt WW, Rieck CE (1981) Inhibition of chlorophyll accumulation by glyphosate. Weed Sci 29(4):513–516

Kurchak ON, Provorov NA, Simarov BV (2001) Plasmid pSym1-32 of *Rhizobium leguminosarumbv. viceae* controlling nitrogen fixation activity, effectiveness of Symbiosis, competitiveness, and acid tolerance. Russ J Genet 37(9):1025–1031

Kelly SJ, Muszyński A, Kawaharada Y, Hubber AM, Sullivan JT, Sandal N, Carlson RW, Stougaard J, Ronson CW (2013) Conditional requirement for exopolysaccharide in the Mesorhizobium–Lotus symbiosis. Mol Plant Microbe Interact 26(3):319–329

Laus MC, Logman TJ, Lamers GE, Van Brussel AA, Carlson RW, Kijne JW (2006) A novel polar surface polysaccharide from Rhizobium leguminosarum binds host plant lectin. Mol Microbiol 59:1704–1713

Laus MC, Van Brussel AA, Kijne JW (2005) Role of cellulose fibrils and exopolysaccharides of Rhizobium leguminosarum in attachment to and infection of Vicia sativa root hairs. Mol Plant Microbe Interact 18(6):533–538

Lee TT (1984) Release of lateral buds from apical dominance by glyphosate in soybean and pea seedlings. J Plant Growth Regul 3(1-4):227

Leigh JA, Signer ER, Walker GC (1985) Exopolysaccharide-deficient mutants of Rhizobium meliloti that form ineffective nodules. Proc Natl Acad Sci 82(18):6231–6235

Lloret J, Wulff BB, Rubio JM, Downie JA, Bonilla I, Rivilla R (1998) Exopolysaccharide II production is regulated by salt in the halotolerant strain Rhizobium meliloti EFB1. Appl Environ Microbiol 64:1024–1028

Lu YL, Chen WF, Wang ET, Guan SH, Yan XR, Chen WX (2009) Genetic diversity and biogeography of rhizobia associated with Caragana species in three ecological regions of China. Syst Appl Microbiol 32(5):351–361

Mallik MA, Tesfai K (1984) Pesticidal effect on soybean-rhizobia symbiosis. Plant Soil 85 (1):33–41

Mashhady AS, Salem SH, Barakah FN, Heggo AM (1998) Effect of salinity on survival and symbiotic performance between *Rhizobium meliloti* and *Medicago sativa l.* in Saudi Arabian soils. Arid Land Res Manag 12(1):3–14

Matamoros MA, Baird LM, Escuredo PR, Dalton DA, Minchin FR, Iturbe-Ormaetxe I, Rubio MC, Moran JF, Gordon AJ, Becana M (1999) Stress-induced legume root nodule senescence, physiological, biochemical and structural alterations. Plant Physiol 121:97–111

Minchin FR, Pate JS (1975) Effects of water, aeration, and salt regime on nitrogen fixation in a nodulated legume definition of an optimum root environment. J Exp Bot 26(1):60–69

Minchin FR, James EK, Becana M (2008) Oxygen diffusion, production of reactive oxygen and nitrogen species, and antioxidants in legume nodules in nitrogen-fixing leguminous symbioses. Springer, Dordrecht, pp 321–362

Miransari M, Smith DL (2009) Alleviating salt stress on soybean (*Glycine max* (L.) Merr.) – *Bradyrhizobium japonicum* symbiosis, using signal molecule genistein. Eur J Soil Biol 45 (2):146–152

Mithöfer A (2002) Suppression of plant defence in rhizobia–legume symbiosis. Trends Plant Sci 7 (10):440–444

Mohammad RM, Akhavan-Kharazian M, Campbell WF, Rumbaugh MD (1991) Identification of salt-and drought-tolerant *Rhizobium meliloti* L. strains. Plant Soil 134(2):271–276

Naya L, Ladrera R, Ramos J, González EM, Arrese-Igor C, Minchin FR, Becana M (2007) The response of carbon metabolism and antioxidant defenses of alfalfa nodules to drought stress and to the subsequent recovery of plants. Plant Physiol 144(2):1104–1114

Nichols PG, Craig AD, Rogers ME, Albertsen TO, Miller SM, McClements DR, Hughes SJ, D'antuono MF, Dear BS (2008) Production and persistence of annual pasture legumes at five saline sites in southern Australia. Aus J Exp Agric 48(4):518–535

Oldroyd GE, Downie JA (2008) Coordinating nodule morphogenesis with rhizobial infection in legumes. Annu Rev Plant Biol 59:519–546

Orchard VA, Cook FJ (1983) Relationship between soil respiration and soil moisture. Soil Biol Biochem 15(4):447–453

O'Neill MA, Darvill AG, Albersheim P (1991) The degree of esterification and points of substitution by O-acetyl and O-(3-hydroxybutanoyl) groups in the acidic extracellular polysaccharides secreted by Rhizobium leguminosarum biovars viciae, trifolii, and phaseoli are not related to host range. J Biol Chem 266:9549–9555

Pandey S, Singh DK (2004) Total bacterial and fungal population after chlorpyrifos and quinalphos treatments in groundnut (*Arachis hypogaea L.*) soil. Chemosphere 55(2):197–205

Parker MB, Dowler CC (1976) Effects of nitrogen with trifluralin and vernolate on soybeans. Weed Sci 24(1):131–133

Pena-Cabriales JJ, Castellanos JZ (1993) Effects of water stress on N2 fixation and grain yield of *Phaseolus vulgaris L* in enhancement of biological nitrogen fixation of common bean in Latin America. Springer, Dordrecht, pp 151–155

Phadtare S, Yamanaka K, Inouye M (2000) The cold shock response. In: Storz G, Hengge-Aronis R (eds) Bacterial stress response. ASM Press, Washington, DC, pp 33–45

Pimratch S, Jogloy S, Vorasoot N, Toomsan B, Patanothai A, Holbrook CC (2008) Relationship between biomass production and nitrogen fixation under drought-stress conditions in peanut genotypes with different levels of drought resistance. J Agron Crop Sci 194(1):15–25

Porcel R, Barea JM, Ruiz-Lozano JM (2003) Antioxidant activities in mycorrhizal soybean plants under drought stress and their possible relationship to the process of nodule senescence. New Phytol 157(1):135–143

Priefer UB, Aurag J, Boesten B, Bouhmouch I, Defez R, Filali-Maltouf A, Miklis M, Moawad H, Mouhsine B, Prell J, Schlüter A (2001) Characterisation of *Phaseolus* symbionts isolated from Mediterranean soils and analysis of genetic factors related to pH tolerance. J Biotechnol 91 (2–3):223–236

Pucciariello C, Boscari A, Tagliani A, Brouquisse R, Perata P (2019) Exploring legume-rhizobia symbiotic models for waterlogging tolerance. Front Plant Sci 10:578

Rai R (1983) The salt tolerance of rhizobium strains and lentil genotypes and the effect of salinity on aspects of symbiotic N-fixation. J Agric Sci 100(1):81–86

Ramos ML, Parsons R, Sprent JI, James EK (2003) Effect of water stress on nitrogen fixation and nodule structure of common bean. Pesq Agrop Brasileira 38(3):339–347

Rhizopoulou S, Psaras GK (2003) Development and structure of drought-tolerant leaves of the Mediterranean shrub *Capparis spinosa L.* Ann Bot 92(3):377–383

Rosas SB, Palacios S, Correa NS (1996) Growth and nodulation of *Cyamopsis tetragonoloba L.* (guar) under conditions of salinity. Int J Exp Bot 58:107–114

Rajwar A, Sahgal M, Johri BN (2013) Legume–rhizobia symbiosis and interactions in agroecosystems, Plant microbe symbiosis: fundamentals and advances. Springer, New Delhi, pp 233–265

Romdhane SB, Trabelsi M, Aouani ME, de Lajudie P, Mhamdi R (2009) The diversity of rhizobia nodulating Chickpea (Cicer arietinum) under water deficiency as a source of more efficient inoculants. Soil Biol Biochem 41:2568–2572

Sadowsky MJ (2005) Soil stress factors influencing symbiotic nitrogen fixation in agriculture, forestry, ecology, and the environment. Springer, Dordrecht, pp 89–112

Salah IB, Albacete A, Andújar CM, Haouala R, Labidi N, Zribi F, Martinez V, Pérez-Alfocea F (2009) Abdelly C (2009) response of nitrogen fixation in relation to nodule carbohydrate metabolism in *Medicago ciliaris* lines subjected to salt stress. J Plant Physiol 166(5):477–488

Santos R, Hérouart D, Sigaud S, Touati D, Puppo A (2001) Oxidative burst in alfalfa-Sinorhizobium meliloti symbiotic interaction. Mol Plan Microbe Interact 14(1):86–89

Seghers D, Verthé K, Reheul D, Bulcke R, Siciliano SD, Verstraete W, Top EM (2003) Effect of long-term herbicide applications on the bacterial community structure and function in an agricultural soil. FEMS Microbiol Ecol 46(2):139–146

Serraj R, Drevon JJ (1998) Effects of salinity and nitrogen source on growth and nitrogen fixation in alfalfa. J Plant Nutr 21(9):1805–1818

Shaw SL, Long SR (2003) Nod factor inhibition of reactive oxygen efflux in a host legume. Plant Physiol 132(4):2196–2204

Skorupska A, Janczarek M, Marczak M, Mazur A, Król J (2006) Rhizobial exopolysaccharides: genetic control and symbiotic functions. Microb Cell Fact 5:1–19

Soria RD, Correa NS, Rosas SB (1996) Effect of water stress on the guar-*Bradyrhizobium* system using PEG 6000. Phyton 58:97–106

Soussi M, Ocana A, Lluch C (1998) Effects of salt stress on growth, photosynthesis and nitrogen fixation in chick-pea (*Cicer arietinum* L). J Exp Bot 49(325):1329–1337

Sung FJ (1993) Waterlogging effect on nodule nitrogenase and leaf nitrate reductase activities in soybean. Field Crops Res 35(3):183–189

Tiaz L, Zeiger E (2002) Plant physiology. Sinauer Associates, Sunderland, pp 591–621

Tilman D (1999) Global environmental impacts of agricultural expansion: the need for sustainable and efficient practices. Proc Natl Acad Sci 96(11):5995–6000

Tiwari RP, Reeve WG, Dilworthan MJ, Glenn AR (1996) An essential role for actA in acid tolerance of *Rhizobium melilotix*. Microbiol 142(3):601–610

Türkan I, Demiral T (2009) Recent developments in understanding salinity tolerance. Environ Exp Bot 67(1):2–9

Turner NC, Wright GC, Siddique KH (2001) Adaptation of grain legumes (pulses) to water-limited environments. Adv Agron 71:194–233

Vriezen JA, De Bruijn FJ, Nüsslein K (2007) Responses of rhizobia to desiccation in relation to osmotic stress, oxygen, and temperature. Appl Environ Microbiol 73(11):3451–3459

Velázquez E, Martínez-Romero E, Rodríguez-Navarro DM, Trujillo ME, Daza A, Mateos PE, Martínez- Molina E, van Berkum P (2001) Characterization of rhizobial isolates of Phaseolus vulgaris by staircase electrophoresis of low-molecular weight RNA. Appl Environ Microbiol 67:1008–1010

Vineusa P, Léon-Barrios M, Silva C, Willems A, Jabaro-Lorenzo A, Pérez-Galdona R, Werner D, Martínez-Romero E (2005) Bradyrhizobium canariense sp. nov., an acid-tolerant endosymbiont that nodulates endemic genistoid legumes (Papilionoideae: Genisteae) from Canary Islands, along with Bradyrhizobium japonicum bv. genistearum, Bradyrhizobium genospecies alpha and Bradyrhizobium genospecies beta. Int J Syst Evol Microbiol 55:569–575

Wadisirisuk P, Danso SK, Hardarson G, Bowen GD (1989) Influence of *Bradyrhizobium japonicum* location and movement on nodulation and nitrogen fixation in soybeans. Appl Environ Microbiol 55(7):1711–1716

Weisz PR, Denison RF, Sinclair TR (1985) Response to drought stress of nitrogen fixation (acetylene reduction) rates by field-grown soybeans. Plant Physiol 78(3):525–530

Worrall VS, Roughley RJ (1976) The effect of moisture stress on infection of *Trifolium subterraneum L. by Rhizobium trifolii* Dang. J Exp Bot 27(6):1233–1241

Wang ET, Rogel MA, Garcia-de los Santos A, Martinez-Romero J, Cevallos MA, Martínez-Romero E (1999) Rhizobium etli bv. mimosae, a novel biovar isolated from Mimosa affinis. Int J Syst Bacteriol 49:1479–1491

Williams A, Wilkinson A, Krehenbrink M, Russo DM, Zorreguieta A, Downie JA (2008) Glucomannan-mediated attachment of Rhizobium leguminosarum to pea root hairs is required for competitive nodule infection. J Bacteriol 190:4706–4715

Yelton MM, Yang SS, Edie SA, Lim ST (1983) Characterization of an effective salt-tolerant, fast-growing strain of *Rhizobium japonicum*. Microbiology 129(5):1537–1547

Yura T (2000) The heat shock response: regulation and function. In: Bacterial stress responses. ACM Press, Washington, DC, pp 3–18

Zablotowicz RM, Eskew DL, Focht DD (1978) Denitrification in *Rhizobium*. Can J Microbiol 24 (6):757–760

Zahran HH (1991) Conditions for successful rhizobium-legume symbiosis in saline environments. Biol Fertil Soils 12(1):73–80

Zahran HH (1999) Rhizobium legume symbiosis and nitrogen fixation under severe conditions and in arid climate. Microbiol Mol Biol Rev 63:968–989

Zahran HH, Sprent JI (1986) Effects of sodium chloride and polyethylene glycol on root-hair infection and nodulation of *Vicia faba L* plants by *Rhizobium leguminosarum*. Planta 167 (3):303–309

Zhang F, Lynch DH, Smith DL (1995) Impact of low root temperatures in soybean [*Glycine max*. (L.) Merr.] on nodulation and nitrogen fixation. Environ Exp Bot 35(3):279–285

Zheng C, Jiang D, Liu F, Dai T, Jing Q, Cao W (2009) Effects of salt and waterlogging stresses and their combination on leaf photosynthesis, chloroplast ATP synthesis, and antioxidant capacity in wheat. Plant Sci 176(4):575–582

Zurayk R, Adlan M, Baalbaki R, Saxena MC (1998) Interactive effects of salinity and biological nitrogen fixation on chickpea (*Cicer arietinum*) growth. J Agron Crop Sci 180(4):249–258
Zahran HH (2001) Rhizobia from wild legumes: diversity, taxonomy, ecology, nitrogen fixation and biotechnology. J Biotechnol 91(2-3):143–153
Zevenhuizen LPTM (1986) Selective synthesis of polysaccharides by Rhizobium trifolii strain TA-1. FEMS microbiology letters 35(1):43–47

Chapter 14
Archaeal Symbiosis for Plant Health and Soil Fertility

Ranjith Sellappan, Senthamilselvi Dhandapani, Anandakumar Selvaraj, and Kalaiselvi Thangavel

Abstract In plant microbes play an indispensable role for their growth and immunity. Microbes are capable of colonizing the plant rhizosphere, phyllosphere and internal part of the plant (endophytes). Plant growth-promoting microorganisms (PGPM) like bacteria, fungi and archaea improve the plant growth through nutritional (nutrient solubilisation, phytohormones production) and non-nutritional (induce defence) mode. Archaea are an important group of microbes which have distinct character and disseminated not only in an extreme environment, they have a broad range of habitat. Plant growth-promoting archaea (PGPA) play a pivotal role in biogeochemical cycle and making accessible of important nutrients like C, N, S and P to the plants through fixation, solubilisation. It also helps in plant growth by phytohormones and siderophore production. Archaea elicit a defence response of the plant against both biotic and abiotic stress through inducing systemic resistance (ISR) of the plants. Soil the substrate for plants, contains huge microbial (bacteria, fungi, archaea) population which improves the soil health through nutrient cycle. So archaea are an important group of microorganism having special attention in agricultural production. However, archaea occupy 20% of world biomass and play important role in plant growth similar to bacteria and fungi it not well studied. Understanding archaeal plant interaction is necessary for improving the plants grown in different environmental conditions. Only a few studies focused on archaeal plant interaction. This chapter briefs the immense role of archaea in plant development and soil fertility improvement.

Keywords Archaea · Plant growth-promoting archaea · Soil health · Induced systemic resistance

---

R. Sellappan (✉) · S. Dhandapani · A. Selvaraj · K. Thangavel
Department of Agricultural Microbiology, Tamil Nadu Agricultural University, Coimbatore, Tamil Nadu, India

N. Shrivastava et al. (eds.), *Symbiotic Soil Microorganisms*, Soil Biology 60,
https://doi.org/10.1007/978-3-030-51916-2_14

14.1 Introduction

Archaea are single-celled prokaryotic organisms that have a distinct character from bacteria. Archaea are extremophiles because of their inhabitants in varied stress environmental conditions like high temperature, pH, salinity, anoxic condition and terrestrial soil (Flemming and Wuertz 2019). Most culturable and well investigated archaeal species are phylum Euryarchaeota and Crenarchaeota and the other phylum Nanoarchaeota, Thaumarchaeota, Aigarchaeota and Korarchaeota have also been reported (Huber et al. 2002; Baker et al. 2006). For a long time archaea consider as the organisms which grow only in the extreme environment, but at the end of the twentieth century, it was found that the extremophiles (archaea) also present in non-extreme conditions (Reitschuler et al. 2016) using metagenomics approaches. They are abundantly present in marine, freshwater, normal terrestrial soil.

Soil the hotspot of the microorganisms, their population varies based on the soil conditions. Plant roots have an intimate association with numerous soil microorganisms including bacteria, archaea and fungi. These microbes modulate plant health in both beneficial and detrimental manner. Among these microbe plant growth-promoting microorganisms play a pivotal role in plant health through nutritional and non-nutritional mode. Microbes not only present in the root, it also presents as epiphytic and endophytic in the plant. Archaea are the third-largest domain after bacteria and fungi in soil (Hassani et al. 2018). Plant growth-promoting archaea play the major role in biogeochemical cycling and protection against pathogen in plants. The role of plant growth-promoting bacteria and fungi for plant development was studied, but the role of archaea in plant growth promotion was not fully studied. Archaea represent an important component in plant microbiome, but their impact on the host is still unclear. This chapter discusses the important role of soil archaea in soil fertility improvement and plant growth and development.

14.2 Distribution of Archaea in Soil

Archaea exist ubiquitously and as a foremost portion of global biomes like marine, sediments, freshwater, soil and hydrothermal vents (Lee et al. 2019) may contribute up to 20% of earth's biomass (Yadav et al. 2017). After the recognition of the archaeal domain (Woese et al. 1991) understanding of archaeal diversity, biology and ecology were not fully understood. Euryarchaeota and Crenarchaeota the two major cultivated archaea defined by Carl woese. Of these, Crenarchaeotaes are majorly distributed and known to dwell environments like hot springs, geysers and ocean vents which are inhospitable to many other organisms (Yadav et al. 2017). Commonly Crenarchaeota phylum is abundant in cultivated soil (Bates et al. 2011), but the structure of the soil community and functional diversity are remain poorly understood.

14.3 Plant Archaeal Microbiome

Distributions and ecological assortment of soil archaeal communities in agricultural ecosystem are essential for increasing the crop productivity and soil fertility. Archaeal microbiome varies depending on different soil conditions and plant genotype (Müller et al. 2015). Based on the 16s rRNA Crenarchaeota and Euryarchaeota retrieved from the soil (Borneman and Triplett 1997). In a later analysis of seven diverse soil, Crenarchaeota was found abundant (0.5–3%) in bulk sandy soil samples, but *Festuca ovina* grown in same soil only 0.16% archaea were retrieved (Buée et al. 2009). Lee et al. (2019) reported the archaea in different rhizocompartment of tomato plant root and the dominant archaeal phylum was Thaumarchaeota followed by Euryarchaeota. These phyla also dominated in *Jatropha curcas* (Dubey et al. 2016) rhizosphere. Archaeal population is more in mycorrhizosphere using glycogen as food sources (Taffner et al. 2018). Phyllosphere microbes present in surface of plant the leaves; plays an important role in plant growth and health. Occurrences of the phyllospheric archaeal community were low compare to rhizosphere (Taffner et al. 2019; Bringel and Couée 2015). In rice plant, methanotrophs are abundant in phyllosphere region (Knief et al. 2012). Archaea associated with arugula phyllosphere are Thaumarchaeota and Euryarchaeota groups. Endophytes reside inside the plant which improves the plant growth and health. Plant endophytic archaeal taxa belong to phyla Crenarchaeota, Thaumarchaeota and Euryarchaeota (Hassani et al. 2018) but their function in plants remain to be clarified. Wassermann et al. (2019) isolated the endophytic archaea in alpine plant seeds. It is dominated by Thaumarchaeota and Crenarchaeota, but Euryarchaeota is less represented. Colonization levels of archaea in the rhizosphere and endosphere are high compared to phyllosphere (Taffner et al. 2019).

14.4 Role of Archaea in Biogeochemical Cycling

Microorganisms play the important role in biogeochemical cycle, which improves the availability of like N, P, S and K to plant for growth and increase soil fertility. Microbial inoculum improves the plant growth in nutrient-poor soil and it also improves the soil nutrient status. An archaea domain serves as enhancing agricultural production because of their different habitats and plays the major role in nutrient cycling.

14.4.1 Carbon Cycling

Carbon is the main building block of the plant metabolism. Catabolic degradation of organic substrates by chemoorganotrophic microbes results in the production of CO_2

as the main end product and absence of terminal electron acceptor results partially oxidized compounds. Numerous archaea are growing organotrophically in both aerobic and anaerobic conditions. Archaea play the major role in carbon mineralization process major groups are Euryarchaeota and Crenarchaeota.

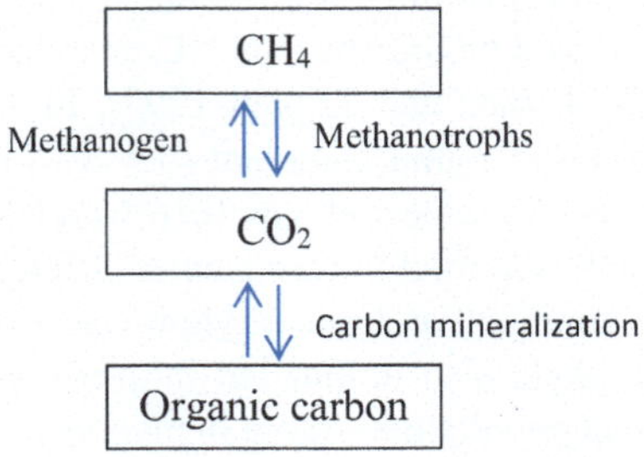

After the mineralization process, CO_2 converts into methane by methanogenic archaea. Methanogenic archaea (*Methanobacteria* spp., *Methanococci* spp., etc.) in the carbon cycle contribute immensely to the decomposition of organic matter by removing hydrogen. Euryarchaeota colonizes the rice roots, which utilize the carbon of plant and produce methane (Buée et al. 2009). Methylotrophic archaea (*Methanosarcinales* spp., *Methanoplasmatales* spp., etc.) convert this methane into CO_2 by anaerobic oxidation. Methanogens and anaerobic methane oxidation are two processes of global importance that are performed exclusively by archaea.

14.4.2 Nitrogen Cycle

Most of the organisms uptake the nitrogen in the form of NH_3 (inorganic form) or nitrate (NO_3) /organic forms. The atmospheric nitrogen enters into the biogeochemical cycle by a group of bacteria/archaea as ammonia (NH_3) followed by nitrification and denitrification process occurs. Hitherto we recognized only lithotrophic group of bacteria involved in nitrogen fixation. But recent discoveries of methanogens are recognized as the broadened nitrogen fixer in archaea. Diazotrophic methanogens belong to three major taxonomic classes, i.e. *Methanobacteria*, *Methanococci* and *Methanomicrobia.* These methanogenic archaea are ubiquitous in aerated soils and become active under anoxic conditions (Angel et al. 2012). Ammonium oxidizing archaea (AOA) belong to the Thaumarchaeota (Santoro and Casciotti 2011) which oxidize the ammonia and convert it into nitrite then nitrate by bacteria. Only few cultivable archaea are recognized as the denitrifiers, for example, *Pyrobaculum aerophilum.*

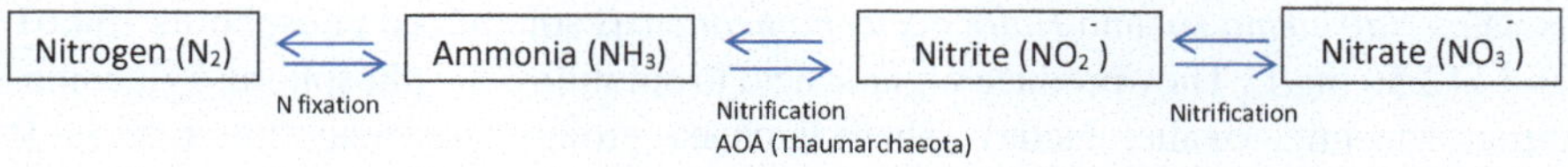

14.4.3 Sulphur Cycle

In plant sulphur presents as amino acids like cysteine, methionine and some substance like thiamine, biotin. It also indirectly influences the plant chlorophyll and nodulation in legumes. Bacteria, archaea, fungi involve in sulphur cycling. Elemental sulphur may respire with H_2 or organic compounds as electron donors. These metabolisms are performed by Crenarchaeota and Thermoplasmata groups (Dubey et al. 2016).

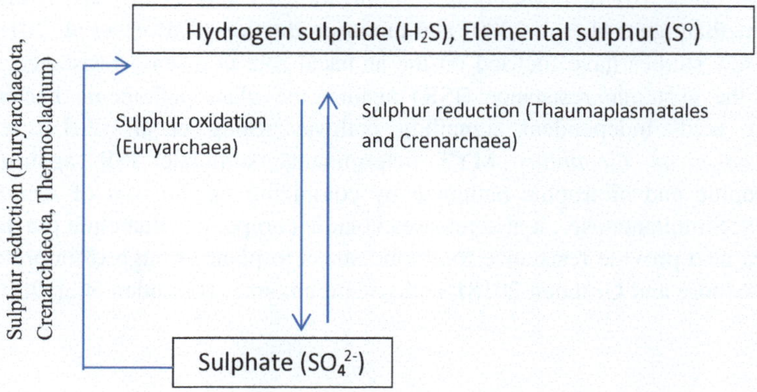

Dimethyl sulphide (DMS) is one of the sulphur-based compound on sulphur cycle which is used for cloud formation have an advantage in reducing the global warming.

14.4.4 Phosphorous Cycle

Phosphorous (P) is one of the essential elements after nitrogen for plants. On the average content of P in soil is 0.05%, but only 0.1% of these available for plants. This problem is mitigated by phosphorus solubilizing microorganisms. They are enormous soil bacteria, fungi are identified as phosphate solubilizer but in the case of archaea only the meagre report available (Alori et al. 2017; Yadav et al. 2017). Yadav et al. (2015) identified the P solubilizing halophilic archaea from plant Abutilon, Cenchrus, Dichanthium, Sporobolous and *Suaeda nudiflora*. Among the isolates *Natrinema* sp. and *Halococcus hamelinensis* solubilized phosphorus 134.61 and 112.56 mg/L. They produce organic acid to solubilize the phosphorus (gluconic, citric, succinic, oxalic, lactic). These archaea groups may play the role in P solubilisation for crops growing in saline environment.

14.5 Archaea in Plant Health Improvement

Plant growth-promoting rhizobacteria (PGPR) improves the plant growth through nutritional and non-nutritional mode. PGPR improves plant growth by nutrient availability and elicits defence of plants against the pathogen and herbivores were well studied. Similar to PGPR, archaea also improve the plant growth and immunity. Hence it is referred to as a plant growth-promoting archaea (PGPA). Several studies reported that archaea improve the plant growth through siderophore production (Dignam et al. 2018), phosphorus solubilization (Yadav et al. 2015), nitrogen-fixing methanogens (Leigh 2000) and auxin production (Taffner et al. 2018). But only a few studies have focused on the archaeal role in disease resistance. PGPA induce the systemic resistance (ISR) against the plant pathogenic bacteria via salicylic acid- independent signalling pathway. Song et al. (2019) reported *Nitrosocosmicus oleophilus* MY3 determinants stimulate ISR against both necrotrophic and biotrophic pathogen by colonizing on the root of *Arabidopsis thaliana*. Simultaneously, it also releases volatile compounds that elicit the defence. Archaea also provide resistance to abiotic stress to plant through osmoprotectants (Smith-Moore and Grunden 2018), antioxidant enzymes (Grunden et al. 2005).

14.6 Conclusion

Archaea contribute nutrient cycle and offered various biotic and abiotic stress tolerances to plant. Because of its uniqueness, archaea adopt several ecosystems having special attention in agriculture, industry and medicinal application. Archaea are still an understudied area of plant microbiome and their role in plant growth and immunity. Understanding the plant archaeal interaction paves the way for new insight in plant grown in different environment conditions and their health improvement.

Acknowledgements Thank to Department of Agricultural Microbiology, Tamil Nadu Agricultural University, Coimbatore.

References

Alori ET, Glick BR, Babalola OO (2017) Microbial phosphorus solubilization and its potential for use in sustainable agriculture. Front Microbiol 2:8–971

Angel R, Claus P, Conrad R (2012) Methanogenic archaea are globally ubiquitous in aerated soils and become active under wet anoxic conditions. ISME J 6(4):847–862

Baker BJ, Tyson GW, Webb RI, Flanagan J, Hugenholtz P, Allen EE, Banfield JF (2006) Lineages of acidophilic archaea revealed by community genomic analysis. Science 22:314–341

Bates ST, Berg-Lyons D, Caporaso JG, Walters WA, Knight R, Fierer N (2011) Examining the global distribution of dominant archaeal populations in soil. ISME J 5(5):908–917

Borneman J, Triplett EW (1997) Molecular microbial diversity in soils from eastern Amazonia: evidence for unusual microorganisms and microbial population shifts associated with deforestation. Appl Environ Microbiol 63(7):2647–2653

Bringel F, Couée I (2015) Pivotal roles of phyllosphere microorganisms at the interface between plant functioning and atmospheric trace gas dynamics. Front Microbiol 22:6–486

Buée M, De Boer W, Martin F, Van Overbeek L, Jurkevitch E (2009) The rhizosphere zoo: an overview of plant-associated communities of microorganisms, including phages, bacteria, archaea, and fungi, and of some of their structuring factors. Plant Soil 321(2):189–212

Dignam BE, O'Callaghan M, Condron LM, Kowalchuk GA, Van Nostrand JD, Zhou J, Wakelin SA (2018) Effect of land use and soil organic matter quality on the structure and function of microbial communities in pastoral soils: implications for disease suppression. PLoS One 13(5): e0196581

Dubey G, Kollah B, Gour VK, Shukla AK, Mohanty SR (2016) Diversity of bacteria and archaea in the rhizosphere of bioenergy crop *Jatropha curcas*. 3 Biotech 6(2):257

Flemming HC, Wuertz S (2019) Bacteria and archaea on earth and their abundance in biofilms. Nat Rev Microbiol 17(4):247–260

Grunden AM, Jenney FE, Ma K, Ji M, Weinberg MV, Adams MW (2005) In vitro reconstitution of an NADPH-dependent superoxide reduction pathway from Pyrococcus furiosus. Appl Environ Microbiol 71(3):1522–1530

Hassani MA, Durán P, Hacquard S (2018) Microbial interactions within the plant holobiont. Microbiome 6(1):58

Huber H, Hohn MJ, Rachel R, Fuchs T, Wimmer VC, Stetter KO (2002) A new phylum of Archaea represented by a nanosized hyperthermophilic symbiont. Nature 417(6884):63–67

Knief C, Delmotte N, Chaffron S, Stark M, Innerebner G, Wassmann R, Von Mering C, Vorholt JA (2012) Metaproteogenomic analysis of microbial communities in the phyllosphere and rhizosphere of rice. ISME J 6(7):1378–1390

Lee SA, Kim Y, Kim JM, Chu B, Joa JH, Sang MK, Song J, Weon HY (2019) A preliminary examination of bacterial, archaeal, and fungal communities inhabiting different rhizocompartments of tomato plants under real-world environments. Sci Rep 9(1):1–5

Leigh JA (2000) Nitrogen fixation in methanogens: the archaeal perspective. Curr Issues Mol Biol 2:125–131

Müller H, Berg C, Landa BB, Auerbach A, Moissl-Eichinger C, Berg G (2015 Mar) Plant genotype-specific archaeal and bacterial endophytes but similar Bacillus antagonists colonize Mediterranean olive trees. Front Microbiol 3(6):138

Reitschuler C, Hofmann K, Illmer P (2016) Abundances, diversity and seasonality of (non-extremophilic) Archaea in alpine freshwaters. Antonie Van Leeuwenhoek 109(6):855–868

Santoro AE, Casciotti KL (2011) Enrichment and characterization of ammonia-oxidizing archaea from the open ocean: phylogeny, physiology and stable isotope fractionation. ISME J 5 (11):1796–1808

Smith-Moore CM, Grunden AM (2018) Bacteria and archaea as the sources of traits for enhanced plant phenotypes. Biotechnol Adv 36(7):1900–1916

Song GC, Im H, Jung J, Lee S, Jung M-Y, Rhee S-K, Ryu C-M (2019) Plant growth-promoting archaea trigger induced systemic resistance in against and. Environ Microbiol 21(3):940–948

Taffner J, Erlacher A, Bragina A, Berg C, Moissl-Eichinger C, Berg G (2018) What is the role of Archaea in plants? New insights from the vegetation of alpine bogs. mSphere 3(3). https://doi.org/10.1128/mSphere.00122-18

Taffner J, Cernava T, Erlacher A, Berg G (2019) Novel insights into plant-associated archaea and their functioning in arugula (*Eruca sativa* Mill.). J Adv Res 30:16–25

Wassermann B, Cernava T, Müller H, Berg C, Berg G (2019) Seeds of native alpine plants host unique microbial communities embedded in cross-kingdom networks. Microbiome 7(1):108

Woese CR, Achenbach L, Rouviere P, Mandelco L (1991) Archaeal phylogeny: reexamination of the phylogenetic position of Archaeoglohus fulgidus in light of certain composition-induced artifacts. Syst Appl Microbiol 14(4):364–371

Yadav AN, Sharma D, Gulati S, Singh S, Dey R, Pal KK, Kaushik R, Saxena AK (2015) Haloarchaea endowed with phosphorus solubilization attribute implicated in phosphorus cycle. Sci Rep 5(1):12293

Yadav AN, Verma P, Kaushik R, Dhaliwal HS, Saxena AK (2017) Archaea endowed with plant growth promoting attributes. EC Microbiol 8(6):294–298

Chapter 15
Microbial Symbionts of Aquatic Plants

Tejaswini Dash, Klaus-J. Appenroth, and K. Sowjanya Sree

Abstract Aquatic plants have been used over time in several ways which include use as food and feed supplement, for bioenergy production, for phytoremediation and wastewater treatment, to name a few. Like in the terrestrial ecosystem, interactions between microorganisms and plants are diverse in an aquatic ecosystem. In the present chapter, one of the positive interactions, symbiosis between the microorganisms and aquatic plants has been discussed focusing on how this interaction can prove to be beneficial to aquatic plants. Some of the features like biofiltration, nitrogen fixation, phytohormone production, and bioenergy production from aquatic plants and the role of microbial symbionts in these processes have been discussed. One of the re-emerging model plants, duckweeds have been considered as an example in order to describe the abovesaid. Lemnaceae (duckweeds) constituting the fastest-growing flowering plants have gained attention in terms of their practical applications because of their huge potential for biomass production. As a future prospect, sustainable ways of use of aquatic plants and their biomass production should be a prime focus in aquatic research.

Keywords Aquatic plants · Bioenergy production · Biofiltration · Duckweeds · Lemnaceae · Microbial symbionts

T. Dash · K. S. Sree (✉)
Department of Environmental Science, Central University of Kerala, Periye, India
e-mail: ksowsree9@cukerala.ac.in

K.-J. Appenroth
Matthias Schleiden Institute – Plant Physiology, Friedrich Schiller University of Jena, Jena, Germany

N. Shrivastava et al. (eds.), *Symbiotic Soil Microorganisms*, Soil Biology 60,
https://doi.org/10.1007/978-3-030-51916-2_15

15.1 Introduction

Symbiosis is a positive ecological interaction between biotic components of an ecosystem, in which all the interacting partners benefit from this relationship. Such relationships can exist in both terrestrial (Unnikumar et al. 2013) as well as aquatic (Appenroth et al. 2016) ecosystems. In the present chapter, we will focus on this symbiotic relationship between microorganisms and plants in a freshwater aquatic ecosystem and how this interaction proves to be beneficial to the plant community. The multitude of organisms living in either lotic or lentic ecosystems, like any other ecosystem, need to adapt themselves to the existing environmental conditions. So also the aquatic plants, which are the primary producers of an aquatic ecosystem. These adaptations can either be through changes or modifications in morphology, anatomy, physiology or can be through their positive and negative interactions with the rest of the biota in their environment. The plants belonging to several different groups of lower plants as well as angiosperms have adapted to these aquatic environments. Some of the examples are water hyacinth (*Eichhornia crassipes*), Indian lotus (*Nelumbo nucifera*), water spinach (*Ipomoea aquatica*), water mimosa (*Neptunia oleracea*), mosquito fern (*Azolla pinnata*), water lettuce (*Pistia stratiotes*), and duckweeds (Lemnaceae, e.g., *Wolffia globosa*). The aquatic plant community is also diverse in its ability to inhabit different strata of the water body. They can be free-floating, submerged, or rooted. Equally diverse are the microbes of the aquatic ecosystem. They belong to several different groups of microorganisms and can differ in their capacity to interact with the plant body, e.g., ectophytes and endophytes. The positive interaction between these two groups of biotic community has been tapped by several fields of applied biology.

15.2 Uses of Aquatic Plants

Several uses of aquatic plants and their economic benefits have been reviewed by Wersal and Madsen (2012). They not only provide ecosystem supporting services but also such of the benefits that are being utilized by human beings.

Aquatic plants can uptake nutrients from contaminated water and consequently purify the water (Zimmels et al. 2004). This phenomenon has been put to practical applications as treatment of wastewater by phytoremediation is an environment friendly approach (Ziegler et al. 2016, 2017, 2019; Calado et al. 2019). Various aquatic plants have been used for remediation of wastewater contaminated with several industrial pollutants. Heavy metal removal by *E. crassipes, Hydrilla, Jussiaea repens, Lemna minor, Pistia stratiotes*, and *Trapa natans* was studied and *L. minor* was found efficient for remediating the particular wastewater (Mishra et al. 2013). In other respect, halophytic plants like *Salicornia europaea*, *Salsola crassa,* and *Bienertia cycloptera* have been investigated for salt removal capacity (Farzi et al. 2017). Some other halophytic species like *Atriplex barclayana* have been found to act as biofilters to remove nutrients from wastewater (Brown et al. 1999).

Duckweeds are some of the well-studied aquatic plants. Duckweeds include the smallest and fastest-growing free-floating flowering monocotyledonous plants (Sree et al. 2015a, b; Ziegler et al. 2015) that have the capacity to grow at a rapid rate in nutrient-rich waters. These plants belonging to the family *Lemnaceae* are classified into five genera, which are *Spirodela, Landoltia, Lemna, Wolffia, and Wolffiella,* and 36 species (Bog et al. 2019). These plants are distributed throughout the world except the polar regions and deserts. Different species of duckweeds have been used for the treatment of municipal, industrial, domestic wastewater (Oron 1994), secondary effluents (Sutton and Ornes 1975), aquaculture wastewater (Porath and Pollock 1982), and Swine wastewater (Mohedano et al. 2014). *Lemna minor has* been studied for removal of heavy metals by Kara (2004), Bianconi et al. (2013), Chaudhuri et al. (2014), and Sree et al. (2015c) and has been reviewed by Ziegler et al. (2016, 2017, 2019).

Different duckweed species viz., *Spirodela polyrhiza, Landoltia punctata, Lemna minor, Lemna gibba, Wolffiella hyalina, and Wolffia microscopica* were investigated for their nutritional value and were found suitable for human consumption (Appenroth et al. 2017, 2018). The duckweed plant extracts also did not show any cytotoxicity and antiproliferative effect on human cell lines (Sree et al. 2019). Owing to their nutritional quality, duckweeds can also be utilized as fish feed (Paul et al. 2013) and animal feed (Culley et al. 1981).

From ancient times, plants have been used as good source of medicines (Shakya 2016). Some of the aquatic plants used for medicine include *Typha domingensis, Ipomoea aquatica*, and *Marsilea minuta* (Panda and Mishra 2011).

Antioxidants play an important role in defense mechanism in our body against oxidative stress. Plants have the ability to synthesize antioxidants, which are capable of reducing the level of Reactive Oxygen Species (Kasote et al. 2015). Gulcin et al. (2010) studied the antioxidant, antibacterial, and anticandidal activities of the duckweed *Lemna minor* against 21 different bacteria and fungi and confirmed the antibacterial, antifungal activities of *Lemna minor,* which is a source of antioxidants, food, medicines, and pharmaceuticals. Zinc supplement to *Spirodela polyrhiza* increased the antioxidant activity and ROS scavenging activity (Upadhyay and Panda 2009). Further, Tipnee et al. (2017) found antioxidant, anti-inflammatory properties in another duckweed, *Wolffia globosa*.

15.3 Microbial Symbionts

Positive interaction of microorganisms with plants not only helps to promote plant growth but also plays a vital role in nutrient cycling (Ishizawa et al. 2020). There are three regions in plants where microorganisms can associate and interact with them. These are phyllosphere, rhizosphere, and endosphere (Compant et al. 2019). Depending on the habit and habitat of the aquatic plant species, the exposure of these regions to different abiotic factors such as temperature, moisture, and radiation will vary. The scale of these abiotic factors may affect the diversity of microorganisms associated with a specific sphere. These microorganisms use carbon, nitrogen, carbohydrate, and ammonium as source of energy (Knief et al. 2012). The microbes that are found in the endosphere region, e.g., inside the root tissue are considered as endophytes. Some of these associated microorganisms are beneficial in nature such as growth-promoting microorganisms (Appenroth et al. 2016; Ishizawa et al. 2020). Some endophytic bacteria have the capacity to produce plant growth hormones such as auxins and some plant growth-promoting rhizobacteria also defend the plants from their pathogens (Sturz et al. 2000).

15.4 Benefits of Microbial Symbiosis in Aquatic Plants

In this mutually beneficial interaction, as implied by the term mutual, both the aquatic plants as well as the associated microorganisms gain advantages from each others' company. In the present chapter, we will focus on the benefits gained by the aquatic plants from this symbiotic relationship. In the following, we have mainly taken the example of the symbiotic relationship between microorganisms and members of the aquatic plant family Lemnaceae (duckweeds). Whereever appropriate we have also included details on the investigations on other aquatic plants.

There are several benefits of this microbial association to the aquatic plants, a few are detailed below.

15.4.1 Biofiltration

Plant-associated microorganisms enhance the uptake capacity of plants and help in removal of nutrients, heavy metals, xenobiotics, and so on from a water body. Hence, this partnership can be made use of in wastewater remediation. Apart from studies in duckweeds (see below), investigations have been carried out on several other aquatic plants. Investigations showed that microbes like *Nitrobacteria irancium* associated with *E. crassipes* help in removal of chromium and zinc (Abou-Shanab et al. 2007).

As already mentioned in the previous section, duckweeds as biofilters are promising candidates for use in wastewater remediation especially because of their extremely high growth rates (Sree et al. 2015a; Ziegler et al. 2015) and by their capacity to interact with a huge diversity of microorganisms (Acosta et al. 2020; Appenroth et al. 2016; Ishizawa et al. 2017a, 2019a). Deeper understanding of the molecular mechanism of plant–microbe interaction is meanwhile facilitated by the availability of genome sequences of several plant growth-promoting and inhibiting bacteria (Sugawara et al. 2015; Ishizawa et al. 2017b, 2018). We present here the effects of duckweed-bacteria interactions on removal of nutrients, heavy metals, and organic xenobiotics.

15.4.1.1 Removal of Nutrients from Eutrophic Wastewater

Excessive use of fertilizers in modern agriculture results finally in eutrophication of surface water bodies, which is a worldwide problem. Because of the very high growth rates of duckweed by vegetative propagation and by the additional effect of plant growth-promoting bacteria, duckweed-bacteria systems show excellent performance in the uptake of excess nutrients from wastewater. The advantage of the system lies in the point that the huge amount of biomass generated this way can be harvested from the surface of the water body and can be utilized for several practical applications depending on the quality of the wastewater (Appenroth et al. 2015).

Bacteria associated with duckweeds play an important role in nutrient removal from water bodies. Bacterial growth was found to be a key factor for the degradation of organic matter (Körner and Vermat 1998). Nitrogen assimilation was especially well investigated (review in Appenroth et al. 2016). Duckweed provides additional adherence space for bacteria both via roots as well as the submerged plant body as shown in the rootless duckweed species of the genus *Wolffia*. Moreover, duckweed provides oxygen for the bacterial nitrogen assimilation (Appenroth et al. 2016). Ying-ru et al. (2013) showed that a strain of *Pseudomonas* species found in the rhizosphere of *L. minor* helps in removal of nitrogen together with promotion of plant growth.

Root exudates from some of the duckweed species like *S. polyrhiza and L. minor* showed the presence of *Pseudomonas fluorescens* (Lu et al. 2014). However, Ardiansyah and Fotedar (2016) found an abundance of heterotrophic bacteria like Gram-negative cocci and Gram-negative bacilli associated with *L. minor*. Also, surface area of a plant influences the growth and structure of bacterial communities (Leonard et al. 2000). More recently, Shen et al. (2019) showed that the association of heterotrophic nitrifying bacteria like *Acinetobacter* sp. with *L. gibba* enhanced the rate of removal of ammonium and total nitrogen from the water body.

In terms of the practical application of a specific duckweed-bacteria association that would have been studied in detail in a laboratory, a certain problem exists concerning the stability of the association under non-axenic conditions, i.e., when other bacteria or microbes present in the environment can compete for a possible association with the surface of duckweed. In this respect, very recently, Ishizawa et al. (2020) suggested a two-step cultivation process with the plant growth-promoting bacterium, *Acinetobacter calcoaceticus* P23 comprising a "colonization step" and a subsequent "mass cultivation step," for increasing the nutrient removal capacity, thereby enhancing the biomass production.

15.4.1.2 Removal of Heavy Metals

The interaction of bacteria and duckweed was investigated hoping that also the uptake of heavy metals by duckweed could be stimulated by certain bacteria and that this will enhance the phytoremediation potential of duckweeds. Related experiments were carried out with cadmium (Cd) and chromate (Cr (VI)). It turned out that, in the toxic concentration range, the uptake of these two heavy metals was not enhanced, but was rather diminished. Several mechanisms might be responsible for these effects but in summary, it can be concluded that the associated bacteria rather play a role in phytoprotection than in enhanced phytoremediation potential of duckweed (Appenroth et al. 2016). The growth promotion activity of *Aquitalea magnusonii* strain H3 on *L. minor* increased remarkably when the plant was subjected to copper (Cu) and zinc (Zn) stresses, which itself inhibited the duckweed growth by ca. 40%. This plant growth-promoting bacterium also enhanced the capacity of the duckweed to accumulate and tolerate the two heavy metals (Ishizawa et al. 2019b). In another study, oxidation of As(III) by bacteria from phyllosphere of *Wolffia australiana* was reported (Xie et al. 2014).

15.4.1.3 Removal of Organic Xenobiotic

Wastewater is often contaminated by dissolved organic material like herbicides, pesticides, antibiotics, or other drugs because of their illegal delivery into the environment. As early as 1985, Pignatello et al. (1985) showed that *L. minor* contributed to removal of pentachlorophenol from wastewater. These authors also showed that not only duckweed but also associated bacteria were responsible for the degradation of this phenolic compound. A phenol degrading bacterium, *Acinetobacter calcoaceticus* associated with *Lemna aequinoctialis* was isolated by Yamanga et al. (2010). Muerdter and LeFeyre (2019) recently reported transformation of the insecticides Imidacloprid and Thiacloprid (Neonicotinoids) into their metabolites by an unsterile culture of *Lemna* sp. After surface sterilization, the metabolites could not be observed demonstrating the role of duckweed-microbe interactions in this transformation reaction. The interaction of rhizospheric bacteria with *S. polyrhiza* for degradation of organic compounds has been investigated by several groups. This association has been found to be capable of degrading surfactants (Mori et al. 2005) and degrading a variety of aromatic compounds (Toyama et al. 2006; Hoang et al. 2010). Strains of *Pseudomonas* and *Cupriavidus* were isolated from the rhizosphere region of *S. polyrhiza* and were found to be able to degrade 3-Nitrophenol which is also an aromatic compound (Kristanti et al. 2012). Bacterium, *Sphingobium fuliginis* associated with *S. polyrhiza* was found capable of degrading 4-tert-butylphenol and using it as a sole carbon and energy source (Ogata et al. 2013).

In the last 20 years, investigations concerning the degradation, removal, and uptake of a broad range of organic xenobiotics has increased to a large degree

(Appenroth et al. 2016; Ziegler et al. 2016, 2017). In several cases, the effects of interactions between duckweeds and different bacteria were demonstrated. More investigations are required before large-scale biotechnological applications can be expected. Ziegler et al. (2019) pointed to the opportunity to even identify the nature of contamination by biotests with duckweed species.

15.4.2 Nitrogen Fixation

Besides assimilation of nitrate and ammonium, interaction of duckweed with nitrogen-fixing bacteria has also been reported, although the mechanism of interaction is still unknown. Nitrogen-fixing heterotrophic bacteria have been found to be associated with duckweed mat that can recycle nitrogen produced by the denitrifiers to produce organic compounds (Zuberer 1982). Nitrogen-fixing free-living bacteria have also been isolated from three aquatic plant species *Azolla filiculoides, Lemna gibba, and Ricciocarpus natans* (Quisehuatl-Tepexicuapan et al. 2016).

15.4.3 Production of Phytohormones

The role of auxin in plant–microbe interaction has been a topic of discussion (Spaepen and Vanderleyden 2011). Gilbert et al. (2018) isolated endophytic bacteria from surface-sterilized duckweed species and found that they are capable of producing indole-3-acetic acid and other indole related compounds. However, the role of these indole related molecules in the association of duckweeds and the bacteria producing them needs to be more clearly established. It is interesting to note that the same study had found a correlation between the duckweed species and the indole related molecules produced by the associated bacteria. Extrapolating this finding to other metabolites and functions, this might have an impact on the use of a particular duckweed species for a particular practical application.

15.4.4 Bioenergy Production

Fuel produced from plant biomass is an alternative source of energy that is safe for the environment. Water hyacinth as a candidate for bioethanol production has been investigated by several researchers (Hronich et al. 2008; Wang and Calderon 2012). Methane producing capacity of eight aquatic plants such as *Salvinia molesta, Hydrilla verticillata, Nymphaea stellate, Azolla pinnata, Ceratopteris sp., Scirpus sp., Cyperus sp., Utricularia reticulata* was investigated by Abbasi et al. (1990) and found that *Salvinia molesta* has the highest yield.

The plant growth-promoting capacity of the microbes can play a role in enhancing the yield of the plants suitable for biofuel production. Mishima et al. (2008) compared the effect of two microbes *Saccharomyces cervisiae* and *Escherichia coli* on ethanol production from water hyacinth and water lettuce. It was found that *E. coli* had more ethanol-producing capacity than *S. cervisiae*. Duckweeds have also been investigated for use as potential feedstock for biofuels such as bioethanol (Chen et al. 2012; Cui et al 2011; Xu et al. 2011; Soda et al. 2015) and biobutanol (Li et al. 2012) production. These studies might owe to the fact that duckweeds include some of the fastest-growing plants which can result in huge amounts of biomass at a fast rate. Xiao et al. (2013) found that a duckweed species, *Landoltia punctata* is a suitable crop for bioenergy production as it was found to have a high-starch accumulation capacity. Another study was conducted by Xiu et al. (2010) for bio-oil production from duckweeds. *Lemna minor* was found to be a potential crop plant for biogas production (Muradov et al. 2012). Further, ethanol and methane production capacity were investigated in four duckweed species *S. polyrhiza, L. minor, L. gibba, and L. punctata* after being grown in effluent from municipal and swine wastewater. It was found that, out of the four, *S. polyrhiza* had highest nitrogen removal capacity as well as highest ethanol and methane production capacity (Toyama et al. 2018). Biomass of *L. minor* produced from swine lagoon wastewater was found to be capable of ethanol production using yeast (Ge et al. 2012). Biomass of duckweed obtained from wastewater points toward the conditions where duckweeds are not growing in isolation but are in association with other beings of the aquatic ecosystem especially pointing toward a closer association between duckweeds and microbes.

15.5 Conclusion

The symbiosis between microorganisms and aquatic plants has a significant role in enhancing the different capacities of aquatic plants. When tapped in a sustainable manner, the aquatic plant resources can prove to be beneficial with several alternatives for practical applications. The enormous growth of different aquatic plants in the nutrient-rich wastewaters has been considered for a long time as a sign of bad health of the ecosystem. However, the huge biomass thus produced can be harvested and efficiently put to use for several applications like biogas, biofuel, or biofertilizer production or bioplastic production based on the quality of wastewater on which the plants had grown. Regular harvest of biomass will ensure the safe health of the ecosystem and will also be a bioresource. In this sustainable use of aquatic plants, microorganisms, especially symbionts of these aquatic plants play a crucial role in supporting and enhancing several of the processes as described in this chapter.

References

Abbasi SA, Nipaney PC, Schaumberg GD (1990) Bioenergy potential of eight common aquatic weeds. Biol Wastes 34:359–366

Abou-Shanab RAI, Angle JS, Van-Berkum P (2007) Chromate tolerant bacteria for enhanced metal uptake by *Eichhornia crassipes*. Int J Phytoremediation 9:91–105

Acosta K, Xu J, Gilbert S, Denison E, Brinkman T, Lebeis S, Lam E (2020) Duckweed hosts a taxonomically similar bacterial assemblage as the terrestrial leaf microbiome. PLoS One 15: e0228560

Appenroth KJ, Sree KS, Fakhoorian T, Lam E (2015) Resurgence of duckweed research and applications: report from the 3rd international duckweed conference. Plant Mol Biol 89:647–654

Appenroth KJ, Ziegler P, Sree KS (2016) Duckweed as a model organism for investigating plant–microbe interactions in an aquatic environment and its applications. Endocytobiosis Cell Res 27:94–106

Appenroth KJ, Sree KS, Boehm V, Hammann S, Vetter W, Leiterer M, Jahreis G (2017) Nutritional value of duckweeds (*Lemnaceae*) as human food. Food Chem 217:266–273

Appenroth KJ, Sree KS, Bog M, Ecker J, Seeliger C, Böhm V, Lorkowski S, Sommer K, Vetter W, Tolzin-Banasch K, Kirmse R, Leiterer M, Dawczynski C, Liebisch G, Jahreis G (2018) Nutritional value of the duckweed species of the genus *Wolffia* (Lemnaceae) as human food. Front Chem 6:483

Ardiansyah A, Fotedar R (2016) The abundance and diversity of heterotrophic bacteria as function of harvesting frequency of duckweed (*Lemna minor L.*) in recirculating aquaculture systems. Lett Appl Microbiol 63:53–59

Bianconi D, Pietrini F, Massacci A, Iannelli MA (2013) Uptake of cadmium by *Lemna minor*, a (hyper?-) accumulator plant involved in phytoremediation applications. E3S Web Conf 1:13002. https://doi.org/10.1051/e3sconf/20130113002

Bog M, Appenroth KJ, Sree KS (2019) Duckweed (Lemnaceae). Its molecular taxonomy. Front Sustain Food Syst 3(117):1–7

Brown JJ, Glenn EP, Fitzsimmons KM, Smith SE (1999) Halophytes for the treatment of saline aquaculture effluent. Aquaculture 175:255–268

Calado SLM, Esterhuizen-Londt M, Assis HCS, Pflugmacher S (2019) Phytoremediation: green technology for the removal of mixed contaminants of a water supply reservoir. Int J Phytoremediation 21:1–9

Chaudhuri D, Majumber A, Misra AK, Bandyopadhyay K (2014) Cadmium removal by *Lemna minor* and *Spirodela polyrhiza*. Int J Phytoremediation 16:1119–1132

Chen Q, Jin Y, Zhang G, Fang Y, Xiao Y, Zhao H (2012) Improving production of bioethanol from duckweed (*Landoltia punctata*) by pectinase pretreatment. Energies 5:3019–3032

Compant S, Samad A, Faist H, Sessitsch A (2019) A review on the plant microbiome: ecology, functions and emerging trends in microbial application. J Adv Res 19:29–37

Cui W, Xu J, Cheng J, Stomp A (2011) Starch accumulation in duckweed for bioethanol production. Biol Eng 3:187–197

Culley DD, Rojimankova KM, Kuet J, Fry JB (1981) Production of chemical quality and use of duckweeds (*Lemnaceae*) in aquaculture. Waste management and animal feeds. J World Aquacult Soc 12:27–49

Farzi A, Borghei SM, Vossoughi M (2017) The use of halophytic plants for salt phytoremediation in constructed wetlands. Int J Phytoremediation 19(7):643–650

Ge XM, Zhang NN, Phillips GC, Xu JF (2012) Growing *Lemna minor* in agricultural wastewater and converting the duckweed biomass to ethanol. Bioresour Technol 124:485–488

Gilbert S, Xu J, Acosta K, Poulev A, Lebesis S, Lam E (2018) Bacterial production of indole related compounds reveals their role in association between duckweeds and endophytes. Front Chem 6 (265):1–14

Gulcin I, Kirecci E, Akkemik E, Topal F, Hisar O (2010) Antioxidant, antibacterial, and anticandidal activities of an aquatic plant duckweed (*Lemna minor*) L. Lemnaceae. Turk J Biol 34:175–188

Hoang H, Yu N, Toyma T, Inoue D, Sei K, Ike M (2010) Accelerated degradation of a variety of aromatic compounds by *Spirodela polyrhiza*-bacterial associations and contribution of root exudates released from *S. polyrhiza*. J Environ Sci 22:494–499

Hronich JE, Martin L, Plawsky J, Bungay HR (2008) Potential of *Eichhornia crassipes* for biomass refining. J Ind Microbiol Biotechnol 35:393–402

Ishizawa H, Kuroda M, Morikawa M, Ike M (2017a) Evaluation of environmental bacterial communities as a factor affecting the growth of duckweed *Lemna minor*. Biotechnol Biofuels 10. https://doi.org/10.1186/s13068-017-0746-8

Ishizawa H, Kuroda M, Ike M (2017b) Draft genome sequence of *Aquitalea magnusonii* strain H3, a plant growth-promoting bacterium of duckweed (*Lemna minor*). Microbiol Resour Announc 5. https://doi.org/10.1128/genomeA.00812-17

Ishizawa H, Kuroda M, Inoue D, Ike M (2018) Complete genome sequences of two plant growth-inhibiting bacteria, *Acinetobacter ursingii* M3 and *Asticcacaulis excentricus* M6, isolated from duckweed (*Lemna minor*). Microbiol Resour Announc 7. https://doi.org/10.1128/MRA.01092-18

Ishizawa H, Kuroda M, Inoue K, Inoue D, Morikawa M, Ike M (2019a) Colonization and competition dynamics of plant growth-promoting/inhibiting bacteria in the phytosphere of the duckweed *Lemna minor*. Microb Ecol 77:440–450. https://doi.org/10.1007/s00248-018-1306-x

Ishizawa H, Tada M, Kuroda M, Inoue D, Ike M (2019b) Performance of plant growth-promoting bacterium of duckweed under different kinds of abiotic stress factors. Biocatal Agric Biotechnol 19. https://doi.org/10.1016/j.bcab.2019.101146

Ishizawa H, Ogata Y, Hachiya Y, Tokura KI, Kuroda M, Inoue D, Toyama T, Tanaka Y, Mori K, Morikawa M, Ike M (2020) Enhanced biomass production and nutrient removal capacity of duckweed via two-step cultivation process with a plant growth-promoting bacterium, *Acinetobacter calcoaceticus* P23. Chemosphere 238:124682

Kara Y (2004) Bioaccumulation of copper from contaminated wastewater by using *Lemna minor* (aquatic green plant). Environ Contamin Toxicol 72:467–471

Kasote DM, Katyare SS, Hegde MV, Bae H (2015) Significance of antioxidant potential of plants and its relevance to therapeutic applications. Int J Biol Sci 11:982–991

Knief C, Delmotte N, Chaffron S, Stark M, Innerebner G, Wassmann R, Mering C, Vorholt JA (2012) Metaproteogenomic analysis of microbial communities in the phyllosphere and rhizosphere of rice. ISME J 6:1378–1390

Körner S, Vermat JE (1998) The relative importance of *Lemna gibba* L., bacteria and algae for the nitrogen and phosphorus removal in duckweed-covered domestic wastewater. Water Res 32:3651–3661

Kristanti RA, Kanbe M, Hadibarta T, Toyama T, Tanka Y, Mori K (2012) Isolation and characterization of 3-nitrophenol-degrading bacteria associated with rhizosphere of *Spirodela polyrhiza*. Environ Sci Pollut Res 19:1852–1858

Leonard N, Blancheton JP, Guiraud JP (2000) Populations of heterotrophic bacteria in an experimental recirculation aquaculture system. Aquacult Eng 22:109–120

Li X, Jin Y, Gao X, Zhang G, Zhao H (2012) Fermentation method of high ratios of biobutanol with *Landoltia punctata*. China Brewing 31:85–88

Lu Y, Zhou Y, Nakai S, Hosomi M, Zhang H, Kronzucker HJ, Shi W (2014) Stimulation of nitrogen removal in the rhizosphere of aquatic duckweed by root exudate components. Planta 239:591–603

Mishima D, Kuniki M, Sei K, Soda S, Ike M, Fujita M (2008) Ethanol production from candidate energy crops: water hyacinth (*Eichhornia crassipes)* and water lettuce (*Pistia stratiotes L.*). Bioresource Technology 99:2495

Mishra S, Mohanty M, Pradhan C, Patra HK, Das R, Sahoo S (2013) Physico-chemical assessment of paper mill effluent and its heavy metal remediation using aquatic macrophytes—a case study at JK Paper mill, Rayagada, India. Environ Monit Assess 185:4347–4359

Mohedano RA, Costa RHR, Hofmann SM, Belli P (2014) Using full-scale duckweed ponds as the finish stage for swine waste treatment with a focus on organic matter degradation. Water Sci Technol 69:2147–2154

Mori K, Toyama T, Sei K (2005) Surfactants degrading activities in the rhizosphere of giant duckweed ("*Spirodela polyrrhiza*"). Jpn J Water Treat Biol 41:129–140

Muerdter CP, LeFeyre GH (2019) Synergistic *Lemna* duckweed and microbial transformation of Imidacloprid and Thiacloprid Neonicotinoids. Environ Sci Technol Lett 6:761–767

Muradov N, Fidalgo B, Gujar A, Garceau N, Raissi T (2012) Production and characterization of *Lemna minor* bio-char and its catalytic application for biogas reforming. Biomass Bioenergy 42:123–131

Ogata Y, Toyama T, Wang NYX, Sei K, Ike M (2013) Occurrence of 4-tert-butyl phenol (4-tert-BP) bio degradation in aquatic sample caused by the presence of *Spirodela polyrhiza* and isolation of a 4-tert-BP utilizing bacterium. Biodegradation 24:191–202

Oron G (1994) Duckweed culture for wastewater renovation and biomass production. Agric Water Manag 26:27–40

Panda A, Mishra MK (2011) Ethnomedicinal survey of some wetland plants of South Orissa and their conservation. Indian J Tradit Knowl 10:296–303

Paul M, Barman AC, Siddque MF, Bhadra A, Rahaman MS (2013) Effect of use of duckweed powder as a fish feed on polyculture of carps. J Environ Sci Nat Resour 6:213–216

Pignatello JJ, Johnson LK, Martinson MM, Carlson RE, Crawford RL (1985) Response of the microflora in outdoor experimental streams to pentachlorophenol: compartmental contributions. Appl Environ Microbiol 50:127–132

Porath D, Pollock J (1982) Ammonia stripping by duckweed and its feasibility in circulating aquaculture. Aqu Bot 13:125–131

Quisehuatl-Tepexicuapan E, Ferrera-Cerrato R, Silva-Rojas HV, Rodriguez-Zaragoza S, Alarcón A, Almaraz-Suárez JJ (2016) Free-living culturable bacteria and protozoa from the rhizoplanes of three floating aquatic plant species. Plant Biosyst 150:855–865

Shakya AK (2016) Medicinal plants: future source of new drugs. Int J Herbal Med 4:59–64

Shen M, Yin Z, Qia D, Zhao Q, Kang Y (2019) Combination of heterotrophic nitrifying bacterium and duckweed (*Lemna gibba* L.) enhances ammonium nitrogen removal efficiency in aquaculture water via mutual growth promotion. J Gen Appl Microbiol 65:151–160

Soda S, Ohchi T, Piradee J, Takai Y, Ike M (2015) Duckweed biomass as a renewable biorefinery feedstock: ethanol and succinate production from *Wolffia globosa*. Biomass Bioenergy 81:364–368

Spaepen S, Vanderleyden J (2011) Auxin and plant-microbe interactions. Cold Spring Harb Perspect Biol 3:a001438. https://doi.org/10.1101/cshperspect.a001438

Sree KS, Sadakaran S, Appenroth K-J (2015a) How fast can angiosperms grow? Species and clonal diversity of growth rates in the genus *Wolffia* (Lemnaceae). Acta Physiol Plant 37:204

Sree KS, Maheshwari SC, Boka K, Khurana JP, Keresztes A, Appenroth KJ (2015b) The duckweed *Wolffia microscopica*: a unique aquatic monocot. Flora 210:31–39

Sree KS, Keresztes A, Muller-Roeber B, Brandt R, Eberius M, Fischer W, Appenroth KJ (2015c) Phytotoxicity of cobalt ions on the duckweed *Lemna minor* - morphology, ion uptake, and starch accumulation. Chemosphere 131:149–156

Sree KS, Dahse H-M, Chandran JN, Schneider B, Jahreis G, Appenroth KJ (2019) Duckweed for human nutrition: no cytotoxic and no anti-proliferative effects on human cell lines. Plant Foods Hum Nutr 74:223–224

Sturz AV, Christie BR, Nowak J (2000) Bacterial endophytes potential role in developing sustainable systems of crop production. Crit Rev Plant Sci 19:1–30. https://doi.org/10.1080/07352680091139169

Sugawara M, Hosoyama A, Yamazoe A, Morikawa M (2015) Draft genome sequence of *Acinetobacter calcoaceticus* strain P23, a plant growth-promoting bacterium of duckweed. Microbiol Resour Announc 3. https://doi.org/10.1128/genomeA.00026-14

Sutton D, Ornes WH (1975) Phosphorus removal from static sewage effluent using duckweed and algal ponds. Water Sci Technol 42:363–370

Tipnee S, Jutiviboonsuk A, Wongtrakul P (2017) The biodiversity of active compounds in *Wolffia globosa* extract for an alternative source of bioactive substances. Cosmetics 4:53

Toyama T, Yu N, Kumada H, Sei K, Ike M, Fujita M (2006) Accelerated aromatic compounds degradation in aquatic environment by use of interaction between *Spirodela polyrhiza* and bacteria in its rhizosphere. J Biosci Bioeng 101:346–353

Toyama T, Hanaoka T, Tanaka Y, Morikawa M, Mori K (2018) Comprehensive evaluation of nitrogen removal rate and biomass ethanol and meth, ethanol and methane yields by combination of four major duckweeds and three type of wastewater effluent. Bioresour Technol 250:464–473

Unnikumar KR, Sree KS, Varma A (2013) *Piriformospora indica*: a versatile root endophytic symbiont. Symbiosis 60:107–113

Upadhyay R, Panda SK (2009) Zinc reduces copper toxicity induced oxidative stress by promoting antioxidant defense in freshly grown aquatic duckweed *Spirodela polyrhiza*. J Hazard Mater 175:1081–1084

Wang Z, Calderon MM (2012) Environmental and economic analysis of application of water hyacinth for eutrophic water treatment coupled with biogas production. J Environ Manag 110:246–253

Wersal RM, Madsen JD (2012) Aquatic plants their uses and risks: a review of the global status of aquatic plants. International plant protection convention. FAO, Italy, Rome

Xiao Y, Fang Y, Jin Y, Zhang G, Zhao H (2013) Culturing duckweed in the field for starch accumulation. Ind Crop Prod 48:183–190

Xie W-Y, Su JQ, Zu YG (2014) Arsenite oxidation by the phyllosphere bacterial community associated with *Wolffia australiana*. Environ Sci Technol 48:9668–9674

Xiu SN, Shahbazi A, Croonenberghs J, Wang LJ (2010) Oil production from duckweed by thermochemical liquefaction. Energy Sources A Recov Util Environ 32:1293

Xu JL, Cui WH, Cheng JJ, Stomp AM (2011) Production of high-starch duckweed and its conversion to bioethanol. Biosyst Eng 110:67–72

Yamanga F, Washio K, Morikawa M (2010) Sustainable biodegradation of phenol by *Acinetobacter calcoaceticus p23* isolated from the rhizosphere of duckweed *Lemna aoukikusa*. Environ Sci Technol 44:6470–6474

Ying-ru Z, Yu-fang L, Hai-lin Z, Wei-ming S (2013) Aerobic denitrifying characteristics of duckweed rhizospheric bacterium RWX3. Afr J Microbiol Res 7:211–219

Ziegler P, Adelmann K, Zimmer S, Schmidt C, Appenroth K-J (2015) Relative in vitro growth rates of duckweeds (Lemnaceae) – the most rapidly growing higher plants. Plant Biol 17(Suppl. 1):33–41

Ziegler P, Sree KS, Appenroth KJ (2016) Duckweeds for water remediation and toxicity testing. Toxicol Environ Chem 98:1127–1154

Ziegler P, Sree KS, Appenroth KJ (2017) The uses of duckweed in relation to water remediation. Desalin Water Treat 63:327–342

Ziegler P, Sree KS, Appenroth KJ (2019) Duckweed biomarkers for identifying toxic water contaminants? Environ Sci Pollut Res 26:14797–14822

Zimmels Y, Kirzhner F, Roitman S (2004) Use of naturally growing aquatic plants for wastewater purification. Water Environ Res 76:220–229

Zuberer DA (1982) Nitrogen fixation (acetylene reduction) associated with duckweed (Lemnaceae) mats. Appl Environ Microbiol 43:823–828

Chapter 16
Rhizobium Presence and Functions in Microbiomes of Non-leguminous Plants

Alexandra Díez-Méndez and Esther Menéndez

Abstract The genus *Rhizobium* is well known in the context of its interaction with leguminous plants. The symbiosis *Rhizobium*-legume constitutes a significant source of ammonia in the biosphere. *Rhizobium* species have been studied and applied as biofertilizers for decades in legumes and nonlegumes, due to the potential as N-fixer and plant growth promoter. Since its discovery, conventional culture-dependent techniques were used to isolate *Rhizobium* members from their natural niche, the nodule, and their identification was routinely performed via 16S rRNA gene and different housekeeping genes. Biotechnological advances based on the use of omics-based technologies showed that species belonging to the genus *Rhizobium* are keystone taxa in several diverse environments, such as forests, agricultural land, Arctic, and Antarctic ecosystems, contaminated soils and plant-associated microbiota. In this chapter, we will summarize the advances in the study of the *Rhizobium* genus, from culturomics strategies to modern omics methodologies, mostly based on next-generation sequencing approaches. These cutting-edge molecular approaches are fundamental in the study of the behavior of *Rhizobium* species in their interaction with Non-leguminous plants, supporting their potential as an ecological alternative to chemical fertilizers in the battle against Climatic Change.

Keywords Rhizobia · Plant growth promotion · Plant microbiome · Biofertilizers · Non-legumes

---

A. Díez-Méndez
Universidad Católica de Ávila, Ávila, Spain
e-mail: alexandra.diez@ucavila.es

E. Menéndez (✉)
Mediterranean Institute for Agriculture, Environment and Development (MED), Institute for Advanced Studies and Research (IIFA), Universidade de Évora, Évora, Portugal
e-mail: esthermenendez@uevora.pt

N. Shrivastava et al. (eds.), *Symbiotic Soil Microorganisms*, Soil Biology 60,
https://doi.org/10.1007/978-3-030-51916-2_16

16.1 The Genus *Rhizobium* at a Glance

The genus *Rhizobium* as known nowadays was firstly described 131 years ago, in 1889 when B. Frank reclassified the *Bacillus radicicola* as named by Beijerinck in 1888 onto the name *Rhizobium leguminosarum*. The name of the genus comes from "Gr. fem. *n. rhiza, a root; Gr. masc. n. bios, life; N.L. neut. n. Rhizobium, that which lives in a root.*" The genus *Rhizobium* is the type genus of the family Rhizobiaceae, which was described by Conn in 1938, and this family belongs to the order Rhizobiales, named many years later, in 2006 by Kuykendall. The species *Rhizobium leguminosarum* is the type species of the genus *Rhizobium*, being USDA2370^{T} its type strain.

According to the List of Prokaryotic names with Standing in Nomenclature—LPSN (Parte 2018; accessed for the last time March 22, 2020) at the time of writing, there are 89 validly published species within the genus *Rhizobium*. In the last 10 years, there has been a boom in describing new species belonging to this genus (Table 16.1) and interestingly, the reclassification of some of the species into newly described genera, such as *Neorhizobium* (Mousavi et al. 2014) and *Pararhizobium* (Mousavi et al. 2015), is becoming more frequent due to the use of the NGS tools (Ormeño-Orrillo et al. 2015; Checcucci et al. 2019; González et al. 2019).

The members of the genus *Rhizobium* were traditionally isolated from legume nodules; nevertheless, in the last years, some reports showed that other sources can be considered for the isolation of *Rhizobium*. Table 16.1 shows in bold characters the newly described species originally isolated from non-leguminous plants. Moreover, the surface of minerals, contaminated soils, and water are also sources of new species of this genus.

It is implicit that the definition of "rhizobium" is based on the ability to elicit the formation of root nodules in leguminous plants and the latter action of fixing nitrogen within these newly formed structures. This symbiotic relationship has been extensively studied (Oldroyd et al. 2011; Wang et al. 2018; Taylor et al. 2020). However, *Rhizobium* has many features apart from the classical function in Nitrogen Fixation, this group of bacteria interacts also with other plants than legumes and has plant growth promotion and biocontrol functions. Member of this genus is able to interact and produce benefits in a wide range of crops, such as the most important grain food crops, including rice, maize, wheat, and barley; vegetable crops, including lettuce, carrots, potatoes, and spinach; or bioenergy crops, such as canola, sunflower, or switchgrass. The functions and properties that *Rhizobium* exert in Non-leguminous plants, especially the ones related to plant growth promotion, are extensively reviewed in the last years (Vargas et al. 2017; Nag et al. 2019; Yoneyama et al. 2019; Velázquez et al. 2019; Mahmud et al. 2020). The involvement of *Rhizobium* in biocontrol actions against pathogens an alleviation of stresses is a not well-studied characteristic; however, there are some studies showing the potential of this feature; for example, Bellabarba et al. (2019) reviewed the importance of plant growth-promoting rhizobia in the alleviation of salt and osmotic stresses as well as in contaminated soils. Volpiano et al. (2019) also reviewed the

Table 16.1 New species of *Rhizobium* validly described in the last 10 years (2010–2020)

Name of the species	Isolated from	References
Rhizobium acidisoli	*Phaseolus vulgaris*	Roman-Ponce et al. (2016)
Rhizobium aegyptiacum	*Trifolium alexandrinum*	Shamseldin et al. (2016)
Rhizobium aethiopicum	*Phaseolus vulgaris*	Aserse et al. (2017)
Rhizobium aggregatum	Hexachlorocyclohexane dump site	(Hirsch and Müller 1986) Kaur et al. (2011)
Rhizobium altiplani	*Mimosa pudica*	Barauna et al. (2016)
Rhizobium alvei	Freshwater river	Sheu et al. (2015)
Rhizobium anhuiense	*Vicia faba* and *Pisum sativum*	Zhang et al. (2015a)
Rhizobium aquaticum	Water of a crater lake	Mathe et al. (2018)
Rhizobium arenae	Sand	Zhang et al. (2017)
Rhizobium azibense	*Phaseolus vulgaris*	Mnasri et al. (2014)
Rhizobium azooxidifex	Arable soils	Behrendt et al. (2016)
Rhizobium bangladeshense	*Lens culinaris*	Rashid et al. (2015)
Rhizobium binae	*Lens culinaris*	Rashid et al. (2015)
Rhizobium calliandrae	*Calliandra grandiflora*	Rincon-Rosales et al. (2013)
Rhizobium capsici	*Capsicum annuum* var. Grossum	Lin et al. (2015)
Rhizobium cauense	*Kummerowia stipulacea*	Liu et al. (2012)
Rhizobium chutanense	*Phaseolus vulgaris*	Huo et al. (2019)
Rhizobium ecuadorense	*Phaseolus vulgaris*	Ribeiro et al. (2015)
Rhizobium endolithicum	Beach sand	Parag et al. (2013)
Rhizobium endophyticum	*Phaseolus vulgaris*	López-López et al. (2010)
Rhizobium esperanzae	*Phaseolus vulgaris*	Cordeiro et al. (2017)
Rhizobium favelukesii	*Medicago sativa*	Torres Tejerizo et al. (2016)
Rhizobium flavum	Soil	Gu et al. (2014)
Rhizobium freirei	*Phaseolus vulgaris*	Dall'Agnol et al. (2013)
Rhizobium gei	***Geum aleppicum***	**Shi et al.** (2016)
Rhizobium grahamii	*Dalea leporina, Leucaena leucocephala* and *Clitoria ternatea*	Lopez-Lopez et al. (2012)
Rhizobium halophytocola	***Rosa rugosa***	**Bibi et al.** (2012)
Rhizobium helianthi	***Helianthus annuum***	**Wei et al.** (2015)
Rhizobium ipomoeae	***Ipomoeba aquatica***	**Sheu et al.** (2016)
Rhizobium jaguaris	*Calliandra grandiflora*	Rincon-Rosales et al. (2013)
Rhizobium laguerreae	*Vicia faba*	Saidi et al. (2014)
Rhizobium lemnae	***Lemna aequinoctialis***	**Kittiwongwattana and Thawai** (2014)
Rhizobium lentis	*Lens culinaris*	Rashid et al. (2015)
Rhizobium leucaenae	*Leucaena leucocephala*	Ribeiro et al. (2012)
Rhizobium marinum	Seawater	Liu et al. (2015)

(continued)

Table 16.1 (continued)

Name of the species	Isolated from	References
Rhizobium mayense	*Calliandra grandiflora*	Rincon-Rosales et al. (2013)
Rhizobium mesoamericanum	*Phaseolus vulgaris*, siratro, cowpea, and *Mimosa pudica*	Lopez-Lopez et al. (2012)
Rhizobium metallidurans	*Anthyllis vulneraria*	Grison et al. (2015)
Rhizobium naphthalenivorans	Sediment of a polychlorinated-dioxin-transforming microcosm	Kaiya et al. (2012)
Rhizobium oryziradicisoryziradicis	***Oryza sativa***	Zhao et al. (2017)
Rhizobium oryziradicis	*Arachis hypogaea*	corrig. Khalid et al. (2015)
Rhizobium paknamense	***Lemna aequinoctialis***	**Kittiwongwattana and Thawai** (2013)
Rhizobium paranaense	*Phaseolus vulgaris*	Dall'Agnol et al. (2014)
Rhizobium petrolearium	Oil-contaminated soil	Zhang et al. (2012)
Rhizobium populi	***Populus euphratica***	**Rozahon et al.** (2014)
Rhizobium puerariae	*Pueraria candollei* var. candollei	Boonsnongcheep et al. (2016)
Rhizobium rhizoryzae	***Oryza sativa***	**Zhang et al.** (2014)
Rhizobium rosettiformans	Hexachlorocyclohexane dump site	Kaur et al. (2011)
Rhizobium smilacinae	***Smilacina japonica***	**Zhang et al.** (2015a, b)
Rhizobium soli	Soil	Yoon et al. (2010)
Rhizobium sophorae	*Sophora flavescens*	Jiao et al. (2015)
Rhizobium sophoriradicis	*Sophora flavescens*	Jiao et al. (2015)
Rhizobium sphaerophysae	*Sphaerophysa salsula*	Xu et al. (2011)
Rhizobium straminoryzae	***Oryza sativa***	**Lin et al.** (2014)
Rhizobium subbaraonis	Beach sand	Ramana et al. (2013)
Rhizobium taibaishanense	*Kummerowia striata*	Yao et al. (2012)
Rhizobium tarimense	River soil	Turdahon et al. (2013)
Rhizobium tubonense	*Oxytropis glabra*	Zhang et al. (2011)
Rhizobium tumorigenes	***Cane gall tumors on thornless blackberry***	**Kuzmanovic et al.** (2018)
Rhizobium vallis	*Phaseolus vulgaris, Mimosa pudica* and *Indigofera spicata*	Wang et al. (2011)
Rhizobium vignae	Nodules of multiple legume species	Ren et al. (2011)
Rhizobium viscosum	Soil	(Gasdorf et al. 1965) Flores-Felix et al. (2017)

(continued)

Table 16.1 (continued)

Name of the species	Isolated from	References
Rhizobium wenxiniae	***Zea mays***	**Gao et al.** (2017)
Rhizobium wuzhouense	***Oryza officinalis***	**Yuan et al.** (2018)
Rhizobium yantingense	Surfaces of weathered rock (purple siltstone)	Chen et al. (2015)
Rhizobium zeae	***Zea mays***	**Celador-Lera et al.** (2017)

Rhizobium species originally isolated from Non-leguminous plants are marked in bold

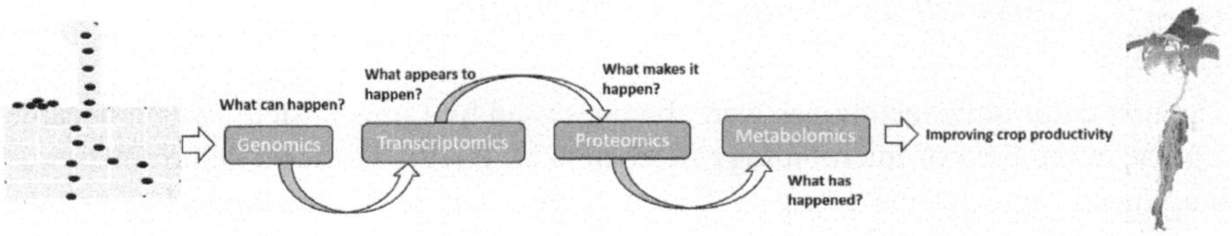

Fig. 16.1 Strategies based on current omics tools that are useful to study of the belowground bacteria–plant–environment interactions

mechanisms and the efficacy of *Rhizobium* and related genera on biocontrol of several pathogen-caused diseases affecting all kinds of plants.

Fortunately, nowadays, there is a plethora of multidisciplinary approaches that can be applied to elucidate aspects that remain unclear (diCenzo et al. 2019), above all the aspects and features attributed to this interesting genus.

16.2 The Study of *Rhizobium* in the Era of the Omics

Omics-based technology is a holistic approach to understand biological systems using predictable tools to know their properties and functions. These omics strategies describe the cell, tissue, and organism features around the dogma central due to the identification of genes (genomics), mRNA (transcriptomics), proteins (proteomics), and metabolites (metabolomics) (Arivaradarajan and Misra 2019). Due to the immense potential of omics approaches and the current competitive prices, scientists from many disciplines began to use them as their primary resource (López-Mondéjar et al. 2017; Menéndez and Paço 2020). In the pursue for a sustainable agriculture, these omics-based technologies have become essential tools to face agricultural challenges (Fig. 16.1).

It is well known that Plant Probiotic Bacteria (PPBs) could be applied as biofertilizers or biopesticides (Menendez and Garcia-Fraile 2017). There is a trend to integrate them as an ecological innovation to face climate change and improving crop productivity without compromising agriculture, which become more sustainable (Compant et al. 2019; Menéndez and Paço 2020). Thus, for the development of eco-friendly agriculture based on the application of safety bacteria, such as those belonging to the genus *Rhizobium*, omics—strategies will be essential tools to improve our knowledge on many features and in the interaction among the inoculant bacteria, the environment, and the host plant. Omics-based techniques should be coordinated with the classical techniques developed to allow the cultivation of microorganisms. This knowledge will help in the selection of a correct inoculum or consortium for each crop.

16.2.1 Classical Techniques: Culturomics

The first culturomic techniques were the ones used by Louis Pasteur, which became a pillar in the field of microbiology (Sarhan et al. 2019). The term culturomics was introduced latter by the group of Didier Raoult and Jean-Christophe Lagier as a strategy to bring more bacterial isolates into cultivation in the laboratories from environmental microbiomes (Lagier et al. 2018). In combination with these culturing approaches, PCR amplification of the ubiquitous 16S ribosomal RNA (rRNA) has been used commonly to identify bacterial isolates (Turner et al. 2013). Classical classification of *Rhizobium* species is based on PCR amplifications of 16S rRNA gene and different housekeeping genes, such as *recA* and *atpD* (Gaunt et al. 2001), amongst others. Nowadays and despite the use of the genome sequences in taxonomical classification, those identification techniques are essential to identify microorganisms, and various reports still endorse these methodologies for the classification of the members of the genus *Rhizobium* (García-Fraile et al. 2007; Kaur et al. 2011; López-López et al. 2010; Rincon-Rosales et al. 2013; Zhang et al. 2012). Moreover, other techniques became complementary to the identification of species of *Rhizobium*, such as the MALDI-TOF mass spectrometry (MALDI-TOF MS). For example, a recent study revealed the usefulness of a combination between MALDI-TOF MS and classical PCR amplification for the taxonomic identification of isolates of *Rhizobium laguerreae* from nodules of *Phaseolus vulgaris* and *Lens culinaris* (Flores-Félix et al. 2019).

The obtention of pure cultures and their correct identification are two essential concerns in the design of biofertilizers (Menéndez and Paço 2020). The members of the genus *Rhizobium* are easy to culture and to identify as well as it is a proven Plant Probiotic Bacteria, which is very interesting for the design of biofertilizers. The most used culture media for the growth of *Rhizobium* is the Yeast Mannitol Agar (YMA; Vincent 1970) and/or the Tryptone Yeast agar (TY; Beringer 1974).

Despite its importance, the "culturomics" has many limitations being the most import one the still limited capacity for the cultivation of some bacterial taxa. Nowadays, the culturome (strains that we are able to culture in laboratory conditions) does not represent the total microbiome (Martiny 2019; Steen et al. 2019). Thus, current studies are focused on the improvement of a variety of factors involved in microbial growth such as culturing conditions (Light/dark, anaerobic/aerobic, proper temperatures), times of incubation, growth media (providing environmental and nutritional conditions), according to the bacteria nutritional requirements (Sarhan et al. 2019; Compant et al. 2019). These studies propose to combine plant-based culture media plus multiple growth factors such as vitamins, amino acids, etc. to improve isolation techniques of plant probiotics, such as *Rhizobium* species (Sarhan et al. 2019).

16.2.2 Genomics

The term genomics was introduced by Tom Roderick in 1986 when mapping the human genes and referred to study of the structure, function, and interaction of the genome, i.e. examine a complete organism set of DNAs, including both non-coding and protein-coding genes (Womack 2019). However, the interest in these techniques has increased in recent years since many disciplines might benefit from their use. Since Plant Probiotics have demonstrated their role supporting benefits effects for plant growth and health, there is an urgent need to bring microbial innovations into agriculture practices (Compant et al. 2019). Thus, the selection of appropriate microbiota inoculants is an urgent need to face current challenges in crop production.

In this sense, Biological Nitrogen Fixation (BNF) is an ecologically and agriculturally process performed by two prokaryotic groups. Among them, the fixation of N_2 gas into ammonia by root nodule bacteria (called rhizobia) is an efficient process in terms of supplying nitrogen to the plant (diCenzo et al. 2019) For decades, researchers have generated massive data to increase crop yield using rhizobial species based on culture-dependent approaches. However, in the first decade of the XXI century, genomic approaches become increasingly, and rhizobial genomes are becoming to be sequenced, according to the proposed minimal standards of the taxonomy of prokaryotes and for the description of new rhizobia and agrobacteria (Chun et al. 2018; De Lajudie et al. 2019). These strategies have resulted in an increased trend on genome sequencing projects by different methodologies to the present. Genomes Online Database (GOLD) is a World Wide Web resource, powered by DOE Joint Genome Institute, for comprehensive access to information regarding genome and metagenome sequencing projects, and their associated metadata around the world (https://gold.jgi.doe.gov/), which showed the increased data from genome sequencing projects over time (Fig. 16.2).

Public databases as NCBI (National Center for Biotechnology Information) and GOLD show an increasing number of reports showing the genome sequences of strains belonging to the genus *Rhizobium*. At the time of writing, there are 691 (NCBI) and 521 (JGI GOLD) sequenced genomes from different species in both databases. All these data are improved due to technological advances, which led to the development of the second-generation sequencing (SGS) technologies improving limitation from Sanger sequencing, such as read lengths, assembly, and determination of complex genomics regions, methylation detection, and isoform

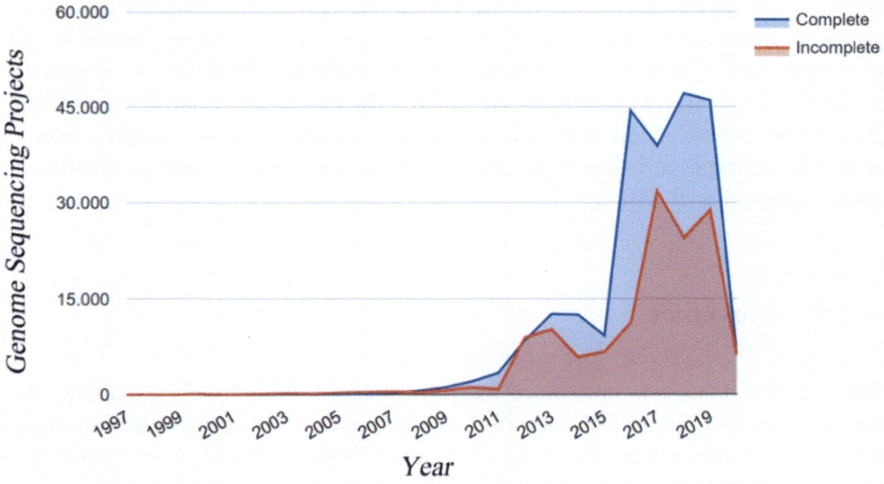

Fig. 16.2 Genome Sequencing Project from the last decades (modified from GOLD Database)

detection. Nowadays, this methodology has evolved using the next-generation platforms such as Illumina, Roche 454, Ion Torrent, and PacBio (Rhoads and Au 2015; Smits 2019).

Recent studies based on genomics research have demonstrated that the genus *Rhizobium* forms part of the core microbiome of many plants (Bulgarelli et al. 2015; Yeoh et al. 2017; Oberholster et al. 2018; Pérez-Jaramillo et al. 2019). Besides, NGS platforms such as Illumina and PacBio are nowadays indispensable in the classification via genomes of the members of the genus *Rhizobium* (Ormeño-Orrillo et al. 2015; González et al. 2019). Some studies already report the use of PacBio (longer reads than Illumina add more possibilities of complete the genome) to generate genomes of novel species, such as *Rhizobium jaguaris* CCGE525^T isolated from *Calliandra grandiflora* nodules (Servín-Garcidueñas et al. 2019) or completing genomes sequence such as *Rhizobium* sp. strain 11515TR isolated from tomato rhizosphere cultivated in the Philippines (Montecillo et al. 2018) and novel non-nodulating *Rhizobium* species isolated from *Agave americana* (Ruíz-Valdiviezo et al. 2017), amongst others.

16.2.3 Proteomics

Proteins are essential molecules that have direct involvement in cellular function. The term "proteome" was introduced by Marc Wilkins in 1994 (Wasinger et al. 1995), defining it as the study of the entire range of proteins in a single cell. The "proteomics" is focused on the identification and characterization of proteins to understand physiological events (Zhao and Lin 2014).

Plant proteomics will be a crucial research tool in developing new technologies to further improve agricultural production and its sustainability. Focusing on the *Rhizobium*-legume symbiosis perspective, there are two types of approaches involving proteomics: (i) the study of the legume root proteome after the inoculation with their respective microsymbiont and, (ii) the studies focused on the *Rhizobium* partner (Larrainzar and Wienkoop 2017).

Regarding the first approach, proteomics methodologies are focused on the identification of proteins in the early stages of symbiosis. Morris and Djordjevic (2001) focused on roots proteome of clover cultivar Woogenellup, and the results showed that ethylene levels are upregulated during the early stages of infection (48 h) by the inoculation of the bacterial strain ANU843 identified as *R. leguminosarum* bv trifolii, but that does not turn in the induction of common pathogenesis mechanisms. However, inoculation with a rhizobial non-nodulating bacterial as the strain ANU743 produced aberrant nodules. This phenotype may be related to the induction of alpha-fucosidase by the bacterial strain on the roots of this clover.

On the other hand, Irar et al. (2014) reported a proteomic approach relating to nodule response to drought on *Pisum sativum*. Pea plants were inoculated with a strain of *R. leguminosarum* and grown under "normal well-irrigated" conditions, and the other was subjected to drought issues. The results showed a total of 18 proteins expressed during drought stress: 11 are encoded by *Rhizobium leguminosarum* and seven nodule proteins encoded by *Pisum sativum*. These proteins are related to flavonoid metabolism, sulfur metabolism, and RNA-binding proteins. All these data provide new targets to improve legumes tolerance to drought.

In the second approach, proteomics strategies have been used to identify specific proteins during the symbiosis comparing free-living *Rhizobium* cells vs bacteroids, their nitrogen fixation symbiotic forms. Despite the importance of these approaches, authors have focused their study on model organisms, such as *Sinorhizobium* or *Bradyrhizobium* species. Tullio et al. (2019) showed that this omic-based methodology is useful to identify proteins related to tolerance to environmental stress, such as soil acidity, using the bacterial strain *Rhizobium freirei* PRF81. Thus, further studies should be performed to understand the proteomic side of the genus *Rhizobium*.

16.2.4 Metabolomics

Metabolome includes a wide range of molecules such as amino acids, peptides, carbohydrates, nucleic acids, organic acids, polyphenols, alkaloids, minerals, etc.; all chemical compounds in a cell, which are metabolized or synthesized (Pandey and Dubey 2019).

In the last decades, the genus *Rhizobium* has been tested for growth promotion of Non-leguminous plants, such as maize, rice, oat, and others cereals (Mia and Shamsuddin 2010; Nag et al. 2019). Thus, in the biofertilizer-technology, *Rhizobium* species are becoming sustainable tools for agriculture of nonlegumes and legumes by substituting agrochemicals gradually. Therefore, the identification of their metabolome is needed because they are the key in the symbiotic relationship, rhizosphere, soil niche, and communications with other organisms (Jacoby et al. 2018).

A recent study aimed to identify the exo-metabolomes produced by *Rhizobium etli* CFN42^T, *Rhizobium leucaenae* CFN299^T, *Rhizobium tropici* CIAT899^T, *Rhizobium phaseoli* Ch24-10 from free-living culture by nuclear magnetic resonance (Montes-Grajales et al. 2019). The exometabolomic profile was carried out growing cultures using minimal medium containing sucrose and glutamate. The results show that there is a different pattern among the bacterial strains, being acetone and C4-dicarboxylate the only compounds secreted by all of them. It has been observed that this amino acid stimulates nitrogen fixation of bacteroids isolated from soybean root nodules (Kouchi et al. 1991). Ornithine was only detected in the culture supernatant of *R. tropici* CIAT 899^T. It has been observed that this compound is involved in symbiotic efficiency and resistance to stress conditions, such as acidity (Rojas-Jiménez et al. 2005; Vences-Guzmán et al. 2011).

There are few studies about the *Rhizobium* metabolomes; therefore, we should highlight the necessity of more works on this, aiming to achieve more knowledge on the role of the metabolome of *Rhizobium* species in improving crop performance.

16.3 *Rhizobium* and co.: Interactions with Non-leguminous Plants

It is well known that the Green Revolution improved crops yield saving people from starvation. The objectives of this revolution were focused on applying a massive amount of chemical fertilizers for growing a single crop. However, global crop production cause not only adverse effects on the environment, such as climate change, pollute fresh and marine water, loss of biodiversity, and alterations of biogeochemical cycles, but also harmful impacts on human health (García-Fraile et al. 2017). Although, agriculture is an important pillar that maintenances society staple food, there are a billion people who still suffer inadequate food supplies and have unhealthy diets (Tilman and Clark 2014). Thus, EAT-Lancet Commission revealed that global food systems need a transformation to focus mainly on environmental sustainability of food production and the health consequences of final consumption (Willett et al. 2019).

Plant probiotics are eco-friendly tools that help face agriculture challenge. However, before using a biofertilizer, it is necessary to know the effects of the microorganisms on the environment and human health (García-Fraile et al. 2012; Menéndez and Paço 2020). For that reason, there is a plethora of studies of new ecological alternatives based on Plant Probiotics. Amongst them, the genus *Rhizobium* represents a group of interest not only for their biosafety features after decades of legume inoculations (Bashan 1998; Bhattacharjee et al. 2008) but also because of their potential to improve the performance of Non-leguminous crops (Celador-Lera et al. 2017; Menendez and Garcia-Fraile 2017; Nag et al. 2019).

Different reports have been shown that these probiotic bacteria have diverse direct and indirect plant growth promotion (PGP) mechanisms in vitro, such as phosphate solubilization, siderophores, or the production of phytohormones as acid indole acetic IAA (Table 16.2). These potential abilities result in the better development of cereal and horticulture crops (Table 16.2). For example, the improvement of grain yield and plant growth in *Oryza sativa* produced by the strain E11 identified as *Rhizobium leguminosarum* bv *trifolii* under gnotobiotic and field conditions (Yanni et al. 2001). Few years later, a report by Hafeez et al. (2004) showed that the same strain (E11) was able to improve dry root weight and root length of cotton (*Gossypium hirsutum)* under growth room conditions. Moreover, the same strain has been used with excellent results in rice crops grown under field conditions (Yanni and Dazzo 2010; Jha et al. 2020). These pieces of evidence may be related to the bacterial ability to synthesize phytohormones.

Other example is the enhance in production of a variety of crops belonging to the genus *Brassica* by two *Rhizobium* species. First, seed inoculations of *B. campestris* and *B. napus* with *R. leguminosarum* VF39SM improved early seedling root growth under chamber conditions (Noel et al. 1996). Second, *R. rubi* increased chlorophyll content, quality, and plant weight under field conditions (Yildirim et al. 2011). Both strains showed PGP in vitro mechanisms associated with the production of phytohormones like IAA and/or cytokinin and resulted in beneficial to *Brassica* crops.

Root colonization is an essential step of a productive relationship between the macro and microsymbiont (Compant et al. 2019). For this attachment, bacteria must produce a variety of surface polysaccharides, being cellulose a synthesized widespread among the genus *Rhizobium* and the principal component of their biofilms (Robledo et al. 2012). In this sense, some reports showed the behavior of GFP-tagged strains on the surface of the roots by using fluorescence microscopy, such as spinach (Jiménez-Gómez et al. 2018), strawberry (Flores-Félix et al. 2015), tomato, and pepper (García-Fraile et al. 2012) and carrots and lettuce (Flores-Félix et al. 2013). Besides, all these bacterial strains showed in vitro PGP, which result in the improvement of plant growth increasing different parameters as nutritional values, number of flowers, chlorophyll content, and so on (Table 16.1). Interestingly, the work presented by Marks et al. (2015) suggest that the presence of *Rhizobium* itself might be not necessary in some occasions. The addition of LCOs extracted from *R. tropici* CIAT 899 to an *Azospirillum*-based inoculant, to the seed or by spraying, increases several parameters in maize cultivated on Brazilian soils.

However, there is enough evidence to claim that the best performance *Rhizobium* inoculants are those able to enter and colonize internal tissues and persist there. This should be addressed with microscopy or other approaches (Romano et al. 2020). Thus, plant growth promotion (PGP) *Rhizobium* can be applied in agriculture because of their capability to confer beneficial effects on plant growth also, their abilities to colonize the rhizosphere and the roots efficiently and their safety for human and the environments.

Table 16.2 *Rhizobium* species described as PGPB of non-leguminous plants

Species	Strain	In vitro PGP mechanisms	Nonlegume host	Parameters improved	References
Rhizobium sp.	Br5	IAA	*Gossypium hirsutum*	Plant height Boll weight and number of bolls per plant Seed cotton yield	Qureshi et al. (2019)
Rhizobium sp.	nd	IAA Siderophore production	*Eruca sativa*	Root system	Rubio-Canalejas et al. (2016)
Rhizobium sp.	YAS34	Exopolysaccharide production	*Heliant hus annum*	Growth promotion N uptake	Alami et al. (2000)
Rhizobium sp.	PEPV13	IAA Siderophore production Phosphate solubilization	*Dianthus caryophyllus*	Plant growth in the first stages of development	Menéndez et al. (2016)
R. leguminosarum bv. phaseoli	LR30/MR2	IAA Catalase Exopolysaccharide under simulated drought conditions	*Triticum aestivum*	Plant growth Biomass Drought stress	Hussain et al. (2014)
R. leguminosarum bv viciae	USDA2370^T	Phytohormones and increased colonization	*Oryza sativa*	SDW Plant height Shoot N Grain yield	Chi et al. (2005)
R. leguminosarum bv. trifolii	IAT168	Probably phytohormones	*Triticum aestivum*	Shoot dry matter Grain yield	Hilali et al. (2001)
R. leguminosarum	*VF39SM*	*IAA.* *Cytokinin.*	*Brassica campestris cv. Tobin*	Root length. Seedling growth.	Noel et al. (1996)
			Brassica napus cv. Westar		
R. leguminosarum bv phaseoli	P31	Phosphate solubilization. IAA	*Zea mays*	Plant height	Chabot et al. (1996)
R. leguminosarum bv phaseoli	R1	Phosphate solubilization. IAA	*Lactuca sativa*	Dry weight. Increased lettuce P concentration.	

R. leguminosarum bv trifolii	PEPT01	IAA. Root colonization.	*Solanum lycopersicum* var. cerasiformein and *Capsicum annuum*	Dry weight. Fresh weight. Number of flowers. Nutritional value (P, K, and Mg).	García-Fraile et al. (2012)
R. leguminosarum bv trifolii	E11	Phosphate solubilization. IAA	*Oryza sativa* L.	Grain yield. Plant growth.	Yanni et al. (2001)
R. leguminosarum bv trifolii	Consortia 2/4 strains	Phytohormones	*Oryza sativa* L.	Straw biomass. Reduced N-fertilizers	Yanni and Dazzo (2010)
R. leguminosarum sv. trifolii	TV-13	IAA	*Lactuca sativa*	Progressive damages in seed vigor and growth	Schlindwein et al. (2008)
R. leguminosarum sv. phaseoli/viciae	R1-R5	IAA	Wheat greenhouse	Growth NPK uptake	Etesami et al. (2009)
R. leguminosarum bv trifolii	E11	IAA	*Gossypium hirsutum L*	Dry root weight. Root length.	Hafeez et al. (2004)
R. leguminosarum	PEPV16	Siderophores. IAA. Phosphate solubilization. Root colonization. Cellulose.	*Fragraria* × *ananassa*	Number of stolons, flowers, and fruits. Nutritional value (Fe, Zn, Mn y Mo). More citric acid and malic acid.	Flores-Félix et al. (2015)
			Daucus carota *Lactuca sativa*	Dry weight. Nutritional values (N, P, Ca, K, S).	Flores-Félix et al. (2013)
R. laguerreae	PEPV40	Siderophores. IAA. Phosphate solubilization. Root colonization. Cellulose	*Spinacia oleracea*	Number of leaves. Leaf size. Fresh weight. Dry weight. Chlorophyll content. Nutritional value (N, P, K, Mg, Ca)	Jiménez-Gómez et al. (2018)

(continued)

Table 16.2 (continued)

Species	Strain	In vitro PGP mechanisms	Nonlegume host	Parameters improved	References
R. etli	CFN42	Not data	*Zea mays*	Root weight. Shoot weight and length.	Gutiérrez-Zamora and Martínez-Romero (2001)
R. rubi	Not data	Phosphate solubilization.	*Brassica oleracea* L., var. italica	Nutritional value Mn, Fe, Cu, Ca. Increase yield. Plant weight, Chlorophyll content.	Yildirim et al. (2011)

16.4 Finding *Rhizobium* in the Microbiomes associated with nonlegumes

The members of the order Rhizobiales and the family Rhizobiaceae are composed of diazotrophic bacteria usually living in the soils associated with plant roots. There are some exceptions, such as *Pseudorhizobium pelagicum* R1-200B4^{T}, which was isolated from the Mediterranean Sea (Kimes et al. 2015), *Rhizobium daejeonense* L61^{T}, which was isolated from a cyanide treatment bioreactor (Quan et al. 2005) or *Rhizobium alvei* TNR-22^{T}, which was isolated from a freshwater river (Sheu et al. 2015), amongst others. Recent studies based on computational inference (omics-based) showed that the order Rhizobiales is a keystone taxon in several environments, such as forests, agricultural land, Arctic, and Antarctic ecosystems, contaminated soils and plant-associated microbiota (Banerjee et al. 2018; LeBlanc and Crouch 2019). More specifically, some of these works identify the genus *Rhizobium* as a keystone taxon in core microbiomes of several plant crops rhizospheres, such as tropical crops, sunflower, and sorghum (Bulgarelli et al. 2015; Yeoh et al. 2017; Oberholster et al. 2018), as well as their well-known presence and functions on legume nodule microbiome (Velázquez et al. 2017; Martínez-Hidalgo and Hirsch 2017; Cardoso et al. 2018; Muresu et al. 2019; Velázquez et al. 2019; Zheng et al. 2020). Other genera belonging to the order Rhizobiales and closely related to the genus *Rhizobium*, such as *Agrobacterium, Bradyrhizobium,* and *Devosia*, were found forming part of the maize rhizospheric core microbiome in long- term assays (Walters et al. 2018). According to Walters et al. (2018), members of the family Rhizobiaceae and some other Rhizobiales members appeared to form part of the heritable fraction of maize rhizosphere microbiome.

Interestingly, over the last years, many reports and studies have been published about the presence of *Rhizobium* and related genera in the rhizosphere, endosphere, and phyllosphere of Non-leguminous crops. This is the result of the high interest in the exploration of the crop microbiomes with the aim of finding native rhizobial and non-rhizobial bacteria that might be endophytes to produce benefits in those crops and also, to be friendly with the microbiomes (Menéndez and Paço 2020).

Rhizobium, which fixes Nitrogen within legume nodules, and other endophytic diazotrophs were usually found in Non-leguminous crops (Yoneyama et al. 2017, 2019). Using *nifH* gene amplification and cloning (not HiSeq/MiSeq or pyrosequencing, but somehow comparable) from various, some works reported the existence of members of the genus *Rhizobium* in several plant structures and compartments, such as *Rhizobium* sp. in roots and stems of maize plants grown in fields (Roesch et al. 2008), *R. etli* in roots of one cultivar of sorghum grown with low and high doses of nitrogen fertilizers (Rodrigues Coelho et al. 2008), *R. leguminosarum* in sweet potato tubers (Terakado-Tonooka et al. 2008), *R. helanshanense* in roots and shoots of switchgrass (Bahulikar et al. 2014) and *R. daejeonense* in stems and roots of sugarcane cultivated in Japan and Brazil (Thaweenut et al. 2011).

Using NGS techniques, Lay et al. (2018) performed a comparative analysis between rhizosphere and endosphere of canola, pea, and wheat grown in Canadian prairies. *R. leguminosarum* was found in different relative abundancies in the endospheres and rhizospheres of the three crops.; however, related members of the family Rhizobiaceae, such as *Agrobacterium* sp. was only found associated with endosphere of canola and wheat, but not pea (Lay et al. 2018).

Moreover, Essel et al. (2019), in a study focused on the selection of adequate agronomic practices for agricultural soils, found that *Rhizobium* is only present in the rhizospheric soils of wheat and pea cultivated in rotation and not in the bulk soils. This finding represents that *Rhizobium* is more abundant in the soils closely attached to the roots, exposing the specialized function of the genus Rhizobium in its interaction with crops.

Jha et al. (2020) revealed *Rhizobium* as a dominant OTU amongst other diazotrophs in rice fields. Other related OTUs, such as unclassified Rhizobiales and unclassified Rhizobiaceae, as well as other OTUs of other rhizobia were also found with less predominance. The addition of a strain of *Rhizobium leguminosarum* as inoculant with or without a low dose of urea fertilizers reduced the OTU richness; *Rhizobium* continued as relevant OTU but other OTUs belonging to Alpha-Proteobacteria are less abundant. Nevertheless, the beneficial effects produced by inoculation and inoculation+lowNdose on the rice growth and yield upon inoculation is the proof of *Rhizobium* addition is beneficial for this crop, suggesting that the communities are not negatively affected, at least regarding the plant.

Most of the studied crops are those cereals or similar crops with agro-economic importance. Nevertheless, in the last years, some research has been performed in microbiomes of vegetable plants, trees, and shrubs, all of them also with agro-economic importance. Members of the genus *Rhizobium* and related genera are also present in those microbiomes, due to their importance in plant growth promotion and biocontrol. For example, *Rhizobium* spp. were detected in bulk and rhizospheric soils of cucumber plants (Jia et al. 2019). Using DGGE and not amplicon sequencing or metagenomics, Marasco et al. (2013) found various *Rhizobium* species in grapevine roots, both in the rhizosphere and in the inner tissues.

Remarkably, members of the complex *Allorhizobium–Rhizobium/Pararhizobium–Rhizobium* were found only in *Xylella*-infected and Xylella-uninfected olive trees of the variety "Leccino" (tolerant to *Xylella* infection). They were found by NGS in the phyllosphere and endosphere of leaves and branches (Vergine et al. 2020). Rhizobium was found only in the resistant and not in the susceptible cultivar, suggesting that this taxon might exert some roles related to this cultivar resistance to pathogens.

16.5 Concluding Remarks

Rhizobium species are essential in the development of ecological agriculture not only for its ability to fix nitrogen but also its qualities as plant probiotic bacteria for different agronomic crops. Moreover, their easy growth under laboratory conditions makes them an ecological alternative to chemicals fertilizers.

Molecular approaches based on "-omics" revealed their selves as essential tools for the study of this particular genus. Overall, all these techniques showed that members of the genus *Rhizobium* and related ones are always present in the tissues or the rhizospheres of the Non-leguminous crops, suggesting their functions are essential for the plants and the environment.

To understand their role in those diverse ecosystems where it has been found, further studies are necessary with a combination of methodologies, improving the techniques of isolation as well as the molecular strategies. In this way, the design of biofertilizers based on *Rhizobium* strains alone or in combination with other beneficial bacteria will be more effective from the economic point of view as well as sustainable with our agricultural systems.

Acknowledgments EM acknowledges the FCT research contract from the Individual Call to Scientific Employment Stimulus 2017 (CEECIND/00270/2017). This work is funded by National Funds through FCT—Foundation for Science and Technology under the Project UIDB/05183/2020.

References

Alami Y, Achouak W, Marol C, Heulin T (2000) Rhizosphere soil aggregation and plant growth promotion of sunflowers by an exopolysaccharide-producing *Rhizobium* sp. strain isolated from sunflower roots. Appl Environ Microbiol 66:3393–3398

Arivaradarajan P, Misra G (2019) Omics approaches, technologies and applications, Omics approaches, technologies and applications: integrative approaches for understanding OMICS data. Springer, New York. https://doi.org/10.1007/978-981-13-2925-8

Aserse AA, Woyke T, Kyrpides NC, Whitman WB, Lindstrom K (2017) Draft genome sequence of type strain HBR26^T and description of *Rhizobium aethiopicum* sp. nov. Stand Genomic Sci 12:14

Bahulikar RA, Torres-Jerez I, Worley E, Craven K, Udvardi MK (2014) Diversity of nitrogen-fixing bacteria associated with switchgrass in the native tallgrass prairie of northern Oklahoma. Appl Environ Microbiol 80(18):5636–5643

Banerjee S, Schlaeppi K, van der Heijden MG (2018) Keystone taxa as drivers of microbiome structure and functioning. Nat Rev Microbiol 16(9):567–576

Barauna AC, Rouws LF, Simoes-Araujo JL, Dos Reis Junior FB, Iannetta PP, Maluk M, Goi SR, Reis VM, James EK, Zilli JE (2016) *Rhizobium altiplani* sp. nov., isolated from effective nodules on *Mimosa pudica* growing in untypically alkaline soil in Central Brazil. Int J Syst Evol Microbiol 66:4118–4124

Bashan Y (1998) Inoculants of plant growth-promoting bacteria for use in agriculture. Biotechnol Adv 1:729–770. https://doi.org/10.1016/S0734-9750(98)00003-2

Behrendt U, Kampfer P, Glaeser SP, Augustin J, Ulrich A (2016) Characterization of the N2O-producing soil bacterium *Rhizobium azooxidifex* sp. nov. Int J Syst Evol Microbiol 66:2354–2361

Bellabarba A, Fagorzi C, diCenzo GC, Pini F, Viti C, Checcucci A (2019) Deciphering the symbiotic plant microbiome: translating the Most recent discoveries on rhizobia for the improvement of agricultural practices in metal-contaminated and high saline lands. Agronomy 9(9):529

Beringer JE (1974) R factor transfer in *Rhizobium leguminosarum*. Microbiology 84(1):188–198

Bhattacharjee RB, Singh A, Mukhopadhyay SN (2008) Use of nitrogen-fixing bacteria as biofertiliser for non-legumes: prospects and challenges. Appl Microbiol Biotechnol. https://doi.org/10.1007/s00253-008-1567-2

Bibi F, Chung EJ, Khan A, Jeon CO, Chung YR (2012) *Rhizobium halophytocola* sp. nov., isolated from the root of a coastal dune plant. Int J Syst Evol Microbiol 62:1997–2003

Boonsnongcheep P, Prathanturarug S, Takahashi Y, Matsumoto A (2016) *Rhizobium puerariae* sp. nov., an endophytic bacterium from the root nodules of the medicinal plant *Pueraria candollei* var. candollei. Int J Syst Evol Microbiol 66:1236–1241

Bulgarelli D, Garrido-Oter R, Münch PC, Weiman A, Dröge J, Pan Y, McHardy AC, Schulze-Lefert P (2015) Structure and function of the bacterial root microbiota in wild and domesticated barley. Cell Host Microbe 17(3):392–403

Cardoso P, Alves A, Silveira P, Sá C, Fidalgo C, Freitas R, Figueira E (2018) Bacteria from nodules of wild legume species: phylogenetic diversity, plant growth promotion abilities and osmotolerance. Sci Total Environ 645:1094–1102

Celador-Lera L, Menendez E, Peix A, Igual JM, Velazquez E, Rivas R (2017) *Rhizobium zeae* sp. nov., isolated from maize (*Zea mays* L.) roots. Int J Syst Evol Microbiol 67:2306–2311

Chabot R, Antoun H, Cescas MP (1996) Growth promotion of maize and lettuce by phosphate-solubilizing *Rhizobium leguminosarum* biovar. Phaseoli. Plant Soil 184:311–321. https://doi.org/10.1007/BF00010460

Checcucci A, Perrin E, Bazzicalupo M, Mengoni A (2019) Genomic diversity and evolution of rhizobia. In: Das S, Dash HR (eds) Microbial diversity in the genomic era. Elsevier, Amsterdam, the Netherlands, pp 37–46

Chen W, Sheng XF, He LY, Huang Z (2015) *Rhizobium yantingense* sp. nov., a mineral-weathering bacterium. Int J Syst Evol Microbiol 65:412–417

Chi F, Shen SH, Cheng HP, Jing YX, Yanni YG, Dazzo FB (2005) Ascending migration of endophytic rhizobia, from roots to leaves, inside rice plants and assessment of benefits to rice growth physiology. Appl Environ Microbiol 71(11):7271–7278

Chun J, Oren A, Ventosa A, Christensen H, Arahal DR, da Costa MS, Rooney AP, Yi H, Xu XW, De Meyer S, Trujillo ME (2018) Proposed minimal standards for the use of genome data for the taxonomy of prokaryotes. Int J Syst Evol Microbiol 68:461–466

Compant S, Samad A, Faist H, Sessitsch A (2019) A review on the plant microbiome: ecology, functions and emerging trends in microbial application. J Adv Res 19:20–37

Cordeiro AB, Ribeiro RA, Helene LCF, Hungria M (2017) *Rhizobium esperanzae* sp. nov., a N_2-fixing root symbiont of *Phaseolus vulgaris* from Mexican soils. Int J Syst Evol Microbiol 67:3937–3945

Dall'Agnol RF, Ribeiro RA, Ormeno-Orrillo E, Rogel MA, Delamuta JR, Andrade DS, Martinez-Romero E, Hungria M (2013) *Rhizobium freirei* sp. nov., a symbiont of *Phaseolus vulgaris* that is very effective at fixing nitrogen. Int J Syst Evol Microbiol 63:4167–4173

Dall'Agnol RF, Ribeiro RA, Delamuta JR, Ormeno-Orrillo E, Rogel MA, Andrade DS, Martinez-Romero E, Hungria M (2014) *Rhizobium paranaense* sp. nov., an effective N_2-fixing symbiont of common bean (*Phaseolus vulgaris* L.) with broad geographical distribution in Brazil. Int J Syst Evol Microbiol 64:3222–3229

De Lajudie PM, Andrews M, Ardley J, Eardly B, Jumas-Bilak E, Kuzmanović N, Lassalle F, Lindström K, Mhamdi R, Martínez-Romero E, Moulin L (2019) Minimal standards for the description of new genera and species of rhizobia and agrobacteria. Int J Syst Evol Microbiol 69 (7):1852–1863

diCenzo GC, Zamani M, Checcucci A, Fondi M, Griffitts JS, Finan TM, Mengoni A (2019) Multidisciplinary approaches for studying rhizobium–legume symbioses. Can J Microbiol 65 (1):1–33

Essel E, Xie J, Deng C, Peng Z, Wang J, Shen J, Xie J, Coulter JA, Li L (2019) Bacterial and fungal diversity in rhizosphere and bulk soil under different long-term tillage and cereal/legume rotation. Soil Tillage Res 19:104302. https://doi.org/10.1016/j.still.2019.104302

Etesami H, Alikhani HA, Jadidi M, Aliakbari A (2009) Effect of superior IAA producing rhizobia on N, P, K uptake by wheat grown under greenhouse condition. World Appl Sci J 6 (12):1629–1633

Flores-Félix JD, Menéndez E, Rivera LP, Marcos-García M, Martínez-Hidalgo P, Mateos PF, Martínez-Molina E, Velázquez E, García-Fraile P, Rivas R (2013) Use of *Rhizobium leguminosarum* as a potential biofertilizer for *Lactuca sativa* and *Daucus carota* crops. J Plant Nutr Soil Sci 176:876–882. https://doi.org/10.1002/jpln.201300116

Flores-Félix JD, Marcos-García M, Silva LR, Menéndez E, Martínez-Molina E, Mateos PF, Velázquez E, García-Fraile P, Andrade P, Rivas R (2015) *Rhizobium* as plant probiotic for strawberry production under microcosm conditions. Symbiosis 67:25–32. https://doi.org/10.1007/s13199-015-0373-8

Flores-Felix JD, Ramirez-Bahena MH, Salazar S, Peix A, Velazquez E (2017) Reclassification of *Arthrobacter viscosus* as *Rhizobium viscosum* comb. nov. Int J Syst Evol Microbiol 67:1789–1792

Flores-Félix JD, Sánchez-Juanes F, García-Fraile P, Valverde A, Mateos PF, Gónzalez-Buitrago JM, Velázquez E, Rivas R (2019) *Phaseolus vulgaris* is nodulated by the symbiovar viciae of several genospecies of *Rhizobium laguerreae* complex in a Spanish region where *Lens culinaris* is the traditionally cultivated legume. Syst Appl Microbiol 42:240–247. https://doi.org/10.1016/j.syapm.2018.10.009

Gao JL, Sun P, Wang XM, Lv FY, Mao XJ, Sun JG (2017) *Rhizobium wenxiniae* sp. nov., an endophytic bacterium isolated from maize root. Int J Syst Evol Microbiol 67:2798–2803

García-Fraile P, Rivas RL, Willems A, Peix A, Martens M, Martínez-Molina E, Mateos PF, Velázquez E, Es E (2007) *Rhizobium cellulosilyticum* sp. nov., isolated from sawdust of *Populus alba*. Int J Syst Evol Microbiol. https://doi.org/10.1099/ijs.0.64680-0

García-Fraile P, Carro L, Robledo M, Ramírez-Bahena MH, Flores-Félix JD, Velazquez E (2012) *Rhizobium* promotes non-legumes growth and quality in several production steps: towards a biofertilization of edible raw vegetables healthy for humans. PLoS One 7:e38122. https://doi.org/10.1371/journal.pone.0038122

García-Fraile P, Menéndez E, Celador-Lera L, Díez-Méndez A, Jiménez-Gómez A, Marcos-García M, Cruz-González XA, Martínez-Hidalgo P, Mateos PF, Rivas R (2017) Bacterial probiotics: a truly green revolution. In: Probiotics and plant health. Springer, Singapore, pp 131–162. https://doi.org/10.1007/978-981-10-3473-2_6

Gasdorf HJ, Benedict RG, Cadmus MG, Anderson RF, Jacksonv RW (1965) Polymer-producing species of *Arthrobacter*. J Bacteriol 90(1):147–150

Gaunt MW, Turner SL, Rigottier-Gois L, Lloyd-Macgilp SA, Young JPW (2001) Phylogenies of atp D and recA support the small subunit rRNA-based classification of rhizobia. Int J Syst Evol Microbiol 51:2037–2048. https://doi.org/10.1099/00207713-51-6-2037

González V, Santamaría RI, Bustos P, Pérez-Carrascal OM, Vinuesa P, Juárez S, Romero D (2019) Phylogenomic rhizobium species are structured by a continuum of diversity and genomic clusters. Front Microbiol 10:910

Grison CM, Jackson S, Merlot S, Dobson A, Grison C (2015) *Rhizobium metallidurans* sp. nov., a symbiotic heavy metal resistant bacterium isolated from the Anthyllis vulneraria Zn-hyperaccumulator. Int J Syst Evol Microbiol 65:1525–1530

Gu T, Sun LN, Zhang J, Sui XH, Li SP (2014) *Rhizobium flavum* sp. nov., a triazophos-degrading bacterium isolated from soil under the long-term application of triazophos. Int J Syst Evol Microbiol 64:2017–2022

Gutiérrez-Zamora M, Martínez-Romero E (2001) Natural endophytic association between *Rhizobium etli* and maize (*Zea mays* L.). J Biotechnol 91:117–126. https://doi.org/10.1016/S0168-1656(01)00332-7

Hafeez FY, Safdar ME, Chaudhry AU, Malik KA (2004) Rhizobial inoculation improves seedling emergence, nutrient uptake and growth of cotton. Aust J Exp Agric 44:617–622. https://doi.org/10.1071/EA03074

Hilali A, Prévost D, Broughton WJ, Antoun H (2001) Effets de l'inoculation avec des souches de *Rhizobium leguminosarum* biovar trifolii sur la croissance du blé dans deux sols du Maroc. Can J Microbiol 47(6):590–593

Huo Y, Tong W, Wang J, Wang F, Bai W, Wang E, Shi P, Chen W, Wei G (2019) *Rhizobium chutanense* sp. nov., isolated from root nodules of *Phaseolus vulgaris* in China. Int J Syst Evol Microbiol 69:2049–2056

Hussain MB, Zahir ZA, Asghar HN, Asgher M (2014) Can catalase and exopolysaccharides producing rhizobia ameliorate drought stress in wheat? Int J Agric Biol 16:3–13

Irar S, González EM, Arrese-Igor C, Marino D (2014) A proteomic approach reveals new actors of nodule response to drought in split-root grown pea plants. Physiol Plant 152:634–645. https://doi.org/10.1111/ppl.12214

Jacoby RP, Martyn A, Kopriva S (2018) Exometabolomic profiling of bacterial strains as cultivated using *Arabidopsis* root extract as the sole carbon source. Mol Plant Microbe Interact 31:803–813. https://doi.org/10.1094/MPMI-10-17-0253-R

Jha PN, Gomaa A, Yanni YG et al (2020) Alterations in the Endophyte-enriched root-associated microbiome of rice receiving growth-promoting treatments of urea fertilizer and *Rhizobium* biofertilizer. Microb Ecol 79:367–382. https://doi.org/10.1007/s00248-019-01406-7

Jia HT, Liu JY, Shi YI, Li DL, Wu FZ, Zhou XG (2019) Characterization of cucumber rhizosphere bacterial community with high-throughput amplicon sequencing. Allelopathy J 47(1):103–112

Jiao YS, Yan H, Ji ZJ, Liu YH, Sui XH, Wang ET, Guo BL, Chen WX, Chen WF (2015) *Rhizobium sophorae* sp. nov. and *Rhizobium sophoriradicis* sp. nov., nitrogen-fixing rhizobial symbionts of the medicinal legume *Sophora flavescens.* Int J Syst Evol Microbiol 65:497–503

Jiménez-Gómez A, Flores-Félix JD, García-Fraile P, Mateos PF, Menéndez E, Velázquez E, Rivas R (2018) Probiotic activities of *Rhizobium laguerreae* on growth and quality of spinach. Sci Rep 8:295. https://doi.org/10.1038/s41598-017-18632-z

Kaiya S, Rubaba O, Yoshida N, Yamada T, Hiraishi A (2012) Characterization of *Rhizobium naphthalenivorans* sp. nov. with special emphasis on aromatic compound degradation and multilocus sequence analysis of housekeeping genes. J Gen Appl Microbiol 58:211–224

Kaur J, Verma M, Lal R (2011) *Rhizobium rosettiformans* sp. nov., isolated from a hexachlorocyclohexane dump site, and reclassification of *Blastobacter aggregatus* Hirsch and Muller 1986 as *Rhizobium aggregatum* comb. nov. Int J Syst Evol Microbiol 61:1218–1225. https://doi.org/10.1099/ijs.0.017491-0

Khalid R, Zhang YJ, Ali S, Sui XH, Zhang XX, Amara U, Chen WX, Hayat R (2015) *Rhizobium pakistanensis* sp. nov., isolated from groundnut (*Arachis hypogaea*) nodules grown in rainfed Pothwar, Pakistan. Anton Van Leeuwenhoek 107:281–290

Kimes NE, López-Pérez M, Flores-Félix JD, Ramírez-Bahena MH, Igual JM, Peix A, Rodriguez-Valera F, Velázquez E (2015) *Pseudorhizobium pelagicum* gen. nov., sp. nov. isolated from a pelagic Mediterranean zone. Syst Appl Microbiol 38(5):293–299

Kittiwongwattana C, Thawai C (2013) *Rhizobium paknamense* sp. nov., isolated from lesser duckweeds (*Lemna aequinoctialis*). Int J Syst Evol Microbiol 63:3823–3828

Kittiwongwattana C, Thawai C (2014) Rhizobium lemnae sp. nov., a bacterial endophyte of lemna aequinoctialis. Int J Syst Evol Microbiol 64(7):2455–2460. https://doi.org/10.1099/ijs.0.061622-0

Kouchi H, Fukai K, Kihara A (1991) Metabolism of glutamate and aspartate in bacteroids isolated from soybean root nodules. J Gen Microbiol 137:2901–2910. https://doi.org/10.1099/00221287-137-12-2901

Kuzmanovic N, Smalla K, Gronow S, Pulawska J (2018) *Rhizobium tumorigenes* sp. nov., a novel plant tumorigenic bacterium isolated from cane gall tumors on thornless blackberry. Sci Rep 8:9051

Lagier JC, Dubourg G, Million M, Cadoret F, Bilen M, Fenollar F, Levasseur A, Rolain JM, Fournier PE, Raoult D (2018) Culturing the human microbiota and culturomics. Nat Rev Microbiol 16(9):540–550

Larrainzar E, Wienkoop S (2017) A proteomic view on the role of legume symbiotic interactions. Front Plant Sci. https://doi.org/10.3389/fpls.2017.01267

Lay CY, Bell TH, Hamel C, Harker KN, Mohr R, Greer CW, Yergeau É, St-Arnaud M (2018) Canola root–associated microbiomes in the Canadian prairies. Front Microbiol 9:1188

LeBlanc N, Crouch JA (2019) Prokaryotic taxa play keystone roles in the soil microbiome associated with woody perennial plants in the genus *Buxus*. Ecol Evol 9:11102–11111. https://doi.org/10.1002/ece3.5614

Lin SY, Hsu YH, Liu YC, Hung MH, Hameed A, Lai WA, Yen WS, Young CC (2014) *Rhizobium straminoryzae* sp. nov., isolated from the surface of rice straw. Int J Syst Evol Microbiol 64:2962–2968

Lin SY, Hung MH, Hameed A, Liu YC, Hsu YH, Wen CZ, Arun AB, Busse HJ, Glaeser SP, Kampfer P et al (2015) *Rhizobium capsici* sp. nov., isolated from root tumor of a green bell pepper (*Capsicum annuum* var. grossum) plant. Antonie Van Leeuwenhoek 107:773–784

Liu TY, Li YJ, Liu XX, Sui XH, Zhang XX, Wang ET, Chen WX, Chen WF, Pulawska J (2012) *Rhizobium cauense* sp. nov., isolated from root nodules of the herbaceous legume *Kummerowia stipulacea* grown in campus lawn soil. Syst Appl Microbiol 35:415–420

Liu Y, Wang RP, Ren C, Lai QL, Zeng RY (2015) *Rhizobium marinum* sp. nov., a malachite-green-tolerant bacterium isolated from seawater. Int J Syst Evol Microbiol 65:4449–4454

López-López A, Rogel MA, Ormeño-Orrillo E, Martínez-Romero J, Martínez-Romero E (2010) *Phaseolus vulgaris* seed-borne endophytic community with novel bacterial species such as *Rhizobium endophyticum* sp. nov. Syst Appl Microbiol 33:322–327. https://doi.org/10.1016/J.SYAPM.2010.07.005

Lopez-Lopez A, Rogel-Hernandez MA, Barois I, Ortiz Ceballos AI, Martinez J, Ormeno-Orrillo E, Martinez-Romero E (2012) *Rhizobium grahamii* sp. nov., from nodules of *Dalea leporina, Leucaena leucocephala* and *Clitoria ternatea*, and *Rhizobium mesoamericanum* sp. nov., from nodules of *Phaseolus vulgaris*, siratro, cowpea and *Mimosa pudica*. Int J Syst Evol Microbiol 62:2264–2271

López-Mondéjar R, Kostovˇcík M, Lladó S, Carro L, García-Fraile P (2017) Exploring the plant microbiome through multi-omics approaches. In: Probiotics in agroecosystem. Springer, Singapore, pp 233–268

Mahmud K, Makaju S, Ibrahim R, Missaoui A (2020) Current progress in nitrogen fixing plants and microbiome research. Plan Theory 9(1):97

Marasco R, Rolli E, Vigani G, Borin S, Sorlini C, Ouzari H, Zocchi G, Daffonchio D (2013) Are drought-resistance promoting bacteria cross-compatible with different plant models? Plant Signal Behavior 8(10):e26741

Marks BB, Megías M, Ollero FJ, Nogueira MA, Araujo RS, Hungria M (2015) Maize growth promotion by inoculation with *Azospirillum brasilense* and metabolites of *Rhizobium tropici* enriched on lipo-chitooligosaccharides (LCOs). AMB Exp 5(1):71

Martínez-Hidalgo P, Hirsch AM (2017) The nodule microbiome: N_2-fixing rhizobia do not live alone. Phytobiomes 1(2):70–82

Martiny AC (2019) High proportions of bacteria are culturable across major biomes. ISME J 4:2125–2128

Mathe I, Toth E, Mentes A, Szabo A, Marialigeti K, Schumann P, Felfoldi T (2018) A new *Rhizobium* species isolated from the water of a crater lake, description of *Rhizobium aquaticum* sp. nov. Antonie Van Leeuwenhoek 111:2175–2183

Menendez E, Garcia-Fraile P (2017) Plant probiotic bacteria: solutions to feed the world. AIMS Microbiol 3(3):502

Menéndez E, Paço A (2020) Is the application of plant probiotic bacterial consortia always beneficial for plants? Exploring synergies between rhizobial and non-rhizobial bacteria and their effects on agro-economically valuable crops. Life 10(3):24

Menéndez E, Escribano-Viana R, Flores-Félix JD, Mateos PF, Rivas R (2016) Rhizobial biofertilizers for ornamental plants. In: Biological nitrogen fixation and beneficial plant-microbe interaction. Springer, Cham, pp 13–21

Mia MAB, Shamsuddin ZH (2010) Rhizobium as a crop enhancer and biofertilizer for increased cereal production greater production of cereals brings forth higher production cost and pollutes the soil environment due to excessive use of chemical fertilizers. Afr J Biotechnol 9:6001–6009

Mnasri B, Liu TY, Saidi S, Chen WF, Chen WX, Zhang XX, Mhamdi R (2014) *Rhizobium azibense* sp. nov., a nitrogen fixing bacterium isolated from root-nodules of Phaseolus vulgaris. Int J Syst Evol Microbiol 64:1501–1506

Montecillo AD, Raymundo AK, Papa IA, Aquino GMB, Rosana ARR (2018) Complete genome sequence of *Rhizobium* sp. strain 11515TR, isolated from tomato rhizosphere in the Philippines. Microbiol Resour Announc 7(7):e00903–e00918

Montes-Grajales D, Esturau-Escofet N, Esquivel B, Martinez-Romero E (2019) Exo-metabolites of *Phaseolus vulgaris*-nodulating rhizobial strains. Metabolites 9:105. https://doi.org/10.3390/metabo9060105

Morris AC, Djordjevic MA (2001) Proteome analysis of cultivar-specific interactions between *Rhizobium leguminosarum* biovar trifolii and subterranean clover cultivar Woogenellup. Electrophoresis 22:586–598. https://doi.org/10.1002/1522-2683

Mousavi SA, Osterman J, Wahlberg N, Nesme X, Lavire C, Vial L, Paulin L, de Lajudie P, Lindstrom K (2014) Phylogeny of the *Rhizobium-Allorhizobium-Agrobacterium* clade supports the delineation of *Neorhizobium* gen. nov. Syst Appl Microbiol 37:208–215

Mousavi SA, Willems A, Nesme X, de Lajudie P, Lindstrom K (2015) Revised phylogeny of Rhizobiaceae: proposal of the delineation of *Pararhizobium* gen. nov., and 13 new species combinations. Syst Appl Microbiol 38:84–90

Muresu R, Porceddu A, Sulas L, Squartini A (2019) Nodule-associated microbiome diversity in wild populations of *Sulla coronaria* reveals clues on the relative importance of culturable rhizobial symbionts and co-infecting endophytes. Microbiol Res 221:10–14

Nag P, Shriti S, Das S (2019) Microbiological strategies for enhancing biological nitrogen fixation in non-legumes. J Appl Microbiol. https://doi.org/10.1111/jam.14557

Noel TC, Sheng C, Yost CK, Pharis RP, Hynes MF (1996) *Rhizobium leguminosarum* as a plant growth-promoting rhizobacterium: direct growth promotion of canola and lettuce. Can J Microbiol 42:279–283. https://doi.org/10.1139/m96-040

Oberholster T, Vikram S, Cowan D, Valverde A (2018) Key microbial taxa in the rhizosphere of sorghum and sunflower grown in crop rotation. Sci Total Environ 624:530–539. https://doi.org/10.1016/j.scitotenv.2017.12.170

Oldroyd GE, Murray JD, Poole PS, Downie JA (2011) The rules of engagement in the legume-rhizobial symbiosis. Ann Rev Genet 45:119–144

Ormeño-Orrillo E, Servín-Garcidueñas LE, Rogel MA, González V, Peralta H, Mora J, Martínez-Romero E (2015) Taxonomy of rhizobia and agrobacteria from the Rhizobiaceae family in light of genomics. Syst Appl Microbiol 38(4):287–291

Pandey P, Dubey RS (2019) Metal toxicity in Rice and strategies for improving stress tolerance. In: Advances in rice research for abiotic stress tolerance. Elsevier, New York, pp 313–339. https://doi.org/10.1016/b978-0-12-814332-2.00015-0

Parag B, Sasikala C, Ramana CV (2013) Molecular and culture dependent characterization of endolithic bacteria in two beach sand samples and description of *Rhizobium endolithicum* sp. nov. Antonie Van Leeuwenhoek 104:1235–1244

Parte AC (2018) LPSN—list of prokaryotic names with standing in nomenclature (bacterio.net), 20 years on. Int J Syst Evol Microbiol 68:1825–1829. https://doi.org/10.1099/ijsem.0.002786

Pérez-Jaramillo JE, de Hollander M, Ramírez CA, Mendes R, Raaijmakers JM, Carrión VJ (2019) Deciphering rhizosphere microbiome assembly of wild and modern common bean (*Phaseolus vulgaris*) in native and agricultural soils from Colombia. Microbiome 7(1):1–16

Quan ZX, Bae HS, Baek JH, Chen WF, Im WT, Lee ST (2005) *Rhizobium daejeonense* sp. nov. isolated from a cyanide treatment bioreactor. Int J Syst Evol Microbiol 55(Pt 6):2543–2549

Qureshi M, Shahzad H, Saeed M, Ullah S, Ali M, Mujeeb F, Anjum M (2019) Relative potential of rhizobium species to enhance the growth and yield attributes of cotton (*Gossypium hirsutum* L.). Eur J Soil Sci 8(2):159–166. https://doi.org/10.18393/ejss.544747

Ramana CV, Parag B, Girija KR, Ram BR, Ramana VV, Sasikala C (2013) *Rhizobium subbaraonis* sp. nov., an endolithic bacterium isolated from beach sand. Int J Syst Evol Microbiol 63:581–585

Rashid MH, Young JP, Everall I, Clercx P, Willems A, Santhosh Braun M, Wink M (2015) Average nucleotide identity of genome sequences supports the description of *Rhizobium lentis* sp. nov., *Rhizobium bangladeshense* sp. nov. and *Rhizobium binae* sp. nov. from lentil (*Lens culinaris*) nodules. Int J Syst Evol Microbiol 65:3037–3045

Ren Da W, Chen WF, Sui XH, Wang ET, Chen WX (2011) *Rhizobium vignae* sp. nov., a symbiotic bacterium isolated from multiple legume species. Int J Syst Evol Microbiol 61:580–586

Rhoads A, Au KF (2015) PacBio sequencing and its applications. Genomics Proteomics Bioinformatics 13(5):278–289

Ribeiro RA, Rogel MA, Lopez-Lopez A, Ormeno-Orrillo E, Barcellos FG, Martinez J, Thompson FL, Martinez-Romero E, Hungria M (2012) Reclassification of *Rhizobium tropici* type A strains as *Rhizobium leucaenae* sp. nov. Int J Syst Evol Microbiol 62:1179–1184

Ribeiro RA, Martins TB, Ormeno-Orrillo E, Marcon Delamuta JR, Rogel MA, Martinez-Romero E, Hungria M (2015) *Rhizobium ecuadorense* sp. nov., an indigenous N_2-fixing symbiont of the Ecuadorian common bean (*Phaseolus vulgaris* L.) genetic pool. Int J Syst Evol Microbiol 65:3162–3169

Rincon-Rosales R, Villalobos-Escobedo JM, Rogel MA, Martinez J, Ormeno-Orrillo E, Martinez-Romero E (2013) *Rhizobium calliandrae* sp. nov., *Rhizobium mayense* sp. nov. and *Rhizobium jaguaris* sp. nov., rhizobial species nodulating the medicinal legume *Calliandra grandiflora*. Int J Syst Evol Microbiol 63:3423–3429. https://doi.org/10.1099/ijs.0.048249-0

Robledo M, Rivera L, Jiménez-Zurdo JI, Rivas R, Dazzo F, Velázquez E, Martínez-Molina E, Hirsch AM, Mateos PF (2012) Role of rhizobium endoglucanase CelC2 in cellulose biosynthesis and biofilm formation on plant roots and abiotic surfaces. Microb Cell Fact 11(1):125

Rodrigues Coelho MR, De Vos M, Carneiro NP, Marriel IE, Paiva E, Seldin L (2008) Diversity of nifH gene pools in the rhizosphere of two cultivars of sorghum (*Sorghum bicolor*) treated with contrasting levels of nitrogen fertilizer. FEMS Microbiol Lett 279(1):15–22

Roesch LFW, Camargo FA, Bento FM, Triplett EW (2008) Biodiversity of diazotrophic bacteria within the soil, root and stem of field-grown maize. Plant Soil 302(1–2):91–104

Rojas-Jiménez K, Sohlenkamp C, Geiger O, Martínez-Romero E, Werner D, Vinuesa P (2005) A CIC chloride channel homolog and ornithine-containing membrane lipids of *Rhizobium tropici* CIAT899 are involved in symbiotic efficiency and acid tolerance. Mol. Plant Microbe Interact 18:1175–1185. https://doi.org/10.1094/MPMI-18-1175

Romano I, Ventorino V, Pepe O (2020) Effectiveness of plant beneficial microbes: overview of the methodological approaches for the assessment of root colonization and persistence. Front Plant Sci 11:6. https://doi.org/10.3389/fpls.2020.00006

Roman-Ponce B, Jing Zhang Y, Soledad Vasquez-Murrieta M, Hua Sui X, Feng Chen W, Carlos Alberto Padilla J, Wu Guo X, Lian Gao J, Yan J, Hong Wei G et al (2016) *Rhizobium acidisoli* sp. nov., isolated from root nodules of *Phaseolus vulgaris* in acid soils. Int J Syst Evol Microbiol 66:398–406

Rozahon M, Ismayil N, Hamood B, Erkin R, Abdurahman M, Mamtimin H, Abdukerim M, Lal R, Rahman E (2014) *Rhizobium populi* sp. nov., an endophytic bacterium isolated from *Populus euphratica*. Int J Syst Evol Microbiol 64:3215–3221

Rubio-Canalejas A, Celador-Lera L, Cruz-González X, Menéndez E, Rivas R (2016) *Rhizobium* as potential biofertilizer of *Eruca sativa*. In: Biological nitrogen fixation and beneficial plant-microbe interaction. Springer, Cham, pp 213–220

Ruíz-Valdiviezo VM, Rogel-Hernandez MA, Guerrero G, Rincón-Molina CI, García-Perez LG, Gutiérrez-Micel FA, Villalobos-Maldonado JJ, López-López A, Martinez-Romero E, Rincón-Rosales R (2017) Complete genome sequence of a novel nonnodulating rhizobium species isolated from *Agave americana* L. rhizosphere. Genome Announc 5(46):e01280–e01217

Saidi S, Ramirez-Bahena MH, Santillana N, Zuniga D, Alvarez-Martinez E, Peix A, Mhamdi R, Velazquez E (2014) *Rhizobium laguerreae* sp. nov. nodulates *Vicia faba* on several continents. Int J Syst Evol Microbiol 64:242–247

Sarhan MS, Hamza MA, Youssef HH, Patz S, Becker M, ElSawey H, Nemr R, Daanaa HS, Mourad EF, Morsi AT, Abdelfadeel MR (2019) Culturomics of the plant prokaryotic microbiome and the dawn of plant-based culture media – a review. J Adv Res 19:15–27

Schlindwein G, Vargas LK, Lisboa BB, Azambuja AC, Granada C, Eichelberger G, Naiana C, Prates F, Stumpf R (2008) Influência da inoculação de rizóbios sobre a germinação e o vigor de plântulas de alface. Ciência Rural 38(3):658–664. https://doi.org/10.1590/S0103-84782008000300010

Servín-Garcidueñas LE, Guerrero G, Rogel-Hernández MA, Martínez-Romero E (2019) Genome sequence of *Rhizobium jaguaris* CCGE525T, a strain isolated from *Calliandra grandiflora* nodules from a rain forest in Mexico. Microbiol Resour Announc 8(9): e01584–e01518

Shamseldin A, Carro L, Peix A, Velazquez E, Moawad H, Sadowsky MJ (2016) The symbiovar trifolii of *Rhizobium bangladeshense* and *Rhizobium aegyptiacum* sp. nov. nodulate *Trifolium alexandrinum* in Egypt. Syst Appl Microbiol 39:275–279

Sheu SY, Huang HW, Young CC, Chen WM (2015) *Rhizobium alvei* sp. nov., isolated from a freshwater river. Int J Syst Evol Microbiol 65:472–478

Sheu SY, Chen ZH, Young CC, Chen WM (2016) *Rhizobium ipomoeae* sp. nov., isolated from a water convolvulus field. Int J Syst Evol Microbiol 66:1633–1640

Shi X, Li C, Zhao L, Si M, Zhu L, Xin K, Chen C, Wang Y, Shen X, Zhang L (2016) *Rhizobium gei* sp. nov., a bacterial endophyte of *Geum aleppicum*. Int J Syst Evol Microbiol 66:4282–4288

Smits TH (2019) The importance of genome sequence quality to microbial comparative genomics. BMC Genomics 20(1):662

Steen AD, Crits-Christoph A, Carini P, DeAngelis KM, Fierer N, Lloyd KG, Thrash JC (2019) High proportions of bacteria and archaea across most biomes remain uncultured. ISME J 13:3126–3130

Taylor BN, Simms EL, Komatsu KJ (2020) More than a functional group: diversity within the legume–rhizobia mutualism and its relationship with ecosystem function. Diversity 12(2):50

Terakado-Tonooka J, Ohwaki Y, Yamakawa H, Tanaka F, Yoneyama T, Fujihara S (2008) Expressed nifH genes of endophytic bacteria detected in field-grown sweet potatoes (*Ipomoea batatas* L.). Microbes Environ 23(1):89–93

Thaweenut N, Hachisuka Y, Ando S, Yanagisawa S, Yoneyama T (2011) Two seasons' study on *nifH* gene expression and nitrogen fixation by diazotrophic endophytes in sugarcane (*Saccharum* spp. hybrids): expression of *nifH* genes similar to those of rhizobia. Plant Soil 338(1–2):435–449

Tilman D, Clark M (2014) Global diets link environmental sustainability and human health. Nature 515. https://doi.org/10.1038/nature13959

Torres Tejerizo G, Rogel MA, Ormeno-Orrillo E, Althabegoiti MJ, Nilsson JF, Niehaus K, Schluter A, Puhler A, Del Papa MF, Lagares A et al (2016) *Rhizobium favelukesii* sp. nov., isolated from the root nodules of alfalfa (*Medicago sativa* L). Int J Syst Evol Microbiol 66:4451–4457

Tullio LD, Gomes DF, Silva LP, Hungria M, da Silva Batista JS (2019) Proteomic analysis of *Rhizobium freirei* PRF 81 T reveals the key role of central metabolic pathways in acid tolerance. Appl Soil Ecol 135:98–103. https://doi.org/10.1016/j.apsoil.2018.11.014

Turdahon M, Osman G, Hamdun M, Yusuf K, Abdurehim Z, Abaydulla G, Abdukerim M, Fang C, Rahman E (2013) *Rhizobium tarimense* sp. nov., isolated from soil in the ancient Khiyik River. Int J Syst Evol Microbiol 63:2424–2429

Turner TR, James EK, Poole PS (2013) The plant microbiome. Genome Biol 14:1–10. https://doi.org/10.1186/gb-2013-14-6-209

Vargas LK, Volpiano CG, Lisboa BB, Giongo A, Beneduzi A, Passaglia LMP (2017) Potential of rhizobia as plant growth-promoting rhizobacteria. In: Microbes for legume improvement. Springer, Cham, pp 153–174

Velázquez E, Carro L, Flores-Félix JD, Martínez-Hidalgo P, Menéndez E, Ramírez-Bahena MH, .. Peix A (2017) The legume nodule microbiome: a source of plant growth-promoting bacteria. In Probiotics and plant health, 41-70. Springer, Singapore

Velázquez E, Carro L, Flores-Félix JD, Menéndez E, Ramírez-Bahena MH, Peix A (2019) Bacteria-inducing legume nodules involved in the improvement of plant growth, health and nutrition. In: Microbiome in plant health and disease. Springer, Singapore, pp 79–104

Vences-Guzmán MÁ, Guan Z, Ormeño-Orrillo E, González-Silva N, López-Lara IM, Martínez-Romero E, Geiger O, Sohlenkamp C (2011) Hydroxylated ornithine lipids increase stress tolerance in *Rhizobium tropici* CIAT899. Mol Microbiol 79:1496–1514. https://doi.org/10.1111/j.1365-2958.2011.07535.x

Vergine M, Meyer JB, Cardinale M, Sabella E, Hartmann M, Cherubini P, De Bellis L, Luvisi A (2020) The *Xylella fastidiosa*-resistant olive cultivar "Leccino" has stable endophytic microbiota during the olive quick decline syndrome (OQDS). Pathogens 9(1):35

Vincent JM (1970) The cultivation, isolation and maintenance of rhizobia. In: A manual for the practical study of the root-nodule bacteria. Blackwell, Oxford, pp 1–13

Volpiano CG, Lisboa BB, Granada CE, São José JFB, de Oliveira AMR, Beneduzi A et al (2019) Rhizobia for biological control of plant diseases. In: Microbiome in plant health and disease. Springer, Singapore, pp 315–336

Walters WA, Jin X, Youngblut N, Wallace JG, Sutter J, Zhang W, González-Peña A, Peiffer J, Koren O, Shi Q, Knight R (2018) Large-scale replicated field study of maize rhizosphere identifies heritable microbes. Proc Natl Acad Sci 115(28):7368–7373

Wang F, Wang ET, Wu LJ, Sui XH, Li YJ, Chen WX (2011) *Rhizobium vallis* sp. nov., isolated from nodules of three leguminous species. Int J Syst Evol Microbiol 61:2582–2588

Wang J, Andersen SU, Ratet P (2018) Molecular and cellular mechanisms of the legume-rhizobia symbiosis. Front Plant Sci 9:1839

Wasinger VC, Cordwell SJ, Cerpa-Poljak A, Yan JX, Gooley AA, Wilkins MR, Duncan MW, Harris R, Williams KL, Humphrey-Smith I (1995) Large-scale amino-acid analysis for proteome studies. Electrophoresis 1995:16

Wei X, Yan S, Li D, Pang H, Li Y, Zhang J (2015) *Rhizobium helianthi* sp. nov., isolated from the rhizosphere of sunflower. Int J Syst Evol Microbiol 65:4455–4460

Willett W, Rockström J, Loken B, Springmann M et al (2019) Food in the anthropocene: the EAT–Lancet commission on healthy diets from sustainable food systems. Lancet 393:447–492. https://doi.org/10.1016/S0140-6736(18)31788-4

Womack JE (2019) Mapping genes is good for you. Ann Rev Anim Biosci 7:1–16

Xu L, Shi JF, Zhao P, Chen WM, Qin W, Tang M, Wei GH (2011) *Rhizobium sphaerophysae* sp. nov., a novel species isolated from root nodules of *Sphaerophysa salsula* in China. Antonie Van Leeuwenhoek 99:845–854

Yanni YG, Dazzo FB (2010) Enhancement of rice production using endophytic strains of *Rhizobium leguminosarum* bv. trifolii in extensive field inoculation trials within the Egypt Nile delta. Plant Soil 336(1–2):129–142. https://doi.org/10.1007/s11104-010-0454-7

Yanni YG, Rizk RY, El-Fattah FKA, Squartini A et al (2001) The beneficial plant growth-promoting association of *Rhizobium leguminosarum* bv. trifolii with rice roots. Funct Plant Biol 28:845. https://doi.org/10.1071/PP01069

Yao LJ, Shen YY, Zhan JP, Xu W, Cui GL, Wei GH (2012) *Rhizobium taibaishanense* sp. nov., isolated from a root nodule of *Kummerowia striata*. Int J Syst Evol Microbiol 62:335–341

Yeoh YK, Dennis PG, Paungfoo-Lonhienne C, Weber L, Brackin R, Ragan MA, Schmidt S, Hugenholtz P (2017) Evolutionary conservation of a core root microbiome across plant phyla along a tropical soil chronosequence. Nat Commun 8(1):1–9

Yildirim E, Karlidag H, Turan M, Dursun A, Goktepe F (2011) Promotion of broccoli by plant growth promoting rhizobacteria. Hort Sci 46:932–936

Yoneyama T, Terakado-Tonooka J, Minamisawa K (2017) Exploration of bacterial N_2-fixation systems in association with soil-grown sugarcane, sweet potato, and paddy rice: a review and synthesis. Soil Sci Plant Nutr 63(6):578–590

Yoneyama T, Terakado-Tonooka J, Bao Z, Minamisawa K (2019) Molecular analyses of the distribution and function of diazotrophic rhizobia and Methanotrophs in the tissues and rhizosphere of non-leguminous plants. Plants 8(10):408. https://doi.org/10.3390/plants8100408

Yoon JH, Kang SJ, Yi HS, Oh TK, Ryu CM (2010) *Rhizobium soli* sp. nov., isolated from soil. Int J Syst Evol Microbiol 60(6):1387–1393

Yuan T, Liu L, Huang S, Taher AH, Tan Z, Wu G, Peng G (2018) *Rhizobium wuzhouense* sp. nov., isolated from roots of *Oryza officinalis*. Int J Syst Evol Microbiol 68:2918–2923

Zhang RJ, Hou BC, Wang ET, Li Y, Zhang XX, Chen WX (2011) *Rhizobium tubonense* sp. nov., isolated from root nodules of *Oxytropis glabra*. Int J Syst Evol Microbiol 61(3):512–517

Zhang X, Li B, Wang H, Sui X, Ma X, Hong Q, Jiang R (2012) *Rhizobium petrolearium* sp. nov., isolated from oil-contaminated soil. Int J Syst Evol Microbiol 62:1871–1876. https://doi.org/10.1099/ijs.0.026880-0

Zhang YJ, Zheng WT, Everall I, Young JP, Zhang XX, Tian CF, Sui XH, Wang ET, Chen WX (2015a) *Rhizobium anhuiense* sp. nov., isolated from effective nodules of *Vicia faba* and *Pisum sativum*. Int J Syst Evol Microbiol 65:2960–2967

Zhang XX, Gao JS, Cao YH, Sheirdil RA, Wang XC, Zhang L (2015b) *Rhizobium oryzicola* sp. nov., potential plant-growth-promoting endophytic bacteria isolated from rice roots. Int J Syst Evol Microbiol 65:2931–2936

Zhang S, Yang S, Chen W, Chen Y, Zhang M, Zhou X, Fan G, Feng FY (2017) *Rhizobium arenae* sp. nov., isolated from the sand of Desert Mu Us, China. Int J Syst Evol Microbiol 67:2098–2103

Zhao YY, Lin RC (2014) UPLC–MSE application in disease biomarker discovery: the discoveries in proteomics to metabolomics. Chem Biol Interact 215:7–16

Zhao JJ, Zhang J, Sun L, Zhang RJ, Zhang CW, Yin HQ, Zhang XX (2017) *Rhizobium oryziradicis* sp. nov., isolated from rice roots. Int J Syst Evol Microbiol 67:963–968

Zhang XX, Tang X, Sheirdil RA, Sun L, Ma XT (2014) *Rhizobium rhizoryzae* sp. nov., isolated from rice roots. Int J Syst Evol Microbiol 64:1373–1377

Zheng Y, Liang J, Zhao DL, Meng C, Xu ZC, Xie ZH, Zhang CS (2020) The root nodule microbiome of cultivated and wild halophytic legumes showed similar diversity but distinct community structure in Yellow River Delta saline soils. Microorganisms 8(2):207

Part III
Insect–Fungus Mutualism

Chapter 17
Symbiotic Harmony Between Insects and Fungi: A Mutualistic Approach

Saraswathy Nagendran, Surendra S. Agrawal, and Sheba Abraham

Abstract Microbes are known to interact with a variety of organisms belonging to different classes, genera, or species through their own diverse and specific pathways and mechanisms. Such an interaction, which exists between microbes and herbivores like insects, has become a topic of great importance for researchers far and wide. Since such interfaces occur in the form of mutual interactions, which in turn leads to the participating organisms achieving rich and important advantages that are necessary for their survival and development. Much of the research on reciprocal interactions between insects and microbes have focused on bacterial associations with insects, more or less ignoring the fact that interactions between insects and fungi are equally important which usually follow the same mechanisms and pathways as associations between insects and Bacteria.

This chapter deliberates the various aspects and properties of fungal interactions with mushroom growers such as leaf-cutters (Attina ants), termites, and ambrosia beetles. These interactions are based on a complex and interesting evolutionary line, which finally introduces the concept of mutuality into this insect community. The benefits of these interactions range from nutrition to the spore dispersal of fungi as well as protection from predators and competitors. The interaction between yeast and insects has also been discussed in ample detail, with our focus mainly on the areas in which each participant in the interaction benefits.

Keywords Fungi · Insects · Beetles · Ants · Termites · Yeast · Mutualism

S. Nagendran (✉) · S. S. Agrawal · S. Abraham
Shobhaben Pratapbhai Patel School of Pharmacy & Technology Management, SVKM's NMIMS, Mumbai, India

N. Shrivastava et al. (eds.), *Symbiotic Soil Microorganisms*, Soil Biology 60,
https://doi.org/10.1007/978-3-030-51916-2_17

17.1 Introduction

Nature has its own way of astonishing us at the intricate details found in each of its naturally occurring phenomena, as well as the beauty of the complexity that revolves around it. The harmony of the ecological interactions between different organisms of different species and the balance that it offers for the environment as a whole is such that, ecologists all over the world always leave with awe and wonder at the perfection of nature's creativity.

Ecological interactions, which indicate that certain organisms are primarily dependent on one another for their survival, are classified according to whether these interactions have neutral, advantageous, or harmful effects for one of the interacting organisms. Symbiotic interactions contribute to a main category of such interactions in which both interacting organisms are closely related. Depending on whether the interaction partners produce a positive, negative, or neutral effect, symbiotic relationships are divided into mutuality, parasitism, or commensalism (Molles 2015; Smith et al. 2012).

Mutualism as a positive category of symbiotic relationships is examined in more detail and investigated, which brings us closer to the complexity of mutual interactions between different species or organisms that cannot be taxonomically related, but thrive together and contribute to mutual survival in a competent and challenging environment. Although there are mutual relationships between almost all organisms on earth, including humans, this chapter mainly focuses on the characteristics and evolutionary history of such an interaction between different types of fungi and insects.

So what exactly is understood when the term "mutualism" is being referred to? Mutualism involves an interaction between two kinds of organisms, a host and a symbiont, belonging to different class, species, or taxa. It is different from other types of interactions as mentioned in Fig. 17.1, wherein both the organisms are equally benefitted from the interaction, in contrast to commensalism or parasitism in which either of the organisms gets no benefit, or they are caused harm by the other. In short, mutualism is a reciprocatory positive interaction between a pair of organisms (Lu et al. 2016). For example, the interaction between ants and aphids in which the ants rear aphids for their source of food which the aphids produce and in turn the aphids are protected by the ants from potential predators. Another example of mutualism, within the aquatic kingdom is that of clownfish and sea anemone. The clownfish lives within the poisonous tentacles of the anemone, being immune to its poison and act as a bait for luring other small sea animals close to the vicinity of the anemone's tentacles. The association of sharks with remora fish is another of the many wonders of mutualistic interaction in the sea world. Last, but not the least, we

Fig. 17.1 Different Types of Mutual Associations

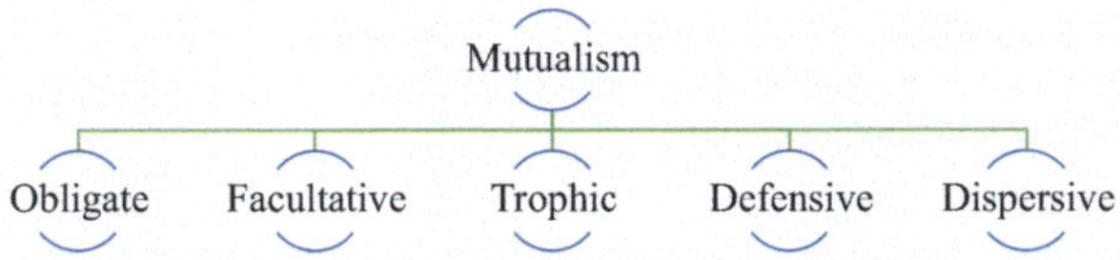

humans ourselves survive on the basis of a mutualistic relationship with microbes such as bacteria which thrive well in the intestinal microenvironment depending on the host's nutrition, and at the same time see to it that we remain protected against the harmful effects of ingested pathogenic microbes (Roossinck 2008).

17.2 Fungi and the Ecosystem

It is a well-known fact that many species belonging to kingdom fungi live in symbiotic associations with lower organisms such as cyanobacteria as well as higher organisms belonging to kingdom plants or animals. While some species may have a parasitic relationship with the associated organism wherein they derive food and thereby have harmful effects on the host, it is found that some other species survive through a mutually beneficial exchange of services between them and their associated partner. One such relationship that observed, is between fungi and plant roots. Mycorrhiza is the term used to refer to such an alliance in which the fungus is benefitted by the food accessed from the plant and in return the fungi mycelia help to absorb water and aid nutrition thereby providing nourishment to the plant.

The mutual association between fungi and photosynthetic organisms, usually cyanobacteria or green algae, is called lichen. The fungus grows around its host and absorbs food prepared by photosynthesis in exchange for water and nutrients (https://www.Ck12.Org/Biology/Fungi-Symbiosis/Lesson/Symbiotic-Relationships-Of-Fungi-Bio/).

When it concerns microbe-insect mutualism, much importance is directed towards exploring associations within the bacterial society, more or less, limited importance being given to the fungal associations that are equally common and important. In contrast to bacterial mutualism, fungal associations with insects are mostly facultative and horizontal, and it is reported that the microbial cells live extracellularly in hemolymph, fat bodies, or other specialized structures of the associated insects (Klepzig et al. 2009). A well-characterized and general example of such a relationship is that between leaf-cutter ants and fungi, in which the ants cultivate the fungi to gain access to their source of nourishment. A similar kind of mutualistic interaction is seen between certain species of bark beetles and fungi too. Other examples of insect–fungi mutualism which may not be too well known include that of wood wasp *Sirex noctilio* and wood decay fungus *Amylostereum areolatum.* The fungus contains wood degrading enzymes which enable the wasps to degrade and colonize tree barks (Nielsen et al. 2009; Kukor and Martin 1983; Talbot 1977) Another example, out of the several species that can be quoted in this context, is that of *Epichloe* species of fungi and *Botanophila* flies. The flies consume the fungal spores and cause deposition of their fecal matter onto unfertilized stroma rendering cross-pollination (Bultman et al. 1998; Bultman and Leuchtmann 2008). Fungal mutualists are now also reported to play a pivotal role as producers of chemical indicators for insect communication. The fungi *Pichia pinus* and *Hansenula capsulate* produce are capable of converting cis and trans verbenol to verbenone which acts as an anti-aggregation pheromone for bark beetles (Hunt and Borden 1990; Leufvén and Nehls 1986).

This chapter also focuses the attention on the evolutionary history as well as on the features that include the mutual relationship between fungi and insects such as leaf-cutting ants, ambrosia beetles, termites, and arthropods.

17.3 The Evolutionary Antiquity of Fungus-Farming Insects

Extensive work and research in this area have proposed that the evolution of an obligate mutualism proceeds primarily through five stages: (a) consistent and extended contact (b) avoidance of lethal harm during contact (c) coadaptation leading to reduced virulence and increased tolerance (d) further co-adaptation leading to dependence or interdependence (e) still further coadaptation leading to permanence and stability in the association. These stages give an overview of the progress of any agonistic or antagonistic interaction to stable mutualism (Taylor and FJR 1983).

Since time immemorial, even when the idea of cultivating plants for food to sustain oneself dawned in man's mind, the lineages of three insects namely, termites, ants, and beetles, rose to ecological importance by evolving into fungus cultivating farmers. Since they were completely dependent on the cultivated produce as the sole source of their nutrition several tasks partitioned societies all playing their own individual role in raising of food for their kind and thus in the process, became important players of the ecosystem (Wilson 1971). These insects, mostly being pests and detested by the human population, were put into the task of being exterminated but only recently it had come to the knowledge of the evolutionary glory that runs through their veins.

The transit of termite, ants, and beetles to the arena of fungi culture follows different evolutionary pathways. In termites, fungi probably had been an important source of nutrition derivation before they turned to cultivation. Many non-farming termites are known to feed on fungus infested woods and the termite fungi culture is now believed to be an expansion of such feeding habits of the ancestors (Batra 1979; Rouland-Lefèvre et al. 2006). Studies reveal that out of the 2600 species of identified termites, about 330 species survive in obligate association with a fungus basidiomycete genus, *Termitomyces*. *Termitomyces* are grown in subterranean combs within the heart of termite mound nest (Batra 1979; Abe et al. 2000a). Consumption of the fungal spores and deposition of fecal pellets consisting of the consumed fungal spores and plant forage within the comb serves as the seed for the growth of new fungi cultivar (Mueller and Gerardo 2002). Such a fungi culture practice also serves as a boon for obtaining the genetic material of the fungus from termites, sidestepping the tedious task of nest excavation. To reassemble the evolutionary history, researchers first compared the DNA of *Termitomyces* with that of non-domesticated fungi, and secondly, comparison was drawn between the DNA of termite farmers and non-farming relatives (Aanen et al. 2002). Studies conducted revealed that the fungus cultivation began in termites with the *Termitomyces* species of cultivars which eventually differentiated into several other cultivar species each almost exclusively associated with a particular termite group. Moreover, each of these cultivar groups has been found to be exchanged between the termite lineages within these group of termites. This led to the adaptation of termite groups to specific fungus cultivars at the same time fungus cultivars have evolved to adapt and survive only in association with certain farming groups (Mueller and Gerardo 2002).

A striking evolutionary parallel can be drawn when comparing the emergence of termite farmers with that of beetle and ant cultivators. In contrast to the ancestral feeding habits of termites that evolved into termite fungi culture, an ancestral vectoring system is responsible for the evolution of beetles into fungus cultivators. Whereas in ants it is still not clear whether fungus cultivation emerged from ancestral mycophagy or ancestral vectoring (Seifert et al. 1993; Mueller et al. 2001). Whatever may be the reason, it is now known that ants started growing fungus (mostly basidiomycetes) in their backyard about 50–60 million years ago (Mueller et al. 2001) and since then till date, roughly about 200 species of fungus-growing ants have emerged (Schultz and Meier 1995). Ants raise their fungus cultivation in subterranean chambers providing manure to their growth in the form of vegetable debris, or in the case of leaf-cutter ants, leaf fragments from live plants. The leaf-cutter ant's sustenance is completely dependent on the fungi they grow and hence, form an obligatory association with their fungal partners (Mueller et al. 2001). As in the case of ants, certain species of beetles commonly known as ambrosia beetles are found in obligatory association with fungi as they grow them as their primary source of food and to derive important nutrition in order to complete their life cycle (Farrell et al. 2001; Batra 1966). The fungus provides nutrition to most of the beetle developmental stages while the insects carry the fungal spores along with them, infecting new trees (Harrington 2005; Paine et al. 1997). The fungi also produce that degrade the indigestible wood into nutritious matter for the insects (Valiev et al. 2009). In stark contrast to termite and ant fungi culture which arose just once in each group and later led to diversification, studies suggest multiple origin of fungus cultivating habit in ambrosia beetles which arose at least seven times giving rise to sheer diversity of beetle species with respect to their feeding habits (Farrell 1998). No records of an evolutionary reversal to a non-fungus-cultivating pattern of life in any of these nine known, independently evolved farmer lineages have been found, suggesting a similar trend to that of humans where transition to way of living depending on agriculture has contributed to a radical and irreversible change that probably has bridled subsequent evolution (Diamond and Renfrew 1997).

17.4 Mutualism Between Leaf-Cutter Ants and Fungi

Leaf-cutter ants are considered major players of the ecosystem owing to the significant effects they elicit on local flora (Wirth et al. 2003), seedling recruitment (Costa et al. 2008), distribution of soil nutrients (Sternberg et al. 2007) and human agriculture (Cherrett and Peregrine 1976). These species of ants have known to be in association with fungi species since as long as 40–50 million years which it is believed, originated on the Amazon basins (Schultz and Brady 2008). Fungus-growing ants (*Myrmicinae*:Attini) cultivated fungus for their young ones using organic debris such as dead insect parts and feces as manure for their garden (Weber 1972; Mueller and Wcislo 1998; Mueller et al. 2005). A significant behavioral change in the ancestors *Acromyrmex* and *Atta* in replacing leaf fragments and other plant parts as compost for their fungus gardens instead of the usual organic debris contributed majorly to the deviation in the evolutionary pattern of the species (Schultz and Brady 2008; Hölldobler and Wilson 2010). Such a change in the pattern of cultivation contributed to an astounding increase in colony size, social structure, and ecological footprint. Leaf-cutter ants have now emerged to be the most dominant and diverse in Neotropical ecosystems, harvesting about 2–17% of the foliar biomass of annual leaf production of the forest and savanna woody plants (Wirth et al. 2003; Costa et al. 2008; Weber 1972; Hölldobler and Wilson 2010; Wheeler 1907).

Although the most common associate of attine ants remains to be basidiomycetes fungus, a larger population of the species is involved in farming lepiotaceous fungus of the genus *Leucoagaricus* while a smaller population are farmers of a distantly related pterulaceous group of fungi (Herz et al. 2007; Chapela et al. 1994). The leaf-cutter ants are precise and definite about the plant species, the individual plant, and the leaves within the plant that they cut. Factors that steer the assumption that leaf-cutter ants prefer relatively easy to cut, less defended leaves with high nutritional values include selection of younger leaves than older ones, woody rather than herbaceous and light demanding rather than shade-tolerant species (Villesen et al. 2004; Blanton and Ewel 1985; Coley and Barone 1996; Farji-Brener 2001). The worker ants are involved in cutting the leaves, carrying the fragments to the nest, cleaning, and processing them to form suitable for the fungus to grow and thrive in underground chambers inside the nest (Schultz and Brady 2008). In return, the fungus cultivar partially degrades the leaf material which serves as a source of nutrition to the ant colonies and their developing larvae. Thus the mutualistic interaction endures on the basis of services offered by ants ranging from weeding and grooming to the disposal of various antimicrobial compounds while reaping the benefits of a healthy fungus cultivar (Wirth et al. 2003; Schultz and Brady 2008; Barke et al. 2010; Currie and Stuart 2001, Fernández-Marín et al. 2006, 2009).

Fungus gardens are principally composed of only one fungal mutualist (Aylward et al. 2012) and factors that contribute to the low diversity of fungi in the fungus gardens include cautious cleaning by ants to maintain the hygiene of the fungal cultivars. Three main hygienic practices have been observed in fungus-growing ants which include: (a) weeding removal of dead fungal debris (b) fungus grooming-removal of spores of foreign fungi (Wirth et al. 2003) (c) application of fecal droplets to the fungal matrix. The fecal droplets of some species if attine ant is reported to contain fungal chitinases and lignocellulases which contributes to both plant biomass degradation as well as eradication of fungal pathogens (Rønhede et al. 2004, 2008; Martin et al. 1973).

Apart from the predominant fungus mutualist, microbes such as *Actinobacterium* of the genus *Pseudonocardia*, that produces antibiotic against fungus attacking parasites (e.g., *Escovopsis*), have been found in association with fungus gardens (Currie 2001; Currie et al. 1999a, b). Many of the bacterial species isolated from these ecosystems have been established to play a major role in not just antibiotic mediated extermination of pathogens but also in nutrient biosynthesis. One particular study carried out on *Klebsiella* and *Pantoea* species of nitrogen-fixing bacteria isolated from leaf-cutter ant nests proves the significance of these bacteria in fixing nitrogen and thus being important nutritional players in the ant–fungi ecosystem (Pinto-Tomás et al. 2009; Hölldobler and Wilson 2009).

The central role of fungus gardens cultivated by ants is the conversion of plant biomass into useful compounds important for the nutrition of the ants. The integrated biomass of fungus gardens includes a rich source of cellulose, hemicellulose, lignin, protein, simple sugars, and other compounds. In gardens of higher attine ants, these compounds are converted to hyphal swellings known as gongylidia, rich in lipids, carbohydrates, and other nutrients (Mueller et al. 2001; Martin et al. 1969). Gongylidia serves as an important food source for the entire colony and is an exclusive nutrient source for the developing larvae and brood (Hölldobler and Wilson 2009; Weber 1966; Nygaard et al. 2011; Suen et al. 2011).

17.5 Adaptation of the Ant Genome

The sequenced genomes of the leaf-cutter ants *Atta cephalotes* and *Achromyrmex echinatior have* opened up a sea of knowledge about the symbiotic association of these species. Obligate dependence of ants on their fungi associate have led to reductions at the genomic level (Suen et al. 2011; International Aphid Genomics Consortium 2010). On examination, genomes of attine ants were found to be deficient in the levels of amino acid arginine in comparison to other non-farming ant genomes, pointing to the fact that the fungal cultivars provide the required arginine thereby reducing the need for the particular pathway. This hypothesis is supported by the evidence obtained from previous studies of the documented compounds found in *Atta colombica* cultivar which showed presence of arginine (Abe et al. 2000b; Johnson et al. 1981). The *A. cephalotes* genome was also found devoid of the hexamerin gene responsible for amino acid sequestration during larval development. Serine protease was another compound found in significantly low amount, and as with arginine and hexamerin, is hypothesized to be provided by the fungal cultivar (Abe et al. 2000b). These data indicate that over the years, having established a mutualistic relationship with fungi, leaf-cutter ants have lost their ability and capacity to acquire nutrients on their own.

17.6 Fungus Cultivating Termite Species

Symbiotic relationship with a variety of organisms such as protists, methanogenic *Archaea,* and bacteria have always been a major player in the evolution of termites (Martin 1992). However, as on date only a single *Termitidae* sub-family, *Macrotermitinae* is known to have evolved into a mutualistic association with fungi of the genus *Termitomyces* (Abe et al. 2000c) and are a predominant component of the termite species in the African and Indomalayan region (Kambhampati and Eggleton 2000). The fungus is grown on a specialized structure known as fungus comb in the termite nest, maintained by the termites by the continual addition of new predigested plant debris while concurrent degradation of the old material is carried out (Kirk et al. 2001). The *Macrotermitinae* is divided into 11 genera with approximately 330 species and roughly about 40 species involved in *Termitomyces* symbiosis have been identified (Kambhampati and Eggleton 2000; Kirk et al. 2001). They play a significant role in litter removal by rummaging through dead wood, dead grass, and dung of herbivorous mammals. The workers collect substrate, chew them into very small fragments, maintain constant conditions in the nest for growth of the fungus cultivar, prevent the growth of potential competitors and thus in this way not just disperses the fungal spores, but also ensures their healthy growth inside the nest (Ausat et al. 1960).

The mature combs of *Macrotermitinae* which nests the cultivar is a firm intricate structure housing sparse growth of mycelium and populous small white spores called "mycotetes" that are accumulations of conidiophores and conidia, the asexual reproductive fruiting bodies of fungi. The combs are constructed of plant material that has been comminuted by chewing and passing through the workers gut. The cell wall polysaccharides undergoes insignificant breakdown and transformation during its transit through the digestive tract, as indicated by the presence of intact cells and cell walls, high cellulose content of the comb, and negligible difference in the cellulose-to-lignin ratio of the comb to the rummaged plant debris (Sands 1956).

The termite's diet consists of both the fungus combs and the mycotetes, both performing an integral role in providing a source of excellent nourishment to the termites (Nygaard et al. 2011; Ausat et al. 1960) due to the presence of elevated nitrogen content constituting a range of 5.7–7.9%. Additionally, the fungal enzymes released from the mycotetes after ingestion combines with the enzymes present in the insect gut rendering high catalytic activity that contributes to cellulose and hemicellulose digestion of the plant material in the *Macrotermes* species (Sands 1956; Abo-Khatwa 1978; Martin and Martin 1978; Rouland et al. 1988). Due to the availability of a better and richer source of nutrition, the termites no longer depend on woodchips or filter paper to derive nutrition, unless in cases when they are destitute of all food obtained through the fungal symbiont (Martin 1987; Wood and Thomas 1989).

It has been observed that the fruiting bodies of Termitomyces are always found in association with the termite nests indicating that within the basidiomycetes (Rouland-Lefèvre 2000; Rouland-Lefèvre and Bignell 2001) mutualistic symbiosis traces back to a single evolutionary origin and no record of reverting back to the nonsymbiotic way of living has been reported (Rouland-Lefevre et al. 2002; Moncalvo et al. 2000, 2002). The patterns of cospeciation and specificity have been found to be steady and undeviating with transmission of fungal symbionts from host to host outside the vertical host lineage (Frank 1996). Horizontal symbiont transmission is the pattern observed in many species of *Macrotermitinae* associated with formation of fruiting bodies in its fungal symbiont. However, for *Microtermes* and species of *M.bellicosus*, clonal uniparental transmission has been observed. The female in *Microtermes* and the male in *M.bellicosus* takes up the task of transmitting the fungus in the absence of sexual fruiting bodies (Darlington 1994).

17.7 Mutualistic Association of Beetle With Fungi

The bark and ambrosia beetles are either considered as two families *Scolytidae* and *Platypodidae* within the weevils (*Curculionoidea*) or as sub-families of *Curculionoidea* (Hsiau and Harrington 2003; Vega 2014). Ambrosia beetles are derived from bark beetles that colonize and consume phloem, which is considered to be more nutritious than wood and are mostly found in association with its fungal symbiont Ascomycotan fungi, the level of association ranging from facultative to obligative. Bark beetles, similar to ambrosia beetles, have been found mostly in association with Ascomycotan and hardly ever in association with basidiomycotan fungi. The variability of their association is reported to range from being facultative to obligate mutualists (Harrington 2005; Vega 2014; Li et al. 2016).

Ambrosiodmus is a genus consisting of over 80 species within the largest group of ambrosia beetles, *Xyleborini* (Wood 1982, 1992; Hopkins 1915; Batra 1985). Ambrosia beetles are known to have evolved into symbiotic fungi culture after at least 11 subsequent evolutionary patterns (Wood 1982). About 3200 species of ambrosia beetles are known to be fungus farmers of the genus *Ambrosia*. The most prominent ambrosia feeding genera include *Xyleborus, Trypodendron, Gnathotrichus, and Anisandrous,* belonging to family *Scolytidae* and are widely distributed across the temperate and tropical regions (Beaver et al. 1989). The fungus cultivar, responsible for providing nutrition to the beetles during their period of dormancy and inactiveness as well as during the stages of active growth and development, is grown and sheltered by beetles in specialized storage organs known as *mycangia*. Mycangia, also known as mycetangia are ectodermal glandular pockets of beetles where ambrosia fungi are stored and where they grow and multiply (Hulcr and Cognato 2010). Mycangia are known to have evolved in two ways, as pocket like dilation of cuticle or as newly developed hollow glands from glands that previously produced defensive compounds and oils in order to prevent excessive flow of sap (Vega and Blackwell 2005). Unlike other species of ambrosia beetles that mostly colonize dead but relatively fresh wood tissues, ambrosia beetles are capable of infesting wood throughout the period of its decay, including the final stages of decay when the xylem is inhabited and colonized by other competitive wood-rotting fungi (Beaver et al. 1989). In contrast to the fungus-farming behavior of the *Ambrosiodmus*, its phylogenetically related sister *Ambrosiophilus* is known to rely on mycoclepty (fungus theft) to obtain nourishment, having lost the ability to culture their own fungal farms (Vega and Blackwell 2005).

The fungi inoculum carried inside mycangia gets discharged into the beetle tunnels on the bark while burrowing or during oviposition. The nitrogenous waste eliminated by the beetles serve as fertile manure providing nutrient-rich medium for the fungus to grow and thrive. The beetle larvae in turn are dependent on the fungus to derive nourishment. Prior to several days before the eggs are hatched, the fungus weakens the wood elements aiding the larvae in excavation as well as broadening of tunnels. Such a mutualistic interaction has its own importance in the ecosystem as many nonspecific wood-decaying fungi in association with insects such as beetles have been found to augment and accelerate the process of wood decaying, degradation, and nutrient recycling in the forest ecosystem (Beaver et al. 1989).

The mycangial secretions of ambrosia beetles are responsible for morphological characteristics and biology of the ambrosia fungi. Conidia and other reproductive spores of these fungi develop into the ambrosia stage within the mycangia of the beetles. Mycangial secretions found to contain a rich store of oils and proteins, serve as a source of nourishment and preservation of the fungal inoculum. Mycangial contents of other species of ambrosia beetles have also reported to contain compounds such as amino acids, fat, and proteins, confirmed through thin layer chromatographic techniques (Hulcr and Cognato 2010).

17.8 Fungi and Insect Mutualistic Association

"Yeast" is used to refer to a fungal growth form consisting of only a single cell and lacking specialized sex organs, thereby reproducing through the process of budding via sexual spores from somatic cells, which are not enclosed within fruiting bodies (Vega and Blackwell 2005; Vega et al. 2008; Kurtzman et al. 2011). This group is considered to be omnipresent and found to occupy a variety of ecological niches in both terrestrial and aquatic ecosystems (Kurtzman and Fell 2006). About 1500 species of yeasts have been identified, mostly which belong to the phyla Ascomycota and Basidiomycota (Urubschurov and Janczyk 2011). Almost 700 species from about 93 genera under the class Saccharomycetes of Ascomycota are referred to as "true yeasts" (Batra 1979). Genera belonging to this class include *Candida, Kluyveromyces, Metschnikowia, Pichia,* and *Saccharomyces. Fellomyces*, *Tremella, Ustilago*, and *Cystofilobasidium* are some examples of fungi that belong to the class

Basidiomycota (Urubschurov and Janczyk 2011; Landell et al. 2009; Suh et al. 2004; Fell et al. 1999; Gibson and Hunter 2005).

Most of the species of yeast discovered to be in mutualistic relationship with insects is found to dwell within the gastrointestinal tract of the host and retrieved from faeces, ovipositors, or other specialized organs of the insects, thus directing a hypothesis of their facultative relationship with the host (Boekhout 2005).

A vast range of variability among the hosts and their habitats have led to the rise of unknown number of new species of yeasts, that almost correspond to the total number of organisms they are found to be in association with since each host is assumed to carry their own particular and specific yeast partner. An example of this is the wide range of yeast species inhabiting the guts of *Erotylidae* and *Tenebrionidae* families of beetles, the number of yeast species being equal to the number of beetle species under each family (Sung-Oui et al. 2005).

Although most often "true yeasts" are involved in symbiosis with insects, studies report that a separate group of fungal endosymbionts known as "yeast like symbionts" (YLS) also exist in association with insects as shown in Fig. 17.2 (Suh et al. 2001, 2004). This group is supposedly believed to have evolved from ascocarpic ascomycetes, especially from the subphylum *Pezizomycotina* (Gibson and Hunter 2010). Notwithstanding their ability to form beneficial association with insects, they are known to be phylogenetically more related to Hypocreales than to Saccharomycetales (true yeasts). This relation directs the attention to an interesting area of evolutionary mutualism, since *Hypocreales* belong to the family of

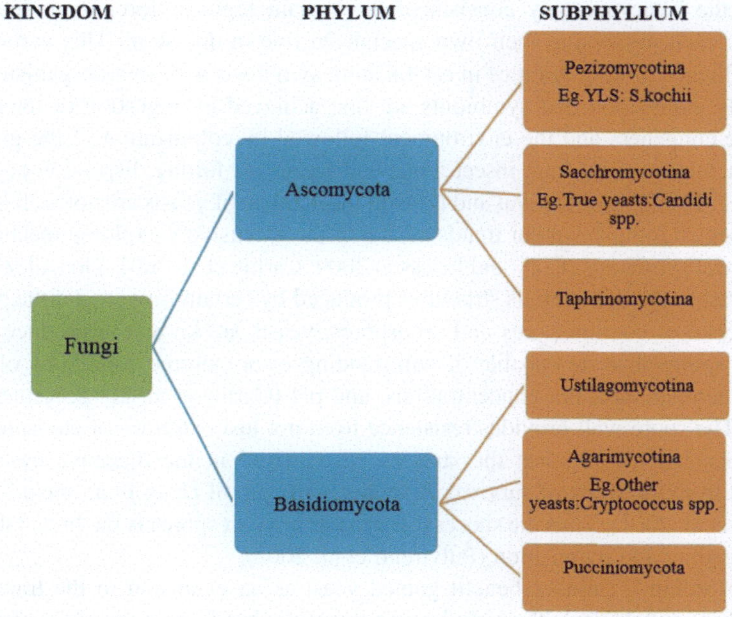

Fig. 17.2 Taxonomic Classification of Yeast and Yeast-like Symbionts [adapted from Suh et al. (2004) and (Suh et al. (2001)]

Clavicipitaceae which consists of entomopathogenic fungi making it quite evident that YLS is comparatively more related to entomopathogenic fungi than to yeasts (Fukatsu and Ishikawa 1996). This places an evidence of two separate pathways of fungi evolution-mutualistic YLS having their source of origin in insect pathogenic microbes whereas other related yeasts apparently having evolved from commensal interactions (Noda and Kodama 1996). However, the variability in interactions between YLS and insects are sparser than yeast-insect interactions. As per studies and researches to date, examples of mutualistic insect-YLS interactions have been found mostly in anobiid beetles and also in some planthopper and aphid species (Sasaki et al. 1996; Douglas and Smith 1989; Engel and Moran 2013).

17.9 Services Offered and Benefits Gained

Though much speculation still exists on the benefits involved in insect–fungi mutualism, studies till date have reported that the major benefits that fungi reap out such a relationship include protection and dispersal of their spores as well as provisions for outbreeding (Kurtzman et al. 2011; Coluccio et al. 2008).

Although insect digestive tracts are considered as one of the highly vulnerable areas for microbial; colonization, few microbes have developed resistance to withstand the harsh gut environment and are successfully passed on to host congeners. Although design of the digestive tract varies among different insects and insect orders, the basic anatomy consists of three main regions: foregut, midgut, and hindgut, each possessing their own specialized role in digestion. This variation in turn facilitates distinct abilities in establishing symbiosis with microorganisms. It is generally understood that symbionts are first achieved by ingestion or interaction with the congeners and the environment followed by colonization of the guts and subsequently released from insect molts and faeces for further dispersion of microbial cells. Though the survival and transfer mechanism of yeasts are not well known, it is expected to have similar trends as that of the extensively explored mechanisms of bacterial symbionts (Cory and Ericsson 2009; Carlile et al. 2001; Gonzalez 2014). In contrast to the filamentous structures produced by certain fungi to aid the process of dispersal of conidiospores and ascospores, yeasts are known to produce reproductive spores that are capable of withstanding even extreme conditions of stress such as temperature, salt concentrations, and pH (Carlile et al. 2001; Reuter et al. 2007). The spore wall provides resistance from not just extremes of environmental conditions but also against the stresses encountered in the digestive system, as inferred from the study of survival of spores in the gut of *Drosophila melanogaster* (Carlile et al. 2001). Thus the successful transfer of yeast spores is facilitated through insect feeding and oviposition (Pulvirenti et al. 2002).

Outbreeding is another benefit gained yeast as an extension to the transit and survival of sexual spores through the unwelcoming, harsh environment in the insect gut. Outbreeding is an important concept in order to maintain genetic variation among the descendants to facilitate adaptation and thus evolution. Unfavorable, harsh conditions mostly contribute to the reproduction in yeasts through the formation of diploid vegetative cells that can undergo meiosis to give rise to a tetrad of haploid spores that can germinate, mate, and restore the diploid state. Most often, it has also come to notice that the four spores contained within the same capsule may undergo selfing or inbreeding. This process is prevented by the gut enzymes that break open the capsule to release the spores, thus inhibiting inbreeding and assisting the phenomena of outbreeding (Guzmán et al. 2013; Lachance and Bowles 2002). A study conducted by making use of genetically marked strains of *S.cerevisiae* divided into two groups, one in contact with fruit flies and another in no contact with insects, demonstrated a significant increase in the number of heterozygotes with insect association than in yeast that was exposed to non-insect mediated mating (Guzmán et al. 2013). An example of yeast receiving benefits in all three forms—protection, facilitation outbreeding, and spore dispersal—is *Metschnikowia* species in association with pollinating insects of the orders *Diptera, Coleoptera,* and *Hymenoptera.* The yeast is found particularly in the nectar of flowers thus explaining their association with insect pollinators (Lachance et al. 2001, 2003; Janson et al. 2008; Bismanis 1976).

The benefits achieved by insects as a result of this interaction include nutrition source, detoxication from harmful substances, protection from biotic stress as well as an aid for chemical communication (Noda and Kodama 1996; Jurzitza 1970).

The role played by yeast mutualists in providing nourishment to insects is interpreted from the fact that insect performance and development decreases in the absence of their yeast associate (Kurtzman et al. 2011; Kurtzman and Fell 2006). Yeast cells form an excellent source of nitrogen, containing about 7.5–8.5% of nitrogen by weight, apart from other essential nutrients such as vitamin B3 and B5, proteins, trace metals and amino acids which are broken down and absorbed by simple digestion in the insect gut (96–97,108). Insects such as Pseodococcus *citri*, and wood-boring cerambycids, *Leptura,* and *Rhagium* are reported to depend on yeasts for their dietary source of nitrogen (Starmer and Aberdeen 1990; Noda and Koizumi 2003). The rich source of nitrogen, lipids, and vitamins provide major nutritional support to Drosophila flies especially during the stages of egg maturation and larval development (Shen and Dowd 1991a). Similar to the nutritional role played by yeasts, YLS are also reported to play a pivotal role in being nutritional hub to their host. *Symbiotaphrina* and anobiid beetles are supplied by nutrients like nitrogen, sterols, vitamins, and essential amino acids by YLS (Starmer and Aberdeen 1990; Noda and Koizumi 2003). Similarly important intermediate precursors for ergosterol biosynthesis in rice planthoppers are also provided by the yeast-like symbionts (Shen and Dowd 1991b).

The wide range of enzymes produced by insect associated yeasts includes exoproteases and peptidases (involved in protein degradation), lipases (for digestion of fatty acids), and hydrolytic enzymes involved in sugar degradation (Chararas et al. 1983). These enzymes play a role in conversion of complex molecules and polysaccharides to simple compounds like glucose or sugars that are easily and directly absorbed by the insects. This role is however mainly attributed to the enzymes produced by YLS, which release digestive enzymes into their surroundings for colonization and spreading to new areas, in contrast to true yeasts, which are more sessile and are not known to release digestive enzymes unless they are trapped in their own erosion zones (Noda and Kodama 1996). An exception to this nature is demonstrated by true yeasts belonging to the genus Candida which possess the ability to degrade wood components such as cellulose, pectin, and glucosides (Listemann 1988).

The process of detoxification by yeasts have proved that mutualism between insects and fungi is a detail of much significance in herbivory. Detoxification in this context relates to making certain nutrients available by neutralizing or decomposing the toxins, thus aiding the process of digestion. In some cases, detoxification converts these nutrients to more polar forms that can be easily removed from the host's digestive system (Kurtzman et al. 2011; Kurtzman and Fell 2006; Cory and Ericsson 2009). An example of detoxification by yeasts is clearly elaborated in the YLS, *S.kochii* which is found in association with the beetle *L.serricorne (*Kurtzman et al. 2011*;* Kurtzman and Fell 2006*)*. *S.kochii* is reported to have the potential to detoxify a variety of plant allelochemicals, metal toxins, insecticides, and herbicides in addition to producing detoxifying enzymes such as aromatic ester hydrolase, glucosidase, phosphatase, and glutathione transferase that converts toxic chemicals to important carbon sources (Chararas et al. 1983).

Biotic hazards faced by herbivores include competitors, parasites, predators and plant chemical defenses. Plant diseases could have a negative impact on insects therefore insect–yeast mutualism aides the availability of safe food sources as yeasts have demonstrated various protection mechanism of plant tissues from infesting pathogens (Pulvirenti et al. 2002; Listemann 1988). Yeasts could play a role in limiting the presence of other fungi or microbes inside plant tissues thus favoring the growth and development of associated insects. *Metschnikowia* species decrease the prevalence of molds inside apples which indirectly correspond to lower mortality and larval development time for *Cydia pomonella* (Witzgall et al. 2012). These results are a clear illustration that yeasts not only performs the role as nutrition provider to insects, but also ensures that opportunistic pathogens and microbes that might hinder the development of both participants of symbiosis are kept at bay (Pulvirenti et al. 2002).

17.10 Conclusion

Researches and studies conducted so far have pointed to the fact that fungus farmers had not always been involved in mutualistic relations with fungi, but it is a phenomenon that has evolved over a period of nearly 50 million years due to variations and adaptations amongst various species of organisms. It has manifested its effects even at the genome level as in case of attine ants whose genome is deficient of genes responsible for the production of several important enzymes since these are now easily available in ready form from the fungi in association. Yeast and yeast-like organisms have developed resistant mechanisms to survive in the harsh dynamic environment of insect gut which in turn facilitates spore disposal through feces and insect oviposition. The mutualistic evolution has come a long a way in developing mechanisms for the survival and development of both insects and fungi alike. It is a demonstration of the potential of organisms to develop adaptations in the face of unfavorable surroundings and environment giving rise to variations and modifications within different species.

References

Aanen DK, Eggleton P, Rouland-Lefevre C, Guldberg-Frøslev T, Rosendahl S, Boomsma JJ (2002) The evolution of fungus-growing termites and their mutualistic fungal symbionts. Proc Natl Acad Sci 99(23):14887–14892

Abe T, Bignell DE, Higashi M, Higashi T, Abe Y (eds) (2000a) Termites: evolution, sociality, symbioses, ecology. Springer, Dordrecht, pp 335–336

Abe T, Bignell DE, Higashi M, Higashi T, Abe Y (eds) (2000b) Termites: evolution, sociality, symbioses, ecology. Springer, Dordrecht, pp 189–208

Abe T, Bignell DE, Higashi M, Higashi T, Abe Y (eds) (2000c) Termites: evolution, sociality, symbioses, ecology. Springer, Dordrecht, pp 289–306

Abo-Khatwa N (1978) Cellulase of fungus-growing termites: a new hypothesis on its origin. Experientia 34(5):559–560

Ausat A, Cheema PS, Koshi T, Perti SL, Ranganathan SK (1960, October) Laboratory culturing of termites. In: Termites in the humid tropics, proceedings of the New Delhi symposium. UNES CO, Paris, pp 121–125

Aylward FO, Currie CR, Suen G (2012) The evolutionary innovation of nutritional symbioses in leaf-cutter ants. Insects 3(1):41–61

Barke J, Seipke RF, Grüschow S, Heavens D, Drou N, Bibb MJ, Goss RJ, Douglas WY, Hutchings MI (2010) A mixed community of actinomycetes produce multiple antibiotics for the fungus farming ant *Acromyrmex octospinosus*. BMC Biol 8(1):109

Batra LR (1966) Ambrosia fungi: extent of specificity to ambrosia beetles. Science 153 (3732):193–195

Batra LR (1979) The mutualistic fungi of Xyleborini beetles. In: Batra LR (ed) Insect-fungus symbiosis. Allanheld, Osmun & Co, Montclair, pp 259–265

Batra LR (1985) Ambrosia beetles and their associated fungi: research trends and techniques. Proc Plant Sci 94(2–3):137–148

Beaver RA, Wilding N, Collins N, Hammond P, Webber J (1989) Insect-fungus relationships in the bark and ambrosia beetles. In: Insect-fungus interactions. Academic Press, New York, pp 121–143

Bismanis JE (1976) Endosymbionts of Sitodrepa panicea. Can J Microbiol 22(10):1415–1424

Blanton CM, Ewel JJ (1985) Leaf-cutting ant herbivory in successional and agricultural tropical ecosystems. Ecology 66(3):861–869

Boekhout T (2005) Gut feeling for yeasts. Nature 434(7032):449–451

Bultman TL, Leuchtmann A (2008) Biology of the Epichloë–Botanophila interaction: an intriguing association between fungi and insects. Fungal Biol Rev 22(3–4):131–138

Bultman TL, White JF Jr, Bowdish TI, Welch AM (1998) A new kind of mutualism between fungi and insects. Mycol Res 102(2):235–238

Carlile MJ, Watkinson SC, Gooday GW (2001) The fungi. Gulf Professional Publishing, Houston

Chapela IH, Rehner SA, Schultz TR, Mueller UG (1994) Evolutionary history of the symbiosis between fungus-growing ants and their fungi. Science 266(5191):1691–1694

Chararas C, Pignal MC, Vodjdani G, Bourgeay-Causse M (1983) Glycosidases and B group vitamins produced by six yeast strains from the digestive tract of *Phoracantha semipunctata* larvae and their role in the insect development. Mycopathologia 83(1):9–15
Cherrett JM, Peregrine DJ (1976) A review of the status of leaf-cutting ants and their control. Ann Appl Biol 84:124–128
Coley PD, Barone JA (1996) Herbivory and plant defenses in tropical forests. Annu Rev Ecol Syst 27(1):305–335
Coluccio AE, Rodriguez RK, Kernan MJ, Neiman AM (2008) The yeast spore wall enables spores to survive passage through the digestive tract of Drosophila. PLoS One 3(8):e2873
Cory JS, Ericsson JD (2009) Fungal entomopathogens in a tritrophic context. In the ecology of fungal Entomopathogens. Springer, Dordrecht, pp 75–88
Costa AN, Vasconcelos HL, Vieira-Neto EH, Bruna EM (2008) Do herbivores exert top-down effects in neotropical savannas? Estimates of biomass consumption by leaf-cutter ants. J Veg Sci 19(6):849–854
Currie CR (2001) Prevalence and impact of a virulent parasite on a tripartite mutualism. Oecologia 128(1):99–106
Currie CR, Stuart AE (2001) Weeding and grooming of pathogens in agriculture by ants. Proc R Soc Lond Ser B Biol Sci 268(1471):1033–1039
Currie CR, Scott JA, Summerbell RC, Malloch D (1999a) Fungus-growing ants use antibiotic-producing bacteria to control garden parasites. Nature 398(6729):701–704
Currie CR, Mueller UG, Malloch D (1999b) The agricultural pathology of ant fungus gardens. Proc Natl Acad Sci 96(14):7998–8002
Darlington JPEC (1994) Nutrition and evolution in fungus-growing termites. In: Hunt JH, Nalepa CA (eds) Nourishment and evolution in insect societies. Westview, Boulder, CO, pp 105–130
Diamond J, Renfrew C (1997) Guns, germs, and steel: the fates of human societies. Nature 386 (6623):339–339
Douglas AE, Smith DC (1989) Are endosymbioses mutualistic? Trends Ecol Evol 4(11):350–352
Engel P, Moran NA (2013) The gut microbiota of insects–diversity in structure and function. FEMS Microbiol Rev 37(5):699–735
Farji-Brener AG (2001) Why are leaf-cutting ants more common in early secondary forests than in old-growth tropical forests? An evaluation of the palatable forage hypothesis. Oikos 92 (1):169–177
Farrell BD (1998) "Inordinate fondness" explained: why are there so many beetles? Science 281 (5376):555–559
Farrell BD, Sequeira AS, O'Meara BC, Normark BB, Chung JH, Jordal BH (2001) The evolution of agriculture in beetles (Curculionidae: Scolytinae and Platypodinae). Evolution 55 (10):2011–2027
Fell JW, Roeijmans H, Boekhout T (1999) Cystofilobasidiales, a new order of basidiomycetous yeasts. Int J Syst Evol Microbiol 49(2):907–913
Fernández-Marín H, Zimmerman JK, Rehner SA, Wcislo WT (2006) Active use of the metapleural glands by ants in controlling fungal infection. Proc R Soc B Biol Sci 273(1594):1689–1695
Fernández-Marín H, Zimmerman JK, Nash DR, Boomsma JJ, Wcislo WT (2009) Reduced biological control and enhanced chemical pest management in the evolution of fungus farming in ants. Proc R Soc B Biol Sci 276(1665):2263–2269
Frank SA (1996) Host–symbiont conflict over the mixing of symbiotic lineages. Proc R Soc Lond Ser B Biol Sci 263(1368):339–344
Fukatsu T, Ishikawa H (1996) Phylogenetic position of yeast-like symbiont of Hamiltonaphis styraci (Homoptera, Aphididae) based on 18S rDNA sequence. Insect Biochem Mol Biol 26 (4):383–388
Gibson CM, Hunter MS (2005) Reconsideration of the role of yeasts associated with Chrysoperla green lacewings. Biol Control 32(1):57–64
Gibson CM, Hunter MS (2010) Extraordinarily widespread and fantastically complex: comparative biology of endosymbiotic bacterial and fungal mutualists of insects. Ecol Lett 13(2):223–234

Gonzalez F (2014) Symbiosis between yeasts and insects (No. 2014: 3)
Guzmán B, Lachance MA, Herrera CM (2013) Phylogenetic analysis of the angiosperm-floricolous insect–yeast association: have yeast and angiosperm lineages co-diversified? Mol Phylogenet Evol 68(2):161–175
Harrington TC (2005) Ecology and evolution of mycophagous bark beetles and their fungal partners. In: Vega FE, Blackwell M (eds) Ecological and evolutionary advances in insect-fungal associations. Oxford University Press, Oxford, pp 257–291
Herz H, Beyschlag W, Hölldobler B (2007) Herbivory rate of leaf-cutting ants in a tropical moist forest in Panama at the population and ecosystem scales. Biotropica 39(4):482–488
Hölldobler B, Wilson EO (2009) The superorganism: the beauty, elegance, and strangeness of insect societies. W.W. Norton & Company, New York
Hölldobler B, Wilson EO (2010) The leafcutter ants: civilization by instinct. W.W. Norton & Company, New York
Hopkins AD (1915) Classification of the Cryphalinae with descriptions of new genera and species (no. 99-106). US Government Printing Office, Washington, DC
Hsiau PT, Harrington TC (2003) Phylogenetics and adaptations of basidiomycetous fungi fed upon by bark beetles (Coleoptera: Scolytidae). Symbiosis 34(2):111–132
https://www.Ck12.Org/Biology/Fungi-Symbiosis/Lesson/Symbiotic-Relationships-Of-Fungi-Bio/. Accessed Feb 2020
Hulcr J, Cognato AI (2010) Repeated evolution of crop theft in fungus-farming ambrosia beetles. Evolution 64(11):3205–3212
Hunt DWA, Borden JH (1990) Conversion of verbenols to verbenone by yeasts isolated from *Dendroctonus ponderosae* (Coleoptera: Scolytidae). J Chem Ecol 16(4):1385–1397
International Aphid Genomics Consortium (2010) Genome sequence of the pea aphid *Acyrthosiphon pisum*. PLoS Biol 8(2). https://doi.org/10.1371/journal.pbio.1000313.t003
Janson EM, Stireman JO III, Singer MS, Abbot P (2008) Phytophagous insect–microbe mutualisms and adaptive evolutionary diversification. Evolution 62(5):997–1012
Johnson RA, Thomas RJ, Wood TG, Swift MJ (1981) The inoculation of the fungus comb in newly founded colonies of some species of the Macrotermitinae (Isoptera) from Nigeria. J Nat Hist 15 (5):751–756
Jurzitza G (1970) Über Isolierung, Kultur und Taxonomie einiger Anobiidensymbionten (Insecta, Coleoptera). Arch Mikrobiol 72(3):203–222
Kambhampati S, Eggleton P (2000) Taxonomy and phylogeny of termites. In: Termites: evolution, sociality, symbioses, ecology. Springer, Dordrecht, pp 1–24
Kirk P, Cannon PF, David JC, Stalpers JA (eds) (2001) Ainsworth & Bisby's dictionary of the fungi. CABI, Wallingford
Klepzig KD, Adams AS, Handelsman J, Raffa KF (2009) Symbioses: a key driver of insect physiological processes, ecological interactions, evolutionary diversification, and impacts on humans. Environ Entomol 38(1):67–77
Kukor JJ, Martin MM (1983) Acquisition of digestive enzymes by siricid woodwasps from their fungal symbiont. Science 220(4602):1161–1163
Kurtzman CP, Fell JW (2006) Yeast systematics and phylogeny—implications of molecular identification methods for studies in ecology. In: Biodiversity and ecophysiology of yeasts. Springer, Berlin, pp 11–30
Kurtzman CP, Fell JW, Boekhout T (2011) Definition, classification and nomenclature of the yeasts. In: The yeasts. Elsevier, New York, pp 3–5
Lachance MA, Bowles JM (2002) Metschnikowia arizonensis and Metschnikowia dekortorum, two new large-spored yeast species associated with floricolous beetles. FEMS Yeast Res 2(2):81–86
Lachance MA, Bowles JM, Kwon S, Marinoni G, Starmer WT, Janzen DH (2001) Metschnikowia lochheadii and Metschnikowia drosophilae, two new yeast species isolated from insects associated with flowers. Can J Microbiol 47(2):103–109

Lachance MA, Bowles JM, Starmer WT (2003) Metschnikowia santaceciliae, Candida hawaiiana, and Candida kipukae, three new yeast species associated with insects of tropical morning glory. FEMS Yeast Res 3(1):97–103

Landell MF, Inacio J, Fonseca A, Vainstein MH, Valente P (2009) Cryptococcus bromeliarum sp. nov., an orange-coloured basidiomycetous yeast isolated from bromeliads in Brazil. Int J Syst Evol Microbiol 59(4):910–913

Leufvén A, Nehls L (1986) Quantification of different yeasts associated with the bark beetle, Ips typographus, during its attack on a spruce tree. Microb Ecol 12(2):237–243

Li Y, Simmons DR, Bateman CC, Short DP, Kasson MT, Rabaglia RJ, Hulcr J (2016) Correction: new fungus-insect Symbiosis: culturing, molecular, and histological methods determine saprophytic Polyporales Mutualists of Ambrosiodmus Ambrosia beetles. PLoS One 11(1):e0147305

Listemann H (1988) Über die Beziehungen zwischen *Candida pulcherrima* und Laspeyresia pomonella: the interrelation between Candida pulcherrima and Laspeyresia pomonella. Mycoses 31(8):423–425

Lu M, Hulcr J, Sun J (2016) The role of symbiotic microbes in insect invasions. Annu Rev Ecol Evol Syst 47:487–505

Martin MM (1987) Invertebrate-microbial interactions. Ingested fungal enzymes in arthropod biology. Cornell University Press, Ithaca

Martin MM (1992) The evolution of insect-fungus associations: from contact to stable symbiosis. Am Zool 32(4):593–605

Martin MM, Martin JS (1978) Cellulose digestion in the midgut of the fungus-growing termite Macrotermes natalensis: the role of acquired digestive enzymes. Science 199(4336):1453–1455

Martin MM, Carman RM, MacConnell JG (1969) Nutrients derived from the fungus cultured by the fungus-growing ant Atta colombica tonsipes. Ann Entomol Soc Am 62(1):11–13

Martin MM, Gieselmann MJ, Martin JS (1973) Rectal enzymes of attine ants. α-amylase and chitinase. J Insect Physiol 19(7):1409–1416

Molles M (2015) Ecology: concepts and applications, 2nd edn. McGraw-Hill, New York, pp 87–118

Moncalvo JM, Lutzoni FM, Rehner SA, Johnson J, Vilgalys R (2000) Phylogenetic relationships of agaric fungi based on nuclear large subunit ribosomal DNA sequences. Syst Biol 49(2):278–305

Moncalvo JM, Vilgalys R, Redhead SA, Johnson JE, James TY, Aime MC, Hofstetter V, Verduin SJ, Larsson E, Baroni TJ, Thorn RG (2002) One hundred and seventeen clades of euagarics. Mol Phylogenet Evol 23(3):357–400

Mueller UG, Gerardo N (2002) Fungus-farming insects: multiple origins and diverse evolutionary histories. Proc Natl Acad Sci 99(24):15247–15249

Mueller UG, Wcislo WT (1998) Nesting biology of the fungus-growing ant *Cyphomyrmex longiscapus* Weber (Attini, Formicidae). Insect Soc 45(2):181–189

Mueller UG, Schultz TR, Currie CR, Adams RMM, Malloch D (2001) The origin of the attine ant-fungus symbiosis. Q Rev Biol 76:169–197

Mueller UG, Gerardo NM, Aanen DK, Six DL, Schultz TR (2005) The evolution of agriculture in insects. Annu Rev Ecol Evol Syst 36:563–595

Nielsen C, Williams DW, Hajek AE (2009) Putative source of the invasive Sirex noctilio fungal symbiont, Amylostereum areolatum, in the eastern United States and its association with native siricid woodwasps. Mycol Res 113(11):1242–1253

Noda H, Kodama K (1996) Phylogenetic position of yeastlike endosymbionts of anobiid beetles. Appl Environ Microbiol 62(1):162–167

Noda H, Koizumi Y (2003) Sterol biosynthesis by symbiotes: cytochrome P450 sterol C-22 desaturase genes from yeastlike symbiotes of rice planthoppers and anobiid beetles. Insect Biochem Mol Biol 33(6):649–658

Nygaard S, Zhang G, Schiøtt M, Li C, Wurm Y, Hu H, Zhou J, Ji L, Qiu F, Rasmussen M, Pan H (2011) The genome of the leaf-cutting ant Acromyrmex echinatior suggests key adaptations to advanced social life and fungus farming. Genome Res 21(8):1339–1348

Paine TD, Raffa KF, Harrington TC (1997) Interactions among scolytid bark beetles, their associated fungi, and live host conifers. Annu Rev Entomol 42(1):179–206

Pinto-Tomás AA, Anderson MA, Suen G, Stevenson DM, Chu FS, Cleland WW, Weimer PJ, Currie CR (2009) Symbiotic nitrogen fixation in the fungus gardens of leaf-cutter ants. Science 326(5956):1120–1123

Pulvirenti A, Zambonelli C, Todaro A, Giudici P (2002) Interspecific hybridisation by digestive tract of invertebrates as a source of environmental biodiversity within the *Saccharomyces cerevisiae*. Ann Microbiol 52(3):245–256

Reuter M, Bell G, Greig D (2007) Increased outbreeding in yeast in response to dispersal by an insect vector. Curr Biol 17(3):R81–R83

Rodrigues A, Carletti CD, Bueno OC, Pagnocca FC (2008) Leaf-cutting ant faecal fluid and mandibular gland secretion: effects on microfungi spore germination. Braz J Microbiol 39 (1):64–67

Rønhede S, Boomsma JJ, Rosendahl S (2004) Fungal enzymes transferred by leaf-cutting ants in their fungus gardens. Mycol Res 108(1):101–106

Roossinck MJ (2008) Symbiosis, mutualism and symbiogenesis. In: Plant virus evolution. Springer, Berlin, pp 157–164

Rouland C, Civas A, Renoux J, Petek F (1988) Synergistic activities of the enzymes involved in cellulose degradation, purified from Macrotermes mülleri and from its symbiotic fungus Termitomyces sp. Comp Biochem Physiol B 91(3):459–465

Rouland-Lefèvre C (2000) Symbiosis with fungi. In: Termites: evolution, sociality, symbioses, ecology. Springer, Dordrecht, pp 289–306

Rouland-Lefèvre C, Bignell DE (2001) Cultivation of symbiotic fungi by termites of the subfamily Macrotermitinae. In: Symbiosis. Springer, Dordrecht, pp 731–756

Rouland-Lefevre C, Diouf MN, Brauman A, Neyra M (2002) Phylogenetic relationships in Termitomyces (family Agaricaceae) based on the nucleotide sequence of ITS: a first approach to elucidate the evolutionary history of the symbiosis between fungus-growing termites and their fungi. Mol Phylogenet Evol 22(3):423–429

Rouland-Lefèvre C, Inoue T, Johjima T (2006) Termitomyces/termite interactions. In: Intestinal microorganisms of termites and other invertebrates. Springer, Berlin, pp 335–350

Sands WA (1956) Some factors affecting the survival of *Odontotermes badius*. Insect Soc 3 (4):531–536

Sasaki T, Kawamura M, Ishikawa H (1996) Nitrogen recycling in the brown planthopper, *Nilaparvata lugens*: involvement of yeast-like endosymbionts in uric acid metabolism. J Insect Physiol 42(2):125–129

Schultz TR, Brady SG (2008) Major evolutionary transitions in ant agriculture. Proc Natl Acad Sci 105(14):5435–5440

Schultz TR, Meier R (1995) A phylogenetic analysis of the fungus-growing ants (Hymenoptera: Formicidae: Attini) based on morphological characters of the larvae. Syst Entomol 20 (4):337–370

Seifert, K.A., Wingfield, M.J. and Kendrick, W.B., 1993. A nomenclator for described species of Ceratocystis, Ophiostoma, Ceratocystiopsis, Ceratostomella and Sphaeronaemella. 269–287. American Phytopathological Society, St. Paul, Minnesota, USA, pp 269–287

Shen SK, Dowd PF (1991a) Detoxification spectrum of the cigarette beetle symbiont Symbiotaphrina kochii in culture. Entomol Exp Appl 60(1):51–59

Shen SK, Dowd PF (1991b) 1-Naphthyl acetate esterase activity from cultures of the symbiont yeast of the cigarette beetle (Coleoptera: Anobiidae). J Econ Entomol 84(2):402–407

Smith TM, Smith RL, Waters I (2012) Elements of ecology. Benjamin Cummings, San Francisco, pp 112–225

Starmer WT, Aberdeen V (1990) The nutritional importance of pure and mixed cultures of yeasts in the development of *Drosophila mulleri* larvae in Opuntia tissues and its relationship to host plant shifts. In: Ecological and evolutionary genetics of Drosophila. Springer, Boston, MA, pp 145–160

Sternberg LDS, Pinzon MC, Moreira MZ, Moutinho P, Rojas EI, Herre EA (2007) Plants use macronutrients accumulated in leaf-cutting ant nests. Proc R Soc B Biol Sci 274(1608):315–321
Suen G, Teiling C, Li L, Holt C, Abouheif E, Bornberg-Bauer E, Bouffard P, Caldera EJ, Cash E, Cavanaugh A, Denas O (2011) The genome sequence of the leaf-cutter ant *Atta cephalotes* reveals insights into its obligate symbiotic lifestyle. PLoS Genet 7(2):e1002007. https://doi.org/10.1371/journal.pgen.1002007
Suh SO, Noda H, Blackwell M (2001) Insect symbiosis: derivation of yeast-like endosymbionts within an entomopathogenic filamentous lineage. Mol Biol Evol 18(6):995–1000
Suh SO, Gibson CM, Blackwell M (2004) Metschnikowia chrysoperlae sp. nov., Candida picachoensis sp. nov. and Candida pimensis sp. nov., isolated from the green lacewings Chrysoperla comanche and Chrysoperla carnea (Neuroptera: Chrysopidae). Int J Syst Evol Microbiol 54(5):1883–1890
Sung-Oui SUH, McHugh JV, Pollock DD, Blackwell M (2005) The beetle gut: a hyperdiverse source of novel yeasts. Mycol Res 109(3):261–265
Talbot PHB (1977) The Sirex-amylostereum-pinus association. Annu Rev Phytopathol 15 (1):41–54
Taylor FJR, FJR T (1983) Some eco-evolutionary aspects of intracellular symbiosis. In: Jeon KW (ed) Intracellular symbiosis. Academic Press, New York, pp 1–28
Urubschurov V, Janczyk P (2011) Biodiversity of yeasts in the gastrointestinal ecosystem with emphasis on its importance for the host. In: The dynamical processes of biodiversity - Case studies of evolution and spatial distribution. IntechOpen Limited, London, pp 277–302
Valiev A, Ogel ZB, Klepzig KD (2009) Analysis of cellulase and polyphenol oxidase production by southern pine beetle associated fungi. Symbiosis 49(1):37–42
Vega FE, Hofstetter RW (eds) (2014) Bark beetles: biology and ecology of native and invasive species. Academic Press, New York, pp 177–178
Vega FE, Blackwell M (eds) (2005) Insect-fungal associations: ecology and evolution. Oxford University Press, Oxford, pp 211–243
Vega FE, Posada F, Aime MC, Pava-Ripoll M, Infante F, Rehner SA (2008) Entomopathogenic fungal endophytes. Biol Control 46(1):72–82
Villesen P, Mueller UG, Schultz TR, Adams RM, Bouck AC (2004) Evolution of ant-cultivar specialization and cultivar switching in Apterostigma fungus-growing ants. Evolution 58 (10):2252–2265
Weber NA (1966) Fungus-growing ants. Science 153(3736):587–604
Weber NA (1972) Gardening ants: the attines, Memoirs of the American Philosophical Society, vol 92. American Philosophical Society, Philadelphia, pp 1–146
Wheeler WM (1907) The fungus-growing ants of North America. Bull AMNH 23:31
Wilson EO (1971) The insect societies. Belknap, Cambridge, MA
Wirth R, Herz H, Ryel RJ, Beyschlag W, Hölldobler B (2003) Herbivory of leaf cutting ants. A case study in the tropical rainforest of Panama. Springer, Berlin, Germany, pp 187–192
Witzgall P, Proffit M, Rozpedowska E, Becher PG, Andreadis S, Coracini M, Lindblom TU, Ream LJ, Hagman A, Bengtsson M, Kurtzman CP (2012) "This is not an apple"-yeast mutualism in codling moth. J Chem Ecol 38(8):949–957
Wood SL (1982) The bark and ambrosia beetles of north and Central America (Coleoptera: Scolytidae), a taxonomic monograph. Great Basin Nat Memoirs 6:1–1356
Wood SL (1992) A catalog of Scolytidae and Platypodidae (Coleoptera), part 2: taxonomic index. Great Basin Nat Memoirs 13:1–1553
Wood TG, Thomas RJ (1989) The mutualistic association between Macrotermitinae and Termitomyces. In: Insect-fungus interactions. Academic Press, New York, pp 69–92

Chapter 18
Panorama of *Metarhizium*: Host Interaction and Its Uses in Biocontrol and Plant Growth Promotion

Srinivas Patil, Gargi Sarraf, and Amit C. Kharkwal

Abstract Vectors have been wreaking a fatal havoc on mankind by causing diseases in agriculturally important plants and humans. Not only diseases caused by them are a hefty task to deal with, but their increasingly successful survival in human settlements is also a rising concern. The entomopathogenic fungi are considered amongst the first organisms for bio management of agriculturally important pests as they are eco-friendly, economically sustainable, and effective. With this, the collateral need for biocontrol in human disease vectors is also being felt. The first observation of fungi infecting insects was in as early as 900 AD, to the first data published in 1726 about entomopathogenic fungi. *Metarhizium* is a widespread fungus found all over the globe. More than 200 species of insects are infected by the fungus thereby making it one of the most sought biocontrol agents. This chapter gives an understanding of interaction between an arthropod host and entomopathogenic fungi genera *Metarhizium*, description of the host and fungal structure, what are some of the conventional and recent efforts done in order to improve the application strategies and what could be some of the possible uses of *Metarhizium* in enhancing plant health. Some of the plant pests and animal vectors which have been explored as host for *Metarhizium* are also mentioned.

Keywords Entomopathogenic fungi · *Metarhizium* · Insect–fungus interaction · PGP · Arthropod

18.1 Introduction

Metarhizium is distributed very much uniformly across the globe, from arctic to tropics and this feature gives it an edge over the other biocontrol agents. Being a fungus it does not need to be necessarily ingested, mostly all it requires is contact

S. Patil · G. Sarraf · A. C. Kharkwal (✉)
Amity Institute of Microbial Technology, Amity University Noida, Noida, Uttar Pradesh, India
e-mail: ackharkwal@amity.edu

N. Shrivastava et al. (eds.), *Symbiotic Soil Microorganisms*, Soil Biology 60,
https://doi.org/10.1007/978-3-030-51916-2_18

with the host cuticle (Kamareddine 2012) also it is very much prolific in terms of mass production (Rasgon 2011). *Metarhizium* genus comprises of species that have both narrow-spectrum entomopathogenecity and broad-spectrum entomopathogenecity, however, one of the most sought after species is *Metarhizhium anisopliae* (Aw and Hue 2017). *M. anisopliae* is a generalist that has been known to infect more than seven orders (Aw and Hue 2017). Majorly the earlier discovered species are *M. anisopliae, M. truncatum, M. cicadinum, M. brunneum, M. flavoviride, M. taii, M. cylindrosporum,* and *M. viridicolumnare* (Bischoff et al. 2009). However with advent of time several species and improved strains were also added to the list. The fungus is target specific, their generation time is short and it can survive in the environment for long when no host is available due to its ability of resting stage (Sandhu et al. 2012). Entomopathogenic fungi can be instilled in IPM (integrated pest management) as they show synergistic activity to control pestiferous insects in addition to the use of natural predators and other biocontrol agents like parasitoids and innovative approaches can be met in IPM using genomic techniques (Chandler et al. 2011; Erler and Ates 2015).

This fungus has many reasons to be used some of which are: firstly, effective broad-spectrum mortality, Secondly, fast and inexpensive mass production (Scholte et al. 2004). The complexity of using fungal spores as biocontrol agent is that the spores need optimal level of abiotic factors namely temperature, relative humidity, salinity, sunlight, and UV light exposure to break dormancy. Also, it needs contact with the host at all times to germinate which means it requires repeated applications if there are no hosts present for prolonged time and when we use it as a vector control agent it might be harmful to many nontarget insects (Scholte et al. 2004). The toxin destruxin affects structural integrity of cell membrane of host thereby damaging host tissue and it also causes fluid loss (Scholte et al. 2004)

18.2 Hosts

Certain strains of *Metarhizium* have successfully shown significant pathogenicity toward certain human disease vectors and plant pests. However, there have been some studies pertaining to some other strains which do not have a firm conclusion. So all in all there are several hosts for different *Metarhizium* strains. But in this chapter only selectively significant *Metarhizium*—host interactions have been mentioned (Table 18.1). The studies mentioning these interactions were aimed at the possible use of *Metarhizium* as a biocontrol agent (Table 18.2).

18.3 Structure and Mechanism

Although the insect anatomical structure is highly detailed, in this chapter we have elaborated certain portions of the arthropod anatomy, which is relevant in understanding the mode of infection. *Metarhizium* spore (conidia) generally germinates

Table 18.1 *Metarhizium* interaction with human and animal disease vectors

Species/strain	Parasitic on	Disease	References
Metarhizium pingshaense (Met_S26 and Met_S10)	*Anopheles coluzzii*	Malaria	Liao et al. (2017), Bilgo et al. (2018)
Metarhizium anisopliae ICIPE-30	*Anopheles gambiae*	Malaria	Mnyone et al. (2011)
M. anisopliae	*Aedes aegypti*	Yellow fever mosquito, chikungunya, dengue, etc.	Carolino et al. (2014)
M. anisopliae	*Culex quinquefasciatus*	Wuchereria, West Nile Virus, avian pox	Lacey et al. (1988)
M. anisopliae	*Aedes albopictus*	Dengue fever, Chikungunya, also capable of hosting Zika virus and certain nematodes	Scholte et al. (2007)
M. anisopliae	*Rhipicephalus sanguineus*	Canine ehrlichiosis	Kirkland et al. (2004)
	Ixodes scapularis	Lyme disease	
M. anisopliae (Ma959, MaE9 and Ma319)	*Anocentor nitens*	Tropical horse tick	Bittencourt et al. (2000)
M. anisopliae	*Ixodes scapularis*		Kurtti and Keyhani (2008)
	Amblyomma americanum	Rocky Mountain spotted fever (*Rickettsia rickettsiae*)	
M. anisopliae ICIPE 30	*Glossina fuscipes fuscipes* (controls parasite *Trypanosoma congolense*)	Trypanosomiasis	Wamiti et al. (2018)
Metarhizium brunneum ARSEF 4556 (with *Toxorhynchites brevipalpis*)	*Aedes aegypti*	Yellow fever mosquito, chikungunya, dengue, etc.	Alkhaibari et al. (2018)
M. anisopliae	*Rhodnius prolixus*	Chagas disease	Garcia et al. (2016)

when it comes into contact with the host's cuticle and then it outgrows the arthropod's body by draining its nutrition, resulting in the death of the mosquito (Fig. 18.1). The journey of the fungal spore from epicuticle (outermost interaction site) to hemocoel (terminal) witnesses the upregulation and production of many genes and proteins in both host and fungus. Prominent ones have been described in pathogenesis section.

18.3.1 Host Structure

A well-known fact about the insects is that they are devoid of an endoskeleton. It has only an exoskeleton. The integument is the outermost layer divided into epidermis

Table 18.2 *Metarhizium* interaction with various plant pests

Species/strain	Parasitic on (common name)	References
Metarhizium anisopliae	*Polyphylla fullo* (june beetle)	Erler and Ates (2015)
M. anisopliae (*ESALQ1604)*	*Mahanarva* (spittlebugs)	Iwanicki et al. (2019)
M. anisopliae (Metschnikoff) Sorokin variety anisopliae	*Tuta absoluta* (tomato borer); *Aethina tumida* (small hive beetle)	Muerrle et al. (2006), Contreras et al. (2014)
M. *brachyspermum* sp. nov. (Clavicipitaceae)	Elateridae (click beetles)	Yamamoto et al. (2019)
M. anisopliae	*Oryctes rhinoceros* (coconut rhinoceros beetle)	Indriyanti et al. (2017)
M. anisopliae	*Culicoides* spp.	Narladkar et al. (2015)
M. brunneum	*Bactrocera oleae* (olive fly)	Yousef et al. (2018)
M. anisopliae	*Bactrocera cucurbitae* (melon fly)	Sookar et al. (2014)
M. anisopliae (ICIPE 69 and ICIPE 18)	*Zeugodacus cucurbitae* (melon fly)	Onsongo et al. (2019)
M. acridum	*Locusta migratoria manilensis* (oriental migratory locust)	Zhang et al. (2015)
M. anisopliae	*Thaumatotibia leucotreta* (false codling moth)	Mkiga et al. (2020)

and cuticle. The cuticle is a chitinous structure in which the host arthropod body is enclosed. Protein makes 70% of the host cuticle (Charnley 2003). It can be interpreted that the cuticle is segregated into the epicuticle and procuticle. The procuticle is coated with a thin waxy and slimy layer known as the epicuticle. The procuticle is comprised of exocuticle and endocuticle. Mesocuticle is sclerotized and hardened region, which sometimes might be present in between them (Chapman 2012). It is mainly the pro cuticle, which is composed mainly of chitin and several other proteins. As discussed above the pro cuticle has endo- and exocuticle. So the endocuticle is a matrix of chitin intermixed with protein, providing a different property to the pure chitin. The modification of chitin with the protein confers its additional properties. While the exocuticle is mainly composed of sclerotin which essentially is a cross-linked form of certain proteins (Pryor 1940; Li and Ortiz 2014). Not only this sclerotin is present in exoskeleton, but it is distributed among the mouthparts used for biting and the dorsal and ventral sides. The composition of sclerotin varies among the insect hosts and also within the host sclerotin composition varies among different body parts, also certain regions are less sclerotinised certain regions are more sclerotinised (Russell et al. 2016). Beneath the exoskeleton lies the body which is divided into head, thorax, and cuticle (Chapman 2012).

18.3.1.1 Hemolymph

Hemolymph, which is considered as blood of arthropods, flows through sinuses referred to as a hemocoel. Hemocytes are the cells, which get circulated through hemolymph. Hemolymph mainly comprises of water, chlorine, sodium, potassium,

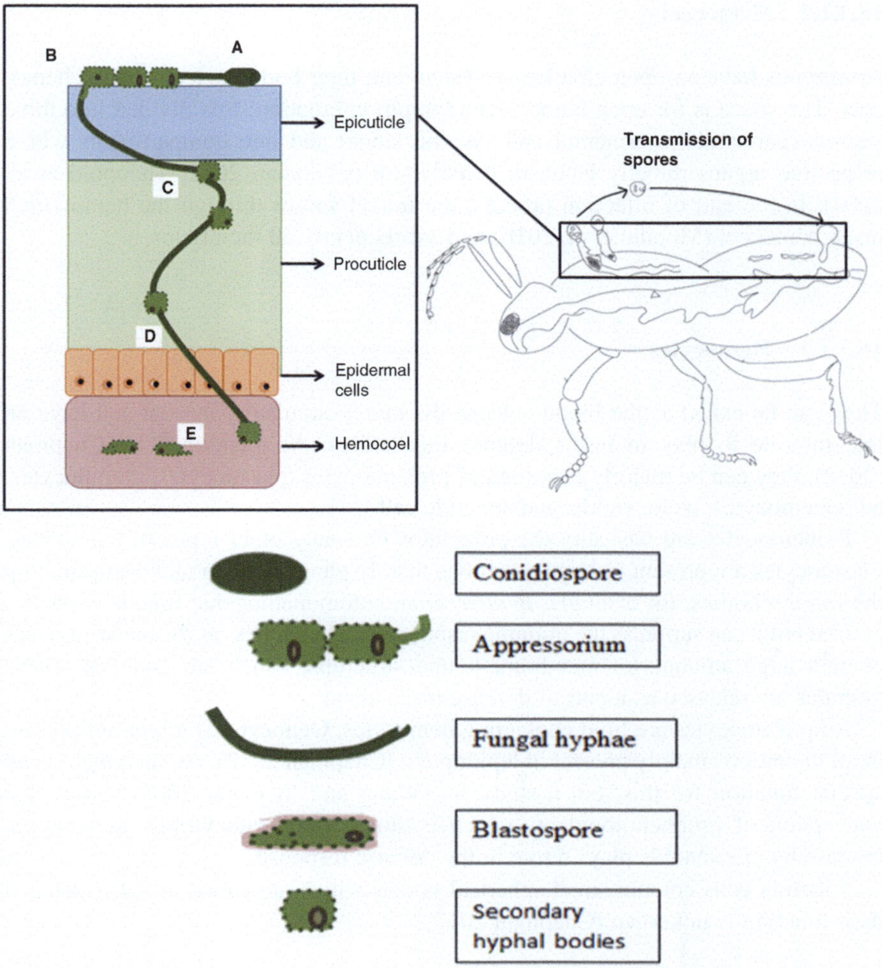

Fig. 18.1 The figure shows the path of fungal spore via the lateral cross-section of anatomical layers of the host and its life cycle. (**A**) conidiospore adhesion, (**B**) appressorium formation and peg penetration, (**C**) hyphal infection progressing in pro cuticle, (**D**) hyphae invading epidermal cells and (**E**) hemocoel colonization blastospores formation. Reference: based on Constanza Mannino et al. (2019)

calcium, magnesium, and biomolecules. Hemocyanin an oxygen-carrying protein is also directly present in hemolymph (Sowers et al. 2006). Significant concentration of free amino acids, presence of glucose, fructose, and sucrose are also found, although concentrations of these substances might vary from stage to stage (Wyatt 1961). As hemolymph is also the site for humoral defense responses for which many proteins are present like antimicrobial peptides, enzymes (later mentioned pathogenesis section), and also in some cases free amino acids like tyrosine aid in a process called humoral encapsulation (Chapman 2012), which contains the foreign body into a thick covering.

18.3.1.2 Hemocoel

Arthropods have an open circulatory system and their body cavity is called hemocoel. The space is for open blood (hemolymph) circulation. It is divided into three sinuses (pericardial, perineural and visceral sinus) and into compartments where respective organs directly bathe in hemolymph (Chapman 2012; Theopold et al. 2004). In the end of infection process, the fungal spores through the hemolymph reach hemocoel (Mondal et al. 2016) and subsequently all the organs.

18.3.1.3 Hemocytes

They can be called as the blood cells of the insect circulatory system and have an integral role to play in insect defense mechanism. As mentioned by Chapman (2012), they can be majorly classified as prohemocytes, plasmocytes, granulocytes, adipohemocytes, oenocytoids, and spherule cells.

Prohemocytes are basically the progenitor for many other types of hemocytes. Plasmocytes are present in large quantities mainly phagocytizing and encapsulating the foreign bodies, for example, *Beauveria* an entomopathogenic fungus is phagocytized but it can suppress the immune response. Granulocytes -as the name suggests contain large amounts of membrane-bound structures which are granules. These granules are released as a part of defense mechanism.

Adipohemocytes are lipid containing hemocytes. Oenocytoids might not present in all the orders, mainly present in lepidoptera (Chapman 2012). As such there is no special function for this, but a study by (Wang and St Leger 2007) shows that conversion of prophenoloxidase (which is stored into oenocytoids) into phenol oxidase by eicosanoids plays a role in the defense response.

Spherule cells contain small spherical bodies which are called as spherules and their function is unknown (Chapman 2012)

18.3.1.4 Fat Bodies

The biomolecules mostly present in the hemolymph are synthesized or stored at some time in the fat bodies. Fat bodies are a tissue-like organization of trophocytes or adipocytes which may be supplemented by the presence of urate cells, tracheal cells. It synthesizes many hemolymph proteins, stores glycogen which is later converted into trehalose a key sugar source in the hemolymph. It also serves as a storage site for most lipids in insects (Chapman 2012).

18.3.2 Fungal Spores

Metarhizium belongs to the Hypocreales order and the family Clavicipitaceae. The peculiar characteristic is that they are widely used insect pathogens. The spores of fungi are reproductive agents adapted for survival during unfavorable conditions and proliferate in a conducive environment. Conidiospores are asexual, exogenic spores formed mitotically. The hyphae are aseptate and the shape of each conidia is ovoid or cylindrical. It forms chains that appear cylindrical column-like or prismatic. It becomes conical at the apex and asci are arranged in dense hymenium (Money 2016; Sinha et al. 2016). The conidia grow on phialides which are whorls of branches ramified from conidiophores (Sinha et al. 2016). Conidiophores which bear conidiospores are translucent in appearance. The conidia can differ from species to species. The conidia can be straight-sided and small as of *M. anisopliae var anisopliae* or another var. *majus* can be large measuring up-to 18 μm. *M. flavoviride* have swollen light green conidia with club-shaped phialides. The rate of growth of conidia may differ too (Glare et al. 1996). The maximum temperature for most *M. anisopliae* isolates is 37 °C. However, there can be variability in thermotolerance (Fernandes et al. 2010).

18.3.2.1 Appressorium as a Structure

Appressorium is a specialized invasive structure which is basically an extension of germ tube to penetrate the host tissues which are intact. Appressorium is one of the salient features of both plant and arthropod pathogenic fungi. A study on in vitro appressorium production (Butt et al. 2016a) states that after germination as soon as the germ tube comes into contact with a hard surface infection structures are produced. As a plant defense response waxes may entrap and inhibit the germination of fungal conidia (Butt et al. 2016a, b). From a general prospect, appressorium can exist in unicellular, multicellular, or simply as a terminal swollen part of germ tube or completely differentiated structure in certain plant pathogens (Liu et al. 2012). Also as explained by Liu et al. (2012), in studies done on *Magnaporthe oryzae* during the development of appressorium—after germination, the conidiospore undergoes a set of events in the cell cycle and cell division. This is followed by appearance of an actomyosin ring which partitions the cell and the structure which will penetrate the host. Autophagy is then initiated in the spore cells which causes the cellular contents to flow inside the appressorium which makes it turgid and enhance its mechanical strength for penetration. Whereas in *Metarhizium*, MPL1 gene which produces perilipin homolog, Ca^{+2}, etc. maintains turgidity, actin cytoskeleton, chitin and dihydroxynaphthalene maintains structural support (Leger et al. 1991b; Gauthier and Keller 2013)

18.3.3 Pathogenesis

Metarhizium spore (conidiospore) generally adheres and germinates when it comes into contact with the host's cuticle and then it outgrows the arthropod's body by draining its nutrition, resulting in the death of the mosquito. Initially it starts with the development of appressorium, which penetrates into the host and then subsequently forms an infection peg, and then when it enters hemolymph, the formation of hyphae takes place, which releases toxins and then leads to death of the host (Scholte et al. 2004). The toxin destruxin affects the structural integrity of cell membrane of host thereby damaging host tissue and it also causes fluid loss (Scholte et al. 2004).

As referred from Aw and Hue (2017), there are six stages in the pathogenesis of *Metarhizium* (Aw and Hue 2017). These are given as below:

18.3.3.1 Adhesion

It is the initial event in which the asexual spores, which are conidia get attached to the cuticle of the host. The *Metarhizium* conidia are surrounded by an outer layer of rodlet cells. These rodlet cells consist of a protein "hydrophobin."

Hydrophobins are cysteine-rich proteins which are present in majority of fungi. These hydrophobins confer an amphipathic nature to rodlets aiding them in attachment to the hydrophobic epicuticle. Hydrophobins also play a major role in reducing the spore wettability thus forming a water-resistant layer (Sunde et al. 2008). When the spores are dispersed aerially they land on the epicuticle and the attachment of spore is due to hydrophobic interactions, electrostatic forces, and interaction of proteins (Aw and Hue 2017). Various external factors affect the attachment such as water, oxygen, nutrients, pH, hydrophobicity of host surface, and environmental conditions. The fungi can have specific requisites to infect restricted hosts (Sandhu et al. 2012).

Another recent study states that Mad 1 and Mad 2 are responsible for anchorage to insects as well as plants, respectively. Mad are *Metarhizium* adhesin like protein (Wang and St Leger 2007).

As explained by Greenfield et al. (2014) the adhesion follows a two-step process. Initially the superficial attachment occurs by electrostatic and hydrophobic forces or by attachment via adhesion proteins. The next step is to release enzymes to facilitate cuticle penetration. Also, the release of hydrolytic enzymes degrades fatty acids and release nutrients, which might aid in germination. The adhesins protein Mad 1 and 2 further strengthen the attachment (Greenfield et al. 2014).

18.3.3.2 Germination

The germination step is initiated by various non-specific exogenous nitrogen and carbon sources (Sandhu et al. 2012; Aw and Hue 2017). A study done by Santi et al. (2010) has reported different enzymes acting on the spore surface proteins. The spore surface proteins have been found to undergo degradation by different proteases.

These degrading enzymes comprise majorly of trehalase, seven different chitinases, two lipolytic enzymes have been detected. The trehalase enzyme ensures steady supply of glucose by breakdown of trehalose. Also phospholipase C which cleaves phospholipids was detected in *M. anisopliae* spores. The spore surface proteins not only have spore proteolytic activities but activities against reactive oxygen species (Santi et al. 2010), which might be produced on the host cuticle as defense response. Expression of Mest1 gene in *M. robertsii* helps aids in fast lipid hydrolysis and germination, on the contrary, *M. acridum* is helped by broadened host range expression of same gene. The expression of particular gene can be specific to particular host (Wang et al. 2011).

During the spore germination the spore absorbs water and nutrition from the host surface by osmosis and develops a germ tube which is an elongated structure (R. Barkai-Golan 2001)

18.3.3.3 Formation of Appressorium

Apart from the general organization and development of appressorium explained above, there are many specific molecular features in *Metarhizium*, which are explained below. Although some of them might be common to certain other entomopathogenic fungi. We can understand appressorium formation given below by bifurcating it into two major points. First is differentiation of germ tube end into appressorium and second maintaining of the turgor pressure for penetration into host cuticle (Fig. 18.2). Expression of ODC1 gene aids in appressorium formation. It encodes for ornithine decarboxylase. As it is known that ornithine decarboxylase enzyme is essential for cell growth as it stabilizes the DNA structure which prevents apoptosis (and supports the excess cell proliferation).The ornithine decarboxylase causes decarboxylation of ornithine which in turn aids in production polyamines which are directly involved in DNA structure stabilization (Pendeville et al. 2001). The expression of this gene is increased during appressorium formation and germination (Pulido et al. 2011). Pmk1 MAP promotes the appressorium maturation (Kershaw and Talbot 2009; Gauthier and Keller 2013). Chitin and dihydroxynaphthalene melanin deposits in appressorium act as structural support against the turgor pressure (Gauthier and Keller 2013). A study done by (Wang and St Leger 2007) shows that *Metarhizium* produces a protein MPL1 which is similar to a mammalian protein perilipin. The study reports that MPL1 confines the lipid molecules into droplets by binding with them similar to perilipin. Phosphorylation

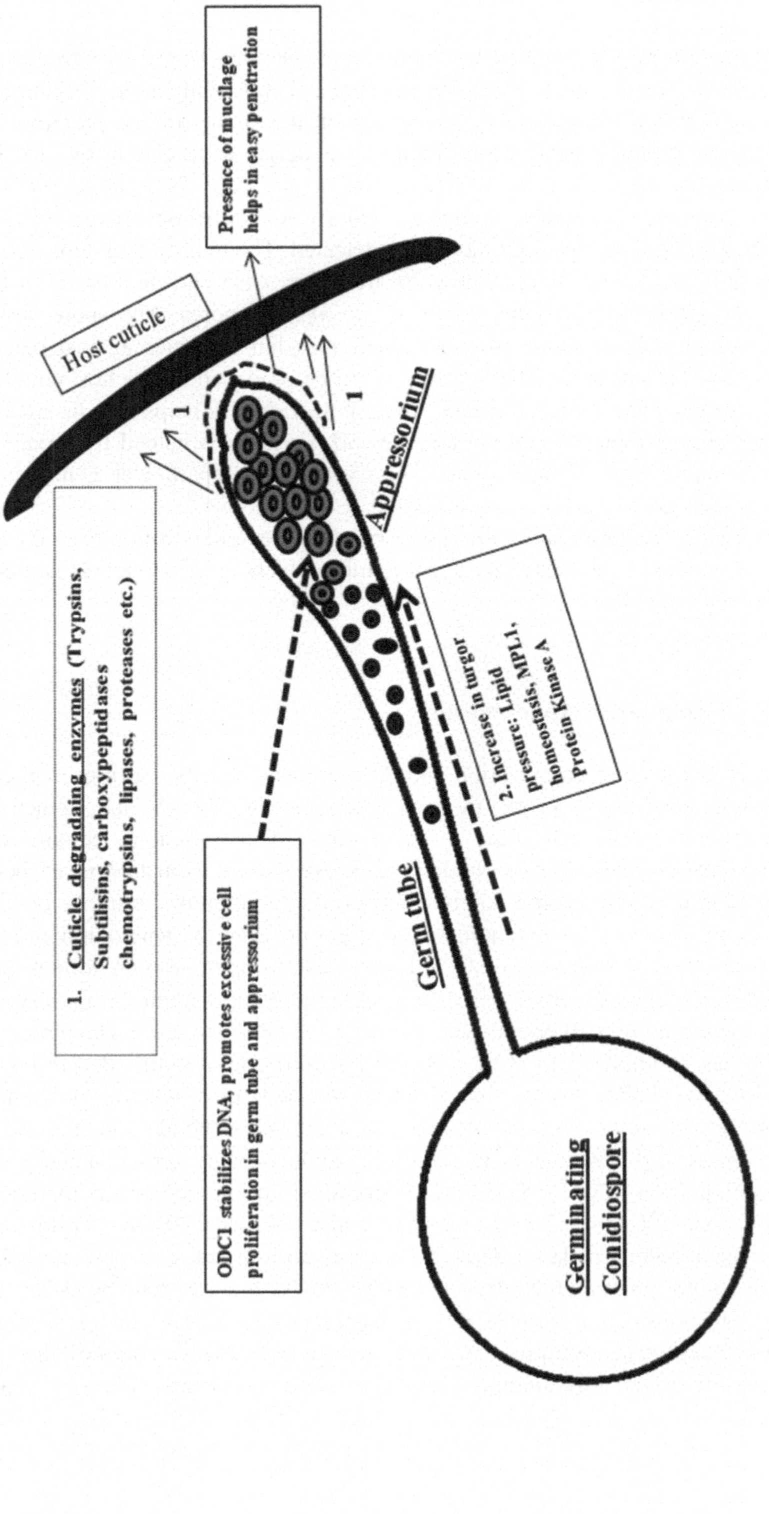

Fig. 18.2 The figure above explains appressorium formation

of this protein by cAMP dependent Protein Kinase A leads to release of the lipids and fatty acids contained in the lipid droplet. This helps in maintaining the turgor pressure, for example, in case of *M. grisea* a lipid breakdown product—glycerol acts as a solute. Its accumulation increases the water uptake (Thines et al. 2000; Wang and St Leger 2007). MAPK—mitogen activates protein kinase—plays a role in appressorium differentiation, Thines et al. (2000) reported that a mutant of *M. grisea* without MAPK did not had lipid mobilization. Aw and Hue (2017) stated about MAPK's intermediate role in adherence and lipid metabolism by regulating their respective genes. Leger et al. (1991a) proposed that Ca^{+2} ions get disrupted in the apical region of hyphae and get redirected to the region of cell enlargement in appressorium. The role of Ca^{+2} is also for actin cytoskeleton maintenance, which helps the appressorium to maintain its structure despite the turgor pressure (Leger et al. 1991b). Zhang et al. mention that chitin synthase MaChsIII, MaChsV, and MaChsVII are also involved in the disruption of host defense responses apart from Appressorium development (Zhang et al. 2019).

18.3.3.4 Penetration

Penetration into the host body involves cuticle breakdown. As cuticle composition varies from host to host as a result the amount and diversity of hydrolytic enzymes released by the fungus also varies. So, different hosts have specificity for different proteins and their concentration (Aw and Hue 2017). Studies by Leger et al. (1991a) have shown the presence of cuticle degrading enzymes in ungerminated conidia of *Metarhizium* namely esterase, chymoelastase protease, and *N*-acetylglucosaminidase. Also, the same study reported that the amount of these enzymes was more on an infected host when compared to in vitro conditions.

Trypsins, subtilisins, carboxypeptidases, and chemotrypsins and other proteases are secreted which degrades the protein part of procuticle of hosts. Pr1 and Pr2 are spore surface proteins responsible for proteolytic activities (Santi et al. 2010). Pr1 is a serine protease which degrades the cuticle by hydrolyzing proteins (Screen et al. 1997). Apart from proteases, chitinases are also involved.

Precisely chitin isoforms are secreted to limit its specificity to the host. Lipases found on conidial surface which interact with the lipid-rich epicuticle release fatty acids which leads to the enhanced hydrophobic interactions between conidia and host (Beys da Silva et al. 2010; Aw and Hue 2017). Also free fatty acid may act as a nutrition source.

18.3.3.5 Colonization

Host hemolymph is colonized next, after breaching the cuticle (Branine et al. 2019). The hemolymph is the site for the host's defense responses. The insect defense system is divided into humoral and cellular responses. The humoral response includes the antimicrobial peptides, reactive oxygen species, etc. The cellular response comprises of encapsulation and phagocytosis by the hemocytes (Lavine and Strand 2002). To counteract the defense response when the encapsulated spores release from the hemocytes, destruxin is released (Aw and Hue 2017).

Destruxins are the insecticidal secondary metabolites responsible for the fungi's virulence (Dornetshuber-Fleiss et al. 2013). The role of destruxins in being virulent to the insect primarily includes (Golo et al. 2014)—suppressing the defense response, hindering the fluid secretion by malpighian tubules thereby interfering with the osmoregulation (James et al. 1993), they are also shown to block vacuolar H^+ ATPases (V-ATPases). V-ATPases via ATP hydrolysis pump the protons into lysosomes, Golgi, late endosomes, and other membrane-bound compartments to maintain the required acidic conditions (Toei et al. 2010) and it also possesses anti-feedant properties (Amiri et al. 1999; Golo et al. 2014).

Presence of catalase, peroxidase (for breakdown of certain reactive oxygen species), genes which code for proteins, which help in breakdown of antimicrobial peptides, MaAC, and certain other genes protect the fungal cells from chemical and physical stresses (Aw and Hue 2017) These stress might arise as a consequence of host immune response. Trehalose makes up the major part of the insect hemolymph. Instant energy to the insects and survival in abiotic stresses are its role. As it is a disaccharide it proves to be the main carbon source for fungal spore growth when it grows in insect hemolymph (Shukla et al. 2015).

Trehalase is also present to hydrolyze the trehalose in the hemolymph into glucose, secreted as extracellular enzyme by the fungi (Xia et al. 2002), Presence of MOS1 gene increases the fungal survival at high osmotic pressure (Wang et al. 2008). All these factors aid in the survival of the fungi in the harsh environment of hemolymph. Inside the hemolymph, some fungal cells proliferate into protoplasts which prevents their recognition by host defense system because the recognition proteins may be present at the cell wall (Mondal et al. 2016).

18.3.3.6 Sporulation

After the fungi have proliferated into host hemolymph and Hemocoel, it feeds on host's nutrients, which leads to the death of the host insect. After this, upon receiving suitable conditions the fungal spore germinates and the hyphae extrude out of the corpse, although spores of *M. anisopliae* have shown to germinate internally in dried corpses (Mondal et al. 2016). Spore formation is an important procedure for the dissemination of fungal diseases. Metalloproteases Mrmep1 and Mrmep2 have been shown to be responsible for the sporulation of *M. robertsii*, (Zhou et al. 2018). The conidia of *M. anisopliae* were present before it fully colonized the host. Mycelium growth is usually observed near the antennae base, on cervix, and its mouthparts (Sun et al. 2002).

18.4 *Metarhizium* Application Methods in Vector Control: A Superfluity

As discussed in this text, it is quintessential need for the fungal spore to come into direct contact with the host in order to infect it so the application strategies must be wisely chosen. Use of *Metarhizium* in agricultural biocontrol procedures mostly comprises of powder-like substances or wettable suspensions as a carrier material but in case of vector biocontrol this entomopathogenic fungi can be inoculated into the breeding grounds, stagnant waters and in some studies have also reported an increase in mortality rate of *Anopheles gambiae* when use of mud panels, polyester netting, and cotton cloth as a holding material was done for the fungus (Aw and Hue 2017). In another study, it was observed that *Metarhizium* was able to interfere with the DDT and permethrin resistance genetically, due to which chances of susceptibility to these insecticides were increased (Farenhorst et al. 2009). Although insecticide-treated net and residual sprays are most popularly used options but apart from them plethora of setups and carrier materials have been experimented with but here we shall discuss some important conventional and recent molecular approaches.

18.4.1 Experimental Huts

A study was carried out in the malaria-endemic region of Tanzania (Mnyone et al. 2012). Local housing huts were designed to contain fungal suspension (which comprised of *B. bassiana* and *M. anisopliae*) infected polyester nets, curtains, bed net strips, panels, and baffles. The results showed a decrease in survival rate of mosquitos and a significant percentage (68%–76%) of mosquitos had fungal growth which shows increased contact surface (Mnyone et al. 2012). It also showed the effect of this strategy on biting behavior and malaria transmission from the mosquito.

18.4.2 Using Paper Substrates as a Resting Material for Fungal Spores

In another study (Farenhorst and Knols 2010) smooth paper pieces were coated with fungal formulations (*Metarhizium anisopliae*) made in low viscous material both manually and mechanically and were used instead of sprays. A stainless steel bar was used to apply this suspension. This method was very effective in standardizing the amount of fungal biomass used and its exposure time to the mosquito.

18.4.3 *Water Storage Pots as a Carrier Material for* Metarhizium

In this project (Farenhorst et al. 2008) African water storage pots were used as wet clay pots were an attractive breeding site for mosquitos. In this study, *Anopheles gambiae* and *Anopheles funestus,* which are an evident malaria vectors, were used. Oil formulated conidia were sprayed uniformly inside the surface of clay pots. The dead mosquito cadavers after the application were studied for fungal growth. The results inferred that clay pots were an attractive sight for mosquito breeding, also these studies (Farenhorst et al. 2008) suggested that in future some other mosquito attractants could also be combined in this clay pot approach.

18.4.4 *Combination of* Metarhizium *with Insecticide-Treated Nets*

With the ease of availability and inexpensive nature, insecticide-treated nets (ITN) have become the mainstay in vector control. But, while combining ITNs with a fungal suspension major focus is to enhance the functional ability of ITNs. A study was done (Hancock 2009) in order to check the performance of fungal suspension intervened ITNs, in this various factors were considered like fungal infection, gonotrophic feeding process, ITNs, etc. then a model was proposed for which mathematical analysis was done. As a result (Hancock 2009) an extensive information was produce which had many suggesting reasons to use fungal biocontrol integrated with ITNs such as high virulence, more probability of fungal infection in host, prolonged period of fungal exposure, etc.

18.4.5 Metarhizium *in Odor Bait Stations (OBS)*

This was a study done by Lwetoijera et al. (2010) where an OBS was used against a malaria vector *Anopheles arabiensis*. OBS are box-like structures made on a wooden frame which is covered with a canvas. The entire device except the floor is covered with a black cloth and inside this device there is a mosquito lure which mainly consists of carboxylic acids, carbon dioxide, and ammonia (Okumu et al. 2010; Lwetoijera et al. 2010). It contains a funnel-like opening and it also has *Metarhizium* conidia treated baffles. The results showed an increase in infection rates and also it is much anticipated because using OBS is much more safe than using fungal suspension in human dwellings as done in experimental huts (Mnyone et al. 2012) and other such methods.

18.4.6 Oil as a Carrier Material

18.4.6.1 Mineral Oil

A study done by (Bukhari et al. 2011) focused on comparing the efficiency of an aqueous substance, a synthetic oil formulation (shellSol T) and dry carrier material (wheat flour, white pepper, and fine bicarbonate particles) as a carrier material of fungal spores in aquatic habitats. The results revealed that Shell Sol T a synthetic oil is an effective spore carrier material in comparison to aqueous and the dry carrier material in both laboratory and field trials. In fact the field trials witnessed a decrease by 39–50 percent of the *Anopheles gambiae* population. In comparison to non-formulated fungal spores where efficacy was very low, Shell Sol T formulated spores showed a promising effect and increased persistence in water. This study gives another perspective—usage of synthetic oils as a carrier material. Also, the oil used here had minimal toxicity to the aquatic habitat as the quantity used here was very less compared to that of the safe limit.

18.4.6.2 Vegetable Oil

This experiment focused to check the viability of *M. flavoviride* spores after storage in different formulations and variable temperature. As a diluent, dedeorized kerosene oil preserved the spores better than Shellsol K, apparently during short duration of storage; however, mean conidial growth was found more in Shellsol K after 32 weeks. Vegetable oils were effective but its efficacy was improved with addition of antioxidant in case of groundnut oil. It was also observed that lower temperature led to increased germination of spores. The addition of silica gels, which aided by drying out spores, showed significant results as well. Still there was requirement for further studies to get the better understanding of storage techniques under variable conditions. (Moore et al. 1995).

18.4.7 Mosquito Landing Boxes (MLBs) for Metarhizium

MLBs are devices that have natural or synthetic human odor as a mosquito attractant. It is a system based on odor baiting technology. These devices are basically a wooden box with solar panels on the top which powers the odor dispenser. Particles of the odor solution lands on the walls of the device which may have fungal spore coating or some insecticide. In a study Lwetoijera et al. (2010) used this device and the walls of the MLB were coated with spores of the entomopathogenic fungi *M. anisoplieae* and vector targeted here was *Anopheles arabiensis.* This trial was conducted in a semi-field system. Separate cups containing larvae were placed near both MLBs and control system and it was supposed that if a mosquito captured from the semi-field system is let into the cups containing larvae and the larvae is not able to survive, and then the mosquito has been contaminated with the fungal spores and vice versa. Many factors like larval mortality, pupation, amount of hyphae growth on cadavers, etc. were assessed to check the level of contamination of the mosquitos with the fungi. The results revealed that 43% of mosquitos were contaminated with the fungal spores compared to 0% in the control. In this study alternatively, a chemical pyriproxyfen was also used in the system. This study inferred that MLBs are an effective tool for delivering fungal biopesticides as a decrease in the survival rate and direct killing of host-seeking mosquitos was witnessed.

18.4.8 **Metarhizium** *in Combination with Phytochemicals*

Use of neem oil as an insecticide is a very popular and old method but despite that today we see very less use of such phytochemical in vector control. A study done by Simone A. Gomes and coworkers (Gomes et al. 2015) focused on the use of neem oil as an adjuvant for the entomopathogenic fungi *Metarhizium*. In their study, they used two systems, one to check the effect of neem oil used solely on *Aedes aegypti* and another one to check the effect of neem oil in combination with *Metarhizium*. Neem oil of variable concentrations was used. Statistical analysis of the survival curve was done. The results revealed that at the concentration of 1×10^8 conidia per ml, a very low survival rate of 12% was observed. Also, later it was suggested that the addition of neem enhanced virulence. This study suggested the use of adjuvants such as phytochemicals along with the fungal biopesticides. In future, many other potential phytochemicals can be used as an adjuvant to variety of fungal biopesticides.

18.4.9 **Metarhizium** *for Chemical Resistant Vector Hosts*

Insecticide resistance is a prominent and emerging problem in the area of vector control. But the use of entomopathogenic fungi for insecticide-resistant vectors is one of the ways to fight insecticide resistance. An interesting study by Blanford et al. (2009), which was done on the vector *Anopheles gambiae* which was insecticide resistant—has interacted with *Metarhizium anisopliae* and another entomopathogenic fungi and results revealed increased susceptibility of the insecticide-resistant strain of host. The resistant strain used here was *Anopheles gambiae s.s.* VKPER, which is a pyrethroid-resistant strain. Also an insecticide susceptible strain *Anopheles gambiae* SKK strain was treated with the same procedure. Various mechanisms were used to deliver the fungal isolates to the mosquito-like formulations in synthetic oils (Kamareddine 2012), dry conidia (Ondiaka et al. 2015), etc.

18.4.10 Delivery System in Agriculture Fields

Apart from the above-given methods, there are some specialized methods that can be instilled for the delivery of fungal spores to the host. The formulations are designed to increase the viability of spores and to expose them lucratively to the host.

18.4.10.1 Kaolin Based

The spores of fungi are mixed with 80% kaolin. Prior to use, it is mixed with water and wetting agent solution. Then solution is sprayed directly onto the plant (Goble et al. 2016).

18.4.10.2 Patty Blend Formulation

Vegetable oil and sugar are mixed with pre-weighed conidiospores. Conidial viability is increased by the addition of Silwet and Saboraud maltose agar. Lactophenol cotton blue is added to stop its germination. The insects are treated with the strip containing serially diluted formulation (Kanga et al. 2010).

18.4.11 Molecular Approaches

The DNA technologies in the new age facilitate the addition of new gene into fungi and perform gene manipulation to increase efficacy of the fungi. Due to various stress condition their efficiency gets decreased but the DNA technology can be useful to improve the ability of virus to sustain in unfavorable conditions and more virulence. By expressing the endogenous proteins in the *Metarhizium* the pathogenesis success rates increases by targeting the cuticle, physiology, and hormones (Lovett and St Leger 2018).

In an experiment performed by Leger et al. (1991b), a genetic modification was done in *Metarhizium* in which more number of copies of *Pr1 gene,* which basically codes for protease that degrades cuticle of host, was inserted. When newly engineered *Metarhizium* was made to infect *Manduca sexta* there was overproduction of gene and activation of phenoloxidase system. The results include reduction in death timing, food consumption, and biological containment of the fungi. One of the reasons for biological containment was due to the accelerated melanization of the host cadaver which in turn provides insufficient substrate source for fungal spores growth (St Leger et al. 1996). Phenoloxidase is activated by prophenoloxidase cascade and provides immunity to the insects and polymerization of the indole group of phenoloxidase leads to the formation of melanin which leads to melanization precisely upon with infection with fungal spores (González-Santoyo and Córdoba-Aguilar 2012; Carolino et al. 2014; Butt et al. 2016b; Zhang et al. 2017).

Peng et al. (2015) performed an experiment in which *ATM1 gene* was overexpressed which codes for trehalase. Trehalase degrades trehalose which provides fungus with the carbon source in hemolymph of insect host. When results were compared to the wild strain of *Metarhizium* the genetically engineered fungi showed more degradation of trehalose and growth enhancement of fungi in host hemolymph.

In an another experiment, scorpion toxin *BjαIT* was used for genetic manipulation of *Metarhizium* the results showed enhanced virulence by the fungi in host infection. The growth of fungal spores on cadaver did not report any drop which might not affect the transmission. Though the yield was reduced but germination and formation of appressorium were the same as the wild-type strain and the lethal dose and lethal time were less to (Peng and Xia 2015).

Transgenic *M. pingshaense* was used to control the insecticide-resistant *Anopheline* mosquitoes. The new genetically modified fungi were hybrid it had voltage-gated calcium blocker with *kappa hexatoxin Hv1a and calcium-activated potassium genes*. The results in labs showed that efficacy was increased and it was able to control the resistant malaria vector. This can be used for on field application in the future (Lovett et al. 2019).

In heat stress conditions hyphal cells may start producing Reactive oxygen species (ROS). Pyruvate acts as ROS scavengers but the rate of formation of pyruvate is slower than the formation of ROS. A transgenic *Metarhizium* was designed so as there is overexpression of genes and therefore increased concentration of pyruvate kinase will be produced. During conidia formation, the pyruvate kinase gets accumulated and this helps conidia to survive during heat stress (Wu et al. 2019).

Genetic modification to create transgenic *Metarhizium* has been successful in many of the cases as mentioned above and showed an increase in virulence as performed by Peng et al. (2015), Lovett et al. (2019) and some (Wu et al. 2019) showed considerable efficacy in lab as well which can further be implemented in field and tested for the outcome. Extensive research in knowing the enzymes, genes, and host immune system in addition to fungi evasion and invasion techniques can help to genetically manipulate the fungus and increase its efficacy.

18.5 Plant Growth Promotion

Mycorrhizae are obligate biotrophs and endophytes aids in improving plant growth and nutrients acquisition from soil to the plants (Karandashov and Bucher 2005; Behie et al. 2017). Unlike mycorrhiza, *Metarhizium* as an endophyte shows no obligatory nature as it can survive in soil freely, as entomopathogens or as saprophytes (Behie et al. 2017). Beyond the activity of being pathogenic to the insects, there is one more benefit which is, plant growth promotion; although this is an area that has not been extensively studied (Canassa et al. 2019). This improves the yield In addition to the pest management of the plants. For sustainable agriculture plant growth promotion coupled with pest inhibiting capabilities can help to discontinue the use of heavy pesticides and fertilizers (Senthil Kumar et al. 2018). *Metarhizium* can induce root hair development, nitrogen translocation, improved absorption of iron, and auxin production (Behie and Bidochka 2014; Moonjely et al. 2019). Colonization of plant tissue makes it an endophytic fungi, which enhance plant biomass as well as it can increase nutrient mobilization and its transfer (Krell et al.

2018). The soil fertility may affect fungal activity and nutrient metabolization. Many studies have been performed but some of them lack a firm conclusion thus it requires further research to be done.

18.5.1 Exchange of Nutrients and Endophytic Nature

Nitrogen is crucial for plant growth and the loss of nitrogen due to insect herbivory leads to deprived nitrogen content available for plants (Behie and Bidochka 2014). Several fungi in symbiotic association with plants transfer nitrogen (Wang et al. 2017a, b). *Metarhizium* has a wide host range and is pervasive worldwide (Hajek and St. Leger 1994). *Laccaria bicolor* transfers the nitrogen derived from insects back to the plant white pine. *Metarhizium* has shown similar results when experimented with the insect *Galleria mellonell* (waxmoth) N^{15} labeled. This was performed on Switchgrass and haricot beans. The *Metarhizium* spp. were able to increase the plant productivity (Behie et al. 2012). In another experiment, five species of *Metarhizium* were tested all expressed positive results. *M. robertsii* was tested on field in natural conditions, showed significant results (Behie and Bidochka 2014). A recent experiment conducted shows that MepC and Mep2 are two ammonium permease genes which have been involved in nitrogen derived from insects and also in colonization process (Moonjely et al. 2019). Research for finding genes responsible for symbiosis of Insect pathogenic fungi can help to understand the functioning elaborately.

A recent study conducted by Behie et al. (2017) gives evidence that *Metarhizium* derives carbon from the plants as much as other endophytes. The carbon translocation can sustain fungi when insect host is absent. Reportedly, when the host was present there was significant increase in carbon transfer to the fungi (Behie et al. 2017). This nature of reciprocation of nutrients helps both plants and fungi.

Plant photosynthates containing carbon were found in the roots of the plants and in the rhizosphere which are utilized by the fungi as substrate. The ^{13}C (CO_2 given to plants had ^{13}C isotope) used was found to be incorporated in the fungi which had been provided by the plants as a symbiotic relationship. In fungi, it was traced to NAG and other Carbon-based components (Behie et al. 2017). The decomposing cadaver when added with *Metarhizium* spores lead to increase in ammonium and nitrate in the soil and spores were able to colonize plants as well that resulted in better plant growth (Kryukov et al. 2019).

Metarhizium has evolved as an endophyte (Moonjely et al. 2016). In an experiment performed by Barelli et al. (2018), the analysis was made between how extensively the *Metarhizium* colonizes the plant roots. For precise detection Plate culture method (c.f.u count) and quantitative PCR, both were done. The results showed that there was fungal colony in rhizosphere, rhizoplane, and within the roots too. Another thing that was noticed that the population of fungi did vary with the number of days post-inoculation (Barelli et al. 2018) but with this experiment it is evident that *Metarhizium* can colonize roots efficiently. Thus there is scope of further investigation, whether this fungi can colonize phyllosphere or not.

18.5.2 Improved Iron Absorption on Calcareous Substrates

In an experiment performed by Raya-Diaz et al. (2017) on sorghum plants, the relationship between plants, entomopathogenic fungi, and soil can be established well. It concluded that the entomopathogenic fungi tested in case of calcareous soil, *M. brunneum* turned out to be the most efficient fungus which lowered the pH of the alkaline soil by releasing the organic acids. The plants suffering from iron chlorosis are iron-deficient plants. This is common concern for the plants growing in calcareous soil either acidic or basic as it leads to the less iron availability in the form plants requires for its uptake (Brown 1956). As tested in vitro, iron oxides changed to dissolved iron forms. The Fe chlorosis symptoms were seen to be assuaged by *M. brunneum* on the sorghum plants which were grown on the artificial substrate that was calcareous. The best method was soil inoculation method (Raya-Diaz et al. 2017). In another experiment performed by Sánchez-Rodríguez et al. (2016) the wheat and sorghum plant showed increase in growth and chlorophyll content by improving the iron bioavailability in soil.

18.5.3 Auxin Formation for Plant Growth

Auxin is a plant growth hormone which influences plant physiology and developmental process to the stimuli sunlight and gravitropism (Bhattacharya 2019; Pandey et al. 2019). In an experiment performed by Liao et al. (2017) vegetative growth of corn plants has shown improvement and the yield was increased. Avirulent *Metarhizium* strain (*Δmcl1*) contributed to the growth of plants as well proving that the plant growth promotion activity is not influenced by entomopathogenic activity. Auxin was produced by fungi, there was increase in formation of leaf collar, foliage biomass, and the length of the stalk where the spores were able to colonize. The culture filtrate even contained auxin which showed positive result (Liao et al. 2014). The growth of plants is due to combined effect of chemicals and auxin (IAA) dependent pathways importantly which is produced by *Metarhizium*; promoting the lateral root and root hair development (Liao et al. 2017).

18.5.4 Proliferation of Plant Cells and Disease Suppression

The *Metarhizium* species can be potential plant endophytes and can live inside plant tissues. In an experiment performed it was noticed that their role can be in increasing the size of stalk, root length, and weight of the root (Mantzoukas et al. 2015; Greenfield et al. 2016). It can cause proliferation of lateral root hair for enhanced plant growth (Sasan and Bidochka 2012). The growth rates differ with respect to the strain used, duration, and the inoculation amount (Jaber and Enkerli 2017). In bean plants, there were significant results to show that it improved the reproductive and vegetative growth also (Canassa et al. 2019).

Jaber (2018) performed an experiment whether *M. brunneum* can be effective for suppression of disease-causing pathogens. The results showed significant decrease in the occurrence of disease, its development, and intensity. The fungus was able to promote shoot and root growth and the weights.

18.6 Conclusion: In the Light of Recent Advances

With the increased understanding of the molecular aspects of *Metarhizium* such as the genes and proteins involved in pathogenesis and their upregulation, secondary metabolites and small molecules, molecular aspects of host immunity, genome-wide studies, etc. have made the scope of research on *Metarhizium* infinitely vast. Donzelli and Krasnoff (2016) state that the recently available genome sequences give many biosynthetic pathways and ability to produce secondary metabolites which surpasses the current knowledge of chemistry. Basically, this study focuses on details of genes involved in the production of secondary metabolites. Also, there have been studies (Brancini et al. 2019) where light has shown to affect the gene expression in *Metarhizium*, after periodic exposure to light upregualtion and down regulation of certain proteins takes place. One such protein is photolyase which is upregulated and is responsible for UV tolerance. The changes concluded in inference that light is involved in stress and signaling. So light might be a factor in controlling the efficacy of *Metarhizium*. Studies by Mukherjee and Vilcinskas (2018) and Hussain (2018) discuss about changes in the gene expression. On recognition of the fungal spore by the host immune system, the host increases the expression of certain antifungal peptides in response to overcome the hostile environment in host, the fungi also increases the expression of certain proteins that destroy these peptides which are epigenetically controlled. This way of the counteracting molecular responses in host and the fungi give further insights into the coevolution process.

Except from entomopathogenecity *Metarhizium* has certain effects on growth promotion in plants. A recent review by Hu and Bidochka (2019) has mentioned that species from *Metarhizium* genera are root endophytes and have a symbiotic relation as they provide insect-derived nitrogen and get photosynthates in return. They have further reviewed the factors governing the rhizospheric interactions. Also, there has been a recent study on the host cadaver decomposition affecting plant growth promotion (Kryukov et al. 2019). The decomposed cadavers contain more ammonia and nitrogen compared to cadavers overgrown by fungus. It was concluded that fungi were unable to sporulate on decomposed cadavers and provided nitrogen faster from the cadavers overgrown by fungus.

So all these facts and studies boil down to some important inferences:

1. Evolution of *Metarhizium defense response*—As discussed in the pathogenesis section there are several lytic enzymes which, facilitate the entry of *Metarhizium* into the host. As stated by Mukherjee and Vilcinskas (2018), host can recognize and counteract these proteins by releasing antimicrobial compounds, proteinase inhibitors, and antifungal compounds. Same study has shown that the expression of chymotrypsin and metalloproteinases by *Metarhizium* can counteract the host defense compounds. An *in vitro* increase in metalloprotease activity was observed in response to the antimicrobial peptides (AMP) like metschnikowia, lysozyme, and proteinase inhibitor. Another significant component of the insect defense system is hemocytes. The presence of destruxin has been mentioned in the text previously. This protein causes actin remodeling, pyknotic nuclei, and blebbing in plasmocytes (Götz et al. 1997). Another toxin cytochalasin is also involved but it is less toxic than destruxin. Both these toxins selectively regulate the expression of IMPI and lysozyme, which are antimicrobial peptides. Also induction of genes involved in epigenetic responses of histone acetylation and deacetylation in *M. robertsii* against an AMP shows, how specific modification in *M. robertsii* at transcriptional level is made to counteract host defense system. This might suggest lesser chances of host gaining resistance against the fungi. (Mukherjee and Vilcinskas 2018)
2. *An ideal biocontrol strategy must focus on enhancing efficacy of both the biological agent and its carrier*—There have been several attempts in creating a recombinant strain of *Metarhizium,* which is more effective by selecting genes like toxin genes as a candidate some of which were discussed in this article. Also carriers with improved efficacy are in a need to be developed considering certain environmental factors which have shown to have a stressful impact on the growth of *Metarhizium.* Study done by Wang et al. (2017a, b) on *Galleria mellonella* and *M. robertsii* mentions an important role of DNA methyltransferases, which are responsible for epigenetic or gene expression control. Here they have shown to have a role in stress tolerance and virulence of the fungi. This shows how transgenic fungi can be effective.
3. *Metarhizium as a complete plant health package*—With the well-established entomopathogenic effects and some endophytic plant growth promotion activities, *Metarhizium* can be used as a holistic supplement. A suggestable effort could be improving *Metarhizium* transgenically with plant growth promotion activity besides its entomopathogenecity. Although all these have been proven experimentally, still there are requirements of field assays. Possessing the knowledge about timely usage of biocontrol agents is very important to avoid emergency pest mitigation; especially when using a fungus, because it is temperature and humidity dependent. All of these characteristics can be brought to better use by focusing on synergistic approach like combining it with bio-fertilizers and using it with other biopesticides. Not only nitrogen translocation but also other mineral utilization by plants can be improved with scientific studies. So as with time,

entomopathogenicty of *Metarhizium* is being explored in newer hosts we are becoming more molecularly aware about it. The confluence of this scientific awareness and industry requirements is where *Metarhizium* promises a vast scope.

References

Alkhaibari AM, Maffeis T, Bull JC, Butt TM (2018) Combined use of the entomopathogenic fungus, Metarhizium brunneum, and the mosquito predator, Toxorhynchites brevipalpis, for control of mosquito larvae: is this a risky biocontrol strategy? J Invertebr Pathol 153:38–50. https://doi.org/10.1016/j.jip.2018.02.003

Amiri B, Ibrahim L, Butt TM (1999) Antifeedant properties of Destruxins and their potential use with the Entomogenous fungus Metarhizium anisopliae for improved control of crucifer pests. Biocontrol Sci Tech 9:487–498. https://doi.org/10.1080/09583159929451

Aw KMS, Hue SM (2017) Mode of Infection of Metarhizium spp. fungus and their potential as biological control agents. J Fungi 3(2):30

Barelli L, Moreira CC, Bidochka MJ (2018) Initial stages of endophytic colonization by metarhizium involves Rhizoplane colonization. Microbiol (United Kingdom). https://doi.org/10.1099/mic.0.000729

Behie SW, Bidochka MJ (2014) Ubiquity of insect-derived nitrogen transfer to plants by endophytic insect-pathogenic fungi: an additional branch of the soil nitrogen cycle. Appl Environ Microbiol 80:1553–1560. https://doi.org/10.1128/AEM.03338-13

Barkai-Golan R (2001) chapter 2 - postharvest disease initiation. In: Barkai-golan (ed) RBT-PD of F and V. Elsevier, Amsterdam, pp 3–24

Behie SW, Zelisko PM, Bidochka MJ (2012) Endophytic insect-parasitic Fungi translocate nitrogen directly from insects to plants. Science 336:1576–1577. https://doi.org/10.1126/science.1222289

Behie SW, Moreira CC, Sementchoukova I et al (2017) Carbon translocation from a plant to an insect-pathogenic endophytic fungus. Nat Commun 8:14245. https://doi.org/10.1038/ncomms14245

Beys da Silva WO, Santi L, Schrank A, Vainstein MH (2010) Metarhizium anisopliae lipolytic activity plays a pivotal role in Rhipicephalus (Boophilus) microplus infection. Fungal Biol 114(1):10–15

Bhattacharya A (2019) Effect of high-temperature stress on the metabolism of plant growth regulators, chapter 6. Academic Press, New York, pp 485–591

Bilgo E, Lovett B, St. Leger RJ et al (2018) Native entomopathogenic Metarhizium spp. from Burkina Faso and their virulence against the malaria vector Anopheles coluzzii and non-target insects. Parasit Vectors 11:209. https://doi.org/10.1186/s13071-018-2796-6

Bischoff JF, Rehner SA, Humber RA (2009) A multilocus phylogeny of the Metarhizium anisopliae lineage. Mycologia 101:512–530. https://doi.org/10.3852/07-202

Bittencourt VREP, Mascarenhas AG, de Menezes GCR, Monteiro SG (2000) In vitro action of Metarhizium anisopliae (Metschnikoff, 1879) Sorokin, 1883 and Beauveria bassiana (Balsamo) Vuillemin, 1912 on eggs of the tick Anocentor nitens (Neummann, 1897) (Acari: Ixodidae). Rev Bras Med Veterinária 22:248–251

Blanford S, Read AF, Thomas MB (2009) Thermal behaviour of Anopheles stephensi in response to infection with malaria and fungal entomopathogens. Malar J 8:72. https://doi.org/10.1186/1475-2875-8-72

Brancini GTP, Ferreira MES, Rangel DEN, Braga GÚL (2019) Combining Transcriptomics and proteomics reveals potential post-transcriptional control of gene expression after light exposure in Metarhizium acridum. G3 (Bethesda) 9:2951–2961. https://doi.org/10.1534/g3.119.400430

Branine M, Bazzicalupo A, Branco S, Hogan DA (2019) Biology and applications of endophytic insect-pathogenic fungi. PLoS Pathog 15(7):e1007831

Brown JC (1956) Iron chlorosis. Annu Rev Plant Physiol 7:171–190. https://doi.org/10.1146/annurev.pp.07.060156.001131

Bukhari T, Takken W, Koenraadt CJM (2011) Development of Metarhizium anisopliae and Beauveria bassiana formulations for control of malaria mosquito larvae. Parasit Vectors 4:23. https://doi.org/10.1186/1756-3305-4-23

Butt TM, Coates CJ, Dubovskiy IM, Ratcliffe NA (2016a) Chapter nine - Entomopathogenic Fungi: new insights into host–pathogen interactions. In: Lovett B, St. Leger RJ (eds) Genetics and molecular biology of Entomopathogenic fungi. Academic Press, New York, pp 307–364

Butt TM, Coates CJ, Dubovskiy IM, Ratcliffe NA (2016b) Entomopathogenic Fungi: new insights into host-pathogen interactions. Adv Genet 94:307–364. https://doi.org/10.1016/bs.adgen.2016.01.006

Canassa F, Tall S, Moral RA et al (2019) Effects of bean seed treatment by the entomopathogenic fungi Metarhizium robertsii and Beauveria bassiana on plant growth, spider mite populations and behavior of predatory mites. Biol Control 132:199–208. https://doi.org/10.1016/J.BIOCONTROL.2019.02.003

Carolino AT, Paula AR, Silva CP et al (2014) Monitoring persistence of the entomopathogenic fungus Metarhizium anisopliae under simulated field conditions with the aim of controlling adult Aedes aegypti (Diptera: Culicidae). Parasit Vectors 7:198. https://doi.org/10.1186/1756-3305-7-198

Chandler D, Bailey AS, Tatchell GM et al (2011) The development, regulation and use of biopesticides for integrated pest management. Philos Trans R Soc Lond Ser B Biol Sci 366:1987–1998. https://doi.org/10.1098/rstb.2010.0390

Chapman RF (2012) The insects: structure and function, 5th edn. Cambridge University Press, Cambridge

Charnley AK (2003) Fungal pathogens of insects: cuticle degrading enzymes and toxins. Academic Press, New York, pp 241–321

Constanza Mannino M, Huarte-Bonnet C, Davyt-Colo B, Pedrini N (2019) Is the insect cuticle the only entry gate for fungal infection? Insights into alternative modes of action of entomopathogenic fungi. J Fungi 5(2):33

Contreras J, Mendoza JE, Martinez-Aguirre MR et al (2014) Efficacy of enthomopathogenic fungus Metarhizium anisopliae against Tuta absoluta (Lepidoptera: Gelechiidae). J Econ Entomol 107:121–124. https://doi.org/10.1603/ec13404

Donzelli BGG, Krasnoff SB (2016) Molecular genetics of secondary chemistry in Metarhizium fungi. Adv Genet 94:365–436. https://doi.org/10.1016/bs.adgen.2016.01.005

Dornetshuber-Fleiss R, Heffeter P, Mohr T et al (2013) Destruxins: fungal-derived cyclohexadepsipeptides with multifaceted anticancer and antiangiogenic activities. Biochem Pharmacol 86:361–377. https://doi.org/10.1016/j.bcp.2013.05.022

Erler F, Ates AO (2015) Potential of two entomopathogenic fungi, Beauveria bassiana and Metarhizium anisopliae (Coleoptera: Scarabaeidae), as biological control agents against the June beetle. J Insect Sci 15:44. https://doi.org/10.1093/jisesa/iev029

Farenhorst M, Knols BGJ (2010) A novel method for standardized application of fungal spore coatings for mosquito exposure bioassays. Malar J 9:27. https://doi.org/10.1186/1475-2875-9-27

Farenhorst M, Farina D, Scholte E-J et al (2008) African water storage pots for the delivery of the entomopathogenic fungus Metarhizium anisopliae to the malaria vectors Anopheles gambiae s.s. and Anopheles funestus. Am J Trop Med Hyg 78:910–916

Farenhorst M, Mouatcho JC, Kikankie CK et al (2009) Fungal infection counters insecticide resistance in African malaria mosquitoes. Proc Natl Acad Sci USA 106:17443–17447. https://doi.org/10.1073/pnas.0908530106

Fernandes EKK, Keyser CA, Chong JP et al (2010) Characterization of Metarhizium species and varieties based on molecular analysis, heat tolerance and cold activity. J Appl Microbiol 108:115–128. https://doi.org/10.1111/j.1365-2672.2009.04422.x

Garcia ARM, de Rocha AP, Moreira CC et al (2016) Screening of fungi for biological control of a Triatomine vector of Chagas disease: temperature and trypanosome infection as factors. PLoS Negl Trop Dis 10:e0005128–e0005128. https://doi.org/10.1371/journal.pntd.0005128

Gauthier GM, Keller NP (2013) Crossover fungal pathogens: the biology and pathogenesis of fungi capable of crossing kingdoms to infect plants and humans. Fungal Genet Biol 61:146–157. https://doi.org/10.1016/J.FGB.2013.08.016

Glare TR, Milner RJ, Beaton CD (1996) Variation in Metarhizium, a genus of fungal pathogens attacking Orthoptera: is Phialide morphology a useful taxonomic criterion? J Orthoptera Res 5:19–27. https://doi.org/10.2307/3503572

Goble TA, Gardescu S, Jackson MA, Hajek AE (2016) Evaluating different carriers of Metarhizium brunneum F52 microsclerotia for control of adult Asian longhorned beetles (Coleoptera: Cerambycidae). Biocontrol Sci Tech 26:1212–1229. https://doi.org/10.1080/09583157.2016.1192103

Golo PS, Gardner DR, Grilley MM et al (2014) Production of destruxins from Metarhizium spp. fungi in artificial medium and in endophytically colonized cowpea plants. PLoS One 9: e104946–e104946. https://doi.org/10.1371/journal.pone.0104946

Gomes SA, Paula AR, Ribeiro A et al (2015) Neem oil increases the efficiency of the entomopathogenic fungus Metarhizium anisopliae for the control of Aedes aegypti (Diptera: Culicidae) larvae. Parasit Vectors 8:669. https://doi.org/10.1186/s13071-015-1280-9

González-Santoyo I, Córdoba-Aguilar A (2012) Phenoloxidase: a key component of the insect immune system. Entomol Exp Appl 142:1–16. https://doi.org/10.1111/j.1570-7458.2011.01187.x

Götz P, Matha V, Vilcinskas A (1997) Effects of the entomopathogenic fungus Metarhizium anisopliae and its secondary metabolites on morphology and cytoskeleton of plasmatocytes isolated from the greater wax moth, galleria mellonella. J Insect Physiol 43(12):1149–1159. https://doi.org/10.1016/s0022-1910(97)00066-8

Greenfield BPJ, Lord AM, Dudley E, Butt TM (2014) Conidia of the insect pathogenic fungus, Metarhizium anisopliae, fail to adhere to mosquito larval cuticle. R Soc Open Sci 1:140193. https://doi.org/10.1098/rsos.140193

Greenfield M, Gómez-Jiménez MI, Ortiz V et al (2016) Beauveria bassiana and Metarhizium anisopliae endophytically colonize cassava roots following soil drench inoculation. Biol Control theory Appl Pest Manag 95:40–48. https://doi.org/10.1016/j.biocontrol.2016.01.002

Hajek AE, St. Leger RJ (1994) Interactions between fungal pathogens and insect hosts. Annu Rev Entomol 39:293–322. https://doi.org/10.1146/annurev.en.39.010194.001453

Hancock PA (2009) Combining fungal biopesticides and insecticide-treated bednets to enhance malaria control. PLoS Comput Biol 5:e1000525–e1000525. https://doi.org/10.1371/journal.pcbi.1000525

Hu S, Bidochka MJ (2019) Root colonization by endophytic insect-pathogenic fungi. J Appl Microbiol. https://doi.org/10.1111/jam.14503

Hussain A (2018) Reprogramming the virulence: insect defense molecules navigating the epigenetic landscape of Metarhizium robertsii. Virulence 9:447–449. https://doi.org/10.1080/21505594.2017.1421828

Indriyanti DR, Widiyaningrum P, Haryuni H et al (2017) Effectiveness of Metarhizium anisopliae and entomopathogenic nematodes to control Oryctes rhinoceros larvae in the rainy season. Pakistan J Biol Sci 20:320–327. https://doi.org/10.3923/pjbs.2017.320.327

Iwanicki NSA, Pereira AA, Botelho ABRZ et al (2019) Monitoring of the field application of Metarhizium anisopliae in Brazil revealed high molecular diversity of Metarhizium spp in insects, soil and sugarcane roots. Sci Rep 9:4443. https://doi.org/10.1038/s41598-019-38594-8

Jaber LR (2018) Seed inoculation with endophytic fungal entomopathogens promotes plant growth and reduces crown and root rot (CRR) caused by Fusarium culmorum in wheat. Planta 248:1525–1535. https://doi.org/10.1007/s00425-018-2991-x

Jaber LR, Enkerli J (2017) Fungal entomopathogens as endophytes: can they promote plant growth? Biocontrol Sci Technol 27:28–41. https://doi.org/10.1080/09583157.2016.1243227

James PJ, Kershaw MJ, Reynolds SE, Charnley AK (1993) Inhibition of desert locust (Schistocerca gregaria) Malpighian tubule fluid secretion by destruxins, cyclic peptide toxins from the insect pathogenic fungus Metarhizium anisopliae. J Insect Physiol 39:797–804. https://doi.org/10.1016/0022-1910(93)90056-W

Kamareddine L (2012) The biological control of the malaria vector. Toxins (Basel) 4:748–767. https://doi.org/10.3390/toxins4090748

Kanga LHB, Adamczyk J, Patt J et al (2010) Development of a user-friendly delivery method for the fungus Metarhizium anisopliae to control the ectoparasitic mite Varroa destructor in honey bee, *Apis mellifera*, colonies. Exp Appl Acarol 52:327–342. https://doi.org/10.1007/s10493-010-9369-5

Karandashov V, Bucher M (2005) Symbiotic phosphate transport in arbuscular mycorrhizas. Trends Plant Sci 10:22–29. https://doi.org/10.1016/j.tplants.2004.12.003

Kershaw MJ, Talbot NJ (2009) Genome-wide functional analysis reveals that infection-associated fungal autophagy is necessary for rice blast disease. Proc Natl Acad Sci 106(37):15967–15972

Kirkland BH, Westwood GS, Keyhani NO (2004) Pathogenicity of Entomopathogenic Fungi Beauveria bassiana and Metarhizium anisopliae to Ixodidae tick species Dermacentor variabilis, Rhipicephalus sanguineus, and Ixodes scapularis. J Med Entomol 41:705–711. https://doi.org/10.1603/0022-2585-41.4.705

Krell V, Unger S, Jakobs-Schoenwandt D, Patel AV (2018) Endophytic Metarhizium brunneum mitigates nutrient deficits in potato and improves plant productivity and vitality. Fungal Ecol 34:43–49. https://doi.org/10.1016/J.FUNECO.2018.04.002

Kryukov VY, Kabilov MR, Smirnova N et al (2019) Bacterial decomposition of insects post-metarhizium infection, possible influence on plant growth. Fungal Biol 123(12):927–935. https://doi.org/10.1016/J.FUNBIO.2019.09.012

Kurtti TJ, Keyhani NO (2008) Intracellular infection of tick cell lines by the entomopathogenic fungus Metarhizium anisopliae. Microbiology 154:1700–1709. https://doi.org/10.1099/mic.0.2008/016667-0

Lacey CM, Lacey LA, Roberts DR (1988) Route of invasion and histopathology of Metarhizium anisopliae in Culex quinquefasciatus. J Invertebr Pathol 52:108–118. https://doi.org/10.1016/0022-2011(88)90109-7

Lavine MD, Strand MR (2002) Insect hemocytes and their role in immunity. Insect Biochem Mol Biol 32:1295–1309. https://doi.org/10.1016/s0965-1748(02)00092-9

Leger RJS, Goettel M, Roberts DW, Staples RC (1991a) Prepenetration events during infection of host cuticle by Metarhizium anisopliae. J Invertebr Pathol 58:168–179. https://doi.org/10.1016/0022-2011(91)90061-T

Leger RJS, Roberts DW, Staples RC (1991b) A model to explain differentiation of appressoria by germlings of Metarhizium anisopliae. J Invertebr Pathol 57:299–310. https://doi.org/10.1016/0022-2011(91)90134-C

Li L, Ortiz C (2014) Pervasive nanoscale deformation twinning as a catalyst for efficient energy dissipation in a bioceramic armour. Nat Mater 13:501–507. https://doi.org/10.1038/nmat3920

Liao X, O'Brien T, Fang W, St Leger R (2014) The plant beneficial effects of Metarhizium species correlate with their association with roots. Appl Microbiol Biotechnol 98. https://doi.org/10.1007/s00253-014-5788-2

Liao X, Lovett B, Fang W, St Leger RJ (2017) Metarhizium robertsii produces indole-3-acetic acid, which promotes root growth in Arabidopsis and enhances virulence to insects. Microbiology 163:980–991. https://doi.org/10.1099/mic.0.000494

Liu X-H, Gao H-M, Xu F et al (2012) Autophagy vitalizes the pathogenicity of pathogenic fungi. Autophagy 8:1415–1425. https://doi.org/10.4161/auto.21274

Lovett B, St Leger RJ (2018) Genetically engineering better fungal biopesticides. Pest Manag Sci 74:781–789. https://doi.org/10.1002/ps.4734

Lovett B, Bilgo E, Millogo SA et al (2019) Transgenic Metarhizium rapidly kills mosquitoes in a malaria-endemic region of Burkina Faso. Science 364:894–897. https://doi.org/10.1126/science.aaw8737

Lwetoijera DW, Sumaye RD, Madumla EP et al (2010) An extra-domiciliary method of delivering entomopathogenic fungus, Metharizium anisopliae IP 46 for controlling adult populations of the malaria vector, Anopheles arabiensis. Parasit Vectors 3:18. https://doi.org/10.1186/1756-3305-3-18

Mantzoukas S, Chondrogiannis C, Grammatikopoulos G (2015) Effects of three endophytic entomopathogens on sweet sorghum and on the larvae of the stalk borer Sesamia nonagrioides. Entomol Exp Appl 154:78–87. https://doi.org/10.1111/eea.12262

Mkiga AM, Mohamed SA, du Plessis H et al (2020) Metarhizium anisopliae and Beauveria bassiana: pathogenicity, horizontal transmission, and their effects on reproductive potential of Thaumatotibia leucotreta (Lepidoptera: Tortricidae). J Econ Entomol 113(2):660–668. https://doi.org/10.1093/jee/toz342

Mnyone LL, Kirby MJ, Mpingwa MW et al (2011) Infection of Anopheles gambiae mosquitoes with entomopathogenic fungi: effect of host age and blood-feeding status. Parasitol Res 108:317–322. https://doi.org/10.1007/s00436-010-2064-y

Mnyone LL, Lyimo IN, Lwetoijera DW et al (2012) Exploiting the behaviour of wild malaria vectors to achieve high infection with fungal biocontrol agents. Malar J 11:87. https://doi.org/10.1186/1475-2875-11-87

Mondal S, Baksi S, Koris A, Vatai G (2016) Journey of enzymes in entomopathogenic fungi. Pacific Sci Rev A Nat Sci Eng 18:85–99. https://doi.org/10.1016/j.psra.2016.10.001

Money NP (2016) Chapter 3 - Spore production, discharge, and dispersal. In: Watkinson SC, Boddy L, Money NP (eds) The fungi, 3rd edn. Academic Press, Boston, pp 67–97

Moonjely S, Barelli L, Bidochka MJ (2016) Insect pathogenic fungi as endophytes. Adv Genet. https://doi.org/10.1016/bs.adgen.2015.12.004

Moonjely S, Zhang X, Fang W, Bidochka MJ (2019) Metarhizium robertsii ammonium permeases (MepC and Mep2) contribute to rhizoplane colonization and modulates the transfer of insect derived nitrogen to plants. PLoS One 14:e0223718. https://doi.org/10.1371/journal.pone.0223718

Moore D, Bateman RP, Carey M, Prior C (1995) Long-term storage of Metarhizium flavoviride conidia in oil formulations for the control of locusts and grasshoppers. Biocontrol Sci Technol 5:193–200. https://doi.org/10.1080/09583159550039918

Muerrle TM, Neumann P, Dames JF et al (2006) Susceptibility of adult Aethina tumida (Coleoptera: Nitidulidae) to entomopathogenic fungi. J Econ Entomol 99:1–6. https://doi.org/10.1093/jee/99.1.1

Mukherjee K, Vilcinskas A (2018) The entomopathogenic fungus Metarhizium robertsii communicates with the insect host galleria mellonella during infection. Virulence 9(1):402–413. https://doi.org/10.1080/21505594.2017.1405190

Narladkar BW, Shivpuje PR, Harke PC (2015) Fungal biological control agents for integrated management of Culicoides spp. (Diptera: Ceratopogonidae) of livestock. Vet World 8:156–163. https://doi.org/10.14202/vetworld.2015.156-163

Okumu F, Biswaro L, Mbeleyela E et al (2010) Using nylon strips to dispense mosquito attractants for sampling the malaria vector Anopheles gambiae s.s. J Med Entomol 47:274–282. https://doi.org/10.1603/me09114

Ondiaka SN, Masinde EW, Koenraadt CJM et al (2015) Effects of fungal infection on feeding and survival of Anopheles gambiae (Diptera: Culicidae) on plant sugars. Parasit Vectors 8:35. https://doi.org/10.1186/s13071-015-0654-3

Onsongo SK, Gichimu BM, Akutse KS et al (2019) Performance of three isolates of Metarhizium Anisopliae and their virulence against Zeugodacus Cucurbitae under different temperature regimes, with global extrapolation of their efficiency. Insects 10:270. https://doi.org/10.3390/insects10090270

Pandey V, Bhatt ID, Nandi SK (2019) Chapter 20 - Role and regulation of auxin signaling in abiotic stress tolerance. In: MIR K, Reddy PS, Ferrante A, Khan N (eds) Plant signaling molecules. Woodhead Publishing, Sawston, pp 319–331

Pendeville H, Carpino N, Marine J-C, Takahashi Y, Muller M, Martial JA, Cleveland JL (2001) The ornithine decarboxylase gene is essential for cell survival during early murine development. Mol Cell Biol 21(19):6549–6558

Peng G, Xia Y (2015) Integration of an insecticidal scorpion toxin (BjαIT) gene into Metarhizium acridum enhances fungal virulence towards Locusta migratoria manilensis. Pest Manag Sci 71:58–64. https://doi.org/10.1002/ps.3762

Peng G, Jin K, Liu Y, Xia Y (2015) Enhancing the utilization of host trehalose by fungal trehalase improves the virulence of fungal insecticide. Appl Microbiol Biotechnol 99:8611–8618. https://doi.org/10.1007/s00253-015-6767-y

Pryor MGM (1940) On the hardening of the cuticle of insects. Proc R Soc Lond Ser B Biol Sci 128:393–407

Pulido JM, Guerrero IP, Martínez IJM, Valadez BC, Guzman JCT, Solis ES, Corona JFG, Schrank A, Bremont FJ, Hernandez AG (2011) Isolation, characterization and expression analysis of the ornithine decarboxylase gene (ODC1) of the entomopathogenic fungus, Metarhizium anisopliae. Microbiol Res 166(6):494–507

Rasgon JL (2011) Using infections to fight infections: paratransgenic fungi can block malaria transmission in mosquitoes. Future Microbiol 6:851–853. https://doi.org/10.2217/fmb.11.71

Raya-Diaz S, Sanchez-Rodriguez AR, Segura-Fernandez JM et al (2017) Entomopathogenic fungi-based mechanisms for improved Fe nutrition in sorghum plants grown on calcareous substrates. PLoS One 12:e0185903. https://doi.org/10.1371/journal.pone.0185903

Russell PJ, Hertz PE, McMillan B (2016) Biology: the dynamic science. Cengage Learning, Belmont, CA

Sánchez-Rodríguez AR, Barrón V, Del Campillo MC, Quesada-Moraga E (2016) The entomopathogenic fungus Metarhizium brunneum: a tool for alleviating Fe chlorosis. Plant Soil 406:295–310. https://doi.org/10.1007/s11104-016-2887-0

Sandhu SS, Sharma AK, Beniwal V et al (2012) Myco-biocontrol of insect pests: factors involved, mechanism, and regulation. J Pathog 2012:126819. https://doi.org/10.1155/2012/126819

Santi L, Beys da Silva WO, Berger M et al (2010) Conidial surface proteins of Metarhizium anisopliae: source of activities related with toxic effects, host penetration and pathogenesis. Toxicon 55:874–880. https://doi.org/10.1016/j.toxicon.2009.12.012

Sasan RK, Bidochka MJ (2012) The insect-pathogenic fungus Metarhizium robertsii (Clavicipitaceae) is also an endophyte that stimulates plant root development. Am J Bot 99:101–107. https://doi.org/10.3732/ajb.1100136

Scholte E-J, Knols BGJ, Samson RA, Takken W (2004) Entomopathogenic fungi for mosquito control: a review. J Insect Sci 4:19. https://doi.org/10.1093/jis/4.1.19

Scholte E-J, Takken W, Knols BGJ (2007) Infection of adult Aedes aegypti and Ae. albopictus mosquitoes with the entomopathogenic fungus Metarhizium anisopliae. Acta Trop 102:151–158. https://doi.org/10.1016/j.actatropica.2007.04.011

Screen S, Bailey A, Charnley K, Cooper R, Clarkson J (1997) Carbon regulation of the cuticle-degrading enzyme PR1 from Metarhizium anisopliae may involve a trans-acting DNA-binding protein CRR1, a functional equivalent of the Aspergillus nidulans CREA protein. Curr Genet 31(6):511–518

Senthil Kumar CM, Jacob TK, Devasahayam S et al (2018) Multifarious plant growth promotion by an entomopathogenic fungus Lecanicillium psalliotae. Microbiol Res 207:153–160. https://doi.org/10.1016/J.MICRES.2017.11.017

Shukla E, Thorat LJ, Nath BB, Gaikwad SM (2015) Insect trehalase: Physiological significance and potential applications. Glycobiology 25(4):357–367

Sinha KK, Choudhary AK, Kumari P (2016) Chapter 15 - Entomopathogenic fungi. In: Omkar BT (ed) Ecofriendly pest management for food security. Academic Press, Cambridge, MA, pp 475–505

Sookar P, Bhagwant S, Allymamod MN (2014) Effect of *Metarhizium anisopliae* on the fertility and fecundity of two species of fruit flies and horizontal transmission of Mycotic infection. J Insect Sci 14:1–12

Sowers AD, Young SP, Grosell M, Browdy CL, Tomasso JR (2006) Hemolymph osmolality and cation concentrations in Litopenaeus vannamei during exposure to artificial sea salt or a mixed-ion solution: relationship to potassium flux. Comp Biochem Physiol A Mol Integr Physiol 145 (2):176–180

St Leger R, Joshi L, Bidochka MJ, Roberts DW (1996) Construction of an improved mycoinsecticide overexpressing a toxic protease. Proc Natl Acad Sci USA 93:6349–6354. https://doi.org/10.1073/pnas.93.13.6349

Sun J, Fuxa JR, Henderson G (2002) Sporulation of Metarhizium anisopliae and Beauveria bassiana on Coptotermes formosanus and in vitro. J Invertebr Pathol 81:78–85. https://doi.org/10.1016/s0022-2011(02)00152-0

Sunde M, Kwan AHY, Templeton MD et al (2008) Structural analysis of hydrophobins. Micron 39:773–784. https://doi.org/10.1016/j.micron.2007.08.003

Theopold U, Schmidt O, Söderhäll K, Dushay MS (2004) Coagulation in arthropods: defence, wound closure and healing. Trends Immunol 25:289–294. https://doi.org/10.1016/j.it.2004.03.004

Thines E, Weber RW, Talbot NJ (2000) MAP kinase and protein kinase A: dependent mobilization of triacylglycerol and glycogen during appressorium turgor generation by Magnaporthe grisea. Plant Cell 12:1703–1718. https://doi.org/10.1105/tpc.12.9.1703

Toei M, Saum R, Forgac M (2010) Regulation and isoform function of the V-ATPases. Biochemistry 49:4715–4723. https://doi.org/10.1021/bi100397s

Wamiti LG, Khamis FM, Abd-Alla AMM et al (2018) Metarhizium anisopliae infection reduces Trypanosoma congolense reproduction in Glossina fuscipes fuscipes and its ability to acquire or transmit the parasite. BMC Microbiol 18:142. https://doi.org/10.1186/s12866-018-1277-6

Wang C, St Leger RJ (2007) The Metarhizium anisopliae Perilipin homolog MPL1 regulates lipid metabolism, appressorial turgor pressure, and virulence. J Biol Chem 282:21110–21115. https://doi.org/10.1074/jbc.M609592200

Wang C, Duan Z, St. Leger RJ (2008) MOS1 Osmosensor of Metarhizium anisopliae Is Required for Adaptation to Insect Host Hemolymph. Eukaryot Cell 7(2):302–309

Wang S, Fang W, Wang C, St Leger RJ (2011) Insertion of an esterase gene into a specific locust pathogen (Metarhizium acridum) enables it to infect caterpillars. PLoS Pathog 7:e1002097–e1002097. https://doi.org/10.1371/journal.ppat.1002097

Wang Y, Wang T, Qiao L, Zhu J, Fan J, Zhang T, Wang Z-X, Li W, Chen A, Huang B (2017a) DNA methyltransferases contribute to the fungal development, stress tolerance and virulence of the entomopathogenic fungus Metarhizium robertsii. Appl Microbiol Biotechnol 101 (10):4215–4226. https://doi.org/10.1007/s00253-017-8197-5

Wang W, Shi J, Xie Q et al (2017b) Nutrient exchange and regulation in arbuscular mycorrhizal symbiosis. Mol Plant 10:1147–1158. https://doi.org/10.1016/J.MOLP.2017.07.012

Wu C, Zhang X, Fang W (2019) Increasing pyruvate concentration enhances conidial thermotolerance in the entomopathogenic fungus Metarhizium robertsii. Front Microbiol 10:519. https://doi.org/10.3389/fmicb.2019.00519

Wyatt GR (1961) The Biochemistry of Insect Hemolymph. Annu Rev Entomol 6(1):75–102

Xia Y, Clarkson JM, Charnley AK (2002) Trehalose-hydrolysing enzymes of Metarhizium anisopliae and their role in pathogenesis of the tobacco hornworm, Manduca sexta. J Invertebr Pathol 80(3):139–147

Yamamoto K, Ohmae M, Orihara T (2019) Metarhizium brachyspermum sp. nov. (Clavicipitaceae), a new species parasitic on Elateridae from Japan. Mycoscience 61(1): 37–42. https://doi.org/10.1016/J.MYC.2019.09.001

Yousef M, Alba-Ramírez C, Garrido Jurado I et al (2018) Metarhizium brunneum (Ascomycota; Hypocreales) treatments targeting olive fly in the soil for sustainable crop production. Front Plant Sci 9:1. https://doi.org/10.3389/fpls.2018.00001

Zhang W, Chen J, Keyhani NO et al (2015) Comparative transcriptomic analysis of immune responses of the migratory locust, Locusta migratoria, to challenge by the fungal insect pathogen, Metarhizium acridum. BMC Genomics 16:867. https://doi.org/10.1186/s12864-015-2089-9

Zhang J, Huang W, Yuan C et al (2017) Prophenoloxidase-mediated ex vivo immunity to delay fungal infection after insect Ecdysis. Front Immunol 8:1445. https://doi.org/10.3389/fimmu.2017.01445

Zhang J, Jiang H, Du Y et al (2019) Members of chitin synthase family in Metarhizium acridum differentially affect fungal growth, stress tolerances, cell wall integrity and virulence. PLoS Pathog 15:e1007964–e1007964. https://doi.org/10.1371/journal.ppat.1007964

Zhou R, Zhou X, Fan A, Wang Z, Huang B (2018) Differential functions of two metalloproteases, Mrmep1 and Mrmep2, in growth, sporulation, cell wall integrity, and virulence in the filamentous fungus Metarhizium robertsii. Front Microbiol 9:1528

Chapter 19
Arbuscular Mycorrhizal Fungi: Potential Plant Protective Agent Against Herbivorous Insect and Its Importance in Sustainable Agriculture

Anandakumar Selvaraj and Kalaiselvi Thangavel

Abstract Wide use of fertilizers and chemicals for food grain production to feed the world population with increasing demand leads to environmental pollution. Alternatively, the use of biological sources such as beneficial microbes to improve crop production as a component of sustainable agriculture production and environmentally friendly. Among them, arbuscular mycorrhizal fungi (AMF) are well-known soil microbe forms a symbiotic association with land plants including agricultural important crops. This beneficial AM fungi improving plant growth, and it also fount to improve resistance capacity of plants against diverse stresses, including herbivorous insect damage through altering the morphological and biochemical traits. In response to herbivore stress, AMF augments plant defense in both constitutive and inducible manner leads to reduce insect damage. AMF induced or primed plant defense mechanisms against herbivorous insect damage have so far underestimated. Therefore, we discuss here an overview of research findings related to AMF induced or priming of immune response in plants against herbivore-induced stress. Acquired mechanisms of plant associated with AMF to protect themselves from pests by altering nutrient availability and physiology. AMF-mediated response of plants to herbivore varied with host plants, AMF species, and degree of colonization, type of pest, and crop management system. With these contexts AMF is could be a good bioprotective agent against pest apart from improving plant growth and this is an integral part of the integrated pest management system for sustainable agriculture production.

Keywords Arbuscular mycorrhizal fungi · Pest · Defense · Sustainable agriculture

---

A. Selvaraj (✉) · K. Thangavel
Department of Agricultural Microbiology, Tamil Nadu Agricultural University, Coimbatore, India

N. Shrivastava et al. (eds.), *Symbiotic Soil Microorganisms*, Soil Biology 60,
https://doi.org/10.1007/978-3-030-51916-2_19

19.1 Introduction

The word mycorrhizae derived from two Greek words, including "mycos" and "rhizos" meaning fungus and root respectively. The successful colonization of mycorrhizae on the roots of plant begins with the signaling from both the partners plant, and arbuscular mycorrhizal fungi (AMF). AMF is a unique soil fungus belongs to the phylum Glomeromycota which forms a symbiotic association with almost 80% of land plants, including agriculturally important crops (Gupta et al. 2019). This plant-associated beneficial biotrophic fungi alters plant root system and improves the uptake of immobile inorganic nutrients, primarily phosphate (P) and other nutrients such as ammonia as well as micronutrients (Mg, Fe, Zn, Cu, Mn, and Al) from the soil solution (Jung et al. 2012; Kumar 2018). Besides, AM fungal hyphal network spread to meters and help the plants to absorb moisture from the surrounding soil. In turn, plants provide around 20% of photosynthetically assimilated carbon (Smith and Read 2008). A plethora of reports confirmed that AMF colonization improves plant growth by increasing the acquisition of nutrients and water from their surrounding (Gill et al. 2013; Seguel et al. 2015; Frew 2019). The effects of AMF colonization on plant growth varied depends on soil fertility, time of AMF inoculation, and degree of fungal colonization on host plant roots (Hart et al. 2018). Many reports are confirmed that crop yield increased by AMF application (Zhang et al. 2019) and it is recognized as an important component of sustainable agriculture production (Ryan and Graham 2018; Rillig et al. 2019). AMF also involved in diverse functions such as soil and plant health improvement (Kumar et al. 2019).

Plants undergo numerous biotic and abiotic stresses and these effects are reduced by AMF colonization through imparting plant defense and tolerance. The plethora of research findings supported that AMF improves plant tolerance against stress like salinity (Rodriguez and Redman 2008), phytopathogens (Slaughter et al. 2012), drought (Ahanger et al. 2014), heavy metals (Salam et al. 2017) and temperature (Schoenherr et al. 2019), etc. In addition, crops are also affected severely by the herbivorous insect to causes transition economic loss in agricultural production. There is a growing interest in developing plant protection measures that are eco-friendly, cost-effective, and sustainable. In this context, it is essential to understand the plant defense mechanisms against insect attack (Gershenzon 2017). It is equally important to understand the adaptation strategies of insects against these plant defensive characteristics to develop sustainable management strategies (Hahn et al. 2019). A research finding confirmed that AMF also found to improve plant resistance or tolerance against herbivorous insect damage by altering morphological and physiological characteristics (Kim and Felton 2013; Schoenherr et al. 2019). The AMF provides tolerance to plants against herbivorous insects through various mechanisms like increasing nutrient uptake, primary and secondary metabolites, antioxidants, volatile organic compounds, and phytohormones production and imparting faster regrowth of tissues (Korpita et al. 2014; Cabral et al. 2018). The effect of AMF on the induction of plant defense against herbivore is studied in many plant species, including strawberry (Gange 2001), big bluestem grass (Miller et al. 2002), lotus (Nishida et al. 2010), ribwort plantain (Wang et al. 2015), faba bean (Cabral et al. 2018), tomato (Formenti and Rasmann 2019), potato (Schoenherr et al. 2019), milkweeds (Meier and Hunter 2019), black gram (Selvaraj et al. 2020), and maize (De Lange et al. 2020), etc. These effects varied on plant and AMF species, degree of AMF colonization, insect pest type (generalist or specific) (Cabral et al. 2018). In this chapter, we discuss the role of AMF in imparting plant defense against herbivorous insects and their importance in sustainable agriculture.

19.2 AMF Primed Plant Defense

AMF symbiosis modulates the plant system in many ways to improve growth and defense against stress by altering plant physiology (Jung et al. 2012). Plants get primed during the establishment of AMF symbiosis by modulating immune responses of plants in both local and systemic manner, in which primed plants also impart efficient activation of defense mechanisms against stress (Jung et al. 2012). This AMF induced response is known as mycorrhiza-induced resistance (MIR), which provides systemic protection against a wide range of attackers and shares characteristics with systemic acquired resistance (SAR) and induced systemic resistance (ISR) (Schoenherr et al. 2019). The MIR protects plants from a wide range of attackers, including pathogens, nematodes, and herbivorous arthropods through SAR and ISR mediated priming by salicylic acid (SA) and jasmonic acid (JA) dependent defenses (Cameron et al. 2013). Herbivore performance was affected either positively (Gange et al. 1999) or negatively (Gange 2001) by AMF colonization, depending on both herbivore and fungal species. In general, mycorrhizal colonization enhances plant resistance to root-feeding insects and generalist herbivores; but it may increase plant susceptibility to sucking insects and specialist herbivores (Hartley and Gange 2009; Pineda et al. 2012). AMF symbioses role in providing enhanced resistance and/or tolerance to plants against pests is remaining elusive and not well studied.

19.3 AMF-Mediated Direct and Indirect Defense Mechanisms of Plants Against Herbivorous Insect

Herbivorous insects fully depend on the energy fixed by plants. However, plants had no chance to escape from attackers, so they must employ other strategies to defend themselves from the deleterious effect of herbivores (Yan and Xie 2015). In response to stress, plants exhibit both constitutive and induced defenses by recognizing stress with a high degree to mount intracellular signaling into appropriate biochemical, physiological, and cellular responses systemically throughout the plant system

(Karban 2020; Wilkinson et al. 2019). During plant–insect interaction they use chemicals as a weapon to overcome each other's ill effect and to bring morphological and biochemical changes in both the partners. Plants first protect themselves from insect damages by bringing changes in the morphological related traits and antioxidant production (Mithöfer and Boland 2012; War et al. 2013). The biochemical mediated defenses developed over a while against insect attack by producing secondary or anti-digestive metabolites, toxic furanocoumarins and amino acids (War et al. 2013) and other related toxic compounds in plants leads to a reduction in feeding capacity (Chen et al. 2012), repelling of herbivorous insects and attraction of insect parasitoids/predators. AMF association affects the development of insects by modifying dietary specialization and feeding mode of the herbivores directly by altering the quality of plants. AMF colonized plants exhibit accumulation of certain defense compounds like phenolics, hormones, and reactive oxygen species quenching enzymes during the interaction, which improves plant defense against herbivore-induced stress (Koricheva et al. 2009; MacLean et al. 2017).

19.4 Direct Defense

19.4.1 AMF Induced Changes in Morphological Characteristics of Plant Against Herbivorous Insect

In the plant system, the constitutive defense provides the first line of defense against stress. Modified structures of plant like thorns, stings, sticky resins, and trichomes provided defense against insects as a type of constitutive defense (Taggar and Gill 2012). All plant parts offer some sort of resistance against herbivory ranging from tissue hardness to highly complex glandular trichomes and spines (Acharya and Bhargava 2008). These special structures act as physical barriers that affect herbivore performance and reduces their severity further. Trichome is an important special plant structure that negatively affects the feeding and ovipositional responses of insect pests (Xing et al. 2017) directly by damaging the insect mechanically and interfere with their movement, thereby reducing their access to leaf epidermis. Glandular trichomes are yet another type of trichomes, which secrete secondary metabolites including flavonoids, terpenoids, and alkaloids that can act as a poison, repellent, or trap insects and other organisms, thus forming a combination of structural and chemical defense (War et al. 2012). AMF colonization exhibited a moderate increase in trichomes density of *Solanum lycopersicum* plants in response to *Spodoptera litura* infestation (Formenti and Rasmann 2019).

Tissue hardiness is another mode of plant defense, which affects the penetration ability of mouthparts of piercing-sucking insects on plant tissues, and also increases mandibular wear in biting-chewing herbivores, thus preventing herbivore feeding (Chaudhary et al. 2018a, b). Hardened leaves reduce the herbivore damage by affecting the palatability and digestibility of the tissues (War et al. 2012). Enhanced leaf toughness by cell wall reinforcement with the deposition of chemicals such as lignin, cellulose, suberin and callose, small organic molecules (phenolics), and even inorganic silica particles provide mechanical resistance to insect feeding and penetration (Keeping and Kvedaras 2008). Kundu et al. (2018) reported that *Spodoptera litura* infestation enhanced lignin content in leaves of *Solanum lycopersicum*. Lignin and phenolic content in cucumber plants increased significantly upon inoculation with the AMF (Chen et al. 2013). Manjarrez et al. (2009) reported that AMF colonization improves the deposition of callose in tomato plants in response to biotic stress.

19.4.2 AMF-Induced Plant Defense by Improving the Nutrient Content

AMF indirectly influences the performance of herbivorous insect through changes in the quality and quantity of plant nutrients, and even affect the behavior of natural enemies of herbivores and plant pollinators (Cardoso Filho et al. 2017). Plant growth and development depend on the available soil nutrient content, which enhances the plant's photosynthetic capacity (Smith and Read 2008), alteration in the production of plant primary and secondary metabolites (Swamy et al. 2016). AMF colonization increases nutrient acquisitions by the plants, thereby have great influences on plant physiology, primary and secondary metabolism, and hormone balance (Cabral et al. 2018). A plethora of reports showed that AMF colonization improves plant growth by enhancing nutrient uptake and by altering plant physiology (Gill et al. 2013; Seguel et al. 2015; Frew 2019). Borowicz (1997) reported that AMF colonization positively affected the growth of *Glycine max* under low phosphorus conditions, which resulted in regulating the performance of *Epilachna varivestis*. These studies support that AMF imparts plant to tolerate herbivore damage by supporting plant growth with nitrogen and phosphorus, which are required to synthesize the defense-related enzymes, secondary metabolites, and modified physical structures (Seguel et al. 2015). Andrade et al. (2013) reported that nicotine contents of the leaves of *Nicotiana tabacum* improved due to AMF colonization. Mycorrhiza colonized *Plantago lanceolata* plants produced diverse carbon-based defensive compounds such as aucubin and catalpol, which decreases feeding and growth characteristic of *Arctia caja* insect (Gange and West 1994). This may evidence that AMF improves plant photosynthetic rates (Miller et al. 2002) and increases the availability of nutrients involved in the synthesis of defense related chemicals. Balestrini et al. (2017) also reported that the AM fungus significantly altered the expression of nutrient transport-related genes in grapevine roots. By contrast, healthy plants rich in nutrients generally favor herbivore growth and get attracted by herbivores (Woods et al. 2004). The AM symbiosis often benefits the host plant in terms of nutrient acquisition and nutritional quality, which can benefit the insects feeding on it (Koricheva et al. 2009); but an increase in nutrient acquisition can also mean greater investment in defenses (Pozo and Azcón-Aguilar 2007). Alterations in plant traits directly affect the herbivory growth by reducing nutrient resources, increasing secondary metabolites, and volatile production. AMF showed a significant impact on plant defense by modulating plant traits directly through improving nutrient requirements, which confers resistance or tolerance against insect damage.

19.4.3 The Antioxidant Activity of Plants Infested with AMF and Herbivore

Plants produce reactive oxygen species (ROS) as a signaling molecule and causative bioactive agent under environmental stress. Herbivore damaged plants accumulate more amount of reactive oxygen species and hydrogen peroxide (H_2O_2) in tissues (Jih et al. 2003) which leads to cellular damage by the degradation of biomolecules like pigments, proteins, lipids, carbohydrates, and DNA. Plants provide the first line of induced cellular defense against herbivore stress by producing antioxidants (Gourlay and Constabel 2019). To mitigate the effect of ROS, plants evolved with antioxidants mediated defense systems to eliminate or detoxify the excess production of reactive oxygen species. Antioxidant systems may be either enzymatic or non- enzymatic mode (Das and Roychoudhury 2014). Enzymatic mode of antioxidants includes superoxide dismutase (SOD), catalase (CAT), guaiacol peroxidase (GPX) and ascorbate peroxidase (APX). The SOD catalyzes the removal of $O_2^{\bullet-}$ by converting it into O_2 and H_2O_2 (Das and Roychoudhury 2014). The peroxidases (POD) and catalases (CAT) catalyze the decomposition of H_2O_2 into H_2O and O_2, thereby eliminate the damage due to excess production of reactive oxygen species, and increase the level of plant resistance to herbivore-induced stress (Ruley et al. 2004). AM fungi enhance the plant antioxidant enzyme activity like superoxide dismutase (SOD), peroxidase (POD), and catalase (CAT) and SA concentration in *Lolium perenne* infested with *Claroideoglomus etunicatum* than non-mycorrhizal plants (Li et al. 2018).

Plants also exhibit a non-enzymatic mode of antioxidant system by producing ascorbic acid (AA), carotenoids, phenolics, flavonoids, and proline. AA is the main metabolite that protects the cell membrane from oxidative damage caused by ROS (Kovalikova et al. 2019). AA is either oxidized into monodehydroascorbate reductase (MDHA) or it reacts with H_2O_2, $OH^{\bullet}$, $O_2^{\bullet-}$, and regenerates α-tocopherol from tocopheroxyl radical thereby protecting the membranes from oxidative damage (Shao et al. 2005). Carotenoids belong to the family of lipophilic antioxidants are localized in the plastids of photosynthetic and non-photosynthetic plant tissues. It exhibits antioxidant activity by protecting the photosynthetic machinery in any one of the four ways, (a) reacting with lipid peroxidation (LPO) products to end the chain reactions, (b) scavenging $^1O_2$ and generating heat as a by-product, (c) preventing the formation of $^1O_2$ by reacting with $^3Chl^*$ and excited chlorophyll (Chl^*), and (d) dissipating the excess excitation energy, via the xanthophyll cycle (Das and

Roychoudhury 2014). Flavonoids are considered as a secondary ROS scavenging system in plants to scavenge the $^1O_2$ and alleviate the damages caused to the outer envelope of the chloroplast membrane (Agati et al. 2012). Proline is one of the important osmolytes that act as a powerful antioxidant to reduce the effect of ROS by scavenging $OH^•$ and $^1O_2$ radicals and can inhibit the damages due to lipid peroxidation (Verbruggen and Hermans 2008). Phenolic compounds act as antioxidants by inactivating lipid free radicals or inhibiting hydroperoxides to break down into free radicals (Bhonwon et al. 2009). AMF colonization reported to enhance the production of flavonoids (Harrison and Dixon 1993), and apocarotenoids (Fester et al. 2002) in various plants.

In response to herbivore attack, plants produce an enhanced level of lipoxygenases (LOXs), polyphenol oxidases (PPOs), peroxidases (PO), and phenylalanine ammonium lyases (PAL) involved in the synthesis of defense metabolites against herbivore damage (Gourlay and Constabel 2019). LOXs are ubiquitous enzymes that play important roles in plants, which involved in the production of fatty acids, jasmonic acid, volatile aldehydes, and oxyacids (Babenko et al. 2017). Phenylalanine ammonium lyase (PAL) is involved in the production of many primary and secondary metabolites of higher plants and regulates lignification during stress conditions (Chen et al. 2009). Polyphenol oxidases (PPOs) are anti-nutritive enzymes that use molecular oxygen to oxidize common phenolic compounds to highly reactive quinones. The quinones further react with amino acids of insects and reduce the amino acid availability (Kundu et al. 2018). PO also plays a vital role in the biosynthesis of lignin and imparting defense against biotic stress by degrading indole acetic acid (IAA) and utilizing H_2O_2 in the process (Das and Roychoudhury 2014). Araji et al. (2014) proved that increase in insect resistance in tomato due to overexpression of PPO genes compared to the mutant. Plants accumulate more of cellulose, lignin, tannins, and silicates, which reduce the palatability or feeding capacity of insects. AMF colonization has been reported to increase the phenolic content and activity of key defense-related enzymes such as lipoxygenases (LOXs) and phenylalanine ammonia lyase in *Piper nigrum* (da Trindade et al. 2019).

19.4.4 AMF Alters Primary Metabolite Production and Allocation of It as Defense Response Against Insect Damage

Synthesis of defense chemicals against herbivorous insect depends on the quantity and quality of primary metabolites accumulation in plants. During defense response, carbon-based materials are converted into nitrogen-based secondary metabolites which affect the quality of feed, and the feeding performance of a herbivore (Pozo and Azcón-Aguilar 2007). AMF colonization altered carbohydrate metabolism in the root system of lychee was reported (Shu et al. 2016). They reported generally increase in C: N ratio of plants during herbivorous, which implies that a reduction in protein, could have significant negative consequences for insects as they are typically limited by protein in their diet and reduction in the nutritional quality of the host plant (Robinson et al. 2012). However, sucking insects which feed on phloem are sometimes able to compensate for this reduction in sap quality, or even benefit from changes in the phloem amino acid profile (Ryalls et al. 2017).

Reallocation of resources is yet another important defense mechanism of plants against insect attack. Upon herbivore attack, plant nutrients are made unavailable to the insect by reallocation of resources from insect infected area to uninfected plant parts or roots (sink) or fungal hyphae in AMF plants. After herbivory passes on all accumulated metabolites are reallocated back to whole plants from the sink. Schultz et al. (2013) proved the translocation of nitrogen from shoots to roots of *Centaurea maculosa* plant upon infection with *Agapeta zoegana* and reallocated back from the root to shoot after removal of insect. Another example is the allocation of sugars from infested green parts into the non-affected roots, as has been shown for *Manduca sexta* infested with *Nicotiana attenuata* plants. Thus, at the necessary time, all rescued material can easily be remobilized and used for building new aboveground organs (Mithöfer and Boland 2012). Foliar arthropod pests might also interact with AMF, as herbivory can alter the allocation of plant photosynthate to the roots (Gange et al. 2007; Machado et al. 2013). Metabolic reprogramming like increased biosynthesis of defense chemicals and relocating of primary metabolites in tissues away from the site of feeding has been noticed in insect attacked plants to affect herbivore performance and growth. Markkola et al. (2004) showed that the increased sink strength of mycorrhizal fungi-colonized roots increased carbon limitation after defoliation. Mycorrhizae could increase shifts in resource allocation to inaccessible tissues of herbivores (Song et al. 2013; Wang et al. 2015).

19.4.5 Herbivore Induced Synthesis of Secondary Metabolites in AMF Associated Plants

AMF association affects the development of insects by modifying dietary specialization and feeding mode of the herbivores directly by altering the quality of plants. AMF colonized plants exhibit accumulation of certain defense compounds like phenolics, hormones, and reactive oxygen species quenching enzymes during the interaction, which improves plant defense against herbivore-induced stress (Koricheva et al. 2009; MacLean et al. 2017). Plants defend themselves by producing secondary metabolites along with primary metabolites, which are toxic to herbivores and act as defense compounds (Wittstock and Gershenzon 2002). These secondary metabolites provide direct defense against herbivorous insects by affecting the growth and development of the pests (Chen et al. 2012; Brunetti et al. 2013). These are targeted against the nervous, digestive, and endocrine organs of herbivores. Plant secondary metabolites can be divided into four chemically distinct groups viz., terpenes, phenolics, nitrogen, and sulfur-containing compounds (Mazid et al. 2011). AMF symbiosis helps to benefit plant health by modulating secondary metabolism and thus potentially fortify both direct and indirect plant defense systems (Borowicz 2013; Jung et al. 2012).

Terpenes are important secondary metabolites that provides insecticidal activity by acting as antifeedants, toxins, or as modifiers of insect development (Lackus et al. 2018). Bennett and Wallsgrove (1994) reported that Azadirachtin is the best insect deterrents terpene and also inhibit egg maturation. Some important terpenoids like gossypol, polygodial, glaucolide-A, pyrethroids, and cucurbitacins are reported as deterrents and toxic to insects (Chaudhary et al. 2018a, b). Some secondary metabolites mainly volatiles provide indirect plant defense by recruiting of natural enemies of insect pests (Yuan et al. 2008). AMF inoculated plants produce more of monoterpenes and sesquiterpenes, which were not found in the control plants (Shrivastava et al. 2015).

Phenolics are another important defense compound in plants act as antioxidants and insect deterrents. Some of the phenolics include coumarin, lignin, flavonoids, isoflavonoids, quinones, and tannins. Quinones formed by oxidation of phenols bind covalently to leaf proteins and inhibit protein digestion in herbivores (Bhonwon et al. 2009). Chickpea infested with *Helicoverpa armigera* showed the production of a greater quantity of isoflavonoids, which deter larval feeding. Salicylates in salix leaves reduce the feeding and growth of polyphagous larvae of *Operophtera brumata* (Lattanzio et al. 2006). AMF (*Gigaspora margarita*) inoculated *Lotus japonicas* plants strongly affects the oviposition of the spider mites by increasing leaf phenolic content (Nishida et al. 2010).

19.4.6 The AMF Associated Plant Produced Anti-Nutritional/ Digestive Proteins Involved in Defense Against Herbivore Performance

Plants can also defend themselves by producing proteins that reduce the nutritive value to the attacking insect or causes physical damage to the insect digestive tract. The major classes of such defense proteins are α-amylase inhibitor, chitinase, proteinase inhibitor, and lectin: α-amylase inhibitor reduces the activity of α-amylase, an enzyme that plays a role in the digestion of starch and glycogen in insects (Sales et al. 2012). Triticale-α amylase inhibitor has a strong inhibitory activity on *Eurygaster integriceps* gut α-amylase (Mehrabadi et al. 2010). Chitinases enzyme degrades the chitin, which is the major component of the insect cuticle and peritrophic membrane (Chandrasekaran et al. 2014). Sharma et al. (2003) reported that the development of Colorado potato beetle is inhibited in transgenic tomato by overexpression of the chitinase gene. Lectin is another entomotoxic protein bind with carbohydrates and glycoproteins to make inhibiting the absorption of the nutrients (Vandenborre et al. 2011). Proteinase inhibitor acts as anti-metabolic proteins, which interfere with the digestive process of insects by reducing the availability of amino acids necessary for growth and development (Habib and Fazili 2007). The AM fungus *Glomus mosseae* promoted expression of serine protease inhibitors, oxygenase D, and allene oxide cyclase genes in tomato (*Solanum lycopersicum*) infested by *Helicoverpa armigera* was reported by Song et al. (2013).

19.5 Indirect Plant Defense Against herbivorous Insect Modulated by AM Fungi Association

In response to environmental stress, plants produce diverse volatile organic compounds (VOCs) as a part of plant protection and signaling molecules to communicate among them. It is also involved in indirect plant defenses against biotic stress, including herbivore attack to serve as an attractant of natural enemies of the attacking insect (De Lange et al. 2020). These volatiles attract both parasitic and predatory insects that are natural enemies of the insect herbivores (Paré and Tumlinson 1999). Volatiles detected in different plants following attacks from different herbivores share notable similarities even though the composition of these volatiles may differ from case to case. Besides feeding, leaf injury caused by caterpillar movement and insect oviposition can also increase volatile emission in plants (Arimura et al. 2005; Hare 2011). The majority of volatiles are from the category of terpenoids, indole (phenylpropanoid), fatty acid derivatives (green leaf volatiles), and nitrogen and sulfur-containing compounds. AM fungal colonization also induces plant defense through jasmonic and salicylic acid signaling pathways, enabling them to achieve compatibility with the plant, and the cocktail of volatile organic compounds (VOCs) released from the leaves (Jung et al. 2012).

Terpenes and terpenoids are the main components of plant volatiles play a significant role in indirect defense by attracting natural enemies of herbivores in various systems (Schuman et al. 2014; Böttger et al. 2018). Chen et al. (2011) reported that terpenes and terpenoids play an important role in plant physiology and defense. Overexpression of terpenes synthase tps10 gene in *Arabidopsis* results in higher emission of sesquiterpenes, and the transgenic *Arabidopsis* plants are more preferred by the parasitic wasps than the wild type (Kappers et al. 2005). AMF also imparts defense by affecting volatile organic compounds (VOCs) emitted by plants (Asensio et al. 2012); these might act as a cue for attracting natural enemies of herbivores (Hunter 2002). The plants colonized with *Glomus intraradices* reported attracting more natural enemies of aphid (Volpe et al. 2018). AMF colonization reported enhanced the production of triterpenoids in various plants (Kapoor et al. 2017). AMF colonization influences plant VOC production influences plant defense against herbivores by attracting predator and parasitoid of insect pests (Meier and Hunter 2019). Maize roots release (E)-β-caryophyllene in response to the attack by the larvae of *Diabrotica virgifera* for attracting *Heterorhabditis megidis* nematodes that feed on the beetle larvae (Howe and Jander 2008). Kessler and Heil (2011) reported that maize seedlings attacked by *Spodoptera exigua* attract parasitoid

Cotesia marginiventris. Plants colonized with *Funneliformis mosseae* showed indirect defense responses through the release of volatiles like β-ocimene and β-caryophyllene, which increased spider mite predation rates in *Phaseolus vulgaris* (Schausberger et al. 2012).

Indole and indole-alkaloid molecules contain nitrogen to attract the natural enemies of herbivorous insects. The effect of indoles on natural enemies of herbivorous insects is very specific to different species under different conditions. Indoles have been detected after an insect attack in many plant species, including cotton (McCall et al. 1994), gerbera (Krips et al. 1999), maize (Zhuang et al. 2012), and lima bean (Maurya et al. 2020). Indoles have also been detected in plants treated with the elicitor volicitin (Frey et al. 2000). Turlings et al. (1995) reported that the blend of indoles and terpenoids released by plants after attack by lepidopteran larvae is attractive to the *Cotesia marginiventris*. In addition to their roles in attracting natural enemies directly indoles also play roles in regulating the emission of other volatiles. In maize, indole emission precedes the release of other volatiles and the release of indoles is essential for priming and to the synthesis of mono- and homoterpenes in the attacked plants, suggesting that indoles may act as potent aerial priming agents that prepare other tissues and neighboring plants for incoming insect attacks (Erb et al. 2015). Mycorrhizal *Solanum lycopersicum* plants are more attractive to the *Aphidius ervi* (parasitoid) than non-mycorrhizal plants (Guerrieri et al. 2004).

Fatty acid derivatives are another common volatiles emitted by plants and they are often referred to as green leaf volatiles because they impart the typical odor of green leaves (Clavijo Mccormick et al. 2014). Multiple pathways can lead to the production of green leaf volatiles and are emitted at elevated levels when leaf tissues are disrupted by the herbivore. Linoleic acid and linolenic acids are unsaturated fatty acids released from the plasma membrane due to cell damage, are oxidized and decarboxylated by hydroperoxide lyases, resulting in the production of volatiles C6-aldehydes, C6-alcohols, and esters such as (*Z*)-3-hexen-1-yl acetate (Feussner and Wasternack 2002). Green leaf volatiles can also be produced by lipoxygenases activity via the oxylipin pathway in many plants (Vincenti et al. 2019). The application of *Manduca sexta* oral secretions to the wounds of wild-type *Nicotiana attenuate* plants leads to a remarkable change in composition resulting in increased foraging efficiency of predators in nature (Allmann and Baldwin 2010). Unlike other HIPVs, green leaf volatiles are released immediately upon herbivore damage. Earlier emission of green leaf volatiles in herbivore-attacked plants can induce the emission of other herbivore-induced plant volatiles and therefore may play a potential role in intra- and interplant signaling (Allmann and Baldwin 2010). Exposure of plants to synthetic green leaf volatiles (GLVs) induces the rapid production of JA and emission of sesquiterpene in maize (Ton et al. 2007), and triggers the emission of local and systemic terpenes in tomato (Farag and Pare 2002). AMF colonized *Leucanthemum vulgare* plants infected with *Chromatomyia syngenesiae* (leaf-miner) attract *Diglyphus isaea* parasitoid (Gange et al. 2003). The HIPV's produced vary according to the AM fungal, plant and herbivore species, the developmental stage and condition of the plants, and the herbivores. These factors also affect the composition of the VOCs emitted from the leaves (Schausberger et al. 2012) that render the infested plant less attractive or even repellent to subsequent herbivores, and attractive to natural enemies of these herbivores, such as parasitoids. It is through these induced changes in plant health and defense responses that mycorrhizal fungi and insect herbivores interact with each other indirectly.

Nitrogen (N) compounds such as nitriles and oximes are among many other volatile compounds that are commonly emitted from herbivore damaged plants (Irmisch et al. 2014). In contrast to the abundant and widespread compounds such as monoterpenes, sesquiterpenes, and green leaf volatiles, nitrogenous volatiles are emitted in minor amounts. However, nitrogenous volatiles do play crucial roles as active gradients of cues for carnivore attraction. The proportion of nitrogenous compounds in a volatile blend can contribute to the specificity of the volatile mix for different plant–carnivore systems (Clavijo Mccormick et al. 2014). In plants, nitrogen-containing volatiles is synthesized from secondary metabolites of glucosinolates (Hopkins et al. 2009) and amino acid derivatives (Irmisch et al. 2014). Some of the nitrogen-containing compounds include aldoximes, nitriles, and benzyl cyanides, which have been detected in the volatile blends of *Populus* sp. in response to *Lymantria dispar* attack (Irmisch et al. 2014). Electrophysiological and behavioral experiments suggest that nitrogenous compounds are key attractants for parasitic Hymenopterans (Clavijo Mccormick et al. 2014). The effect of AMF induced variation in parasitoid behavior fully depends on plant size or volatile emission (Schausberger et al. 2012).

The plant produced volatile compounds prime the plants to resist forthcoming herbivore attack apart from the indirect defense. Defense priming is defined as the enhanced readiness of defense responses (Conrath et al. 2006; Kim and Felton 2013). Primed plants display faster and/or stronger activation of various cellular defense responses to forthcoming stress (Ton et al. 2007; Jung et al. 2009; Slaughter et al. 2012). Plants primed by neighboring plants volatiles and display faster or stronger defense activation and enhanced insect resistance following herbivore attack (Kessler et al. 2006). Anti-herbivore defense response often is induced with greater efficiency in plants that have previously experienced with insect attacks (Karban and Baldwin 1997). Recent studies demonstrate that the primed state of *Arabidopsis thaliana* plants can be transferred to their progeny, conferring better protection from pathogen attack as compared to the descendants of unprimed plants (Slaughter et al. 2012). The primed state in the plant also can be provoked by various natural and synthetic compounds, such as jasmonic acid (JA), salicylic acid (SA), and β-aminobutyric acid (BABA) (Worrall et al. 2012). Tomato plants are grown from JA-treated and BABA treated seeds showed increased resistance against herbivory by spider mites, caterpillars, and aphids, and fungal pathogens (Worrall et al. 2012). Even though many reports displayed that AMF could prime the plant defense against stress in several plants, the underlying mechanisms remain elusive?. In order to feed the growing world population and to achieve global food security under climate change and environmental stress, we need to have a sustainable agriculture that is less impacted by the environmental factors. A sustainable agriculture refers to the key components of agriculture conservation that practices the use of continuous cover crops, no tilling, crop rotations, and environmental friendly. To make use of mycorrhizal technologies for a greener future we need to have a clear understanding of the contribution of mycorrhiza to the causal metabolic pathways of above ground plant parts (Plant–insect interaction) that influence the interaction with rhizosphere for maintaining a sustainable agriculture.

19.6 Conclusion and Future Thrust

AMF symbiosis has an important impact on plant growth by enhancing nutrient availability and also aided in sustaining plant health during herbivore infestation. Mycorrhiza-induced resistance (MIR) in aboveground tissues seems effective in generalist chewing insects. Experimental evidence confirms that this protection is based not only on improved nutrition or local changes within the roots and rhizosphere but that priming of plant immunity plays a major role in MIR. AMF colonization improved the tolerance against insect infestation by modulating morphological and physiological characteristics of the plant by producing defense-related compounds like antioxidant enzymes and metabolites, phytoalexins, herbivore-induced plant volatiles, green leaf volatiles, primary and secondary metabolites, cell wall components, and also improved photosynthetic efficiency of the plants. The vast research indicates that there are changes in the morphological and biochemical behavior of plants in response to insect attack. These changes have been modulated positively to provide defense against insect attack in AM infected plants. It was also understood that insect cues accountable for eliciting defense response in plants vary with the insects, plants, and AMF colonization. These insect cues and plant responses have not been understood fully. Further, the mechanisms of plant defense against insect attacks are only in selected plant species. Although the molecular basis for the regulation of plant defenses and the priming of the plant immune system during mycorrhization remains mostly unknown, a prominent role of jasmonate signaling has been confirmed. Thus, further studies are required to elucidate the insect cues, responses of AMF plants to different insect attack, and mechanisms of resistance or counter adaptation developed by insects. This chapter indicated that AM fungi may be recommended as a potential bioprotective agent against pests during the cultivation of field crops. Moreover, we need to develop strategies to investigate different parameters influencing the effectiveness of mycorrhizal colonization in plants- insect interaction along with the implementation of systemic approaches for crop improvement and integrated pest management (IPM). This approach will be a further step towards sustainable agricultural production.

References

Acharya V, Bhargava M (2008) Morphological basis of resistance in cotton (*Gossypium hirsutum*) against whitefly (*Bemisia tabaci*). Indian J Agr Sci 78:818–820

Agati G, Azzarello E, Pollastri S, Tattini M (2012) Flavonoids as antioxidants in plants: location and functional significance. Plant Sci 196:67–76

Ahanger MA, Tyagi SR, Wani MR, Ahmad P (2014) Drought tolerance: role of organic osmolytes, growth regulators, and mineral nutrients: physiological mechanisms and adaptation strategies in plants under changing environment. Springer, New York, pp 25–55

Allmann S, Baldwin IT (2010) Insects betray themselves in nature to predators by rapid isomerization of green leaf volatiles. Science 329:1075–1078

Andrade S, Malik S, Sawaya A, Bottcher A, Mazzafera P (2013) Association with arbuscular mycorrhizal fungi influences alkaloid synthesis and accumulation in *Catharanthus roseus* and *Nicotiana tabacum* plants. Acta Physiol Plant 35:867–880

Araji S, Grammer TA, Gertzen R, Anderson SD, Mikulic-Petkovsek M, Veberic R, Dandekar AM (2014) Novel roles for the polyphenol oxidase enzyme in secondary metabolism and the regulation of cell death in walnut. Plant Physiol 164:1191–1203

Arimura GI, Kost C, Boland W (2005) Herbivore-induced, indirect plant defences. Biochim Biophys Acta 1734:91–111

Asensio D, Rapparini F, Peñuelas J (2012) AM fungi root colonization increases the production of essential isoprenoids vs. nonessential isoprenoids especially under drought stress conditions or after jasmonic acid application. Phytochemistry 77:149–161

Babenko L, Skaterna T, Kosakivska I (2017) Lipoxygenase activity in ontogenesis of ferns *Salvinia natans* and *Polystichum aculeatum*. Ukr Biochem J 89:5–12

Balestrini R, Salvioli A, Dal Molin A, Novero M, Gabelli G, Paparelli E, Bonfante P (2017) Impact of an arbuscular mycorrhizal fungus versus a mixed microbial inoculum on the transcriptome reprogramming of grapevine roots. Mycorrhiza 27:417–430

Bennett RN, Wallsgrove RM (1994) Secondary metabolites in plant defence mechanisms. New Phytol 127:617–633

Bhonwon A, Stout MJ, Attajarusit J, Tantasawat P (2009) Defensive role of tomato polyphenol oxidases against cotton bollworm (*Helicoverpa armigera*) and beet armyworm (*Spodoptera exigua*). J Chem Ecol 35:28–38

Borowicz VA (1997) A fungal root symbiont modifies plant resistance to an insect herbivore. Oecologia 112:534–542

Borowicz VA (2013) The impact of arbuscular mycorrhizal fungi on plant growth following herbivory: A search for pattern. Acta Oecol 52:1–9

Böttger A, Vothknecht U, Bolle C, Wolf A (2018) Terpenes and Terpenoids. In: Lessons on caffeine, Cannabis & Co. Springer, New York, pp 153–170

Brunetti C, Di Ferdinando M, Fini A, Pollastri S, Tattini M (2013) Flavonoids as antioxidants and developmental regulators: relative significance in plants and humans. Int J Mol Sci 14:3540–3555

Cabral C, Wollenweber B, António C, Rodrigues AM, Ravnskov S (2018) Aphid infestation in the phyllosphere affects primary metabolic profiles in the arbuscular mycorrhizal hyphosphere. Sci Rep 8:1–11

Cameron DD, Neal AL, Wees SC, Ton J (2013) Mycorrhiza-induced resistance: more than the sum of its parts? Trends Plant Sci 18:539–545

Cardoso Filho JA, Sobrinho RR, Pascholati SF (2017) Arbuscular mycorrhizal symbiosis and its role in plant nutrition in sustainable agriculture. In: Agriculturally important microbes for sustainable agriculture. Springer, New York, pp 129–164

Chandrasekaran M, Boughattas S, Hu S, Oh SH, Sa T (2014) A meta-analysis of arbuscular mycorrhizal effects on plants grown under salt stress. Mycorrhiza 24:611–625

Chaudhary A, Bala K, Thakur S, Kamboj R, Dumra N (2018a) Plant defenses against herbivorous insects: a review. IJCS 6:681–688

Chaudhary DK, Dahal RH, Oren A, Kim J (2018b) Proposal of Nemorincola gen. nov. to replace the illegitimate prokaryotic genus name Nemorella (2018). Int J Syst Evol Microbiol 68:1319–1320

Chen X, Koumoutsi A, Scholz R, Schneider K, Vater J, Süssmuth R, Borriss R (2009) Genome analysis of *Bacillus amyloliquefaciens* FZB42 reveals its potential for biocontrol of plant pathogens. J Biotechnol 140:27–37

Chen F, Tholl D, Bohlmann J, Pichersky E (2011) The family of terpene synthases in plants: a mid-size family of genes for specialized metabolism that is highly diversified throughout the kingdom. Plant J 66:212–229

Chen H, Stout MJ, Qian Q, Chen F (2012) Genetic, molecular and genomic basis of rice defense against insects. Crit Rev Plant Sci 31:74–91

Chen D, Zheng S, Shan Y, Taube F, Bai Y (2013) Vertebrate herbivore-induced changes in plants and soils: linkages to ecosystem functioning in a semi-arid steppe. Funct Ecol 27:273–281

Clavijo Mccormick A, Gershenzon J, Unsicker SB (2014) Little peaks with big effects: establishing the role of minor plant volatiles in plant–insect interactions. Plant Cell Environ 37:1836–1844

Conrath U, Beckers GJ, Flors V, García-Agustín P, Jakab G, Mauch F, Pozo MJ (2006) Priming: getting ready for battle. Mol Plant Microbe Interact 19:1062–1071

da Trindade R, Almeida L, Xavier L, Lins AL, Andrade EH, Maia JG, da Silva JK (2019) Arbuscular mycorrhizal fungi colonization promotes changes in the volatile compounds and enzymatic activity of lipoxygenase and phenylalanine ammonia lyase in *Piper nigrum* L.'Bragantina'. Plant 8:442

Das K, Roychoudhury A (2014) Reactive oxygen species (ROS) and response of antioxidants as ROS-scavengers during environmental stress in plants. Front Environ Sci 2:53

De Lange ES, Laplanche D, Guo H, Xu W, Vlimant M, Erb M, Turlings TC (2020) *Spodoptera frugiperda* caterpillars suppress herbivore-induced volatile emissions in maize. J Chem Ecol 46:344–360

Erb M, Veyrat N, Robert CA, Xu H, Frey M, Ton J, Turlings TC (2015) Indole is an essential herbivore-induced volatile priming signal in maize. Nat Commun 6:1–10

Farag MA, Pare PW (2002) C6-green leaf volatiles trigger local and systemic VOC emissions in tomato. Phytochemistry 61:545–554

Fester T, Schmidt D, Lohse S, Walter MH, Giuliano G, Bramley PM, Strack D (2002) Stimulation of carotenoid metabolism in arbuscular mycorrhizal roots. Plant 216:148–154

Feussner I, Wasternack C (2002) The lipoxygenase pathway. Annu Rev Plant Biol 53:275–297

Formenti L, Rasmann S (2019) Mycorrhizal fungi enhance resistance to herbivores in tomato plants with reduced jasmonic acid production. Agron 9:131

Frew A (2019) Arbuscular mycorrhizal fungal diversity increases growth and phosphorus uptake in C3 and C4 crop plants. Soil Biol Biochem 135:248–250

Frey M, Stettner C, Paré PW, Schmelz EA, Tumlinson JH, Gierl A (2000) An herbivore elicitor activates the gene for indole emission in maize. Proc Natl Acad Sci 97:14801–14806

Gange AC (2001) Species-specific responses of a root-and shoot-feeding insect to arbuscular mycorrhizal colonization of its host plant. New Phytol 150:611–618

Gange A, West H (1994) Interactions between arbuscular mycorrhizal fungi and foliar-feeding insects in *Plantago lanceolata* L. New Phytol 128:79–87

Gange AC, Bower E, Brown VK (1999) Positive effects of an arbuscular mycorrhizal fungus on aphid life history traits. Oecologia 120:123–131

Gange AC, Brown VK, Aplin DM (2003) Multitrophic links between arbuscular mycorrhizal fungi and insect parasitoids. Ecol Lett 6:1051–1055

Gange AC, Dey S, Currie AF, Sutton BC (2007) Site-and species-specific differences in endophyte occurrence in two herbaceous plants. J Ecol 95:614–622

Gershenzon J (2017) The cost of plant chemical defense against herbivory: a biochemical perspective. In: Insect-plant interactions. CRC Press, Boca Raton, FL, pp 105–176

Gill SS, Anjum NA, Hasanuzzaman M, Gill R, Trivedi DK, Ahmad I, Tuteja N (2013) Glutathione and glutathione reductase: a boon in disguise for plant abiotic stress defense operations. Plant Physiol Biochem 70:204–212

Gourlay G, Constabel CP (2019) Condensed tannins are inducible antioxidants and protect hybrid poplar against oxidative stress. Tree Physiol 39:345–355

Guerrieri E, Lingua G, Digilio MC, Massa N, Berta G (2004) Do interactions between plant roots and the rhizosphere affect parasitoid behaviour? Ecol Entomol 29:753–756

Gupta MM, Chourasiya D, Sharma MP (2019) Diversity of arbuscular mycorrhizal fungi in relation to sustainable plant production systems. In: Microbial diversity in ecosystem sustainability and biotechnological applications. Springer, New York, pp 167–186

Habib H, Fazili KM (2007) Plant protease inhibitors: a defense strategy in plants. Biotechnol Mol Biol Rev 2:68–85

Hahn PG, Agrawal AA, Sussman KI, Maron JL (2019) Population variation, environmental gradients, and the evolutionary ecology of plant defense against herbivory. Am Nat 193:20–34

Hare JD (2011) Ecological role of volatiles produced by plants in response to damage by herbivorous insects. Annu Rev Entomol 56(1):161–180

Harrison M, Dixon R (1993) Isoflavonoid accumulation and expression of defense gene transcripts during the establishment of vesicular-arbuscular mycorrhizal associations in roots of *Medicago truncatula*. Mol Plant-Microbe Interact 6:643–654

Hart MM, Antunes PM, Chaudhary VB, Abbott LK (2018) Fungal inoculants in the field: is the reward greater than the risk? Funct Ecol 32:126–135

Hartley SE, Gange AC (2009) Impacts of plant symbiotic fungi on insect herbivores: mutualism in a multitrophic context. Annu Rev Entomol 54:323–342

Hopkins RJ, van Dam NM, van Loon JJ (2009) Role of glucosinolates in insect-plant relationships and multitrophic interactions. Annu Rev Entomol 54:57–83

Howe GA, Jander G (2008) Plant immunity to insect herbivores. Annu Rev Plant Biol 59:41–66

Hunter MD (2002) A breath of fresh air: beyond laboratory studies of plant volatile–natural enemy interactions. Agric For Entomol 4:81–86

Irmisch S, Clavijo McCormick A, Günther J, Schmidt A, Boeckler GA, Gershenzon J, Köllner TG (2014) Herbivore-induced poplar cytochrome P450 enzymes of the CYP 71 family convert aldoximes to nitriles which repel a generalist caterpillar. Plant J 80:1095–1107

Jih PJ, Chen YC, Jeng ST (2003) Involvement of hydrogen peroxide and nitric oxide in expression of the ipomoelin gene from sweet potato. Plant Physiol 132:381–389

Jung HW, Tschaplinski TJ, Wang L, Glazebrook J, Greenberg JT (2009) Priming in systemic plant immunity. Science 324:89–91

Jung SC, Martinez-Medina A, Lopez-Raez JA, Pozo MJ (2012) Mycorrhiza-induced resistance and priming of plant defenses. J Chem Ecol 38:651–664

Kapoor R, Anand G, Gupta P, Mandal S (2017) Insight into the mechanisms of enhanced production of valuable terpenoids by arbuscular mycorrhiza. Phytochem Rev 16:677–692

Kappers IF, Aharoni A, Van Herpen TW, Luckerhoff LL, Dicke M, Bouwmeester HJ (2005) Genetic engineering of terpenoid metabolism attracts bodyguards to Arabidopsis. Science 309:2070–2072

Karban R (2020) The ecology and evolution of induced responses to herbivory and how plants perceive risk. Ecol Entomol 45:1–9

Karban RB, Baldwin I (1997) Induced responses to herbivory. University of Chicago Press, Chicago

Keeping MG, Kvedaras OL (2008) Silicon as a plant defence against insect herbivory: response to Massey, Ennos and Hartley. J Anim Ecol 77:631–633

Kessler A, Heil M (2011) The multiple faces of indirect defences and their agents of natural selection. Funct Ecol 25:348–357

Kessler A, Halitschke R, Diezel C, Baldwin IT (2006) Priming of plant defense responses in nature by airborne signaling between *Artemisia tridentata* and *Nicotiana attenuata*. Oecologia 148:280–292

Kim J, Felton GW (2013) Priming of antiherbivore defensive responses in plants. Insect Sci 20:273–285

Koricheva J, Gange AC, Jones T (2009) Effects of mycorrhizal fungi on insect herbivores: a meta-analysis. Ecology 90:2088–2097

Korpita T, Gómez S, Orians CM (2014) Cues from a specialist herbivore increase tolerance to defoliation in tomato. Funct Ecol 28:395–401

Kovalikova Z, Kubes J, Skalicky M, Kuchtickova N, Maskova L, Tuma J, Hejnak V (2019) Changes in content of polyphenols and ascorbic acid in leaves of white cabbage after pest infestation. Molecules 24:2622

Krips O, Willems P, Gols R, Posthumus M, Dicke M (1999) The response of Phytoseiulus persimilis to spider mite-induced volatiles from gerbera: influence of starvation and experience. J Chem Ecol 25:2623–2641

Kumar VV (2018) Biofertilizers and biopesticides in sustainable agriculture. In: Role of rhizospheric microbes in soil. Springer, Singapore, pp 377–398

Kumar D, Priyanka P, Yadav P, Yadav A, Yadav K (2019) Arbuscular mycorrhizal fungi-mediated mycoremediation of saline soil: current knowledge and future prospects recent advancement in white biotechnology through Fungi. Springer, New York, pp 319–348

Kundu A, Mishra S, Vadassery J (2018) *Spodoptera litura*-mediated chemical defense is differentially modulated in older and younger systemic leaves of *Solanum lycopersicum*. Planta 248:981–997

Lackus ND, Lackner S, Gershenzon J, Unsicker SB, Köllner TG (2018) The occurrence and formation of monoterpenes in herbivore-damaged poplar roots. Sci Rep 8:1–13

Lattanzio V, Lattanzio VM, Cardinali A (2006) Role of phenolics in the resistance mechanisms of plants against fungal pathogens and insects. Phytochem Advan Res 661:23–67

Li F, Christensen MJ, Gao P, Li Y, Duan T (2018) An arbuscular mycorrhizal fungus and *Epichloë festucae* var. *lolii* reduce *Bipolaris sorokiniana* disease incidence and improve perennial ryegrass growth. Mycorrhiza 28:159–169

Machado BB, Casanova D, Gonçalves WN, Bruno OM (2013) Partial differential equations and fractal analysis to plant leaf identification. J Phys Conf Ser 410:012066

MacLean AM, Bravo A, Harrison MJ (2017) Plant signaling and metabolic pathways enabling arbuscular mycorrhizal symbiosis. Plant Cell 29:2319–2335

Manjarrez M, Wallwork M, Smith SE, Smith FA, Dickson S (2009) Different arbuscular mycorrhizal fungi induce differences in cellular responses and fungal activity in a mycorrhiza-defective mutant of tomato (rmc). Funct Plant Biol 36:86–96

Markkola A, Kuikka K, Rautio P, Härmä E, Roitto M, Tuomi J (2004) Defoliation increases carbon limitation in ectomycorrhizal symbiosis of *Betula pubescens*. Oecologia 140:234–240

Maurya AK, Patel RC, Frost CJ (2020) Acute toxicity of the plant volatile indole depends on herbivore specialization. J Pest Sci 1–11

Mazid M, Khan T, Mohammad F (2011) Role of secondary metabolites in defense mechanisms of plants. Biol Med 3:232–249

McCall PJ, Turlings TC, Loughrin J, Proveaux AT, Tumlinson JH (1994) Herbivore-induced volatile emissions from cotton (*Gossypium hirsutum* L.) seedlings. J Chem Ecol 20:3039–3050

Mehrabadi M, Bandani AR, Saadati F (2010) Inhibition of Sunn pest, *Eurygaster integriceps*, α-amylases by α-amylase inhibitors (T-αAI) from Triticale. J Insect Sci 10:1

Meier AR, Hunter MD (2019) Mycorrhizae alter constitutive and herbivore-induced volatile emissions by milkweeds. J Chem Ecol 45:610–625

Miller R, Miller S, Jastrow J, Rivetta C (2002) Mycorrhizal mediated feedbacks influence net carbon gain and nutrient uptake in *Andropogon gerardi*. New Phytol 155(1):149–162

Mithöfer A, Boland W (2012) Plant defense against herbivores: chemical aspects. Annu Rev Plant Biol 63:431–450

Nishida T, Katayama N, Izumi N, Ohgushi T (2010) Arbuscular mycorrhizal fungi species-specifically affect induced plant responses to a spider mite. Popul Ecol 52:507–515

Paré PW, Tumlinson JH (1999) Plant volatiles as a defense against insect herbivores. Plant Physiol 121:325–332

Pineda A, Zheng SJ, Van Loon J, Dicke M (2012) Rhizobacteria modify plant–aphid interactions: a case of induced systemic susceptibility. Plant Biol 14:83–90

Pozo MJ, Azcón-Aguilar C (2007) Unraveling mycorrhiza-induced resistance. Curr Opin Plant Biol 10:393–398

Rillig MC, Aguilar-Trigueros CA, Camenzind T, Cavagnaro TR, Degrune F, Hohmann P, Yang G (2019) Why farmers should manage the arbuscular mycorrhizal symbiosis. New Phytol 222:1171–1175

Robinson EA, Ryan GD, Newman JA (2012) A meta-analytical review of the effects of elevated CO2 on plant–arthropod interactions highlights the importance of interacting environmental and biological variables. New Phytol 194:321–336

Rodriguez R, Redman R (2008) More than 400 million years of evolution and some plants still can't make it on their own: plant stress tolerance via fungal symbiosis. J Exp Bot 59:1109–1114

Ruley AT, Sharma NC, Sahi SV (2004) Antioxidant defense in a lead accumulating plant, *Sesbania drummondii*. Plant Physiol Biochem 42:899–906

Ryalls JM, Moore BD, Riegler M, Bromfield LM, Hall AA, Johnson SN (2017) Climate and atmospheric change impacts on sap-feeding herbivores: a mechanistic explanation based on functional groups of primary metabolites. Funct Ecol 31:161–171

Ryan MH, Graham JH (2018) Little evidence that farmers should consider abundance or diversity of arbuscular mycorrhizal fungi when managing crops. New Phytol 220:1092–1107

Salam JA, Hatha MA, Das N (2017) Microbial-enhanced lindane removal by sugarcane (*Saccharum officinarum*) in doped soil-applications in phytoremediation and bioaugmentation. J Environ Manag 193:394–399

Sales PM, Souza PM, Simeoni LA, Magalhães PO, Silveira D (2012) α-Amylase inhibitors: a review of raw material and isolated compounds from plant source. J Pharm Pharm Sci 15:141–183

Schausberger P, Peneder S, Jürschik S, Hoffmann D (2012) Mycorrhiza changes plant volatiles to attract spider mite enemies. Funct Ecol 26:441–449

Schoenherr AP, Rizzo E, Jackson N, Manosalva P, Gomez SK (2019) Mycorrhiza-induced resistance in potato involves priming of defense responses against cabbage Looper (Noctuidae: Lepidoptera). Environ Entomol 48:370–381

Schultz JC, Appel HM, Ferrieri A, Arnold TM (2013) Flexible resource allocation during plant defense responses. Front Plant Sci 4:324

Schuman MC, Palmer-Young EC, Schmidt A, Gershenzon J, Baldwin IT (2014) Ectopic terpene synthase expression enhances sesquiterpene emission in *Nicotiana attenuata* without altering defense or development of transgenic plants or neighbors. Plant Physiol 166(2):779–797

Seguel A, Barea JM, Cornejo P, Borie F (2015) Role of arbuscular mycorrhizal symbiosis in phosphorus-uptake efficiency and aluminium tolerance in barley growing in acid soils. Crop Pasture Sci 66:696–705

Selvaraj A, Thangavel K, Uthandi S (2020) Arbuscular mycorrhizal fungi (*Glomus intraradices*) and diazotrophic bacterium (*Rhizobium* BMBS) primed defense in blackgram against herbivorous insect (*Spodoptera litura*) infestation. Microbiol Res 231:126355

Shao L, Young LT, Wang JF (2005) Chronic treatment with mood stabilizers lithium and valproate prevents excitotoxicity by inhibiting oxidative stress in rat cerebral cortical cells. Biol Psychiatry 58:879–884

Sharma AD, Gill PK, Singh P (2003) RNA isolation from plant tissues rich in polysaccharides. Anal Biochem 314:689–689

Shrivastava G, Ownley BH, Augé RM, Toler H, Dee M, Vu A, Chen F (2015) Colonization by arbuscular mycorrhizal and endophytic fungi enhanced terpene production in tomato plants and their defense against a herbivorous insect. Symbiosis 65:65–74

Shu B, Li W, Liu L, Wei Y, Shi S (2016) Effects of girdling on arbuscular mycorrhizal colonization and root hair development of litchi seedlings. Sci Hortic 210:25–33

Slaughter A, Daniel X, Flors V, Luna E, Hohn B, Mauch-Mani B (2012) Descendants of primed arabidopsis plants exhibit resistance to biotic stress. Plant Physiol 158:835–843

Smith S, Read D (2008) Colonization of roots and anatomy of arbuscular mycorrhiza. In: Mycorrhizal symbiosis. Academic Press, London, pp 42–90

Song YY, Ye M, Li CY, Wang RL, Wei XC, Luo SM, Zeng RS (2013) Priming of anti-herbivore defense in tomato by arbuscular mycorrhizal fungus and involvement of the jasmonate pathway. J Chem Ecol 39:1036–1044

Swamy MK, Akhtar MS, Sinniah UR (2016) Response of PGPR and AM fungi toward growth and secondary metabolite production in medicinal and aromatic plants. In: Plant, soil and microbes. Springer, New York, pp 145–168

Taggar G, Gill R (2012) Preference of whitefly, *Bemisia tabaci*, towards black gram genotypes: role of morphological leaf characteristics. Phytoparasitica 40:461–474

Ton J, D'Alessandro M, Jourdie V, Jakab G, Karlen D, Held M, Turlings TC (2007) Priming by airborne signals boosts direct and indirect resistance in maize. Plant J 49:16–26

Turlings T, Loughrin JH, Mccall PJ, Röse U, Lewis WJ, Tumlinson JH (1995) How caterpillar-damaged plants protect themselves by attracting parasitic wasps. Proc Natl Acad Sci 92:4169–4174

Vandenborre G, Smagghe G, Van Damme EJ (2011) Plant lectins as defense proteins against phytophagous insects. Phytochemistry 72:1538–1550

Verbruggen N, Hermans C (2008) Proline accumulation in plants: a review. Amino Acids 35:753–759

Vincenti S, Mariani M, Alberti JC, Jacopini S, Brunini-Bronzini de Caraffa V, Berti L, Maury J (2019) Biocatalytic synthesis of natural green leaf volatiles using the lipoxygenase metabolic pathway. Catalysts 9:873

Volpe V, Chitarra W, Cascone P, Volpe MG, Bartolini P, Moneti G, Guerrieri E (2018) The association with two different arbuscular mycorrhizal fungi differently affects water stress tolerance in tomato. Front Plant Sci 9:1480

Wang M, Bezemer TM, Van der Putten WH, Biere A (2015) Effects of the timing of herbivory on plant defense induction and insect performance in ribwort plantain (*Plantago lanceolata* L.) depend on plant mycorrhizal status. J Chem Ecol 41:1006–1017

War AR, Paulraj MG, Ahmad T, Buhroo AA, Hussain B, Ignacimuthu S, Sharma HC (2012) Mechanisms of plant defense against insect herbivores. Plant Signal Behav 7:1306–1320

War AR, Paulraj MG, Hussain B, Buhroo AA, Ignacimuthu S, Sharma HC (2013) Effect of plant secondary metabolites on legume pod borer, *Helicoverpa armigera*. J Pest Sci 86:399–408

Wilkinson S, Weston AK, Marks DJ (2019) Stabilising urea amine nitrogen increases potato tuber yield by increasing chlorophyll content, reducing shoot growth rate and increasing biomass partitioning to roots and tubers. Potato Res 1–23. https://doi.org/10.1007/s11540-019-09436-x

Wittstock U, Gershenzon J (2002) Constitutive plant toxins and their role in defense against herbivores and pathogens. Curr Opin Plant Biol 5:300–307

Woods H, Fagan W, Elser J, Harrison J (2004) Allometric and phylogenetic variation in insect phosphorus content. Funct Ecol 18:103–109

Worrall D, Holroyd GH, Moore JP, Glowacz M, Croft P, Taylor J, Roberts MR (2012) Treating seeds with activators of plant defence generates long-lasting priming of resistance to pests and pathogens. New Phytol 193:770–778

Xing Z, Liu Y, Cai W, Huang X, Wu S, Lei Z (2017) Efficiency of trichome-based plant defense in *Phaseolus vulgaris* depends on insect behavior, plant ontogeny, and structure. Front Plant Sci 8:2006

Yan C, Xie D (2015) Jasmonate in plant defence: sentinel or double agent? Plant Biotechnol J 13:1233–1240

Yuan JS, Köllner TG, Wiggins G, Grant J, Degenhardt J, Chen F (2008) Molecular and genomic basis of volatile-mediated indirect defense against insects in rice. Plant J 55:491–503

Zhang S, Lehmann A, Zheng W, You Z, Rillig MC (2019) Arbuscular mycorrhizal fungi increase grain yields: a meta-analysis. New Phytol 222:543–555

Zhuang X, Fiesselmann A, Zhao N, Chen H, Frey M, Chen F (2012) Biosynthesis and emission of insect herbivory-induced volatile indole in rice. Phytochemistry 73:15–22

Chapter 20
Eradication of Malaria by the Mutualistic Interaction Between *Wickerhamomyces anomalus* and *Anopheles* sp.

Arpit Gupta, Arpita Balakrishnan, and Amit C. Kharkwal

Abstract The fatality of malaria is quite prominent worldwide, which is majorly affecting developing nations having subtropical to tropical climate. The disease burden is quite hefty and its eradication is of utmost importance for the economic development of nations. Many chemical control methods have been used in the past decades, but these methods have been turned obsolete due to the development of resistance of the parasite *Plasmodium* and the vector *Anopheles* sp. A new potent method has been developed which focuses on the elimination of the parasite, using the mutualistic interactions between the naturally present microbiota in the mosquito midgut and the mosquito itself; which is termed as symbiotic control. Here we focus on the yeast *Wickerhamomyces anomalus* and elucidate its mutualistic interactions with *Anopheles* sp. which can help eradicate the fatal malarial infection.

Keywords Symbiotic control · *Wickerhamomyces anomalus* · *Pichia anomala* · *Anopheles* sp. · *Plasmodium* sp.

20.1 Introduction

Malaria, a devastatingly fatal epidemic affects millions annually all across the globe. According to the WHO in 2017 itself, nearly half of the world's population was at the risk of malaria, the numbers of the affected were approximately 219 million (World Health Organization 2019). Malaria is quite prominent all around the world but certain hotspots are the hubs for this malicious disease. These hubs are usually confined to tropical and subtropical regions (World Health Organization 2019). Another major factor that affects this disease is the economic development of the region; countries that are underdeveloped or are developing nations are quite prone

A. Gupta · A. Balakrishnan · A. C. Kharkwal (✉)
Amity Institute of Microbial Technology, Amity University Uttar Pradesh, Noida, Uttar Pradesh, India
e-mail: ackharkwal@amity.edu

N. Shrivastava et al. (eds.), *Symbiotic Soil Microorganisms*, Soil Biology 60,
https://doi.org/10.1007/978-3-030-51916-2_20

to the intensive fatality of malaria and the developed countries are usually out of bounds (Gallup and Sachs 2001).

International discussions usually have malaria as a focal point due to its health burden on the social and economic development of nations. Thus, constant steps toward the global eradication of the disease are being taken place, be it preventive measures or treatment options (Alonso et al. 2011). Governments of the affected nations are spreading the word of awareness regarding the precautions that the citizens can follow to minimize the acquirement of the disease. Even these small steps help in decreasing the fatality of the disease. If we shift our prime focus to India, the statistical evidence shows a decrease in the disease burden between 2016 and 2017 with 3 million fewer cases (World Health Organization 2018a). This evidence shows that certain steps toward the prevention of malaria may have helped to invoke this kind of result.

To eliminate the disease, local transmission of the parasite needs to be interrupted and continued control measures need to be implemented to prevent the re-establishment of the disease (World Health Organization 2018b). Eradication, on the other hand, is the complete removal of the overall worldwide incidence of the disease, once this has been achieved intervention methods are no longer required (World Health Organization 2018b).

In 1955, WHO launched the Global Malaria Eradication Programme which utilized two major tools: chloroquine for prophylaxis and prevention and dichloro-diphenyl-trichloroethane (DDT) for vector control (Greenwood et al. 2008). The implementation of these tools had a substantial influence in some areas, which had relatively low transmission rates, such as Sri Lanka and India (Greenwood et al. 2008). Regardless of these achievements, the campaign was eventually unsuccessful due to the emergence of chloroquine-resistant *Plasmodium* strains and DDT-resistant *Anopheles* mosquitoes; another factor which caused the campaign to become unsuccessful was that the campaign never attempted to eradicate malaria in most parts of Africa, where the malarial transmission is intense (Greenwood et al. 2008).

More potent methods and new tools have been deployed and one of the most important ones is being highlighted in this chapter, which is by using microflora present in the mosquito midgut naturally (Drexler et al. 2008), as they are present in the same compartment where the most vulnerable stages of *plasmodium* development occur (Wang et al. 2012). Microbes present naturally may offer opportunities to successfully manipulate the vector competence to reduce their abilities to transmit the human pathogens (Ricci et al. 2012a). Furthermore, there is a growing interest of symbionts in mosquito disease vectors since their manipulation may offer novel control methods that are uniformly defined as Symbiotic Control (Ricci et al. 2012a), which can disrupt mosquito-transmitted pathogens.

A plethora of microorganisms inhabits the midgut that has been known to show symbiotic interactions; such as bacterium *Enterobacter agglomerans* (Ricci et al. 2012b), yeast *Wickerhamomyces anomalus* (Cappelli et al. 2014), fungus *Metarhizium anisopliae* (Fang et al. 2011), and even virus *Densovirus* (Ren et al. 2008). This chapter deals with the symbiotic control using the mutualistic interactions between *Wickerhamomyces anomalus* and *Anopheles stephensi* to eliminate the burden of the fatal disease—Malaria.

20.2 Malaria and Its Mechanisms

Malaria is a widely spread and very significant health problem all around the globe (Mawson 2013). Even though it affects millions around the world, the core regions where it is predominating are places having the necessary climatic conditions required for that disease to thrive (Gallup and Sachs 2001).

Malaria is a mosquito-borne infectious disease that affects a majority of vertebrates and not only human beings. Inside the host body, the target cell (organ) of the causative agent is the liver (hepatic cells), which it reaches through the bloodstream. The causative agent is a parasite that belongs to the *Plasmodium* group. This parasite requires two hosts to complete its life cycle—vertebrate and female *Anopheles* mosquito (Arrow et al. 2004). It usually infects the mosquitos and is transmitted to the vertebrate *via* the bite of the infected mosquito. Other transmission methods may also involve blood transfusions, in which *Plasmodium*-infected blood products are transferred to a healthy host (Arrow et al. 2004).

Particularly in the human host, the clinical manifestations in the beginning stages of the infection are nonspecific and mimic a flu-like syndrome (Bartoloni and Zammarchi 2012). Flu-like syndromes usually elicit fever, headache, and many more common symptoms. An extremely diverse range of severity is observed, which ranges from mild headaches to serious complications which may even lead to death (Bartoloni and Zammarchi 2012). In other vertebrates like mice, the commonly observed symptoms were behavioral and neurological (Basir et al. 2012).

Progression to these complications can be quite rapid, thus patients who are suspected of malaria must be assessed and treated swiftly (Bartoloni and Zammarchi 2012). Severe malaria is a life-threatening disease but is also a curable disease (Bartoloni and Zammarchi 2012). Nonspecific clinical findings are quite common in malaria, this often leads to a false diagnosis of the disease (Bartoloni and Zammarchi 2012). Malaria is sometimes falsely detected as influenza, dengue, typhoid fever, viral hepatitis, gastroenteritis, or encephalitis (Bartoloni and Zammarchi 2012) due to the presence of nonspecific symptoms. This can be quite detrimental for the patient and this may allow the parasite to thrive for a much longer time and may lead to a deadly outcome.

Thus, immediate diagnosis and appropriate treatment are of utmost importance to prevent morbidity and this may also help in averting fatal consequences (Bartoloni and Zammarchi 2012).

20.2.1 *The Causative Agent:* Plasmodium

Plasmodium is an obligate parasitic protozoan. It is from the phylum Apicomplexa, which is a large group of parasitic eukaryotes (Zilversmit and Perkins 2008). It differentiates into a series of morphologically distinct forms in the vertebrate and mosquito hosts (Zilversmit and Perkins 2008) which is highlighted in Fig. 20.1. It alternates between invasive stages (sporozoite, merozoite, and ookinete) and replicative stages (pre-erythrocytic, erythrocytic-schizont, and oocyst) interposed by a single phase of sexual development that facilitates the transmission from the human host to the anopheline vector (Hall et al. 2005).

In the *Plasmodium* group, the main species involved in causing malaria are mentioned below:

- *P. falciparum* is found in tropical and subtropical areas, predominately found in Africa (Global Health 2018). *P. falciparum* can cause severe malaria because it multiplies rapidly in the blood, which may lead to anemia. Infected parasites can also clog small blood vessels (Global Health 2018). Cerebral malaria can occur if the parasite reaches the brain, and as a result, the complications that follow can be lethal (Global Health 2018). The incubation period is approximately—2 weeks (Bartoloni and Zammarchi 2012). This malignant species is known for its high fatality. It mainly infects humans among other vertebrates.
- *P. vivax* is found mostly in Asia, Latin America, and in some parts of Africa. It is the most prevalent species of *Plasmodium* (Global Health 2018). *P. vivax* has the same incubation period as *P. falciparum* (Bartoloni and Zammarchi 2012). *P. vivax* has dormant liver stages called hypnozoites that can activate and invade the blood several months or years after the infection (Global Health 2018).

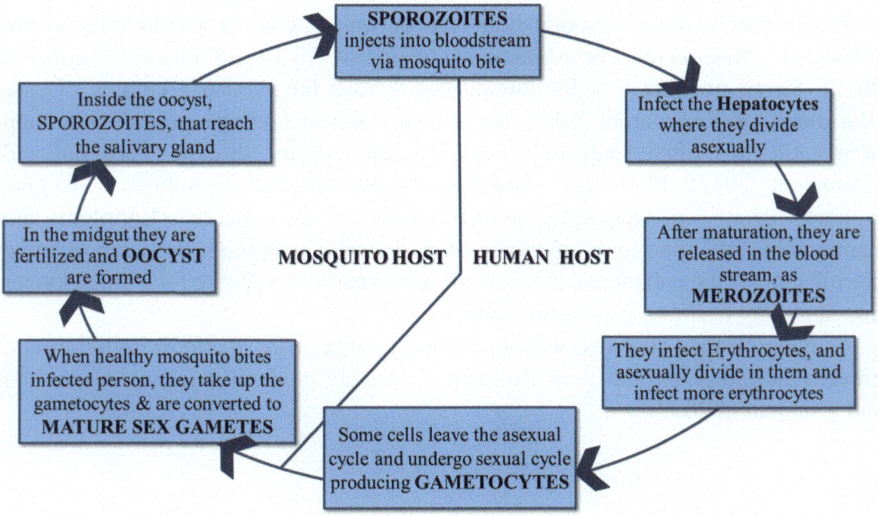

Fig. 20.1 Lifecycle of *Plasmodium* sp.

P. vivax is also rather malignant has come to light by many studies and research in the same matter is being conducted (Bartoloni and Zammarchi 2012).

- *P. ovale* is found mostly in Africa and the islands of the western Pacific. It is biologically and morphologically very similar to *P. vivax* (Global Health 2018). It is rarely very severe, but respiratory distresses have been observed in this case (Lee and Maguire 1999).
- *P. malariae* is found worldwide and is the only human malarial parasitic species that has a three-day cycle (Global Health 2018). *P. malariae* causes chronic infections that can even last a lifetime if left untreated (Global Health 2018). In some chronically infected patients, *P. malariae* can cause grave complications such as nephrotic syndrome (Global Health 2018). It is known to cause the mildest but the most persistent form of malaria (Harinasuta and Bunnang 1988).
- *P. knowlesi* is found throughout Southeast Asia and it has been known to be a significant cause of zoonotic malaria in that region (Global Health 2018). It was also identified as an infectious agent in long-tailed and pig-tailed macaques (Garnham 1966; Singh et al. 2004). *P. knowlesi* has a 24-h replication cycle and so can rapidly progress from an uncomplicated to a severe infection with a high fatality rate (Global Health 2018).
- *P. berghei* is a parasite that causes infection in rodents and it has been noticed in African murine rats, it is the most commonly used model organism for studying severe malaria cases in humans caused by *P. falciparum* (Franke-Fayard et al. 2010). The symptoms shown by this species are to a certain degree comparable to *P. falciparum* infection (Franke-Fayard et al. 2010).
- *P. chabaudi* is a rodent malaria species that cause uncomplicated disease (Wilson et al. 2016). It is usually located in the central regions of Africa (Landau and Killick-Kendrick 1966).

20.2.2 Lifecycle of **Plasmodium**

The malaria parasite, *Plasmodium* develops in both vertebrate and the female *Anopheles* mosquitoes. It requires two hosts to complete its lifecycle. It changes through several stages with different morphological characteristics in both of the hosts (Fig. 20.1).

It starts when an infected female *Anopheles* mosquito bites a healthy vertebrate, injecting *Plasmodium* parasites in the form of sporozoites (Siciliano and Alano 2015) into the bloodstream. The sporozoites move to liver hepatocytes where they divide asexually over the next 7–10 days, causing no symptoms (Siciliano and Alano 2015). After the maturation of the cells in the hepatocytes, they are released from the liver in the form of vesicles into the bloodstream, they are called merozoites (Siciliano and Alano 2015). These merozoites invade red blood cells (erythrocytes) and multiply until lysis of the cell occurs, here asexual reproduction also occurs (Siciliano and Alano 2015). This cycle is repeated and many more blood cells are invaded; Some of the infected blood cells leave the cycle of asexual multiplication and instead multiply into sexual forms of the parasite, called gametocytes (Siciliano and Alano 2015), and these then circulate in the bloodstream.

When a healthy mosquito bites an infected vertebrate, the parasite is transmitted from the vertebrate host to the mosquito. The mosquito ingests these gametocytes, which develop into mature sex cells called gametes (Siciliano and Alano 2015). This maturation occurs in the midgut of the mosquitos, where the fertilization of female gamete takes place and they form oocysts (Siciliano and Alano 2015). Inside the oocyst, thousands of active sporozoites develop, and when it eventually bursts, the release of sporozoites ensues into the body cavity that travels to the mosquito's salivary glands (Siciliano and Alano 2015). And from the salivary gland, the sporozoites get transferred to the vertebrate host (Siciliano and Alano 2015) when this infected mosquito bites, and then the cycle of infection begins again.

20.2.3 The Vector: Anopheles *Mosquito*

The description of the *Anopheles* genus of mosquito was provided in the year 1818 by J.W. Meigen (Meigen 1818). Around 40 species of this *Anopheles* genus are vectors in the malarial infection (Tiwari 2017). They derive blood meal from vertebrates for nurturing their eggs, and sometimes it occurs that the blood being derived is from an infected host and thus the *plasmodium* present in the bloodstream is ingested by the mosquito and hence it becomes a carrier of the disease (Tiwari 2017). The blood meal ingested by the mosquito is the primary link between the vertebrate and the mosquito vector in the parasite's lifecycle (Tiwari 2017).

In Amazon, many species of the *Anopheles* mosquito are present and three species are considered as the primary vectors of malaria, they are namely, *A. darlingi*, *A. albitarsis s.l.*, and *A. aquasalis* (Pimenta et al. 2015). In South America, the major vector is *A. darlingi* and in the Amazonian regions of French Guiana, Bolivia, Guyana, Columbia, Peru, Suriname, and Venezuela the same has been known to be associated with the dynamics of malaria transmission (Zimmerman 1992; Hiwat et al. 2010; Pimenta et al. 2015). In Venezuela *A. albitarsis s.l.* is found (Rubio-Palis et al. 1992; Pimenta et al. 2015) and *A. aquasalisi* is found in Trinidad (Chadee and Kitron 1999; Pimenta et al. 2015) and Venezuela as well (Berti et al. 1993; Pimenta et al. 2015).

Other *Anopheles* species can be occasional malarial vectors due to their geographic distributions, natural infectivity, and their population density (Deane 1986; Zimmerman 1992; Sinka et al. 2010, 2012; Pimenta et al. 2015). *A. nuneztovari s.l.* and *A. triannulatus s.l.* are commonly observed in the Amazon by researchers and they are known to be infected with *P. vivax* and *P. falciparum*, but their role as malaria vectors has yet to be elucidated (de Arruda et al. 1986; de Oliveira-Ferreira et al. 1990; Klein et al. 1991; Tadei and Dutary Thatcher 2000; da Silva-Vasconcelos et al. 2002; Povoa et al. 2003; dos Santos et al. 2005; Póvoa et al. 2006; Galardo et al. 2007; da Rocha et al. 2008; Santos et al. 2009; Pimenta et al. 2015).

In the *Anopheles* mosquito, there are certain defenses against the parasite, the first line of defense or the physical barriers along with the innate immune system poses a significant challenge for parasite development (Pimenta et al. 2015). The limiting of the infection in the salivary glands is majorly done by the innate immune system (Pimenta et al. 2015). For the successful completion of the *Plasmodium* life cycle, it is necessary for the transmission to occur and for the parasite to survive in nature. Thus, there has been enormous pressure on the parasite to evolve means to escape the mosquito's immune defenses (Pimenta et al. 2015). Much more research on this matter needs to be conducted for a better understanding of the role of mosquito's immune response against *Plasmodium* (Pimenta et al. 2015).

20.3 *Wickerhamomyces anomalus*

Wickerhamomyces anomalus is a yeast, which is a robust microorganism and is known for its environmental, industrial, and medical aspects (Walker 2011). A distinguished metabolic and physiological diversity is exhibited by this microorganism (Walker 2011). From an application viewpoint, its role in biotechnology has been of much significance and future research on the same is quite beneficial for better understanding (Walker 2011).

20.3.1 Basic Characteristics and Morphology

W. anomalus is a killer fungus, which is identified by many other names *viz. Pichia anomala, Hansenula anomala, Candida beverwijkiae, Candida pelliculosa,* and commonly termed as non-Saccharomyces wine yeast (Ricci et al. 2011a; Riley et al. 2016). It is a heterothallic yeast, that reproduces by forming one to four hat-shaped ascospores, morphologically it creates highly textured white colored colonies and is classified as a biosafety level-1 organism (Kurtzman 2011; Ricci et al. 2011a; Satora et al. 2014; Landis 2018).

20.3.1.1 Niche

They are found in diverse environments and have also been isolated from various alternate sources exempli gratia soil, plants, cereal grains, maize silage, fruit skin, fruit juices, wine, and food products (like dairy, baked, fermented, high-sugar, and salted foods) (Pfaller et al. 2009; Cappelli et al. 2014; Satora et al. 2014; Riley et al. 2016). These killer toxin-producing yeasts have also been weeded out from contaminated oil, marine environment, wastewater, and also from immunocompromised patients (Cappelli et al. 2014; Riley et al. 2016). They even have been isolated (in trace amounts) from microflora of human skin, throat, and alimentary canal;

likewise, in the gut of insects like beetles, mosquitoes, and flies (Pfaller et al. 2009; Cappelli et al. 2014).

20.3.1.2 Abiding Environment

W. anomalus can survive in highly intolerable (by other microbes) environmental stress and is adaptive to inexcusable growth conditions like varying temperature (3–37 °C), high osmotic pressure, a wide range of pH (2–12), low water activity; simultaneously it possesses the ability to endure in very little or no oxygen environment that is anaerobic conditions, all these qualities make it highly adaptable to a plethora of environments which makes them highly competitive (Cappelli et al. 2014; Satora et al. 2014; Riley et al. 2016).

20.3.1.3 Compounds Secreted and Its Uses

W. anomalus can produce a substance called Killer toxin, which can kill other molds and yeasts, therefore, it can be used as a biocontrol agent and bio preservative in the agro-food sectors (recognized by European Food Safety Authority) (Cappelli et al. 2014; Riley et al. 2016). Other than killer toxin production, it secrets 2-phenyl ethanol, which prevents spore formation in human pathogenic fungus *Aspergillus flavus* and reduces the biosynthesis of aflatoxins (Pretscher et al. 2018).

Moreover, it is one of the strong producers of volatile compounds including Isoamyl acetate (which has a redolence of artificial banana), acetic acid, and ethyl acetate in pure culture (Satora et al. 2014; Landis 2018). Apart from its cytocidal activity, it is used in biofuel production, it is used in wine fermentation due to its ability to produces volatile compounds that contribute to the aroma of wine and some of its molecular products are also used in medicines for humans (Cappelli et al. 2014; Riley et al. 2016).

20.3.2 Killer Toxin

The ability of *W. anomalus* to produce killer toxin is known, but incidences of various types of killer toxins being produced have also been observed, which can be accounted to post-translational modifications; these modifications render peculiar differences in molecular weight among the different types of Killer Toxins (Cecarini et al. 2019). Those yeasts which are known to produce Killer Toxins are immune to their own Killer Toxins but they might be susceptible to toxins produced by other yeasts (Cecarini et al. 2019). Killer Toxins are potent to kill various microbes but what makes them fatal is a prevailing question. These killer toxins are a group of glycoproteins consisting of a major globular core with extensive polar surfaces, 12% Beta sheets, and 32% Helix (4% 3(10) helix and 28% Alpha-helix) (Cappelli et al.

2014; Cecarini et al. 2019). There are different strains of *W. anomalus* which are known for their killer toxin production, some of them are depicted as follows:

W. anomalus HN1-2 isolated from the mangrove ecosystem produces a killer toxin that can kill cells of *Metschnikowia bicuspidata, Candida albicans, Kluyveromyces aestuarii, Pichia guilliermondii, Lodderomyces elongisporus, Yarrowia lipolytica,* and *Saccharomyces cerevisiae,* these are present in natural marine environments (Sun et al. 2012).

W. anomalus-killer toxin a and *W. anomalus*-killer toxin b secreted by *W. anomalus* VKM Y-159 are reported to show cytocidal effects, *W. anomalus*-killer toxin a shows the effect on pathogen *Candida spp.* and several other yeasts while *W. anomalus*-killer toxin b exhibits killer activity on *Candida norvegica* and *Candida alai* (Farkas et al. 2012).

Panomycocin, a potent antifungal agent, is a fatal protein secreted by *W. anomalus* NCYC 343 (Satora et al. 2014).

W. anomalus F17.12 exerts strong anti-plasmodial activities on *Plasmodium berghei* (Cecarini et al. 2019).

20.3.2.1 Conditions for Secretion

Killer phenotype's killer toxin secretion gets enhanced or activated due to stress in the environment which can be experienced due to competition with other microbes for the acquisition of resources; but host body serves as a niche, where yeast can proliferate and utilize a good amount of food thus the secretion of the killer toxin may be hindered (Cappelli et al. 2019). Therefore, in general terms, killer toxin production gets triggered, in the presence of environmental stress and competition.

20.3.2.2 Mechanism of Action

Killer Toxins exert cytocidal effects via a two-step mechanism (Cecarini et al. 2019). The first step involves the binding of Killer toxins with primary receptors, which are usually cell wall carbohydrates (mainly β-1,3 glucans), and these glucans are hydrolyzed by the β-1,3 glucanase activity of the killer toxins into glucose. Later the second step proceeds by translocation of Killer Toxins to the secondary receptors of the plasma membrane, where it ultimately performs cell lysis which leads to death (Cappelli et al. 2014, 2019; Satora et al. 2014; Cecarini et al. 2019) as demonstrated in Fig. 20.2.

20.3.2.3 Validation of the Mechanism of Action

EXGI and EXG2 are the genes coding for β-glucanase synthesis, which are found in the genome of *W. anomalus* and the lack of the *W. anomalus* antimicrobial activity was found to be correlated with coupled or single silencing of these genes (Valzano

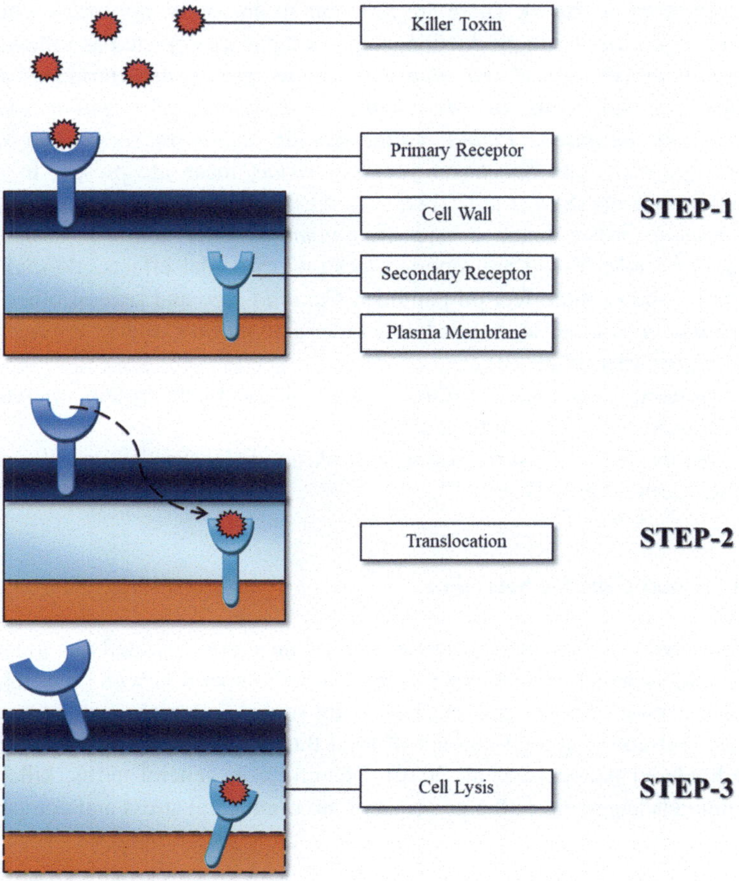

Fig 20.2 Mechanism of action of *Killer Toxins*

et al. 2016). To confirm that the cytocidal effect of Killer Toxins is mediated by β-glucanase activity, Castanospermine (a β-glucanase inhibitor) was induced with Killer Toxins to observe its effect against a parasite and as a consequence, the killer activity of the toxin was reduced to 46.2% from 79.7% in case of Killer Toxins isolated from *W. anomalus* F17.12 strain and in case of the *Killer toxin* isolated from *W. anomalus* ATCC 96603 strain it was reduced to 49.5% from 88.5% (Valzano et al. 2016). So, castanospermine confirmed the Killer Toxins' β-glucanase mediated activity (Valzano et al. 2016).

20.3.3 *The Mutualism of* W. anomalus *and* Anopheles *Mosquito*

W. anomalus a microorganism with high potential is known to form symbiotic relationships with many organisms (Ricci et al. 2011c). Within the insect populations, it has been located within the inner body of *Drosophila* sp. (Ricci et al. 2011c), it has also been identified as a killer yeast, acting against pathogenic fungi in the crab *Portunus trituberculatus* (Wang et al. 2007) and its presence has been observed in many mosquito species (Ricci et al. 2012b) as well.

W. anomalus has been known to form an association with specific species of mosquito hosts (Ricci et al. 2012a); They were detected at all the developmental stages of both malaria (*A. stephensi, A. gambiae*) and dengue mosquito vector species (*Aedes aegypti, Aedes albopictus*) where it localizes in the midgut and reproductive organs (Ricci et al. 2011a, c). These mutualistic interactions shine a light on the possibility that this particular yeast strain is pre-adapted to colonize the mosquito host body not only in the GI tract but it is intimately associated with the mosquito's reproductive systems (Ricci et al. 2011a). This suggests that some beneficial biological properties are present in the yeast to aid its locomotion to different body compartments while escaping the host immune barriers (Ricci et al. 2011a). These host-barrier evasion mechanisms, are yet to be explored.

The combination of the niche preference by *W. anomalus*, together with its capacity to retain growth under variable environmental conditions, are major factors involved in its survival in the midgut and gonads (Ricci et al. 2011a). The presence of *W. anomalus* has been identified through transmission electron microscopic analysis of the laboratory-reared colony of *A. stephensi* (Ricci et al. 2011a). This proved the presence of the yeast in the mosquito species but to confirm symbiotic association much more evidence was collected, which is mentioned as follows:

- *W. anomalus* was identified by molecular and cultivation-based methods.
- *W. anomalus* was PCR-detected in mosquitoes from all development stages.
- *W. anomalus* was detected by using specific fluorescence in-situ hybridization probes in both male and female guts and reproductive systems (Ricci et al. 2011a).

This association between the organisms is quite favorable to both the counterparts; *W. anomalus* is located in certain vital organs, which is a great source of nutrition for the yeast required for its growth and development (Ricci et al. 2011a). Their presence in the gonads could indicate that they are colonizing there to increase fecundity (Ricci et al. 2011c). Interestingly, the insect gonads are characterized by a thick system of tracheal trunks that could represent a means for meeting the oxygen demands (Gibson and Hunter 2005). The localization of *W. anomalus* in the reproductive organs suggests a vertical transmission route through generations; Such a possibility is supported by the occurrence of *W. anomalus* in all life stages of *A. stephensi* (Ricci et al. 2011a). *W. anomalus*-Killer toxin signals were also observed in progeny of mosquitoes fed with *W. anomalus* F17.12 strain, which suggested it could be a consequence of vertical transmission (Cappelli et al. 2014).

On the other hand, *A. stephensi* is attracted to yeasts due to its high CO_2 production levels and the fact that there is a vital requirement of CO_2 in the mosquito body (Smallegange et al. 2010; Blumberg et al. 2015). Yeasts could also take part in driving some behavioral traits of the insects such as oviposition site selection and host-seeking (Phelan and Lin 1991; Ricci et al. 2011c). More possible functional roles of *W. anomalus* are still to be assessed, but potential benefits can be hypothesized. We assume that a certain, protective role against pathogenic fungi is being provided to mosquitoes through the killer toxin production activity of the *W. anomalus* (Polonelli 2000; Passoth et al. 2006). Considering that the marine strains of *W. anomalus* have already been proposed to protect the host crab against pathogens (Wang et al. 2007), such a possibility should be also considered for *A. stephensi*.

In recent years, the relationship between symbionts and mosquitoes has attracted a great deal of attention, mainly for the perspective of exploiting the symbionts for blocking the transmission of parasites, through the production of antagonistic factors (Favia et al. 2008; Moreira et al. 2009). This can help improve the future potential of vector-borne disease control (Bian et al. 2013; Koehler and Kaltenpoth 2013). The novel finding of the yeast from *A. stephensi* displays the presence of a killer strain, which supports the hypothesis of a protective function (Cappelli et al. 2014) and could make the *Plasmodium* vulnerable to the antagonistic actions of the killer toxins (Abraham and Jacobs-Lorena 2004).

20.3.4 Competition

Now, we are aware that *W. anomalus* is present inside the mosquito, not as a transient commensal, but as a symbiont (Ricci et al. 2011a; Wang and Jacobs-Lorena 2013) and that they secrete a Killer Toxin which has been reported to show anti-*Plasmodium* activity on different developmental stages of *Plasmodium spp.* (Cappelli et al. 2019; Cecarini et al. 2019). Thus, serving as a competition to the *Plasmodium spp.* in the mosquito midgut; there are many in vivo and in vitro, experimental researches performed to gather more shreds of evidence to support this hypothesis.

20.3.4.1 In Vivo

Evidence proved that Killer Toxins are secreted by *W. anomalus* F17.12 and *W. anomalus* ATCC 96603 strains in the mosquito's body and its presence was confirmed by a fluorescent signal produced by staining with a monoclonal antibody (mAb) KT4 on the free yeast's cell surface (this result was obtained using immunofluorescence assay), while *W. anomalus* UM3 strain did not reflect any such signals

(Cappelli et al. 2014). By using immunofluorescence assay, it was revealed that Killer Toxin signals were found on the 10th day after the introduction of the yeast, exclusively in the gut but after a time they produced a comparable signal in the gonads as well (Cappelli et al. 2014).

When activated *W. anomalus* F17.12 strain was introduced to *A. stephensi* through diet, it showed up in its midgut interfering with ookinete development of *P. berghei*, while in mice, reduction in parasitemia was observed due to inhibitory action of purified *W. anomalus* F17.12-killer toxin on *P. berghei*'s erythrocytic stages (Cappelli et al. 2019).

Inhibitory effect of *W. anomalus* F17.12 strain on the development of *P. berghei*'s early sporogonic stages in *A. stephensi* was found to be 65.21%, while *W. anomalus* UM3 strain, which is a non-killer toxin-producing strain of *W. anomalus*, exerted 38.02% effect on parasites (Cappelli et al. 2019).

20.3.4.2 In Vitro

Killer Toxins isolated from *W. anomalus* F17.12 strain (obtained from *An. stephensi*) unveiled strong anti-plasmodial activity on *P. berghei*'s early sporogonic stages and found to be lethal due to membrane damage (Cappelli et al. 2019). In another experiment, Killer Toxins' effect in inhibiting plasmodial development was exhibited to be ~90%, it exerted a strong anti-plasmodial response against sporogonic stages of *berghei* species of *Plasmodium* (Adnani et al. 2017). *W. anomalus* F17.12 strain demonstrated the prevention of ookinete development to 40% (Valzano et al. 2016; Cappelli et al. 2019).

It was observed that Killer toxin tends to follow dose-dependent activity, as Killer toxin obtained from both *W. anomalus* ATCC 96603 and *W. anomalus* F17.12 strains exhibited highest inhibition percentage (92.3 and 87.5%, respectively) against the development of *P. berghei*'s sporogonic stages at 100 μg/ml (Valzano et al. 2016). While Lethal Concentration 50 (it is the concentration of Killer toxin at which half of the parasite population is killed) values for *W. anomalus* F17.12 and *W. anomalus* ATCC 96603 strains were found to be 64.4 and 61.3 μg/ml, respectively, which could be the possible equivalent activity of Killer Toxin (Valzano et al. 2016).

20.3.4.3 Impact

After studies, it was found that Killer Toxin strongly inhibits the development of *P. berghei*'s gametocyte stage to ookinete, as the consequence of its interaction with β-glucans as discussed earlier (Valzano et al. 2016). Microscopic observation threw a limelight on some structural/morphological alterations in post-zygotic stages due to its interaction with Killer toxin, these changes include—jagged cell borders, irregular cell-shape, lack of crystalloid assembly, cytoplasmic region's feeble staining and less-defined cytoplasmic granule (Valzano et al. 2016). Using Giemsa staining it was revealed that *W. anomalus* F17.12 strain's Killer Toxins strategy is to hinder the mature ookinete development in the *Plasmodium*'s sporogonic stages (Valzano et al. 2016).

20.4 Malaria Eradication

Eradication of malaria is a topmost priority for any developing nation, due to the fatality that the disease poses on the health and welfare of the citizens. Malaria is a disease that is known to cause a significant health burden on individuals (Alonso et al. 2011) and the need for treatment is required to be promptly provided (Bartoloni and Zammarchi 2012) otherwise if the disease is not dealt with it may showcase some highly complicated clinical manifestations (Bartoloni and Zammarchi 2012).

Certain measures can be taken to limit the pathogenic effect of the disease. Vitamin A supplementation seems to provide a rather positive impact on the disease; a trial conducted on children in the malaria-endemic area of Papua New Guinea led to a 68% decrease in parasite density in those consuming the Vitamin A supplement (Shankar et al. 1999; Mawson 2013).

Global strategies against malaria eradication have been deployed, by the means of rapid application of Dichloro-diphenyl-trichloroethane (DDT) to ensue interruption in transmission around the world regardless of the geographic conditions (World Health Organization 1956). This approach was quite successful in the beginning, but later on, the mosquito species evolved to resist the impact of DDT, rendering it unusable for malaria elimination. Malaria eradication through chemicals has a large impact but constant evolution through time has made the mosquito vectors resistant to the chemicals.

So, an alternative for chemicals was the need of the hour, which could be more potent in the eradication of mosquitos. A much more lethal and effective substitute could be Symbiotic control (Ricci et al. 2012b). Symbiotic Control also referred to as symbiont-based control (Douglas 2007) is a multifaceted approach that employs symbiotic microorganisms, to control insect pests or to render the vector incompetent (Ricci et al. 2012b). This approach was originally termed as paratransgenesis, defined as the modification of the insect phenotype by genetic transformation of its associated microorganisms (Ashburner 1998).

Symbiotic control of Malaria is possible by many microorganisms, like *Pantoea agglomerans* (Bisi and Lampe 2011), *P. berghei* (Cappelli et al. 2014), etc. A proper investigative study of these microorganisms on model organisms, like mice, can aid us to better understand how to curb malaria.

20.4.1 Symbiotic Control of Malaria by Mutualism Between W. anomalus *and* Anopheles *sp.*

Numerous studies and experimental data indicate that *W. anomalus* is found as a symbiont in the gut of *Anopheles* sp. and *Aedes* sp. mosquitoes, while various species of *W. anomalus* are found in reproductive organs of *A.stephensi* and *A. gambiae* (Ignatova et al. 1996; Ricci et al. 2011a, b; Crotti et al. 2012; Wang and Jacobs-Lorena 2013).

Studies indicate that killer toxin produced, elicits antimicrobial properties which are mediated by β-glucanase activity that targets cell wall of yeasts, protozoa, and bacteria; moreover, it is attested to be lethal for some marine microbes as well (Sun et al. 2012; Cappelli et al. 2019).

W. anomalus F17.12 strain is coming into the limelight due to its antimicrobial properties, mainly for inhibition of early sporogonic developmental stages *P. berghei* therefore, it is deemed as a potent symbiotic control of malaria (Cappelli et al. 2014, 2019). *W. anomalus* is a quality presumption of safety status recommended biological agent (Koutsoumanis et al. 2019) but at the same time, it is also known to produce killer toxin which is quite ironical (Sundh and Melin 2011; Cappelli et al. 2019).

Some notable experimental evidence that indicates how *W. anomalus* can be effectively used against malaria in the form of Symbiotic control

- The killer toxin produced by *W. anomalus* F17.12 strain elicited good response against sporogonic stages of *P. berghei* (Cappelli et al. 2019).
- The smell produced by yeast's fermentation acted as a bait for mosquitoes (St Laurent et al. 2016; Cappelli et al. 2019).
- When tested on Murine cell lines, the killer toxin produced by *W. anomalus* F17.12 strains exhibited no harm to vertebrate cells (Cappelli et al. 2019).
- Interaction between *W. anomalus* F17.12 strain and carbohydrates of the parasite cell wall is more productive in the lumen of the midgut in mosquitoes (Cappelli et al. 2019).

An important point to be noted here is, as per Cappelli and colleagues suggestion *W. anomalus* can be easily released in mosquito feeding sites, which implies that it can be ingested by mosquitoes and get into their gut (Cappelli et al. 2019).

20.4.2 Symbiotic Control of Malaria by Other Mutualistic Examples

Many methods of symbiotic control for malaria have been researched and developed in recent years. Other than *W. anomalus,* many other robust microorganisms are also active components in symbiotic control of malaria. The bacteria *Enterobacter agglomerans* present in the midgut of mosquitoes is known to survive in mosquito bodies (Ricci et al. 2012b). *E. agglomerans* is non-pathogenic and it is branded as an excellent candidate for a paratransgenic malaria control strategy (Ricci et al. 2012b).

Moreover, *Pantoea agglomerans*, a bacterial symbiont of *Anopheles* mosquitoes, has been engineered to secrete anti-*Plasmodium* (Bisi and Lampe 2011) effector proteins, thus can also be a potentially useful tool for malaria paratransgenic control (Ricci et al. 2012b). This has been demonstrated by a study that used these engineered *Pantoea agglomerans* strains to see the effect it would have on *Plasmodium,* it gave a result of 98% inhibition of *P. falciparum* in vivo (Hertig and Wolbach 1924).

Densoviruses, capable of infecting and disseminating in *A. gambiae*, has been proposed as a paratransgenic tool for malaria control strategies (Ren et al. 2008). Studies describe the use of the transgenic *Metarhizium anisopliae* fungus to inhibit malaria transmission, which abolishes the parasite development within the mosquito (Fang et al. 2011).

Attempts to modify *Asaia* to produce strains able to express and secrete anti-*Plasmodium* effector molecules to be used in 'malaria transmission-blocking' experiments are being conducted (Ricci et al. 2012b).

20.5 Conclusion

In the current world scenario, when globalization and urbanization are the topmost priorities across the nations, people are enticed toward synthetic or chemical-based drugs, which are quite fast and effective in treating illnesses, but their life-threatening side-effects should not be ignored. Other than this, parasites and even vectors get resistant to it and may cause a situation of quagmire.

To prevent this situation, we should roll toward, more natural ways to combat deadly diseases. There was a time when chloroquine and DDT were the most effective weapon for combating malaria worldwide but after a few years, both the parasite and mosquito vector got resistant to it. Therefore, an alternative against malaria could be *W. anomalus*, a mutualistic yeast which lives in mutual harmony inside the *Anopheles* mosquitoes has unveiled a strong anti-plasmodial activity on *P. berghei*'s sporogonic stages and from gametocyte to ookinete stages via the production of killer toxin.

In our suggestion, spraying of *W. anomalus* in water bodies and environments where the mosquito resides can be beneficial. It could be a wise alternative because it can survive in a highly intolerable environment where many other microbes cannot; and as it is a fermenting yeast the smell produced by it, can act as a bait for the mosquitoes. Furthermore, when it comes to safety, *W. anomalus* is a Biosafety level—1 microorganism, which can kill other yeasts and fungi but not impose any harmful effects on human beings. Due to this, it has been thought to be used as a biocontrol agent in agro-food industries. It has already been found in many organisms such as *Drosophila* sp., and crab *Portunus trituberculatus* where it showed no harm and rather killed the pathogen present in them. When Killer toxin was tested on murine cell lines, it exhibited no harm to vertebrate cells. Also, *W. anomalus* was found to be transmitted vertically in male as well as female mosquitoes of all ages. In our opinion, these reasons are more than satisfactory for *W. anomalus* to be sprayed in water bodies and environments where the mosquito resides to control malarial infections worldwide.

References

Abraham EG, Jacobs-Lorena M (2004) Mosquito midgut barriers to malaria parasite development. Insect Biochem Mol Biol 34:667–671. https://doi.org/10.1016/J.IBMB.2004.03.019

Adnani N, Rajski SR, Bugni TS (2017) Symbiosis-inspired approaches to antibiotic discovery. Nat Prod Rep 34:784–814. https://doi.org/10.1039/c7np00009j

Alonso PL, Brown G, Arevalo-Herrera M et al (2011) A research agenda to underpin malaria eradication. PLoS Med 8:e1000406. https://doi.org/10.1371/journal.pmed.1000406

Arrow K, Panosian C, Gelband H (2004) The parasite, the mosquito, and the disease. In: Saving lives, buying time: economics of malaria drugs in an age of resistance, 6th edn. National Academies Press, Washington, DC

Ashburner M (1998) Prospects for the genetic transformation of arthropods. Insect Mol Biol 7:201–213

Bartoloni A, Zammarchi L (2012) Clinical aspects of uncomplicated and severe malaria. Mediterr J Hematol Infect Dis 4:e2012026. https://doi.org/10.4084/MJHID.2012.026

Basir R, Rahiman SSF, Hasballah K et al (2012) Plasmodium berghei ANKA infection in ICR mice as a model of cerebral malaria. Iran J Parasitol 7:62–74

Berti J, Zimmerman R, Amarista J (1993) Spatial and temporal distribution of anopheline larvae in two malarious areas in Sucre state, Venezuela. Mem Inst Oswaldo Cruz 88:353–362. https://doi.org/10.1590/s0074-02761993000300003

Bian G, Joshi D, Dong Y et al (2013) Wolbachia invades Anopheles stephensi populations and induces refractoriness to plasmodium infection. Science 340(80):748–751

Bisi DC, Lampe DJ (2011) Secretion of anti-plasmodium effector proteins from a natural Pantoea agglomerans isolate by using PelB and HlyA secretion signals. Appl Environ Microbiol 77:4669–4675. https://doi.org/10.1128/AEM.00514-11

Blumberg BJ, Short SM, Dimopoulos G (2015) Employing the mosquito microflora for disease control. Elsevier Inc., Amsterdam

Cappelli A, Ulissi U, Valzano M et al (2014) A Wickerhamomyces anomalus killer strain in the malaria vector anopheles stephensi. PLoS One 9:e95988

Cappelli A, Valzano M, Cecarini V et al (2019) Killer yeasts exert anti-plasmodial activities against the malaria parasite Plasmodium berghei in the vector mosquito Anopheles stephensi and in mice. Parasit Vectors 12:329. https://doi.org/10.1186/s13071-019-3587-4

Cecarini V, Cuccioloni M, Bonfili L et al (2019) Identification of a killer toxin from wickerhamomyces anomalus with beta-glucanase activity. Toxins (Basel) 11:568. https://doi.org/10.3390/toxins11100568

Chadee DD, Kitron U (1999) Spatial and temporal patterns of imported malaria cases and local transmission in Trinidad. Am J Trop Med Hyg 61:513–517. https://doi.org/10.4269/ajtmh.1999.61.513

Crotti E, Balloi A, Hamdi C et al (2012) Microbial symbionts: a resource for the management of insect-related problems. Microb Biotechnol 5:307–317. https://doi.org/10.1111/j.1751-7915.2011.00312.x

da Rocha JAM, de Oliveira SB, Povoa MM et al (2008) Malaria vectors in areas of Plasmodium falciparum epidemic transmission in the Amazon region, Brazil. Am J Trop Med Hyg 78:872–877 da Silva-Vasconcelos A, Kató M, Mourão E et al (2002) Biting indices, host-seeking activity and natural infection rates of anopheline species in Boa Vista, Roraima, Brazil from 1996 to 1998. Mem Inst Oswaldo Cruz 97:151–161

de Arruda M, Carvalho MB, Nussenzweig RS et al (1986) Potential vectors of malaria and their different susceptibility to Plasmodium falciparum and Plasmodium vivax in northern Brazil identified by immunoassay. Am J Trop Med Hyg 35:873–881. https://doi.org/10.4269/ajtmh.1986.35.873

de Oliveira-Ferreira J, Lourenco-de-Oliveira R, Teva A et al (1990) Natural malaria infections in anophelines in Rondonia State, Brazilian Amazon. Am J Trop Med Hyg 43:6–10

Deane LM (1986) Malaria vectors in Brazil. Mem Inst Oswaldo Cruz 81:5–14. https://doi.org/10.1590/S0074-02761986000600002

dos Santos RLC, Sucupira IMC, Lacerda RN d L (2005) Entomological survey and infectivity during malaria outbreak in the Anajas municipality, Para State. Rev Soc Bras Med Trop 38:202–204. https://doi.org/10.1590/s0037-86822005000200018

Douglas AE (2007) Symbiotic microorganisms: untapped resources for insect pest control. Trends Biotechnol 25:338–342. https://doi.org/10.1016/j.tibtech.2007.06.003

Drexler AL, Vodovotz Y, Luckhart S (2008) Plasmodium development in the mosquito: biology bottlenecks and opportunities for mathematical modeling. Trends Parasitol 24:333–336. https://doi.org/10.1016/j.pt.2008.05.005

Fang W, Vega-Rodríguez J, Ghosh A et al (2011) Development of transgenic fungi that kill human malaria parasites in mosquitoes. Science 331(80):1074–1077

Farkas Z, Marki-Zay J, Kucsera J et al (2012) Characterization of two different toxins of Wickerhamomyces anomalus (Pichia anomala) VKM Y-159. Acta Biol Hung 63:277–287. https://doi.org/10.1556/ABiol.63.2012.2.9

Favia G, Ricci I, Marzorati M et al (2008) Bacteria of the genus Asaia: a potential paratransgenic weapon against malaria. Adv Exp Med Biol 627:49–59

Franke-Fayard B, Fonager J, Braks A et al (2010) Sequestration and tissue accumulation of human malaria parasites: can we learn anything from rodent models of malaria? PLoS Pathog 6: e1001032. https://doi.org/10.1371/journal.ppat.1001032

Galardo A, Arruda M, Couto A et al (2007) Malaria vector incrimination in three rural riverine villages in the Brazilian Amazon. Am J Trop Med Hyg 76:461–469

Gallup JL, Sachs JD (2001) The economic burden of malaria. Am J Trop Med Hyg 64:85–96

Garnham P (1966) Malaria parasites and other haemosporidia. Blackwell Science, Oxford

Gibson C, Hunter M (2005) A reconsideration of the role of yeast associated with Chrysoperla lacewings. Biol Control 32:57–64

Global Health (2018) Malaria parasites-about. In: CDC-Cent Dis Control Prev

Greenwood BM, Fidock DA, Kyle DE et al (2008) Malaria: progress, perils, and prospects for eradication find the latest version: review series malaria: progress, perils, and prospects for eradication. J Clin Invest 118:1266–1276. https://doi.org/10.1172/JCI33996.1266

Hall N, Karras M, Raine JD et al (2005) A comprehensive survey of the plasmodium life cycle by genomic, transcriptomic, and proteomic analyses. Science 307(80):82–86

Harinasuta T, Bunnang D (1988) The clinical features of malaria. Malar Princ Pract Malariol:709–734

Hertig M, Wolbach SB (1924) Studies on rickettsia-like micro-organisms in insects. J Med Res 44:329–374.7

Hiwat H, Issaly J, Gaborit P et al (2010) Behavioral heterogeneity of Anopheles darlingi (Diptera: Culicidae) and malaria transmission dynamics along the Maroni River, Suriname, French Guiana. Trans R Soc Trop Med Hyg 104:207–213

Ignatova EA, Nagornaia SS, Povazhnaia TN, Ianishevskaia GS (1996) The yeast flora of blood-sucking mosquitoes. Mikrobiol Z 58:12–15

Klein TA, Lima JB, Tada MS, Miller R (1991) Comparative susceptibility of anopheline mosquitoes in Rondonia, Brazil to infection by Plasmodium vivax. Am J Trop Med Hyg 45:463–470. https://doi.org/10.4269/ajtmh.1991.45.463

Koehler S, Kaltenpoth M (2013) Maternal and environmental effects on symbiont-mediated antimicrobial defense. J Chem Ecol 39:978–988

Koutsoumanis K, Allende A, Alvarez-Ordóñez A et al (2019) Update of the list of QPS-recommended biological agents intentionally added to food or feed as notified to EFSA 10: suitability of taxonomic units notified to EFSA until March 2019. EFSA J 17(7):5753. https://doi.org/10.2903/j.efsa.2019.5753

Kurtzman CP (2011) Wickerhamomyces Kurtzman, Robnett & Basehoar-Powers (2008). In: The yeasts, 5th edn. Elsevier, pp 899–917

Landau I, Killick-Kendrick R (1966) Rodent plasmodia of the République Centrafricaine: the sporogony and tissue stages of Plasmodium chabaudi and P. berghei yoelii. Trans R Soc Trop Med Hyg 60:633–649. https://doi.org/10.1016/0035-9203(66)90010-1

Landis E (2018) Get to know your sourdough yeast – microbial foods. Bread & Grains. MicrobialFood.org

Lee E, Maguire J (1999) Acute pulmonary edema complicating ovale malaria. Clin Infect Dis 29:697–698. https://doi.org/10.1086/598667

Mawson AR (2013) The pathogenesis of malaria: a new perspective. Pathog Glob Health 107:122–129. https://doi.org/10.1179/2047773213Y.0000000084

Meigen JW (1818) Systematische beschreibung der bekannten europäischen zweiflügeligen insekten. FW Forstmann, Aachen

Moreira LA, Iturbe-Ormaetxe I, Jeffery JA et al (2009) A Wolbachia symbiont in Aedes aegypti limits infection with dengue, Chikungunya, and Plasmodium. Cell 139:1268–1278

Passoth V, Fredlund E, Druvefors U, Schnu¨rer J (2006) Biotechnology, physiology and genetics of the yeast Pichia anomala. FEMS Yeast Res 6:3–13

Pfaller MA, Merz WG, Second M (2009) Wickerhamomyces anomalus infections caused by non-Candida, non-Cryptococcus yeasts. In: Clinical mycology, 2nd edn. Elsevier, pp 251–270

Phelan P, Lin H (1991) Chemical characterization of fruit and fungal volatiles attractive to dried-fruit beetle, Carpophilus hemipterus (L.) (Coleoptera: Nitidulidae). J Chem Ecol 17:1253–1272

Pimenta PFP, Orfano AS, Bahia AC et al (2015) An overview of malaria transmission from the perspective of amazon anopheles vectors. Mem Inst Oswaldo Cruz 110:1–25. https://doi.org/10.1590/0074-02760140266

Polonelli L (2000) Human, monoclonal and recombinant candidacidal, pneumocysticidal and mycobactericidal antibodies. Indian J Chest Dis Allied Sci 42:259–263

Povoa MM, Conn JE, Schlichting CD et al (2003) Malaria vectors, epidemiology, and the re-emergence of Anopheles darlingi in Belem, Para, Brazil. J Med Entomol 40:379–386. https://doi.org/10.1603/0022-2585-40.4.379

Póvoa MM, de Souza RTL, da Luz Lacerda RN et al (2006) The importance of Anopheles albitarsis E and An. Darlingi in human malaria transmission in Boa Vista, state of Roraima, Brazil. Mem Inst Oswaldo Cruz 101:163–168. https://doi.org/10.1590/S0074-02762006000200008

Pretscher J, Fischkal T, Branscheidt S et al (2018) Yeasts from different habitats and their potential as biocontrol agents. Fermentation 4(2):31. https://doi.org/10.3390/fermentation4020031

Ren X, Hoiczyk E, Rasgon JL (2008) Viral paratransgenesis in the malaria vector Anopheles gambiae. PLoS Pathog 4:e1000135. https://doi.org/10.1371/journal.ppat.1000135

Ricci I, Damiani C, Scuppa P et al (2011a) The yeast wickerhamomyces anomalus (Pichia anomala) inhabits the midgut and reproductive system of the Asian malaria vector Anopheles stephensi. Environ Microbiol 13:911–921. https://doi.org/10.1111/j.1462-2920.2010.02395.x

Ricci I, Mosca M, Valzano M et al (2011b) Different mosquito species host Wickerhamomyces anomalus (Pichia anomala): perspectives on vector-borne diseases symbiotic control. Antonie van Leeuwenhoek, Int J Gen Mol Microbiol 99:43–50. https://doi.org/10.1007/s10482-010-9532-3

Ricci I, Mosca M, Valzano M, Damiani C (2011c) Different mosquito species host Wickerhamomyces anomalus (Pichia anomala): perspectives on vector-borne diseases symbiotic control. Antonie Van Leeuwenhoek 99:43–50. https://doi.org/10.1007/s10482-010-9532-3

Ricci I, Damiani C, Capone A et al (2012a) Mosquito/microbiota interactions: from complex relationships to biotechnological perspectives. Curr Opin Microbiol 15:1–7. https://doi.org/10.1016/j.mib.2012.03.004

Ricci I, Valzano M, Ulissi U et al (2012b) Symbiotic control of mosquito borne disease. Pathog Glob Health 106:380–385. https://doi.org/10.1179/2047773212Y.0000000051

Riley R, Haridas S, Wolfe KH et al (2016) Comparative genomics of biotechnologically important yeasts. Proc Natl Acad Sci 113:9882 LP–9889887. https://doi.org/10.1073/pnas.1603941113

Rubio-Palis Y, Wirtz RA, Curtis CF (1992) Malaria entomological inoculation rates in western Venezuela. Acta Trop 52:167–174. https://doi.org/10.1016/0001-706x(92)90033-t

Satora P, Tarko T, Sroka P, Blaszczyk U (2014) The influence of Wickerhamomyces anomalus killer yeast on the fermentation and chemical composition of apple wines. FEMS Yeast Res 14:729–740. https://doi.org/10.1111/1567-1364.12159

Shankar AH, Genton B, Semba RD et al (1999) Effect of vitamin a supplementation on morbidity due to Plasmodium falciparum in young children in Papua New Guinea: a randomised trial. Lancet 354:203–209. https://doi.org/10.1016/S0140-6736(98)08293-2

Siciliano G, Alano P (2015) Enlightening the malaria parasite life cycle: bioluminescent Plasmodium in fundamental and applied research. Front Microbiol 6:1–8. https://doi.org/10.3389/fmicb.2015.00391

Singh B, Kim Sung L, Matusop A et al (2004) A large focus of naturally acquired Plasmodium knowlesi infections in human beings. Lancet 363:1017–1024. https://doi.org/10.1016/S0140-6736(04)15836-4

Sinka ME, Rubio-Palis Y, Manguin S et al (2010) The dominant Anopheles vectors of human malaria in the Americas: occurrence data, distribution maps and bionomic precis. Parasit Vectors 3:72. https://doi.org/10.1186/1756-3305-3-72

Sinka ME, Bangs MJ, Manguin S et al (2012) A global map of dominant malaria vectors. Parasit Vectors 5:69. https://doi.org/10.1186/1756-3305-5-69

Smallegange R, Schmied W, van Roey K, Al E (2010) Sugar-fermenting yeast as an organic source of carbon dioxide to attract the malaria mosquito Anopheles gambiae. Malar J 9:292

St Laurent B, Oy K, Miller B et al (2016) Cow-baited tents are highly effective in sampling diverse Anopheles malaria vectors in Cambodia. Malar J 15:440. https://doi.org/10.1186/s12936-016-1488-y

Sun H-Y, Wang K, Chi Z et al (2012) Simultaneous production of single cell protein and killer toxin by Wickerhamomyces anomalus HN1-2 isolated from mangrove ecosystem. Elsevier, Amsterdam

Sundh I, Melin P (2011) Safety and regulation of yeasts used for biocontrol or biopreservation in the food or feed chain. Antonie Van Leeuwenhoek 99:113–119. https://doi.org/10.1007/s10482-010-9528-z

Tadei WP, Dutary Thatcher B (2000) Malaria vectors in the Brazilian Amazon: Anopheles of the subgenus Nyssorhynchus. Rev Inst Med Trop Sao Paulo 42:87–94. https://doi.org/10.1590/S0036-46652000000200005

Tiwari A (2017) Are mosquitoes that carry malaria parasites also affected? In: Sci. ABC

Valzano M, Cecarini V, Cappelli A et al (2016) A yeast strain associated to Anopheles mosquitoes produces a toxin able to kill malaria parasites. Malar J 15:21. https://doi.org/10.1186/s12936-015-1059-7

Walker GM (2011) Pichia anomala: cell physiology and biotechnology relative to other yeasts. Antonie Van Leeuwenhoek 99:25–34. https://doi.org/10.1007/s10482-010-9491-8

Wang S, Jacobs-Lorena M (2013) Genetic approaches to interfere with malaria transmission by vector mosquitoes. Trends Biotechnol 31:185–193. https://doi.org/10.1016/j.tibtech.2013.01.001

Wang X, Chi Z, Yue L et al (2007) A marine killer yeast against the pathogenic yeast strain in crab (Portunus trituberculatus) and an optimization of the toxin production. Microbiol Res (Pavia) 62:77–85

Wang S, Ghosh AK, Bongio N et al (2012) Fighting malaria with engineered symbiotic bacteria from vector mosquitoes. Proc Natl Acad Sci U S A 109:12734–12739. https://doi.org/10.1073/pnas.1204158109

Wilson KD, Stutz SJ, Ochoa LF et al (2016) Behavioural and neurological symptoms accompanied by cellular neuroinflammation in IL-10-deficient mice infected with Plasmodium chabaudi. Malar J 15:1–12. https://doi.org/10.1186/s12936-016-1477-1

World Health Organization (1956) Ninth world health assembly. Intn Organ 10:642–644

World Health Organization (2018a) World malaria report

World Health Organization (2018b) Overview of malaria elimination

World Health Organization (2019) Malaria. WHO

Zilversmit M, Perkins S (2008) Plasmodium: malaria parasite. In: Tree Life Web Proj.

Zimmerman R (1992) Ecology of malaria vectors in the Americas and future direction. Mem Inst Oswaldo Cruz 87:371–383

Part IV
Microbial Symbiosis in Disease and Stress Management

Chapter 21
Halophyte–Endophyte Interactions: Linking Microbiome Community Distribution and Functionality to Salinity

Bliss Ursula Furtado and Katarzyna Hrynkiewicz

Abstract Many plants are unable to adapt to rapid environmental changes (e.g., salinity, drought, or limited nutrients) and may acquire assistance from microbes that have the capacity to increase tolerance of host-plants in stress conditions. By having the right microbes, the plants are more resilient! Such microbes include endophytes that inhabit inner tissues of the plant without causing symptoms of disease in their host. However, this plant–endophytic association exists only when chemical equilibrium is maintained between both, therefore making this mutualistic interaction even more unique. Therefore it is interesting to decode the endophytic community composition in halophytes specifically in the most salt-tolerant halophyte species *Salicornia europaea*, and further determine the factors that could affect this association. Moreover, understanding the endophytes potential plant growth-promoting activities in association with host (*S. europaea*) and non-host plant (non-halophytes) are the focus of this chapter.

Keywords Salicornia europaea · Non-mycorrhizal plant · Salt tolerance · Bacteria · Fungi · Proteobacteria · Ascomycota · Soil salinity · Culture dependent · Metagenomic · Enzyme activity

21.1 Introduction

Soil salinity is among the major abiotic stresses affecting crop production today. It is caused due to climate change, low precipitation, high surface evaporation, weathering of native rocks, improper irrigation management on landscapes, salts used for deicing roads, and poor agricultural practices (Litalien and Zeeb 2019). Soil salinization is characterized by a high concentration of soluble salts, when the

B. U. Furtado · K. Hrynkiewicz (✉)
Department of Microbiology, Faculty of Biological and Veterinary Science, Nicolaus Copernicus University, Torun, Poland
e-mail: hrynk@umk.pl

N. Shrivastava et al. (eds.), *Symbiotic Soil Microorganisms*, Soil Biology 60,
https://doi.org/10.1007/978-3-030-51916-2_21

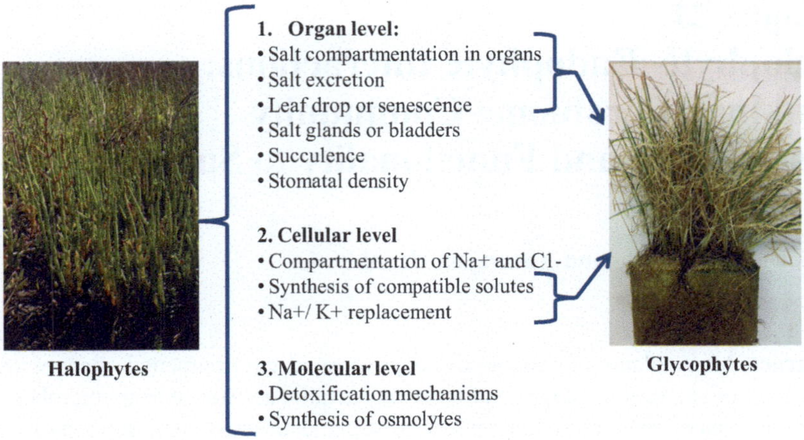

Fig. 21.1 Response of halophytic and glycophytic plants under salinity stress at the (1) organ, (2) cellular, and (3) molecular levels. Halophytes possess all of the mechanisms to combat salinity stress while glycophytes have minimal tolerance to stress *via* organ and cellular responses only and no mechanisms at the molecular level

electrical conductivity (EC_e) in the root zone exceeds 4 dS m^{-1} (approximately 40 mM NaCl) at 25 °C and has exchangeable sodium of 15% and osmotic pressure of approximately 0.2 MPa (reviewed by Munns et al. 2020). The yield of most crop plants is reduced at this EC_e, though many crops exhibit yield reduction at lower EC_e (reviewed by Munns et al. 2020). Salt stress damages various physiological and metabolic processes in plants by inducing osmotic stress and increasing ion toxicity, nutrient deficiency, membrane disruption, and inhibition of metabolic activities, as well as changes in gene expression (Shahzad et al. 2019).

Nevertheless, there are some plants that show no evidence of inhibition in extreme salinity, and normally exhibit a pronounced salt requirement for optimal plant growth (Nikalje et al. 2019). These plants are called halophytes, which are representative vegetation of saline habitats ranging from coastal regions, salt marshes and mudflats, saline depressions, inland deserts and sand dunes and rocky coasts (Nikalje et al. 2019). Most of the halophytes belong to family Amaranthaceae, Plumbaginaceae, Plantaginaceae, Aizoaceae, Poaceae, Brassicaceae, with the Chenopodiaceae being dominant (Slama et al. 2015). The database "eHALOPH" currently identified more than 1500 halophyte species reported from different parts of the world. Clearly, halophytes have evolved a number of adaptive traits expressed at various levels of organization (Fig. 21.1) that includes adjustment of their internal water relations through salt exclusion, succulence, salt-secreting glands and bladders, ion compartmentation in cell vacuoles and accumulation of compatible organic solutes and are distinguishable from glycophytes (salt-sensitive plants) (reviewed by Fan 2020).

It is inevitable that no organism thrives alone, which questions the probable existence and role of the closely associated microbiome present in halophytes and their contribution toward the halophytes' high salt tolerance ability. Endophytic microorganisms ubiquitously occur in all plant species and are important components of the biological diversity. The term "endophyte" can be defined as all organisms that at some point in their life cycle colonize internal plant tissues without causing harm or symptoms of disease in their plant host (Petrini 1991). These include bacteria, actinomycetes, or fungi that colonize the internal plant tissues. Endophytes are particularly fascinating because of their multifaceted lifestyle, i.e., they may exist as either free-living soil microbes or saprobes or pathogens (Tadych et al. 2009). They exhibit variations in their mechanisms of transmission from one plant host to another including strict vertical transmission from maternal plants to seeds, infectious transmission from one host plant to another, or infection by spores from environmental sources like wind, rain, soil, and leaf litter (Tadych et al. 2009). Endophytes are known to share a symbiotic relationship with their plant host, wherein they use the internal environment of the plant as an ecological niche, and in return protect their host from the negative effects of the adversely changing environment (Petrini 1991). Alternatively, some endophytic species could be neutral and do not offer any benefits or neither harm their hosts (Backman and Sikora 2008). With this background, this chapter discusses on the diversity of endophytic bacterial and fungal communities in halophyte *Salicornia europaea* found in different salinity environments. It is a well-known fact, that the true endophyte state in the host plant is established only when the chemical equilibrium between the host and microbe is achieved during their long-term association. During this process of endophyte stabilization in host plant, many environmental factors such as abiotic (e.g., temperature, salinity, soil composition) and biotic (e.g., competition with microbes already present in the host) could alter or redefine the metabolic capacity of endophytes. Hence, we assess on the possibility of these endophytes possessing beneficial traits or properties related to plant growth promotion or stress tolerance in their host further on their resilience when inoculated in other non-host plants. Further, we underlie the importance of understanding the endophytic microbiome in halophytes which will likely pave a new avenue in engineering endophyte-mediated stress tolerance in plants.

21.2 Salicornia a Potential Halophytic Crop Plant

The halophyte chosen for this discussion, i.e., *Salicornia europaea* L. (Amaranthaceae) is a non-mycorrhizal plant (Sonjak et al. 2009) and one of the most salt-accumulating halophyte known. It is a "pioneering plant" in both coastal and inland saline sites, which has generated significant interest as a multi-purpose halophyte (Nikalje et al. 2018). *S. europaea* grows up to a height of 35 cm and is fairly richly branched. They are dark green becoming yellow-green and ultimately flushed pink or red (Ventura and Sagi 2013). They are succulent annuals with extremely reduced leaves and a spike-like terminal inflorescence (Ventura and

Sagi 2013). This genus is widely dispersed in Eurasia, North America, and South Africa, presently comprises around 25–30 species (Ventura and Sagi 2013).

S. europaea is a crop with great commercial value and ecological importance. It is suitable for cultivation as a vegetable in highly saline environments (Ventura and Sagi 2013) and as a source of valuable secondary compounds (Singh et al. 2014). *Salicornia* spp. has been successfully grown in aquaculture systems (Singh et al. 2014). As a salt-accumulating halophyte, as much as 50% of the dry weight of *S. europaea* may be salt ions (Ushakova et al. 2005; Furtado et al. 2019a). This was also observed in the results of Furtado et al. (2019a) where high accumulation of Na^+ in shoots, whereas K^+ and Ca^{2+} levels were higher in roots throughout the sampling period (spring and autumn) in all *S. europaea* samples collected from the two salt-affected sites. Therefore, this species is promising for soil desalination, which is required for the development of agriculture on salty soils and beaches. Moreover, research on halophyte plants is of particular interest today not only for their high salt tolerance and agronomic value but also because it is a non-mycorrhizal plant. This fact makes it interesting to discuss their associated microorganisms (endophytes) that could compensate for the missing symbiotic protection and increase plant growth and fitness particularly under unfavorable conditions. *S. europaea* is especially known to produce secondary compounds (e.g., alkaloids, triterpenoid saponins, and flavonoids, among others) (Isca et al. 2014). Many of these compounds are used as energy sources by endophytic bacteria and fungi suggesting that fluctuations in the proportions and type of compounds during the plant development play an active role in selection for specific bacteria and/or fungi.

21.3 Biodiversity of Endophytic Microbiome in *S. europaea*

The endophytic community in halophytes may be different from those in other plants because salinity acts as an environmental filter. Thus, allowing the survival of selected taxa that can withstand extreme environments. To date, endophyte community diversity investigated in *S. europaea* have been analyzed *via* culture-independent approaches, e.g., sequencing of the 16S rRNA gene and/or the internal transcribed spacer regions (ITS1 and ITS2), 454 pyrosequencing (e.g., Shi et al. 2015; Szymańska et al. 2016; Zhao et al. 2016a; Szymańska et al. 2018; Yamamoto et al. 2018; Hrynkiewicz et al. 2019; Furtado et al. 2019a) or culture-dependent approaches involving procedures for surface sterilization of plant tissue followed by fragmentation and culture of the fragments onto specific agar plates (e.g., Potato Dextrose Agar, R2A media, Czapek dox agar, etc.) amended with antibiotics (e.g., tetracycline, Nystatin, etc.) (e.g., You et al. 2014; Okane and Nakagiri 2015; Park et al. 2016; Zhao et al. 2016b; Furtado et al. 2019b).

21.3.1 *Salicornia* Bacterial Diversity

Prokaryotic endophytes are considered to be diverse comprising of phyla bacteria and a small portion of Archaea. Overall, the bacterial endophytes are found across many phyla, generally the most abundant community is Proteobacteria (group of Gram-negative bacteria), including α, β, and γ-Proteobacteria classes followed by Firmicutes, Actinobacteria, and Bacteriodetes. Using the culture-independent approach, the bacterial community in the rhizosphere soil and endophytes of *S. europaea* growing at the Fukang Desert Ecosystem Observation and Experimental Station (FDEOES) in Xinjiang Province, China were investigated (Shi et al. 2015). Proteobacteria was most dominant phylum in all the samples followed by other large phyla Firmicutes, Bacteroidetes, and Actinobacteria. This study is consistent with results obtained in Fukang, China (Zhao et al. 2016a), Ciechocinek and Inowroclaw, Poland (Szymańska et al. 2018), Lake Notoro in the eastern part of Hokkaido, Japan (Yamamoto et al. 2018), Ciechocinek and Inowroclaw, Poland (Furtado et al. 2019a). Szymańska et al. (2018) suggested that the endophytes representing phyla Proteobacteria and Bacteroidetes predominate in saline environments regardless of the level of salinity in the root zone soil and plant roots. It was also found that bacteria representing Actinobacteria exhibited lower tolerance to salt stress and this phylum exists more frequently in environments with lower levels of salinity (e.g., Shi et al. 2015; Szymańska et al. 2018; Yamamoto et al. 2018).

This is contrary to the results obtained *via* culture-dependent methods in which phylum Firmicutes and Actinobacteria dominate in many reports. Zhao et al. (2016b) isolated endophytic bacteria of *S. europaea* growing at Gurbantünggüt Desert, China. These isolates belonged to phylum Firmicutes and Actinobacteria with 13 different bacterial genera *Arthrobacter*, *Bacillus*, *Brachybacterium*, *Brevibacterium*, *Glycomyces*, *Isoptericola*, *Kocuria*, *Mesorhizobium*, *Pseudomonas*, *Phyllobacterium*, *Planococcus*, *Streptomyces*, and *Variovorax*. Another showed Gram-positive bacteria (phylum Firmicutes and Actinobacteria) dominating in all experimental variants including genera *Bacillus* sp., *Streptomyces* sp., and *Microbacterium* sp. (Szymańska et al. 2016).

Some microbes are ubiquitous and can exist as free-living soil microbes, as epiphytes or as endophytes. For instance, *Bacillus* sp., *Salinicola* sp., *Serratia* sp., *Streptomyces* sp., *Microbacterium* sp. and *Rhodococcus* sp. were reported as endophytes in *S. europaea* shoot and roots collected from two sites in Poland (Furtado et al. 2019b) and similar bacterial diversity was previously obtained in the rhizosphere samples at the same investigated sites (Szymańska et al. 2016). Although another study showed no significant differences in bacterial diversity and richness between the bulk soil, rhizosphere, and the root endosphere for *S. europaea* but the differences were observed at the genera level with the most abundant root endophytes including *Sulfurimonas*, *Coleofasciculus*, and *Aestuariispira* while, *Roseovarius,* and *Halochromatium* were highly abundant in the rhizosphere and seven genera were dominant the *S. europaea* bulk soil: *Thiogranum*, SEEP-SRB1, *Caldithrix*, *Ignavibacterium*, Sva008 sediment group, Candidatus Thiobios, and

Spirochaeta (Yamamoto et al. 2018). Overall, the bacterial endophytic community in *S. europaea* is much greater in comparison to the fungal community according to results obtained by Szymańska et al. (2016) and Furtado et al. (2019b).

21.3.2 Salicornia Fungal Diversity

Although the endophytic microbiome may comprise a small portion of fungi (compared to bacteria) their contribution can be essential in the plants development. Notably, class 2 and class 4 fungal endophytes are commonly found in halophytes, are capable of extensive tissue colonization and are relevant for plant survival in stress habitats (Flowers and Colmer 2015). Moreover, the class 2 fungal endophytes can establish habitat-adapted symbiosis and confer specific stress tolerance to their host and non-host plants in extreme habitats (Flowers and Colmer 2015).

Furtado et al. (2019a) performed metagenomic analysis of *S. europaea* shoots and roots obtained from two different high-salinity environments in Poland and showed that 95% of the fungal reads belonged to phyla Ascomycota. In the next study, the culturable endophytic fungal diversity was investigated by Furtado et al. (2019a) which isolated 320 fungal strains mainly represented phylum Ascomycota (96% of the isolates) from the roots and shoots of *S. europaea* in the same sites. The endophytic fungal strains mainly consisted of the orders: Pleosporales (dominated by *Alternaria* sp. and *Stemphylium* sp.), Eurotiales (mainly *Aspergillus* sp. and *Penicillium* sp.) and Hypocreales (only *Fusarium* sp. and *Trichoderma* sp.). The remaining genera represented the order Dothideales, *Incertae sedis*, Capnodiales, Sordariales, Botryosphaeriales, and Chaetothyriales. Okane and Nakagiri (2015) found *Pleospora* sp. and *Alternaria alternata* were the major endophytes of *S. europaea* roots in the eastern Hokkaido and the Seto Inland Sea (Setouchi) regions in Japan. Endophytic fungal isolates belonging to 9 genera: *Aspergillus*, *Penicillium,* and *Fusarium* were dominantly distributed genera in roots of *S. europaea* native to saltern of the Korea, followed by *Aureobasidium*, *Cladosporium*, *Gibberella*, *Macrophoma*, *Phoma*, *Stemphylium*, and unidentified (Pleosporales), respectively (Park et al. 2016). Alike, Booth et al. (1988) reported a high frequency of *Alternaria* species isolated from *S. europaea* in southern Manitoba and Saskatchewan, Canada. Furthermore, the dominance of *Alternaria* sp. found in *S. europaea* was also previously reported in Canada (Muhsin and Booth 1987), South Korea (You et al. 2014), Japan (Okane and Nakagiri 2015) and Poland (Furtado et al. 2019b). To date, most of the fungal strains (e.g., *Epicoccum* sp., *Alternaria* sp., *Phoma* sp., *Fusarium* sp., *Cladosporium* sp., *Penicillium* sp., *Acremonium* sp., *Lewia* sp., *Pleosporales* sp., *Stemphylium* sp. and *Aspergillus,* etc.) reported in *S. europaea* belong to a group of common plant pathogens and saprobes found in other plants (Okane and Nakagiri 2015; Park et al. 2016; Furtado et al. 2019b). However, colonization of these endophytes in healthy halophyte plant tissue indicates they are not pathogenic and we propose that these fungi may have "co-evolved" in their host. Few studies have discussed the abilities of fungi to switch lifestyles between the endophyte-pathogen (Hyde and Soytong 2008) and endophyte-saprotroph (Promputtha et al. 2010).

In general, very few endophytic fungal strains belonging to the Phylum Basidiomycota have been reported in *S. europaea*, e.g., a study showed 4% of the isolated diversity were categorized in Basidiomycota into four orders; Polyporales (most representative genus being *Peniophora*) and the other strains corresponded to Russulales, Agaricales, and Cantharellales (Furtado et al. 2019b). These findings were parallel to studies using culture-independent techniques (e.g., Furtado et al. 2019a).

21.4 Factors Shaping *S. europaea*–Endophyte Association

The halophyte–endophyte association is dependent on many factors, mainly the host plant lifestyle (i.e., obligate halophyte *S. europaea* in this discussion) and soil characteristics (e.g., salinity) can have a strong effect on the rate of endophyte colonization (Szymańska et al. 2014; Shi et al. 2015; Zhao et al. 2016a; Szymańska et al. 2018; Furtado et al. 2019a, b). Most endophytic communities are strongly influenced by the soil micro-ecological environment as endophytes are recruited from the soil which may act as a filter for microbial species (e.g., Szymańska et al. 2014; Shi et al. 2015). Research has shown that plants growing in soils with high salt content harbor endophytic assemblages that differ significantly from those in other environments. Two saline sites characterized by different salinity and ion composition (anthropogenic salinity site: lower EC_e with Ca^{2+} and natural salinity site: higher ECe with Na^{+}) showed the greatest abundance of bacteria in the rhizosphere samples than that observed in plant roots (Szymańska et al. 2016). They found majority of rhizosphere bacteria belonged to Firmicutes however; this proportion was lower compared to the endophytes (consisted of 50% Firmicutes). This is in line with previous observations by Shi et al. (2015) where a higher total bacterial biomass was measured in soil, followed by roots of *S. europaea* and the bacterial diversity in the endosphere of *S. europaea* was lower than that in the rhizosphere of *S. europaea* (Shi et al. 2015).

In a second survey, Szymańska et al. (2018) analyzed the endophytic bacterial community in *S. europaea* roots from the same test sites using metagenomic approaches. The results revealed that the higher levels of soil salinity did not reduce the composition of endophytic bacterial diversity in roots. However, a distinct taxonomic composition was observed at the two sites which were attributed to the distinct adaptation of halotolerant microorganisms. This was also found in the culturable diversity study where the highest abundance of bacterial endophytes was isolated from the natural saline site compared to the anthropogenic site and frequency of isolation in the culture medium increased with increasing NaCl concentrations (Szymańska et al. 2016). Thus, indicating the presence of a significant number of halophilic bacteria known to survive at high salt concentrations. This difference indicates that the local environment has a complex effect on the bacterial community. Supporting this view, a clear influence of soil properties on the existence of unique bacterial and fungal endophytes of *S. europaea*, specific to the geographical regions in different countries, such as Japan (Okane and Nakagiri 2015), Canada (Muhsin and Booth 1987), China (Zhao et al. 2016a, b), South Korea (You et al. 2014) and Poland (Furtado et al. 2019a, b) was demonstrated.

The importance of acclimatization time in adaptation of microorganisms to salinity was emphasized by Szymańska et al. (2016, 2018) and Furtado et al. (2019a, b) when they found that the site exposed to salinity for a long time had a greater diversity of endophytic bacteria and fungi in native *S. europaea*. They propose that time necessary for shaping the bacterial community at the saline sites was longer in the naturally saline area which existed much longer as compared to the anthropogenic salinity area. A study by Furtado et al. (2019a) investigated the composition of endophytic community in *S. europaea* and observed some bacteria were specific in some samples. Bacterial class Sphingobacteriia were found only in the high-salinity site, while Epsilonproteobacteria were characteristic for the site with lower salinity. Some endophytic genera: *Halomonas*, *Levinella*, *Vibrio*, *Pseudoalteromonas*, and *Leuweenhoekiella* were found exclusively at high-salinity site.

Yamamoto et al. (2018) studied the bacterial diversity and community structure of rhizosphere, root endosphere, and bulk control soil samples in two halophytic plants: *Glaux maritima* and *Salicornia europaea*. Among the *G. maritima* samples, the richness and diversity of bacteria in the rhizosphere were higher than in the root endosphere but were lower in comparison to the bulk soil. In contrast to *S. europaea*, the bulk soil, the rhizosphere, and the root endosphere had similar bacterial richness and diversity (Yamamoto et al. 2018). Another study by Shi et al. (2015) showed the diversity of bacteria was abundant in the rhizosphere soil, while the endophytic diversity was poor in *S. europaea* samples. In conclusion, the variation in endophyte frequencies could be due to differences in host preference rather than environmental factors as the halophyte hosts in the abovementioned studies were at the same location and were investigated simultaneously. This also confirms high endophyte species specificity of *S. europaea*, which depends on the halophytic plant species and the sampling site.

Previous data revealed that bacterial communities in the rhizosphere exhibit greater richness than endophytes in the organs of halophytes. In addition, Momonoki and Kamimura (1994) reported that during the growth period of wild *S. europaea* found around Lake Notoro, the pH and osmotic pressure of the plants increased from 7.6 to 8.8, and from 650 to 2000–2600 mOsm/kg (1 mOsm/kg = 17.02 mm Hg), respectively. This fact is clear that the internal environments of halophytes such as *S. europaea* are likely to be stressful to microbes. Much of the endophytic diversity research is limited to the roots of *S. europaea*. However, some studies have indicated plant organ specificity among endophytes in *S. europaea,* where different bacterial and fungal species inhabited tissues and a few species were overlapping between shoots and roots (Furtado et al. 2019a). At the family level, Halomonadaceae were much more abundant in shoots, whereas Alteromonadaceae, Cellvibrionaceae, Flammeovirgaceae, Rhodobacteraceae, and Saccharospirillaceae were characteristic for roots. *Kushneria* sp. was abundant genus in shoots while *Saccharospirillum* was significantly more common in the roots. In case of fungi, roots at the high-salinity soils were strongly dominated by Pleosporaceae. Family Leptosphaeriaceae, Teratosphaeriaceae, and Didymosphaeriaceae were found exclusively in shoots, and *Paradendryphiella arenariae* was the only species present in all sample types (Furtado et al. 2019a). Moreover, *Paradendryphiella arenariae* (Nicot) Woudenberg and Crous, was previously found in *S. europaea* in Hokkaido (Okane and Nakagiri 2015) and in Canada (as *Dendryphiella arenariae* Nicot) (Booth et al. 1988). In general, the endophytic community (both bacteria and fungi) in *S. europaea* shoots was found to be less diverse than in roots (*S. europaea*). One of the reasons being that the endophyte colonization is dependent on the plant host, that is, imposed by a large salt concentration in this organ (*Salicornia* shoot accumulate salts). Secondly, these differences in endophytic assemblages in different tissue types might be the preferences of individual dominating taxa and may reflect on their capacity for utilizing or surviving within a specific substrate.

Zhao et al. (2016a) observed a marked difference in endophytic bacterial communities from different stages of plant growth. The richest endophytic bacteria diversity of *S. europaea* was detected at the seedling stage, and thereafter, variety of endophytic bacteria declined during flower and fruit setting stage. Phylum Gammaproteobacteria increased during the growing period while Betaproteobacteria decreased. Five genera *Serpens*, *Halomonas*, *Pseudomonas*, *Azomonas*, and *Pantoea* were observed during all growth phases which were suggested as the core-microbiome of *S. europaea*. The fungal diversity in *S. europaea* reported by Furtado et al. (2019b) changed from one growing season to another. The fungal strain from two investigated sites was more diverse (based on genera) in spring (young plants) than in autumn (mature plants) sampling. Certain genera (e.g., *Aureobasidium*, *Cladosporium*, *Epicoccum*, and *Talaromyces*) occurred only in the autumn, while *Neocamarosporium*, *Ascochyta,* and *Acremonium* in spring. The ease of colonization in the young host plant stage and the microorganism's potency to compete for survival in their host may account for this effect.

21.5 *S. europaea* as a Reservoir of Specialized Endophytic Diversity

Soil salinity not only affects microbial community composition and abundance but also affects microbial functions, i.e., enzymatic and metabolic processes. High salinity can reduce the level of respiration, biomass, and activity of microorganisms (Szymańska et al. 2014). For example, the activity of endophytes was found to be higher in sites with lower salinity than at the site with high salinity. Further the endophytes isolated from different samples (different sites and plant organ) although in the same genera *Epiccocum* sp., *Arthrinium* sp., and *Trichoderma* sp. displayed different metabolic activities (Furtado et al. 2019b). Moreover, different microbial species have specific salt requirements and possess varying tolerance levels. Based on their ability to grow in the saline environments, microbes can be grouped as halotolerant (tolerate up to 25% NaCl) and halophilic (require salt for growth) (Sultanpuram and Mothe 2019). Bacterial and fungal endophytes of *S. europaea* are classified halotolerant as experimentally tested by Szymańska et al. (2016, 2018) and Furtado et al. (2019b). The halotolerant microbiome can positively affect the halophyte by providing nutrients (e.g., atmospheric nitrogen fixation, phosphorus solubilization), producing hormones (e.g., 1-aminocyclopropane-1-carboxylic acid (ACC) deaminase and indole-3-acetic acid (IAA)), regulating antioxidant response, synthesizing exopolysaccharides (EPS), maintaining plant defense against biotic stress (e.g., production of antibiotics, competition with pathogens for nutrients, and induction of systemic resistance), accumulating organic solutes such as proline and betaine and increasing soil aggregation (Zhao et al. 2016b; Szymańska et al. 2014, 2016, 2018; Piernik et al. 2017; Hrynkiewicz et al. 2019; Furtado et al. 2019b).

Secretion of phytohormones, particularly indole-3-acetic acid (IAA), increases plant salt tolerance by stimulating root proliferation. IAA-secreting endophytes have been isolated from *S. europaea*, e.g., Szymańska et al. (2016) observed a higher frequency of endophytic bacteria *Serratia marcescens*, *Kushneria marisflavi*, *Microbacterium* sp., *Hymenobacter psychrotolerans* capable of IAA synthesis as compared to rhizosphere bacteria reflecting on direct interaction between the *S. europaea* and endophytes. The endophytic fungi from the roots of *S. europaea* synthesized more IAA compare to endophytes from the shoots (Furtado et al. 2019b).

Bacterial ACC deaminase (1-aminocyclopropane-1-carboxylic acid (ACC)-deaminase) reduces plant ethylene levels and involved in nitrogen fixation, both are considered important mechanisms in bacteria that can promote plant growth, especially under stress conditions (del Carmen Orozco-Mosqueda et al. 2020). ACC deaminase-producing root endophytes isolated from *S. europaea*: *Pseudomonas* sp. I-S-E-12 and *Rhodococcus erythropolis* I-S-E-16 strains showed the presence of a gene encoding dinitrogenase reductase (Szymańska et al. 2016). Similarly, *S. europaea* endophytes with ACC deaminase activity belonging to the genera Arthrobacter, Bacillus, Planococcus, and Variovorax enhanced *in vitro* *S. europaea* seedlings growth in the presence of increasing concentrations of NaCl. In addition, these strains were also able to solubilize phosphate and produce IAA, which stimulated seed germination and increased the rate of root development (Zhao et al. 2016b).

Plant growth-promoting endophytes are able to release siderophores that are involved in biocontrol mechanisms thus preventing the growth of pathogenic micro-organisms in plant hosts (Johnson 2008). High synthesis of siderophores was obtained for bacteria *Streptomyces* sp., *S. griseoplanus,* and *Bacillus thuringiensis* (Szymańska et al. 2016). Siderophores production by endophytic fungi was the most active function reported in *S. europaea* (Furtado et al. 2019b).

Hrynkiewicz et al. (2019) analyzed the endophytic diazotrophs of *S. europaea* which represented a broad range of N_2-fixing bacteria with Actinobacteria dominating at the site characterized by higher salinity, and Proteobacteria at lower salinity. Most of the isolates from Actinobacteria belonged to the genus Curtobacterium (*Curtobacterium* sp., *C. flaccumfaciens*, *C. herbarum*) and Microbacterium (*Microbacterium* sp., *M. kitamiense*, *M. oxydans*), and some of them to Rhodococcus, Mycobacterium, Cellulomonas, Sanguibacter, Clavibacter, Cryocola/Labedella, Frigoribacterium, Agreia, Herbiconiux, and Plantibacter. The presence of the *nif*H gene was also identified in *S. europaea* endophytic bacteria: *Pseudomonas* sp. ISE12 and *R. erythropolis* ISE16 (Szymańska et al. 2016).

Phosphate solubilizers, e.g., *Bacillus endophyticus*, *B. tequilensis*, *Planococcus rifietoensis*, *Variovorax paradoxus*, and *Arthrobacter agilis* were identified in *S. europaea* roots (Zhao et al. 2016b). No activity for phosphate solubilization was observed in fungal endophytes of *S. europaea* (Furtado et al. 2019b).

All fungi possess pathways to biosynthesize polyamines, which are important in restoring cellular homeostasis under stressful conditions (Nikolaou et al. 2009). Furtado et al. (2019b) investigated the fungal culturable diversity and found nearly all of the strains possessed the ability to produce polyamines (90% of isolated strains from shoots and 83% from roots). However, the strains isolated from the *S. europaea* shoots actively produced polyamines that were correlated with the hyper-accumulation of salts in *Salicornia* shoots, which can be more stressful for fungal colonization.

The sulfur-oxidizing genera *Sulfurimonas* and *Halochromatium* were significantly abundant in the root endosphere and rhizosphere in *S. europaea* (Yamamoto et al. 2018). According to a previous report, *Sulfurimonas* is involved in host detoxification by oxidizing sulfide and producing sulfate as an end product, suggesting that the accumulation of these bacteria around the rhizosphere might be critical for the host tolerance of coastal environments (Fahimipour et al. 2017).

Most of the endophytic fungal strains isolated from two saline sites in Poland displayed proteolytic, lipolytic, and chitinolytic activity (Furtado et al. 2019b). On comparing the two sites, the fungal strains obtained from higher salinity site possessed higher cellulolytic, proteolytic, and amylolytic activities. While the strains isolated from lower salinity sites possessed proteolytic, lipolytic, and chitinolytic activities. Fungal strains *Aureobasidium pullulans* and *Sarocladium* sp. displayed high cellulolytic activity (Furtado et al. 2019a, b). Endophytic bacteria *Bacillus baekryungensis*, *Thalassospira permensis*, and *Xanthomonadales* sp. from the same sites showed high activity for hydrolysis of cellulose (Szymańska et al. 2016). High proteolytic activity was exhibited by few fungal strains, e.g., *Sarocladium* sp., *Stereum gausapatum*, *Epicoccum nigrum*, *Epicoccum* sp., *Porostereum spadiceum,* and *Stemphylium* sp. (Furtado et al. 2019b). Endophtyic fungi belonging to *Aureobasidium* sp. tested positive for all the enzymatic activities, while some strains were negative for most of the activities, e.g., *Coprinellus domesticus*, *C. ellisii,* and *Arthrinium arundinis* (Furtado et al. 2019b).

Many pigmented spore- and/or mycelia-forming fungi black fungi or dematiaceous fungi such as the genera *Alternaria*, *Phoma*, *Cladosporium*, *Lewia*, *Pleospora*, *Epicoccum*, *Stemphylium*, *Ascochyta*, *Plenodomus*, *Neocamarosporium*, *Dematiopleospora*, *Aspergillus*, *Penicillium*, *Eurotium*, *Talaromyces*, *Fusarium,* and *Aureobasidium* are frequently isolated from *S. europaea* in salt marsh and desert areas (Okane and Nakagiri 2015; Furtado et al. 2019b). This suggests that pigmented dematiaceous fungi universally inhabit this halophyte and can play an important ecological role in plant stress resistance as they possess some protective substances and are capable of tolerating extreme temperatures, desiccation, and saline environments (Gostinčar et al. 2009).

Endophytes can have myriad effects on host plant fitness, with the outcome of interactions ranging from beneficial to antagonistic. They exhibit a range of symbiotic relationships with their host plant and are well known to contribute to plant fitness, which helps the host to better adapt in stress conditions (Gopi and Jayaprakashvel 2017; Hrynkiewicz et al. 2019; del Carmen Orozco-Mosqueda et al. 2020). Therefore, the application of endophytes in crop productivity has gained importance today. Few studies on the compatibility and role of *S. europaea* endophytes have been established experimentally. For instance, five *S. europaea* strains, namely *Bacillus endophyticus*, *Bacillus tequilensis*, *Planococcus rifietoensis*, *Variovorax paradoxus,* and *Arthrobacter agilis* inoculated in *S. europaea* seeds significantly enhanced seed germination percentage, seedling growth, shoot and root length under salt stress condition (approx. 500 mM NaCl) (Zhao et al. 2016b). A pot experiment by Piernik and co-workers (2017) provided evidence that endophytic bacteria (*Xanthomonadales* sp. CSE-34 and *Pseudomonas* sp. ISE-12) isolated from *S. europaea* shoot and root stimulated the growth of *Beta vulgaris* (cv. Zentaur poly) under salinity conditions (approx. 300 mM NaCl). Szymańska et al. (2019) showed that halotolerant plant growth-promoting endophytic bacteria may have a beneficial effect on the growth and development of *Brassica napus* L. cultivated under salt stress conditions. *Psuedomonas stutzeri* ISE12 stimulated the elongation of roots, hypocotyls, and stems of *B. napus* and decreased the level of oxidative damage to cellular membranes exposed to salt stress. Another study re-inoculated *Staphylococcus* sp. (isolated from *S. europaea*) in *S. europaea* seedlings and found that this strain promoted plant growth and alleviated the negative effects of salinity at 200 mM NaCl in comparison to the non-inoculated plants (Komaresofla et al. 2019). Further, fungal endophytes isolated from *S. europaea* have also demonstrated plant growth-promoting effects in *Lolium perenne* (Furtado et al. 2019b). These fungal strains were selected based on specific properties (positive metabolic activity for IAA, siderophores, polyamines, and enzyme activity for cellulase, protease).

21.6 Conclusions

Through this chapter, we provide an overview of the community diversity and role of the closely associated microbiome, i.e., the endophytes present in halophyte *S. europaea*. The most abundant endophytic community among bacteria is the phylum Proteobacteria, while the majority of the fungal endophytes mainly belong to Ascomycota. Some of the factors responsible for shaping the endophytic diversity in *S. europaea* include soil salinity, acclimatization time for microbes, halophytic plant species, plant organ, and stages of plant growth. The halophytic plant host, e.g., *S. europaea*, filters the specific microbiome. The saline soil harbors a higher microbial diversity in comparison to the endophytic diversity in roots (closer to the soil and the main entry point for endophytes), while the diversity in shoots is very low (the shoot accumulates salt that can be detrimental to the endophytes). Overall, we can infer that increasing salinization may eventually disturb the plant–endophytic association, regardless of the plant host having robust mechanisms to cope with salt stress. Moreover, much of the research on *S. europaea* endophytic microbiome found they are mainly halotolerant and possess traits that contribute toward the host and non-host plant growth and salt tolerance. Application of these endophytes for commercial use requires optimization of endophyte inocula, as questions on their association in natural field conditions, the effect of climate change, and the microbial diversity present in the new site that may affect the endophyte functioning in new host needs investigation.

Conflict of Interests The authors of this chapter declare no conflict of interests.

References

Backman PA, Sikora RA (2008) Endophytes: an emerging tool for biological control. Biol Control 46:1

Booth T, Gorrie S, Muhsin TM (1988) Life strategies among fungal assemblages on *Salicornia europaea* aggregate. Mycologia 80:176–191

del Carmen O-MM, Glick BR, Santoyo G (2020) ACC deaminase in plant growth-promoting bacteria (PGPB): an efficient mechanism to counter salt stress in crops. Microbiol Res 235:126439

Fahimipour AK, Kardish MR, Lang JM, Green JL, Eisen JA, Stachowicz JJ (2017) Global-scale structure of the eelgrass microbiome. Appl Environ Microbiol 83(12):e03391–e03316

Fan C (2020) Genetic mechanisms of salt stress responses in halophytes. Plant Signal Behav 15 (1):1704528

Flowers TJ, Colmer TD (2015) Plant salt tolerance: adaptations in halophytes. Ann Bot 115:327–331

Furtado BU, Gołębiewski M, Skorupa M, Hulisz P, Hrynkiewicz K (2019a) Exploring bacterial and fungal endophytic microbiomes of *Salicornia europaea*. Appl Environ Microbiol 85(13): e00305–e00319

Furtado BU, Szymańska S, Hrynkiewicz K (2019b) A window into fungal endophytism in *Salicornia europaea*: deciphering fungal characteristics as plant growth promoting agents. Plant Soil 445(1–2):577–594

Gopi K, Jayaprakashvel M (2017) Endophytic fungi from halophytes and their antioxidant potential. Res J Pharm Technol 10(11):4105–4107

Gostinčar C, Grube M, De Hoog S, Zalar P, Gunde-Cimerman N (2009) Extremotolerance in fungi: evolution on the edge. FEMS Microbiol Ecol 71(1):2–11

Hrynkiewicz K, Patz S, Ruppel S (2019) *Salicornia europaea* L. as an underutilized saline-tolerant plant inhabited by endophytic diazotrophs. J Adv Res 19:49–56

Hyde KD, Soytong K (2008) The fungal endophyte dilemma. Fungal Divers 33:163–173

Isca VMS, Seca AML, Pinto DCGA, Silva AMS (2014) An overview of Salicornia genus: the phytochemical and pharmacological profile. Research review, Natural products. Daya Publishing House, New Delhi, pp 145–164

Johnson L (2008) Iron and siderophores in fungal–host interactions. Mycol Res 112:170–183

Komaresofla BR, Alikhani HA, Etesami H, Khoshkholgh-Sima NA (2019) Improved growth and salinity tolerance of the halophyte *Salicornia* sp. by co–inoculation with endophytic and rhizosphere bacteria. Appl Soil Ecol 138:160–170

Litalien A, Zeeb B (2019) Curing the earth: a review of anthropogenic soil salinization and plant-based strategies for sustainable mitigation. Sci Total Environ 698:134235

Momonoki YS, Kamimura H (1994) Studies on the mechanism of salt tolerance in *Salicornia europaea*. L.: I. Changes in pH and osmotic pressure in Salicornia plants during the growth period. Jpn J Crop Sci 63(3):518–523

Muhsin TM, Booth T (1987) Fungi associated with halophytes of an inland salt marsh, Manitoba, Canada. Can J Bot 65:1137–1151

Munns R, Passioura JB, Colmer TD, Byrt CS (2020) Osmotic adjustment and energy limitations to plant growth in saline soil. New Phytol 225(3):1091–1096

Nikalje GC, Srivastava AK, Pandey GK, Suprasanna P (2018) Halophytes in biosaline agriculture: mechanism, utilization, and value addition. Land Degrad Dev 29(4):1081–1095

Nikalje GC, Yadav K, Penna S (2019) Halophyte responses and tolerance to abiotic stresses. In: Hasanuzzaman M, Nahar K, Öztürk M (eds) Ecophysiology, abiotic stress responses and utilization of halophytes. Springer, Singapore, pp 1–23

Nikolaou E, Agrafioti I, Stumpf M, Quinn J, Stansfield I, Brown AJ (2009) Phylogenetic diversity of stress signalling pathways in fungi. BMC Evol Biol 9(1):44

Okane I, Nakagiri A (2015) Assemblages of endophytic fungi on *Salicornia europaea* disjunctively distributed in Japan: towards clarification of the ubiquity of fungal endophytes on halophytes and their ecological roles. Curr Sci:62–71

Park JM, Park JH, Kim JG, You YH (2016) Securing and analysis of fungal endophytic diversity from roots of *Salicornia europaea* L. J Life Sci 50(3):89–98

Petrini O (1991) Fungal endophytes of tree leaves. In: Andrews JH, Hirano SS (eds) Microbial ecology of leaves, Brock/Springer Series in Contemporary Bioscience. Springer, New York, NY

Piernik A, Hrynkiewicz K, Wojciechowska A, Szymańska S, Lis MI, Muscolo A (2017) Effect of halotolerant endophytic bacteria isolated from *Salicornia europaea* L. on the growth of fodder beet (*Beta vulgaris* L.) under salt stress. Arch Agron Soil Sci 63(10):1404–1418

Promputtha I, Hyde KD, McKenzie EH, Peberdy JF, Lumyong S (2010) Can leaf degrading enzymes provide evidence that endophytic fungi becoming saprobes? Fungal Divers 41 (1):89–99

Shahzad B, Fahad S, Tanveer M, Saud S, Khan IA (2019) Plant responses and tolerance to salt stress. In: Approaches for enhancing abiotic stress tolerance in plants. Taylor, Francis, pp 61–77

Shi YW, Lou K, Li C, Wang L, Zhao ZY, Zhao S, Tian CY (2015) Illumina-based analysis of bacterial diversity related to halophytes *Salicornia europaea* and *Suaeda aralocaspica*. J Microbiol 53(10):678–685

Singh D, Buhmann AK, Flowers TJ, Seal CE, Papenbrock J (2014) Salicornia as a crop plant in temperate regions: selection of genetically characterized ecotypes and optimization of their cultivation conditions. AoB Plants:6

Slama I, Abdelly C, Bouchereau A, Flowers T, Savour A (2015) Diversity, distribution and roles of osmoprotective compounds accumulated in halophytes under abiotic stress. Ann Bot 115:433–447

Sonjak S, Udovič M, Wraber T, Likar M, Regvar M (2009) Diversity of halophytes and identification of arbuscular mycorrhizal fungi colonising their roots in an abandoned and sustained part of Sečovlje salterns. Soil Biol Biochem 41(9):1847–1856

Sultanpuram VR, Mothe T (2019) Microbial ecology of saline ecosystems. In: Microorganisms in saline environments: strategies and functions. Springer, Cham, pp 39–63

Szymańska S, Piernik A, Baum C, Złoch M, Hrynkiewicz K (2014) Metabolic profiles of microorganisms associated with the halophyte *Salicornia europaea* in soils with different levels of salinity. Ecoscience 21(2):114–122

Szymańska S, Płociniczak T, Piotrowska-Seget Z, Hrynkiewicz K (2016) Endophytic and rhizosphere bacteria associated with the roots of the halophyte *Salicornia europaea* L.–community structure and metabolic potential. Microbiol Res 192:37–51

Szymańska S, Borruso L, Brusetti L, Hulisz P, Furtado B, Hrynkiewicz K (2018) Bacterial microbiome of root-associated endophytes of *Salicornia europaea* in correspondence to different levels of salinity. Environ Sci Pollut Res 25:25420–25431

Szymańska S, Dąbrowska GB, Tyburski J, Niedojadło K, Piernik A, Hrynkiewicz K (2019) Boosting the Brassica napus L. tolerance to salinity by the halotolerant strain Pseudomonas stutzeri ISE12. Environ Exp Bot 163:55–68

Tadych M, White JF, Moselio S (2009) Endophytic microbes. In: Schaechter M (ed) Encyclopedia of microbiology, pp 431–442

Ushakova SA, Kovaleva NP, Gribovskaya IV, Dolgushev VA, Tikhomirova NA (2005) Effect of NaCl concentration on productivity and mineral composition of *Salicornia europaea* as a potential crop for utilization NaCl in LSS. Adv Space Res 36(7):1349–1353

Ventura Y, Sagi M (2013) Halophyte crop cultivation: the case for salicornia and sarcocornia. Environ Exp Bot 92:144–153

Yamamoto K, Shiwa Y, Ishige T, Sakamoto H, Tanaka K, Uchino M, Tanaka N, Oguri S, Saitoh H, Tsushima S (2018) Bacterial diversity associated with the rhizosphere and endosphere of two halophytes: *Glaux maritima* and *Salicornia europaea*. Front Microbiol 9:2878

You Y, Lee M, Kim J (2014) Endophytic fungal diversity isolated from the root of halophytes in Taean peninsula. Korean J Mycol 42:269–275

Zhao S, Zhou N, Zhao ZY, Zhang K, Tian CY (2016a) High-throughput sequencing analysis of the endophytic bacterial diversity and dynamics in roots of the halophyte *Salicornia europaea*. Curr Microbiol 72(5):557–562

Zhao S, Zhou N, Zhao ZY, Zhang K, Wu GH, Tian CY (2016b) Isolation of endophytic plant growth-promoting bacteria associated with the halophyte *Salicornia europaea* and evaluation of their promoting activity under salt stress. Curr Microbiol 73(4):574–581

Chapter 22
Root Endophytic Microbes and Their Potential Applications in Crop Disease Management

Alka Tripathi, Ajit Varma, and Swati Tripathi

Abstract As the increasing incidence of plant yield loss due to pathogen infection as well as abiotic stress is increasing, the need to reduce loss is required more and more. The use of chemical pesticides, insecticide, and fertilizers does improve plant growth and reduce the losses, but the harm posed to the biosphere cannot be neglected and a new problem is posed through bioaccumulation and biomagnification of these chemicals. Thus, plant growth-promoting rhizobacteria (PGPR) and endophytes are studied more concerning their role as a bio protector. These microorganisms alleviate the disease symptoms in many plants and minimize the loss. Their mechanism of action has now been well-studied and their role has been well established. The utilization of endophytes for sustainable agriculture has been looked upon as an alternative. In this chapter, we will see how the yield of crop plant is challenged due to pathogens and the role of endophytes in controlling pathogens as well as growth promotion.

Keywords PGPR · Endophytes · Biological control · Sustainable agriculture · Crop plants

22.1 Introduction

The word Agriculture is originated from Greek αγρός, and *cultūra* meaning "cultivation" or "growing." Agricultural practices are believed to be originated in South-East Asia and eventually reaching the far east with evolving cropping techniques used by Sumerians and Romans like monocropping, organized irrigation, and trade-related farming (Agriculture 2019). During the middle age, North Africa and near east regions develop better irrigation technologies based on hydraulic conductivity and hydrostatic pressure, building dams and reservoirs, location-specific cropping

A. Tripathi · A. Varma · S. Tripathi (✉)
Amity Institute of Microbial Technology, Amity University, Uttar Pradesh, Noida, India
e-mail: stripathi2@amity.edu

N. Shrivastava et al. (eds.), *Symbiotic Soil Microorganisms*, Soil Biology 60,
https://doi.org/10.1007/978-3-030-51916-2_22

manuals, and crop-rotation also lead to increased agriculture productivity (Greatest engineering achievements of the twentieth century 2007). Agriculture in modern days is less dependent on human efficiency and has become more efficient with the rapid rise of modern agricultural tools (World Agricultural Production 2020).

Global production of food grains has reached 2666.5 million metric tons on 57 million square miles arable land being utilized for agriculture (Tracking Productivity: The GAP Index 2018). The USA Department of Agriculture's Economic Research Service (USDA ERS) estimates that TFP growth globally has been rising by an average annual rate of only 1.51% since 2010. According to GHI, global agricultural productivity must increase by a rate of 1.75% to meet productivity demand (U.S. Code 2007). For this purpose, sustainable agriculture must be incorporated in year-round agricultural practices, which can

1. satisfy human food and fiber needs
2. enhance environmental quality and the natural resource based upon which the agricultural economy depends
3. make the most efficient use of non-renewable resources and on-farm resources and integrate, where appropriate, natural biological cycles and controls
4. sustain the economic viability of farm operations and
5. enhance the quality of life for farmers and society as a whole (Barra-Bucarei et al. 2020). Around 375 million people in the world are vegetarians and are entirely dependent on plant-derived nourishment, advocating the adoption of the sustainable agriculture practices.

22.2 Concerns of Economically Important Plants

Majority of world population is dependent on plant-based food products exclusively, while the rest of the population still incorporates plant-based food in their regular diet. Although the productivity of crop plants has been improved significantly with crop management practices and modern mechanized tools, it is not sufficient to meet the demand of splurging population. Biotic stress has been documented as a major constraint for crop production worldwide (Teng and Krupa 1980) and frequency of both biotic and abiotic stress incidences have been increasing at unprecedented scale due to changing climate in past few decades. The direct yield loss in global agricultural productivity caused by pathogens alone ranges from 20 to 40% (Martínez et al. 2003) hampering gross agricultural productivity by more than 10% (Egel and Martyn 2007).

22.3 Crop Pathogens

Despite several morphological and biochemical barriers, which prevents invasion of pathogen in plants, a variety of crop-specific and nonspecific pathogens cause disease across the crop species. For example, *Botrytis cinerea* affects members of the solanaceae family and this pathogen is difficult to control as it invades plants through pruning cuts, flower, and fruit as well as it has several hosts and infection strategies (Mittler 2002). Cucurbitaceae family is affected by a variety of fungus and bacterial pathogen, in which major fungal diseases are Fusarium wilt caused by *Fusarium oxysporum,* damping-off by *Fusarium* spp., Alternaria blight by *Alternaria cucumerina*, powdery mildew by *Erysiphe cichoracearum,* anthracanose by *Colletotrichum orbiculare*, downy mildew by *Pseudoperonospora cubensis*, scab, or gummosis by *Cladosporium cucumerinum*, damping-off of seeds by *Pythium* spp., charcoal rot by *Macrophomina phaseolina*, vine decline by *Monosporascus cannonballus*, Phytophthora blight by *Phytophthora capsica* and major bacterial diseases are angular leaf spot by *Pseudomonas syringae pv. lachrymans*, bacterial leaf spot by *Xanthomonas campestris pv. cucurbitae*, bacterial fruit blotch by *Acidovorax avenae subsp. citrulli*, bacterial rind necrosis by *Erwinia carnegieana* (Winther and Friedman 2008). The crop plants belonging to the family cucurbitaceae are used differently in culinary like salad, cooked food, pickle, etc. and they have a good nutritious value but the loss of crops due to biotic disease is alarming, *Fusarium* spp., alone causes a loss above 70% of the yield in India (Tenberge 1999). Overall, pathogens affect plant fitness, reduce growth, and competitive ability of plants (Burdon et al. 2006). Conventional approaches to reduce disease incidences in agricultural crops include use of different set of pesticides, which are effective to control pathogen spread and contain yield losses. However, indiscriminate use of pesticides has emerged as a greater challenge for ecosystem and human health. Thus, a sustainable approach is warranted to improve crop productivity without using harmful levels of pesticides. One of the promising alternatives that has been documented across the studies is use of certain plant growth-promoting rhizobia (PGPR) and endophytes, which can augment plant defense response to a broad range of pathogens as well as improve growth and productivity of crop plants.

22.4 Endophytes and Their Mechanism of Action

The term "endophyte" is derived from "endon" meaning within, and "phyton" meaning plant. The name was coined by De Bary in 1866 for the organism living inside the plant body, however, the term was precisely defined in terms of their type into bacterial, fungi and actinomyces, or algae (Li et al. 2008). Endophytes reside inside plant tissues without damaging and causing any negative impact resulting in a disease. By assisting in the process like nitrogen fixation, phosphate solubilization, chelation, and production of phytohormones these microbes directly or indirectly

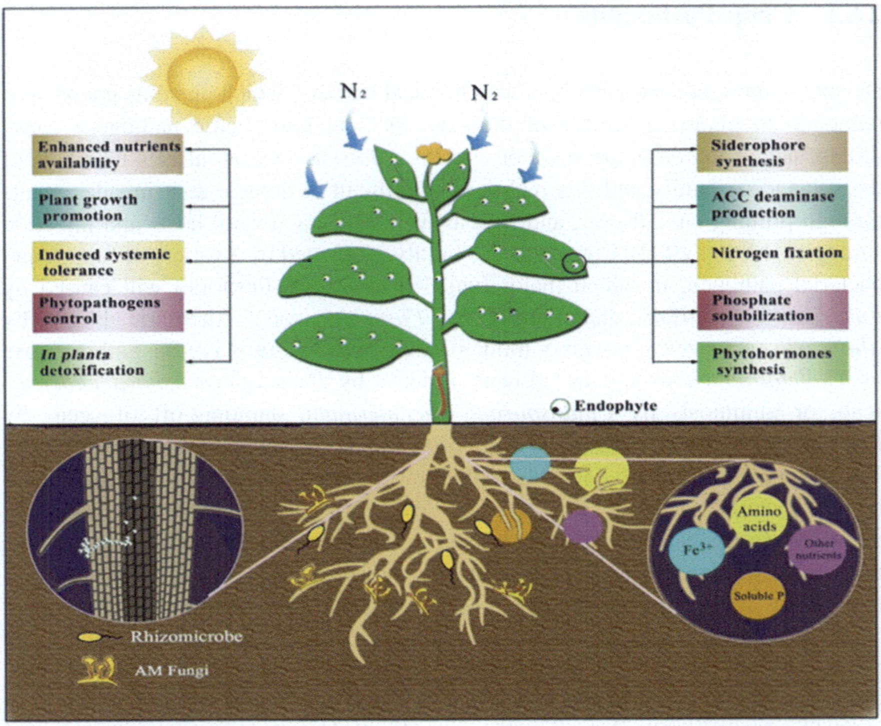

Fig. 22.1 Mechanism of action of endophytes associated with plant (Vendan et al. 2010)

improve plant performance under various biotic and abiotic stresses (Backman and Sikora 2008) (Fig. 22.1). Endophytic fungi and bacteria are found intimately associated with plant rhizospheric region and over 129 different bacterial species have been reported to be isolated from crop plants (Xu et al. 2012; Ryan et al. 2008).

22.4.1 Phytohormone Production

Phytohormones are small organic compounds produced in small quantities in plants that have a role in all aspects of plant growth and development ranging from embryogenesis, reproduction to maturity. Some of these hormones are also known to play a role in disease resistance and prevention. Phytohormones such as auxin, cytokinin, and gibberellic acid are known to induce germination, promote growth, and have roles in various developmental stages of plant (Glick et al. 1998). Conversely, abscisic acid, salicylic acid, jasmonic acid, and ethylene are important hormones that regulate stress-induced responses in plants (Asgher et al. 2015). Endosymbionts, when present in symbiotic association with plants have been shown to enhance various phytohormones and many hormones are produced by these microorganisms, which regulate tolerance to biotic and abiotic stresses. Firmicutes isolated from stem of ginseng are the pronounced producer of IAA

(Vendan et al. 2010). Also, *Acetobacter diazotrophicus* and *Herbaspirillum seropedicae* are known to produce gibberellins on their own through a pathway which is yet to be discovered (Bömke and Tudzynski 2009). Besides phytohormones, there is certain category of the enzymes that are produced by endosymbionts, for example, Cryptocin, which is produced by *Cryptosporiopsis quercina*. This enzyme is a unique tetramic acid that inhibits pathogens like *Pyricularia oryzae* and other plant pathogenic fungi. Other enzymes documented are β-1,3-glucanases, chitinases, and cellulases. These enzymes could function directly against plant pathogens and hydrolyze the cell wall of fungal pathogens and oomycetes. These enzyme are induced in the presence of endosymbionts which act as a bioprotectant for the plants (Li et al. 2018; Sharma et al. 2013). Likewise, endosymbionts stimulate enzymes such as superoxide dismutase (SOD), ascorbate peroxidase (APX), catalase (CAT), and glutathione reductase (GR), which play a vital role for plants to cope up biotic and abiotic stress (Abdelshafy Mohamad et al. 2020). Helmut Baltruschat et al. reported increase in ascorbate enzyme activity in salt sensitive barley co-inoculated with *P. indica;* thus, this endosymbiont directly acts to neutralise oxygen free radicals (Baltruschat et al. 2008).

22.4.2 Nitrogen Fixation

Nitrogen is an essential component of many bio-molecules like amino acid, protein, nucleic acid, enzymes, and hormones. Atmospheric nitrogen cannot be readily utilized by the organisms as such, it has to be converted into ammonia by rhizobacteria, a process known as nitrogen fixation. Nitrogen-fixing bacteria move toward the plant root by chemotaxis and colonize the roots followed by infection as seen in the Rhizobium-cereal model (Khare et al. 2016). The bacteria are mostly endosymbiotic associating with both leguminous and non-leguminous plants. *Frankia* is one of the bacteria, which is well-studied model for the association for actinorhizal plants (Narayan et al. 2017). Besides, *Parasponia andersonii* is the only non-legume that shows nitrogen fixation symbiosis (Waller et al. 2005). Moreover, *Azospirillum* has been shown to improve growth and fitness of crops such as wheat, maize, and rice (Chenniappana et al. 2018).

22.4.3 Phosphate Solubilization

Phosphorous is an inorganic chemical element found in the various form on earth's surface mainly on rocks. The weathering of rocks causes phosphorous leaching to water bodies and this, in turn, is taken directly by plants and indirectly by animals by feeding on plants. Phosphorous forms the component of nucleic acids and various proteins in the form of the disulfide bond. Phosphorous is an essential nutrient for plants for their growth and development. Phosphorous quantity is insufficient in soils of the various area thus farmers have to supply them externally in the form of phosphate fertilizers. The fertilizer is available in the form of superphosphate, which is prepared by the reaction of sulfuric acid with rock phosphate. Conversely, phosphate solubilizing bacteria and fungus such as *Bacillus*, *Pseudomonas*, *Rhizobium*, *Penicillium*, and *Aspergillus* have been documented to solubilize insoluble inorganic phosphate in soil and make it available for plants in the rhizosphere (Yanni et al. 2016). Through phosphate solubilization, endophytes also provide resistance towards different environmental stresses such as biotic stresses. (Li et al. 2018; Salama and Mishricky 1973).

22.4.4 Siderophore Production

Siderophores are low molecular chelator, which assist in transport of iron from the soil microbiome to the rhizobacteria. Strains that have more beneficial traits belonged to the genera *Bacillus*, *Enterobacter*, *Pseudomonas*, *Klebsiella*, and *Microbacterium* are documented to produce 54% of siderophores (Ingram 2011). It has been reported that arbuscular mycorrhizal fungi significantly enhance iron uptake rate in the associated host plants (Lee and George 2005). *Pantoea agglomerans, P. ananatis, P. stewartia, Enterobacter ludwigii, Ralstonia* sp., and *Pseudomonas thivervalensis* are known to be involved in uptake of copper and siderophore production (Zhang et al. 2011). These strains stimulate plant biomass and assist in phytoextraction of metals like Ni, Zn, and Cr. Various studies report that many siderophore producing endophytes exhibit antagonistic activity as reported in case of strains of the genera *Pseudomonas* and *Burkholderia* and two species of Pantoea (*P. ananatis* and *P. agglomerans*) (Yang et al. 2008). Thus, siderophore production is a part of defence strategy against other pathogen as well as mechanism to increase the uptake of mineral like iron.

22.4.5 ACC Deaminase Production

The enzyme 1-Aminocyclopropane-1-carboxylase deaminase was first characterized by Honma and Shimomura in the year 1978 in a soil microorganism and it was known to metabolize ACC into a form which could be metabolized by another microorganism (Hardoim et al. 2008). ACC is an immediate precursor compound in the biosynthesis of ethylene and plays an important role in growth physiology and stress (Ali et al. 2012). The plant growth promoting bacteria employ ACC deaminase to convert ACC into α-ketobutyrate and ammonia, thus reducing the abnormal increase in ethylene level during the stress condition, therefore protecting plants during different biotic and abiotic stresses (Steenhoudt and Vanderleyden 2000).

22.5 Importance in Sustainable Agriculture

Endophytes have different modes of inoculation and transmission. They infect plants directly through roots or transmitted via seeds as reported in maize (Smil 2000) and systemically infect the plant following seed germination or by parasitic competition for infection site as seen in celery yellow disease caused by *F. oxysporum* f.sp. *apii* (Igarashi et al. 2000). Maize plants have also known to be infected by air and rain splash and through insect damage (Dutta et al. 2014). The endophytes may remain dormant or sporulate to benefit the host soon after the infection (Li et al. 2000).

Some of the endophytes and role played by them in agronomically important crops are listed below.

22.5.1 Cereals

Endophytes play an important part in sustainable agriculture by reducing biotic and abiotic stresses and by lessening the harm posed by chemical alternatives. Cereals form a major diet for the majority of world population. Cereal production has to be increased enough to meet the demand of growing population. However, both biotic and abiotic stresses are major constraints causing significant yield losses in cereal crops. Endophytes have been reported to act as bioprotectant in biotic stress, for example, maize seeds when coated with conidia of entomopathogenic fungi offers better protection against *Costelytra giveni* and *F. graminearum* (Quecine et al. 2012). In another example, *P. indica*-inoculated barley plants are more resistant to abiotic and biotic stress (Moreno et al. 2009; Bunbury-Blanchette and Walker 2019). *Streptomyces* sp. Strain DEF09 obtained from wheat root has been documented to act as bioprotectant against *Fusarium* sp. in crops (Egel and Martyn 2007).

22.5.2 Pulses

Pulses are a rich source of protein and nutrients. Pulses increase soil fertility and utilize less water than any other crop plants. Endophytic fungi like *P. indica* decrease disease severity of chickpea by inhibiting its pathogen *B. cinerea* and also promotes healthier growth (Moreno et al. 2009).

22.5.3 Vegetables

Extracts from rhizobacteria like *Streptomyces* sp. suppress the infection of *Alternaria brassicicola* on chinese cabbage seedlings (Gao et al. 2010). Similarly, many antagonistic strains of *Trichoderma* have been isolated which alleviate the disease symptoms as seen in the effect of *T. atroviride* and *T. harzianum* in foot base rot of onion (Feng et al. 2017).

22.5.4 Fruits

Tomato is an economically important crop and is used in various forms in the diet like ketchup, sauce, vegetable gravy, and as a salad. This crop is harmed by a variety of pathogens of which *Fusarium* spp., harms it the most. Endophytes like *Trichoderma, Beauveria bassiana* and *P. indica* are used as bioprotectant against different races of *Fusarium* spp. (Dong et al. 1994; Sefloo et al. 2019).

22.5.5 Sugar and Starches

Sugarcane is prone to many diseases due to its delicate rind and rich source of the nutrient. An endophytic bacterium was seen to colonize and promote the growth of sugarcane as studied by Quecine by inducing synthesis of chitinase and endoglucanase enzymes which protects the plant (Lee and George 2005). Acetobacter, an N_2-fixing endophyte colonizes the intercellular spaces of sugarcane and contributes to high content of N, approximately 180 kg of N required by the crop each season (Vendan et al. 2010).

22.5.6 Spices and Condiments

The PGPR enhance the rhizome yield and curcumin content in turmeric and act as a biocontrol agent against rhizome rot disease of turmeric (Rivas-Franco et al. 2019).

22.6 Conclusion

Endophytic bacteria and fungi are promising to control disease and protection of crops during both biotic and abiotic stress. Agricultural practices involving the use of endophytes is warranted particularly when the world is already facing the impact of changing global climate and ill effects of indiscriminate uses of pesticide, insecticide, and chemical fertilizers. Several endophytic fungal species are now successfully tested and recommended as potential biocontrol agents against a variety of plant pathogens. Conversely, a positive impact of these endophytes in plant growth promotion and improving yield advocate their need in sustainable agricultural practices. On the other hand, some endophytes assist in the process of phytoremediation and bioremediation, thus improving not only agricultural crop production, but also the soil and environment (Fig. 22.1).

References

Abdelshafy Mohamad OA, Ma J-B, Liu Y-H, Zhang D, Hua S, Bhute S, Hedlund BP, Li W-J, Li L (2020) Beneficial Endophytic bacterial populations associated with medicinal plant *Thymus vulgaris* alleviate salt stress and confer resistance to *Fusarium oxysporum*. Front Plant Sci 11:47

Agriculture (2019, February 9) New world encyclopedia. Retrieved 05:03, March 14, 2020 from www.newworldencyclopedia.org/p/index.php?title=Agriculture&oldid=1018022

Ali S, Charles TC, Glick BR (2012) Delay of flower senescence by bacterial endophytes expressing 1-aminocyclopropane-1-carboxylate deaminase. J Appl Microbiol 113:1139–1144

Asgher M, Khan MIR, Anjum NA, Khan NA (2015) Minimising toxicity of cadmium in plants-role of plant growth regulators. Protoplasma 252:399–413

Backman PA, Sikora RA (2008) Endophytes: an emerging tool for biological control. Biol Control 46:1–3

Baltruschat H, Fodor J, Harrach BD, Niemczyk E, Barna B, Gullner G, Janeczko A, Kogel KH, Schäfer P, Schwarczinger I, Zuccaro A, Skoczowski A (2008) Salt tolerance of barley induced by the root endophyte *Piriformospora indica* is associated with a strong increase in antioxidants. New Phytol 180(2):501–510

Barra-Bucarei L, Iglesias AF, González MG et al (2020) Antifungal activity of *Beauveria bassiana* endophyte against *Botrytis cinerea* in two Solanaceae crops. Microorganisms 8(1):65

Bömke C, Tudzynski B (2009) Diversity, regulation, and evolution of the gibberellin biosynthetic pathway in fungi compared to plants and bacteria. Phytochemistry 70(15–16):1876–1893

Bunbury-Blanchette AL, Walker AK (2019) Trichoderma species show biocontrol potential in dual culture and greenhouse bioassays against Fusarium basal rot of onion. Biol Control 130:127-135

Burdon JJ, Thrall PH, Ericson L (2006) The current and future dynamics of disease in plant communities. Annu Rev Phytopathol 44:19–39

Chenniappana C, Narayanasamya M, Daniela GM, Ramarajb GB, Ponnusamyb P, Sekarc J, Vaiyapuri Ramalingamc P (2018) Biocontrol efficiency of native plant growth promoting rhizobacteria against rhizome rot disease of turmeric. Biol Control 129:55–64

Dong Z, Canny MJ, McCully ME, Roboredo MR, Cabadilla CF, Ortega E, Rodes R (1994) A nitrogen-fixing endophyte of sugarcane stems (a new role for the apoplast). Plant Physiol 105 (4):1139–1147

Dutta D, Puzari KC, Gogoi R, Dutta P (2014) Endophytes: exploitation as a tool in plant protection. Braz Arch Biol Technol 57(5):621–629. https://doi.org/10.1590/S1516-8913201402043

Egel DS, Martyn RD (2007) Fusarium wilt of watermelon and other cucurbits. The Plant Health Instructor. https://doi.org/10.1094/PHI-I-2007-0122-01. Updated 2013

Feng N-X, Yu J, Zhao H-M, Cheng Y-T, Mo C-H, Cai Q-Y, Li Y-W, Li H, Wong M (2017) Efficient phytoremediation of organic contaminants in soils using plant–endophyte partnerships. Sci Total Environ 583. https://doi.org/10.1016/j.scitotenv.2017.01.075

Gao FK, Dai CC, Liu XZ (2010) Mechanism of fungal endophytes in plant protection against pathogen. Afr J Microbiol Res 4(13):1346–1351

Glick BR, Penrose DM, Li J (1998) A model for the lowering of plant ethylene concentrations by plant growth-promoting bacteria. J Theor Biol 190:63–68

Greatest engineering achievements of the twentieth century. National Academy of Engineering. Retrieved February 25, 2007

Hardoim PR, Van O, Verbeek LS, Van E, Lsas JD (2008) Properties of bacterial endophytes and their proposed role in plant growth. Trends Microbiol 16:463–471

Honma M, Shimomura T (1978) Metabolism of 1-aminocyclopropane-1-carboxylic acid. Agric Biol Chem 42:1825–1831

Igarashi Y, Iida T, Sasaki Y, Saito N, Yoshida R, Furumai T (2000) Isolation of actinomycetes from live plants and evaluation of antiphytopathogenic activity of their metabolites. Actinomycetologica 16:9–13

Ingram J (2011) A food systems approach to researching food security and its interactions with global environmental change. Food Security 3:417–431

Khare E, Kim KM, Lee K-J (2016) Rice OsPBL1 (*Oryza sativa* Arabidopsis PBS1-LIKE 1) enhanced defense of Arabidopsis against *Pseudomonas syringae* DC3000. Eur J Plant Pathol 146:901–910

Lee YJ, George E (2005) Contributions of mycorrhizal hyphae to the uptake of metal cations by cucumber plants at two levels of phosphorus supply. Plant Soil 278:361–370

Li JY, Strobel GA, Harper JK, Lobkovsky E, Clardy J (2000) Cryptocin, a potent tetramic acid antimycotic from the endophytic fungus *Cryptosporiopsis quercina*. Org Lett 2:767–770

Li J, Zhao GZ, Chen HH, Qin S, Xu LH, Jiang CL, Li WJ (2008) *Rhodococcus cercidiphylli* sp. nov., a new endophytic actinobacterium isolated from a *Cercidiphyllum japonicum* leaf. Syst Appl Microbiol 31:108–113

Li L, Mohamad OAA, Ma J, Friel AD, Su Y, Wang Y et al (2018) Synergistic plant–microbe interactions between endophytic bacterial communities and the medicinal plant *Glycyrrhiza uralensis* F. Antonie Van Leeuwenhoek 111:1735–1748

Martínez R, Aguilar MI, Guirado ML, Álvarez A, Gómez J (2003) First report of fusarium wilt of cucumber caused by *Fusarium oxysporum* in Spain. Plant Pathol 52:410. https://doi.org/10.1046/j.1365-3059.2003.00832.x

Mittler R (2002) Oxidative stress, antioxidants and stress tolerance. Trends Plant Sci 7:405–410

Moreno CA, Castillo F, González A, Bernal D, Jaimes Y, Chaparro M, González C, Rodriguez F, Restrepo S, Cotes AM (2009) Biological and molecular characterization of the response of tomato plants treated with *Trichoderma koningiopsis*. Physiol Mol Plant Pathol 74(2):111–120

Narayan OP, Verma N, Singh AK, Oelmüller R, Kumar M, Prasad D, Kapoor R, Dua M, Johri AK (2017) Antioxidant enzymes in chickpea colonized by Piriformospora indica participate in defense against the pathogen *Botrytis cinerea*. Scientific reports 7: 13553

Quecine MC, Araújo WL, Rossetto PB, Ferreira A, Tsui S, Lacava PT, Mondin M, Azevedo JL, Pizzirani-Kleiner AA (2012) Sugarcane growth promotion by the endophytic bacterium *Pantoea agglomerans* 33.1. Appl Environ Microbiol 78(21):7511–7518

Rivas-Franco F, Hampton JG, Morán-Diez ME, Narciso J, Rostás M, Wessman P, Jackson TA, Glare TR (2019) Effect of coating maize seed with entomopathogenic fungi on plant growth and resistance against *Fusarium graminearum* and *Costelytra giveni*. Biocontrol Sci Tech 29 (9):877–900

Ryan PR, Germaine K, Franks A, Ryan DJ, Dowling DN (2008) Bacterial endophytes: recent developments and applications. FEMS Microbiol Lett 278:1–9

Salama AM, Mishricky AG (1973) Seed transmission of maize wilt fungi with special reference to *Fusarium moniliforme* Sheld. Phytopathology 77:356–362

Sefloo NG, Wieczorek K, Steinkellner S, Hage-Ahmed K (2019) Serendipita species trigger cultivar-specific responses to Fusarium wilt in tomato. Agronomy 9(10):595

Sharma SB, Sayyed RZ, Trivedi MH, Gobi TA (2013) Phosphate solubilizing microbes: sustainable approach for managing phosphorus deficiency in agricultural soils. Springerplus 2:587–600

Smil V (2000) Feeding the world: a challenge for the twenty-first century. The Massachusetts Institute of Technology Press, Cambridge

Steenhoudt O, Vanderleyden J (2000) Azospirillum, a free-living nitrogen-fixing bacterium closely associated with grasses: genetic, biochemical and ecological aspects. FEMS Microbiol Rev 24:487–506

Tenberge KB (1999) Biology and life strategy of ergot fungi. Institut fur Botanik, Munster, Germany

Teng PS, Krupa SV (eds) (1980) Assessment of losses which constrain production and crop improvement in agriculture and forestry. Proceedings of the E. C. Stackman commemorative symposium. University of Minnesota, St. Paul

Tracking Productivity: The GAP Index™ (2018) Global Agricultural Productivity Report® (GAP Report®)

U.S. Code Subchapter I: Findings, purposes, and definitions, U.S. Code, Title 7, Chapter 64-Agricultural research, extension and teaching, Available at GPO Access: http://frwebgate.access.gpo.gov/cgibin/getdoc.cgi?dbname=browse_usc&docid=Cite:+7USC3103 (8/23/2007)

Vendan RT, Yu YJ, Lee SH, Rhee YH (2010) Diversity of endophytic bacteria in ginseng and their potential for plant growth promotion. J Microbiol 48(5):559–565

Waller F, Achatz B, Baltruschat H, Fodor J, Becker K, Fischer M, Heier T, Huckelhoven R, Neumann C, von Wettstein D, Franken P, Kogel K-H (2005) The endophytic fungus *Piriformospora indica* reprograms barley to salt-stress tolerance, disease resistance, and higher yield. PNAS 102(38):13386–13391

Waqas M, Khan AL, Kamran M, Hamayun M, Kang SM, Kim YH, Lee IJ (2012) Endophytic fungi produce gibberellins and indoleacetic acid and promotes host-plant growth during stress. Molecules 17(9):10754–10773

Winther JL, Friedman WE (2008) Arbuscular mycorrhizal associations in Lycopodiaceae. New Phytol 177:790–801

World Agricultural Production. United States Department of Agriculture, Foreign Agricultural Service Circular Series WAP 3-20, March 2020

Xu Z, Gao D, Sung X, Xu Y (2012) A review of endophyte and its use and function. Adv Biomed Eng 8:124–130

Yang J-H, Liu H-X, Zhu G-M, Pan Y-L, Guo J-H (2008) Diversity analysis of antagonists from rice-associated bacteria and their application in biocontrol of rice diseases. J Appl Microbiol 104:91–104

Yanni G, Dazzo FB, Squartini A, Zanardo M, Zidan MI, Elsadany AEY (2016) Assessment of the natural endophytic association between *Rhizobium* and wheat and its ability to increase wheat production in the Nile delta. Plant Soil 407(1–2):367–383

Zhang Y, He L, Chen Z, Wang Q, Qian M, Sheng X (2011) Characterization of ACC deaminase-producing endophytic bacteria isolated from copper-tolerant plants and their potential in promoting the growth and copper accumulation of *Brassica napus*. Chemosphere 83:57–62

Chapter 23
Do Mycorrhizal Fungi Enable Plants to Cope with Abiotic Stresses by Overcoming the Detrimental Effects of Salinity and Improving Drought Tolerance?

I. Ortas, M. Rafique, and F. Ö. Çekiç

Abstract Soil salinization and drought are major and growing ecological problems. They limit the productivity of crop plants cultivated on more than 20% of total agricultural lands worldwide. Global climate changes and sequences of agriculture-related management practices would induce salinity to more than 50% of the arable land by 2050. Excess salt in soil impedes plant photosynthetic processes, seed germination, and root uptake of water and nutrients such as K^+. Under the same soil and climate conditions, water deficiency is also one of the serious limiting factors for plant growth and food security. Application of biological processes such as mycorrhizal fungi as inoculants provide a cost-effective long-term solution for coping with saline and drought conditions. Inoculation of mycorrhizal fungi along with certain microbial strains in salt and drought-affected soils increase root infection. Arbuscular mycorrhizal fungi (AMF) are renowned for effective scavengers of free radicals in soil thereby increasing soil parameters optimal for plant growth. The mechanism to cope with drought stress involves in AMF-enhance drought and salt tolerance through direct water and nutrient uptake via extraradical hyphae, better root system architecture, enhancement of antioxidant defense systems, and greater osmotic adjustment. Mycorrhizal colonization upregulates the expression of chloroplast genes in leaves, and genes encoding membrane transport

I. Ortas (✉)
Faculty of Agriculture, Department of Soil Science and Plant Nutrition, University of Cukurova, Adana, Turkey
e-mail: iortas@cu.edu.tr

M. Rafique
Department of Soil Science, The University of Haripur, Haripur, Khyber Pakhtunkhwa, Pakistan

F. Ö. Çekiç
Faculty of Science and Literature, Aksaray University, Aksaray, Turkey
e-mail: ozlemcekic@aksaray.edu.tr

N. Shrivastava et al. (eds.), *Symbiotic Soil Microorganisms*, Soil Biology 60,
https://doi.org/10.1007/978-3-030-51916-2_23

proteins involved in K^+/Na^+ homeostasis in roots. Mycorrhizal inoculated seedlings exhibit high root salicylic acid concentrations and lower leaf and root jasmonic acid concentrations under salt stress. The AMF improve root hydraulic conductivity as well as the plant water status and tolerance under drought stress. Essential nutrients are also taken up through mycorrhizal hyphae and differences in P and K acquisition, transpiration, and stomatal conductance are related to mycorrhizal efficiencies of different fungi. Indigenous microorganisms may be a promising biological technology to improve plant performance and development and to alleviate salt stress damage.

Keywords Salinity · Drought · Mycorrhizal fungi · Genes · Root architecture · Abiotic stress

23.1 Introduction

Abiotic stresses such as salinity and drought seriously threat the agricultural productivity and food security (Wang et al. 2003). Especially under semiarid and arid climatic conditions climate change increases soil salinization. Currently, over 7% of the Earth's land area is estimated to have saline soils (Ruiz-Lozano et al. 2001). Nearly 20% of the cultivated world's land and half of the irrigated land are affected by high salt concentrations Sudhir and Murthy (2004). Wang et al. (2003) and Porcel et al. (2012) reported that increased salinization of arable land is expected to have devastating global effects, resulting in up to 50% soil salinization by the year of 2050. At present, nearly 5% (77 million hectares) of total cultivated arable land is affected by salinity (Sheng et al. 2008). In the same climatic regions, water scarcity as well as salinity problem poses a serious threat to food security. Especially under arid and semiarid regions, water stress has limited crop productivity (Maggio et al. 2000). Under such semiarid and arid soil regions, salinity problem leads to major constraints on agricultural production. The salt tolerance of a plant is affected by soil, water, plant, and environmental conditions. Plant roots, soil nutrients availability, nutrient absorption capacity, and soil microbial activity especially mycorrhizal fungi significantly affect the plant tolerance to salt and drought stresses.

Arbuscular mycorrhizal fungi (AMF) are present in all kind of ecosystems, regardless of soil type, vegetation, or growing conditions (Mosse et al. 1981). This may lead to indirect effects of the AM association on the plant nutrients availability and uptake (Smith and Read 2008). Their early establishment in the growth process of plants is important. Mycorrhizae have a significant contribution on soil stabilization, as well. If soil erosion is pronounced, the scarcity of microbial propagules in such ecosystems may be a serious handicap to plant establishment and survival. In such cases it may be necessary to inoculate with indigenous fungal species or augment the natural AMF already present within the rhizosphere of leguminous plants.

According to Chitarra et al. (2016), Ruíz-Sánchez et al. (2011) mycorrhizal fungi play an instrumental role in the protection against abiotic stresses such as drought and salt stresses (Naher et al. 2013). Since salt and drought stress factors are serious for food security and sustainable agriculture; it is sound to use plant rhizosphere mechanisms to address the problems. Soil and environmental stress factors affect the efficiency and establishment of mycorrhizae. Mycorrhizae also have many advantages on stress tolerance (Barea et al. 1996) and abiotic stress factors (Swaty et al. 2004). For example, salinity suppressed the growth of AM. Salinity stress significantly reduced the root, stem and leaf dry matter and leaf area due to the direct effects of ion toxicity on plant. However, it has been shown that mycorrhizal colonization significantly improved plant chlorophyll concentration in comparison to the non-salinized and salinized plants (Latef and Chaoxing 2011).

The hyphal networks of AM fungi improve soil particle aggregation thereby they improve the resistance of plant to stress factors, as well. Lehnert et al. (2018) indicated that worldwide in the majority of wheat-growing areas, the incidence of drought stress has increased significantly resulting in a negative impact on plant development and grain yield. In several pot cultures, it has been tested the effects of AM symbiosis on the improvement of drought stress tolerance of wheat plants (Lehnert et al. 2018; Al-Karaki et al. 2004; Al-Karaki 1998; Al-Karaki and Clark 1999). Mycorrhizae can be a strong supporter to help symbiosis needed plantlets. Mycorrhizal fungi seem to act in three ways:

1. Help the plants to attain its best performance
2. Buffering the stress during acclimatization
3. Improve overall plant and soil health

It has been indicated that AMF can promote many aspects of plant life such as plant growth improvement, nutrients uptake, and stress tolerance (Chen et al. 2018). Also, AMF inoculation can increase resistance potential of plant against salt and drought stresses. Mycorrhizal association increases plant tolerance to drought stress as well (Wu et al. 2013). The work of Latef and He (2011) has shown that under several levels of salt application, concentrations of P and K were higher in *Rhizophagus mosseae* inoculated tomato (*Lycopersicon esculentum* L. cv. Zhongzha105) plants when compared with non-AM plants grown under non-saline and saline soil conditions. Usually, under salt stress conditions, AM inoculation reduces the tissue Na^+ concentration. Previously it has been shown that mycorrhizal inoculation significantly affects plant biochemical defense system by enhancement of antioxidant enzyme activities such as superoxide dismutase (SOD), catalase (CAT), peroxidase (POD) and ascorbate peroxidase (APX) in leaves of both salt-affected and non-affected plants. The results of Latef and Chaoxing (2011) have shown that AMF may protect tomato plants against salinity by alleviating salt-induced oxidative stress. Fan and Liu (2011)'s results also have shown that under drought, *G. mosseae* inoculated *Poncirus trifoliata* seedlings exhibited higher level of proline and activities of two antioxidant enzymes, superoxide dismutase (SOD) and peroxidase (POD) as compared to non-inoculated plants. Several studies showed that mycorrhizal inoculation can protect plants against salinity by alleviating the salt-induced oxidative stress. Lower oxidative damage in the mycorrhizal colonized plants may help plants to survive and grow properly. Wu et al. (2010) have shown that *R. mosseae* inoculated trifoliate orange seedlings significantly alleviated the growth reduction of salinity. It seems that mycorrhizae-inoculated citrus seedlings exhibited a more efficient antioxidant defense system, which may provide better protection against salt damage.

Salinity and AMF also had a significant effect on the concentration of phenols and ascorbic acid in the fruits (Grimaldo-Pantoja et al. 2017). Phenols and ascorbic acid may have direct and/or indirect effects on P nutrition (Smith and Read 2008). It has been estimated that approximately 90% of the land plant species roots have symbiosis with mycorrhizal fungi (Gadkar et al. 2001). Despite the low mycorrhizal affinity of the halophytes (Brundrett 1991) mycorrhizae occur under natural saline environmental conditions (Yamato et al. 2008).

It has been found that AMF inoculation improved water relations and alleviated the salt stress of many plants. Also, AMF inoculation provide high resistance to drought through enhanced water uptake (Ruiz-Lozano et al. 2001; Ruiz-Lozano et al. 2006). Under water deficiency since AMF associated plant roots enhanced mineral nutrients especially P, crop productivity is high (Al-Karaki et al. 1998; Marschner and Dell 1994). Also, mycorrhizae-inoculated plants have higher water uptake due to hyphal extraction of soil water and higher root hydraulic conductivity than non-mycorrhizal plants (Auge 2004; Ortuno et al. 2018). All such results suggest that mycorrhization brought biochemical changes helpful in mitigating different stresses experienced by drought and salt factors. The AMF (Smith and Read 2008) and rhizosphere organisms colonization may alleviate drought, salt, and metal stress of plants showing capability in binding heavy metals and mitigate the stress tension.

23.2 Effects of Mycorrhizal Inoculation on Salt and Drought Tolerance

Since the increase in human population has negative effects on land use for food security, soils are under stress. Recently poor soil quality and crop management have negative impact on salt stress, nutrient deficiency, and heavy metal pollution. Many researchers reported that mycorrhizae-inoculated plant species are more tolerant to stress factors such as nutrients concentrations (Zrnic and Siric 2017). Soil and environmental stress factors affect the efficiency and establishment of mycorrhizae. For example, salinity suppressed the growth of AMF. Salinity stress significantly reduced the root, stem and leaf dry matter and leaf area due to direct effects of ion toxicity on plant. However, it has been shown that mycorrhizal colonization significantly improved plant chlorophyll concentration in comparison to the non-salinized and salinized plants (Latef and Chaoxing 2011). The AM fungi also decrease nutrient leaching from the soil, consequently contributing to the retention of nutrients in the soil for saving the chemical fertilizers.

Various ecological approaches such as AM fungi, PGPR, and entophytic bacteria have been conducted. It is well known that a wide range of soil microbes including AM fungi and interaction with endophytic bacteria are able to alleviate soil stresses by (1) enhancing the availability of soil nutrients and water, (2) production of plant hormones, (3) controlling pathogens by producing antibiotics, (4) adjusting and regulating the concentrations of toxic ions, and (5) production of different biochemical compounds to increase plant defense systems (Miransari 2016; Hamilton III and Frank 2001; Miransari 2014). Miransari (2016) indicates that especially under stress, the right combination of AM fungi and the host plant may result in the highest level of efficiency. It is possible under stress conditions that the growth and activity of both AM fungi, other microorganism, and the plant can be adversely affected. It is well known that salt has detrimental effects on spore germination of AMF. For better management of stress factors such as salt on plant growth and plant health, it is possible to isolate the tolerant species of mycorrhizal fungi from the saline soils. Rivero et al. (2018) reported that AMF isolated from the stressful environment was the most effective approach in improving plant tolerance to salt stress. In many Central European soils, AMF spores were isolated from different sodic soils and results showed that up to 80% of all spores from different sites gave single PCR-pattern which closely matched to *R. geosporum* (Bothe 2012).

23.3 Effects of Salinity and Water Stress on Soil Properties and Plant Growth

It is estimated that more than 33% world's irrigated arable land is affected by salinity. Salt stress causes decrease in plant productivity by disrupting the photosynthesis mechanism. Hokmabadi et al. (2005) indicated that relative growth rate (RGR), net assimilation rate (NARw) decreased with increasing salinity level with time for all treatments and rootstocks of pistachio. In general, salt stress is due to accumulation of Na^+ of The intracellular accumulation of Na^+ ions under salt stress conditions alters the ratio of K: Na, which affect the bioenergetic processes of photosynthesis (Sudhir and Murthy 2004). High concentrations of Na^+ and Cl^- accumulation in the root and root cells produce extreme rations of Na^+/Ca^{2+}, Na^+/K^+, $Ca^{2+/}Mg^{2+}$, and Cl^-/NO_3^- which are depressed nutrient–ion activities (Grattan and Grieve 1998). As a result, of nutritional disorders, plants undergo stress conditions. Since mycorrhizae have affected the absorption of other nutrients, they may dilute the effect of Na^+ and Cl^- ions in the root medium. In a screened work with 29 different citrus genotypes and rootstocks, it has been postulated that high concentration of Cl^- and/or Na^+ in the leaves of citrus has been related to disturbances in mineral nutrients, CO_2 gas exchange and water relations (Cimen and Yesiloglu 2016). Navarro et al. (2014) result indicated that response of mycorrhizal inoculation to salt stress was related to the nutritional status and their findings confirm that AM fungi can alter host responses to salinity stress, improving P, K, Fe, and Cu plant nutrition in Cleopatra mandarin plant. Under saline treatments, mycorrhizal inoculation increased root Mg concentration. More Mg concentration may dilute Na^+ concentration in the root medium. The results of Latef and Chaoxing (2011) under nonsaline and saline conditions, Na^+ concentration was lower in AM than non-AM tomato plants. Also, the concentrations of P and K were higher in AM when compared with non-AM tomato plants. According to results of Porras-Soriano et al. (2009a) *R. mosseae* was the most efficient inoculum in reducing the detrimental effects of salinity; it increased shoot growth by 163% and root growth by 295% in the nonsaline medium, and by 239% (shoot) and by 468% (root) under the saline conditions in olive plantlets. Porras-Soriano et al. (2009a) have shown that K content was enhanced under salt conditions by 6.4-fold with *R. mosseae*, 3.4-fold with *R. intraradices*, and 3.7-fold with *R. claroideum* inoculated olive plantlets under nursery conditions.

Mycorrhizal inoculation also decreased the plant shoot/root dry weight ratio as well (Porras-Soriano et al. 2009a). The results of Yang et al. (2014) show that mycorrhizal inoculation significantly increased the root length colonization of mycorrhizal apple plants under high degrees of salinity levels as compared to non-mycorrhizal plants. The work of Al-Khaliel (2010) has shown that under salinity and P deficient soil conditions, mycorrhiza inoculated peanut plants chlorophyll content and leaf water content increased significantly and also salinity tolerance significantly increased. In another pot experiment, Al-Karaki (1998) used two durum wheat genotypes (drought-sensitive and drought-tolerant) under water-stressed and well-watered conditions and the results showed that AM inoculated plants had high shoot and root dry matters under water stress than non-inoculated plants.

23.3.1 Mycorrhizal Fungi: Role in Soil Property Improvement Under Stress Conditions

In order to rehabilitate land sources suffering from soil salinization, useful reclamation programs are undertaken. The AM fungi is suggested as a useful strategy for saline soils (Zhang et al. 2019a). Mycorrhizae can influence soil aggregation by improving structure of the soil (Rillig and Mummey 2006), and they have a direct impact on soil aggregation (Ji et al. 2019). The AM fungi can enhance soil stability by producing hyphal network and glycoproteins (Trouvelot et al. 2014). They can provide a direct link between soil and roots (Lenoir et al. 2016). Therefore, the relationship and feedback between soil structure and mycorrhizae are of special interest. The AM fungi can affect N metabolism of the soil by enhancing the proportion of soil macroaggregate in saline soil. The increase in soil NH_4^+-N by AM fungi can regulate the hyphal growth and AM fungal hyphae can cause a decrease in salt concentration in the hyphosphere (Zhang et al. 2019a). Therefore, AM fungi can be suggested as an important bioindicator for soil quality or soil pollution and a potential application for restoring the degraded ecosystems (Lenoir et al. 2016).

The AM fungi regulate glomalin-related soil protein which is produced by spores and hyphae of AM fungi (Chi et al. 2018; Lovelock et al. 2004). Glomalin and glomalin-related soil protein (GRSP) can lead to an interesting response between soil structure and fungal growth. In poorly aggregated soil, GRSP can be produced in considerable concentrations (Rillig and Mummey 2006; Jones et al. 1997). In a previous study, it was mentioned that spore density, GRSP content, and hyphal length were significantly enhanced by AM fungi under both drought and well-watered conditions, and they strongly suggest dominant role of AM fungi in the management of soil water-stable aggregates by improving soil aggregate stability especially under drought stress (Ji et al. 2019). Likewise, AM fungi can display strongly positive feedback to the conditions by stimulating aggregated soils (Rillig and Mummey 2006).

The GRSP released from mycorrhiza influences the properties of rhizosphere. Previous studies indicate that exogenously applied GRSP could strongly stimulate root morphology and plant growth under drought stress. The GRSP can also modulate the phytohormones especially auxin (IAA), abscisic acid (ABA), and methyl jasmonate (MeJA) concentrations under drought stress. Therefore, the exogenous treatment of GRSPs is suggested as a plant growth regulator for improving drought tolerance (Chi et al. 2018). However, sodic soil can cause a decrease in GRSP concentration. It can be related to the adaptation of the plant-fungi interaction to various environmental conditions (Zhang et al. 2017). The AM fungi symbiosis can be used as an adaptation strategy by enhancing water use efficiency of the plant and it can eliminate the deleterious effects of water stress (Pavithra and Yapa 2018). Moreover, different species of mycorrhizae can stimulate the aggregation of soil to different degrees (Rillig and Mummey 2006). In addition, the physical and chemical properties of the soil effect AM fungi colonization such as low nutrients level and clay concentration are the key factors for colonization (Coutinho et al. 2019). Pesticide residues such as glyphosate in the soil or the plant can cause an inhibition in the mycorrhizal colonization, and a reduction in mycorrhizal symbiosis can be strongly dependent on the soil history (Helander et al. 2018). Therefore, the interaction between AM fungi and plant species should be well evaluated to improve success.

23.3.2 Effects of Salt and Drought Condition on Arbuscular Mycorrhiza Development

The interaction between plants and AM fungi is strongly correlated with the soil properties, and it depends on environmental and atmospheric conditions (Bitterlich et al. 2018; Mirzaei and Moradi 2017). Drought and salinity affect mycorrhizal symbiosis (Füzy et al. 2008). Spore production and hyphal development of AM fungi are negatively affected by water availability. Le Pioufle and Declerck (2018) reported that polyethylene glycol which stimulates osmotic stress in plants decreased AM fungi development. Hyphal growth and spore germination of AM fungi can be reduced in saline environments (Evelin et al. 2009; Juniper and Abbott 2006). The colonization of some AM fungal species is reduced because NaCl impacts directly on AM fungi. The extension of hyphae and spore germination can be inhibited in the presence of NaCl. However, various AM fungal species have different germination abilities under salinity. Spores of some AM fungal species can germinate even under 300 mM NaCl (Juniper and Abbott 2006). Bencherif et al. (2015) demonstrated that number of spores of some AM fungal species isolated from various saline soils could increase under salinity, and this adaptation of AM fungi could be used to restore saline arid lands. Carvalho et al. (2004) mentioned that some AM fungal species could adapt to salt marsh soils. The AM fungi diversity and fungal spore density are also strongly affected by soil nutrient availability and land use practices and agricultural soils (Soka and Ritchie 2018). Therefore, environmental conditions should be well evaluated before the application of AM fungi.

23.3.3 Mycorrhizal Fungi for Salinity Stress Remediation

Mycorrhizae remediation is an important on-site remediation strategy which uses microorganisms and plants for cleaning heavy metals from contaminated environments. As mycorrhizae remediation is a relative low-cost, natural method, it is suggested as a solution for environmental problems. In the mycorrhizae remediation process fungi can sequester or degrade the contaminants because of their mycelium morphology supplying highly extensive and reactive surface (Aroca et al. 2017; Barun Kumar Manjhi et al. 2016). The AM fungi can reduce the transport of heavy metals from roots to shoots (Zhang et al. 2005). Khan (2006) mentioned that AM fungi can act as a bioprotectant, biofertilizer, and biodegrader. The AM fungi can have a significant role in phytoremediation (Wang et al. 2005) and in contaminated soils, phytoremediation can be enhanced by AM fungi inoculation to crops (Khan et al. 2014). The AM fungi have also beneficial role in phytoremediation under drought and salt stresses (Liu et al. 2018a).

Cadmium (Cd) and Nickel (Ni) are important heavy metals, which can enter to food chain via contaminated agricultural products or drinking water. These metals have deleterious effects on human health (Barun Kumar Manjhi et al. 2016). Several studies indicate that heavy metals such as Zn, Cd, As, and Se can be taken up by AM fungi from the environment (Aroca et al. 2017; Giasson et al. 2006). The AM fungi can help to immobilize heavy metals such as Cu, Pb, Zn, and Cd in the roots and alleviate the toxicity of heavy metals (Zhang et al. 2005). The useful effects of AM fungal colonization under metal toxicity are the improved P uptake and decreased Cd, As and Cu concentrations in the shoots. In another study, it was reported that

AM fungi could induce the expression of metallothionein genes under Ni stress and help to alleviate the negative effects of Ni stress (Shabani and Sabzalian 2016). Therefore, AM fungi could have a beneficial impact on the ecological restoration of metalliferous mine areas (Chen et al. 2007).

Basidiomycetes can also degrade persistent organic pollutants, recalcitrant hydrocarbons, such as polyaromatic hydrocarbons (PAHs), aromatic hydrocarbons, halogenated hydrocarbons, and phenols, explosives and dyes. Basidiomycete species can degrade various kinds of hydrocarbons by their essential enzymes (Treu and Falandysz 2017). Boldt-Burisch et al. (2018) mentioned that AM fungal colonization in the roots are not affected by soil hydrocarbons. Therefore; the use of AM fungi symbiosis is suggested in contaminated soils with high efficiency (Zaefarian et al. 2013). However, heavy metal translocation can be affected by the interaction between host plants and different AM fungi isolates (Liang et al. 2009) and the interaction of heavy metal with other metals (Giasson et al. 2006). Therefore, the selection of the species used for bioremediation and the potential interactions with other soil organisms should be well evaluated (Treu and Falandysz 2017).

Barun Kumar Manjhi (Barun Kumar Manjhi et al. 2016) mentioned that AM fungi can be suggested as a filter and AM fungi can inhibit the transport of heavy metals to the plants. The AM fungi can protect the host plants from heavy metals such as Zn, Cd, and Pb toxicity in contaminated soils. The AM fungi can inhibit heavy metals uptake in high concentrations, therefore; they could enhance the plant growth (Liang et al. 2009). In addition, AM fungi can induce heavy metal accumulation in some plant species. The AM fungi may enhance stress tolerance in the contaminated soils via trapping heavy metals in their extraradical hyphae and plant root systems (Carvalho et al. 2006). Roy et al. (2018) reported that AM fungi could be used in the remediation of toxic fly ash by phyto-bio-rhizo-mycoremediation application. The AM fungal species such as *R. tenue*, *R. mosseae,* and *Gigaspora* spp. can defend plants against the deleterious effects of heavy metals (Lal 2002). Giasson et al. (2006) mentioned that plants infected with *R. intradices* can sequester Cd, Zn, As, and Se more than non-mycorrhizal plants. In another study, it was mentioned that *R. mosseae* could lead to high tolerance to heavy metal toxicity (Zhang et al. 2005).

Gai et al. (2018) reported that some species such as *Claroideoglomus claroideum* and *R. etunicatum* tolerated Cd in soil. *R. claroideum* was identified as more tolerant to the toxicity of Cd by measuring root colonization and total extraradical mycelium length. In another study, it was mentioned that *R. intraradices* increased plant growth in Cd contaminated areas (Redon et al. 2008). The AM fungi can also increase As uptake via active arsenite-translocating ATPase. However, the translocation can be altered via AM fungi (Giasson et al. 2006). Glomalin has an important impact on the achievement of mycorrhizoremediation in heavy metal tolerance. Glomalin can isolate heavy metals (Khan 2005). The AM fungi can have an important impact on the accumulation of glomalin-related soil protein, soil organic matter and soil organic carbon and influence the particle-size distribution and aggregate formation in heavy metal contaminated areas (Li et al. 2017). Therefore, the AM fungi are suggested for the recovery of contaminated soils (Abu-Elsaoud et al. 2017; Yang et al. 2017). (Loha et al. 2018) mentioned that a cyclin (SiPHO80) in the protein family could play important roles in homeostasis of inorganic phosphate and regulate the tolerance to heavy metal stress in *Serendipita indica* an osmotolerant AM fungal specie. Therefore, *S. indica* is suggested as a potential biofertilizer. We can also benefit from AM fungi by their role on heavy metal phytoextraction (Wang et al. 2005). Among other metals, Se can be extracted more easily by AM fungi inoculated plants (Giasson et al. 2006). In another study, *R. intraradices* caused an increase in available Cd and reduced Cd contents in leachates. Therefore, AM fungi were suggested as a good approach for the phytoextraction process (Redon et al. 2008).

23.4 Mycorrhizal Inoculation: Effects on Plant Shoot and Root Growth Under Salt Conditions

Salinity affects the plant root and shoot growth. Number of studies have been conducted to evaluate the counter-effect of AM fungi in salinity tolerance implacable to plant growth attributes which include root and shoot growth, chlorophyll content, stomatal conductance, inter and intracellular CO_2 concentration in the plant leaves. Elhindi et al. (2017) conducted a study in sweet basil (*Ocimum basilicum*) plants at three salinity levels which were nonsaline (EC = 0.64 dS m^{-1}), low saline (EC = 5 dS m^{-1}), and highly saline (EC = 10 dS m^{-1}). The AM fungi used in the study were *R. deserticola*. Observed data showed that AM fungi significantly increased dry biomass of plant as a whole but shoot height and their branches were not enhanced significantly. Similarly, another study was conducted on rice plants under salinity stress in the presence of AM fungi (*C. etunicatum*) by Porcel et al. (2016). The study was performed at two salinity levels (75 mM and 150 mM NaCl) where results showed that plants inoculated with AM fungi had more growth and shoot dry weight was increased by 40–62% under nonsaline (75 mM NaCl) conditions. Only increase of 51% was observed under 150 mM NaCl in the presence of AM fungi. It was decreased only by 10% in AM fungi inoculated plants at 150 mM NaCl. Besides that, a 17% reduction in dry weight was observed at 75 mM NaCl non-AM fungi inoculated plants.

In another study, Garg and Bhandari (2016) used silicon nutrition and AM fungi (*F. mosseae*) inoculation to evaluate plant biomass, root to shoot ratio and yield of *Cicer arietinum* L. genotypes under saline (0, 60, 80, and 100 mM NaCl) conditions. Observations showed that salinity significantly reduced the plant dry matter and declined the plant growth in all genotypes. Roots were observed more prone to salinity than leaves resulted in decrease of R/S ratio. Study also concluded that individual application of Si and AM fungi mitigate salinity effects on plant and induce significantly positive changes in plant growth. Moreover, when Si was applied with *F. mosseae*, in saline conditions, plant biomass significantly increased in comparison to Si fertilization only. Additionally, AM fungi directly facilitate in root biomass development and it enhances in comparison to Si application only.

23.5 Mycorrhizal Symbiosis and Mineral Uptake Under Salt and Drought Stress Factors

Several studies have been performed to evaluate the contribution of AM fungi in mitigating salinity and drought effect for better nutrient uptake and plant growth (Hammer et al. 2011; Ruiz-Lozano et al. 2012; Evelin et al. 2012). Studies suggest the mechanism of selective nutrients uptake through AM fungal hyphae into the plant roots which eliminate toxic effects of salt. Moreover, studies showed an increase in K^+ and a decrease in Na^+ concentrations in AM fungi inoculated plants (Evelin et al. 2012; Garg and Manchanda 2009).

Elhindi et al. (2017) grew sweet basil plants in nonsaline ($EC = 0.64$ dS m^{-1}), low saline ($EC = 5$ dS m^{-1}), and high saline soils ($EC = 10$ dS m^{-1}) where observed the mineral nutrients concentration in plant leaves. Results showed that K^+, P, and Ca^{2+} were higher in *G. deserticola* inoculated plants under non-stressed conditions. The content of Na^+ and Cl^- were exceptionally low in sweet basil. Similarly, significantly high amount of K^+, P, and Ca^{2+} was recorded in the sweet basil leaves of AM fungi inoculated plants either NaCl treatment was there or not. Overall increase in NaCl concentration decreased the nutrient content with exception for Na^+ and Cl^- content. The AM fungi significantly enhance leaf-K^+, P, and Ca^{2+} content under salinity stress conditions and reduce Na^+ and Cl^- content. Salinity had significant reducing effect on the ratio between K^+ and Na^+ and between Ca^{2+} and Na^+. There was a significant difference between K^+/Na^+ ratio and Ca^{2+} and Na^+ in AM fungi inoculated plants. In comparison to non-AM fungi inoculated plants, K^+/Na^+ ratio was higher regardless of the salinity strength. Therefore, AM fungi have the capability to reduce the imbalance of ions and their ratios such as Na^+ and Ca^{2+} uptake and K^+/Na^+, Ca^{2+}/Na^+ ratios in saline soil (Kaya et al. 2009).

Contribution of AM fungi in coping with abiotic stress to the associated plant is related to the alterations in hormonal homeostasis where ABA signaling is thoroughly studied (Ruiz-Lozano et al. 2012; Osakabe et al. 2014) (Fig. 23.1). The ABA is a stress hormone, and its production is triggered during environmental stresses such as salinity and drought (Osakabe et al. 2014). Plants adjust their ABA level according to physiological changes, environmental stress, and symbiotic relation. Aroca et al. (2013) endorsed the increase in ABA content when the plant is under stress condition. İn presence of AM fungi, stress is mitigated and ABA level decreases which may suggest that AM fungi inoculated plants are less stressed than non-AM fungi inoculated plants. Moreover, plant growth parameters and plant yield also prove plant fitness.

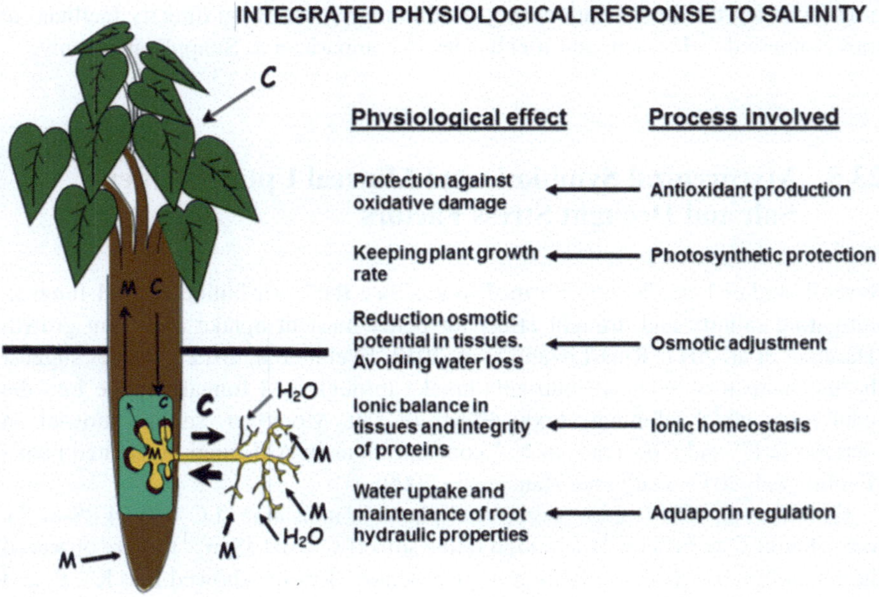

Fig. 23.1 Summary of the main processes by which AM fungi symbiosis can regulate the integrated physiological plant response in order to improve tolerance to salinity. The exchange flux of water, minerals (M), and carbon compounds (C) between the plant and the fungus is also represented. Minerals include nutrients and salt ions present in the soil solution. Adapted from Ruiz-Lozano et al. (2012)

23.5.1 Phosphate Uptake Assisted by the AM Symbiosis Under Salt Stress Conditions

Phosphorus (P) is a major nutrient of interest in symbiotic relationship of plant-AM fungi. It derives basic cellular functions in bioenergetics, metabolites activation, and enzymes regulation during structural formation as nucleic acids (Bucher 2007). Besides that, P is a limiting nutrient for plant productivity because of its immobile nature in soil. Porcel et al. (2016) conducted a study to evaluate the salinity effect on rice plant with non-AM fungi and AM fungi (*C. etunicatum*) application. Results revealed that P concentration was high in root and shoot of AM fungi inoculated plants regardless of the salt concentration. In roots, 175% increase in P concentration on average was recorded, whereas in the shoots, it approached 460% under nonsaline conditions and 190% in highly saline conditions. Indeed, the shoot P concentration decreased in AM plants due to salinity applied, although these plants always maintained higher shoot P concentration than non-AM plants.

23.5.2 Nitrogen Uptake and Transfer at the Mycorrhizal Interface Under Salt Stress Conditions

Nitrogen is a major nutrient required to the plants for their biomass (1–5% of dry weight), biochemical processes, and plant yield. As a significant amount of N requirement to the plant, it is available in the soil as two inorganic forms such as NO_3^- in upland soil and NH_4^+ in flooded soil (Bücking and Kafle 2015). İn AM fungi inoculated soils, N is taken up by three forms such as NH_4^+ (Frey and Schüepp 1993), NO_3^- (Tobar et al. 1994a), and amino acids (Cliquet et al. 1997). Besides that, AM fungi prefer to uptake NH_4^+ for the plant as it need extra energy to reduce NO_3^- into NH_4^+ before conversion into organic compounds (Courty et al. 2015; Nakano et al. 2001) (Fig. 23.2). This mechanism was further supported molecularly, and three mycorrhizal fungal ammonium transporter (AMT) genes such as *GintAMT1*, *2*, and *3* have been identified in *R. irregularis* (Pérez-Tienda et al. 2011; Calabrese et al. 2016). Among them, *GintAMT1* has high affinity for NH_4^+ transporter expressed in cortical cells with arbuscules and extraradical mycelium (ERM). The transcripts of *GintAMT1* could up-regulate in low supply of NH_4^+ particularly in acidic soils (López-Pedrosa et al. 2006). *GintAMT2* expresses in high-P soils and it suggests that more NH_4^+ is transferred (Calabrese et al. 2016). In anaerobic soils, NO_3^- uptake to the plant root is supported by mycorrhizal hyphae coupled to H^+-symport dependent process (Bago et al. 1996). Molecular evidence of NO_3^- absorption was confirmed as NO_3^- transporter, GiNT, from *R. irregularis* was found which is usually present in ERM (Koegel et al. 2015). The fungal GiNT represses in surplus supply of NH_4^+ which shows that upregulation of GiNT and *GintAMT* is dependent on NH_4^+ to NO_3^- ratio in soil.

Besides inorganic form of N, AM fungi can take up certain organic forms of N. Several studies have been conducted with labeled ^{13}C and ^{15}N from organic sources (amino acids) to evaluate the uptake capability of AM fungi. Results showed that only ^{15}N was taken up (Atul-Nayyar et al. 2009; Hodge 2001). Hodge, Campbell (Hodge et al. 2001) noticed 72% capturing from glycine source. An amino acid permease, *GmosAAP1*, involved in transporting amino acid across fungal membrane has been identified in the AM fungus *Funnelliformis mosseae* (Cappellazzo et al. 2008). Multiple transporters are involved in the uptake of amino acids through ERM. Another transporter *RiPTR2* has been identified in *R. irregularis* which is responsible for the transportation of dipeptides (Belmondo et al. 2014).

23.5.3 Water and Potassium Relationship in AM Colonized Plant Under Salt and Drought Conditions

The AM fungi are widely distributed in saline environment. In the presence of NaCl, AM fungi inoculated plant leaves showed high relative water content, moreover, high efficiency of water usage, and reduced water saturation (Sheng et al. 2008). The

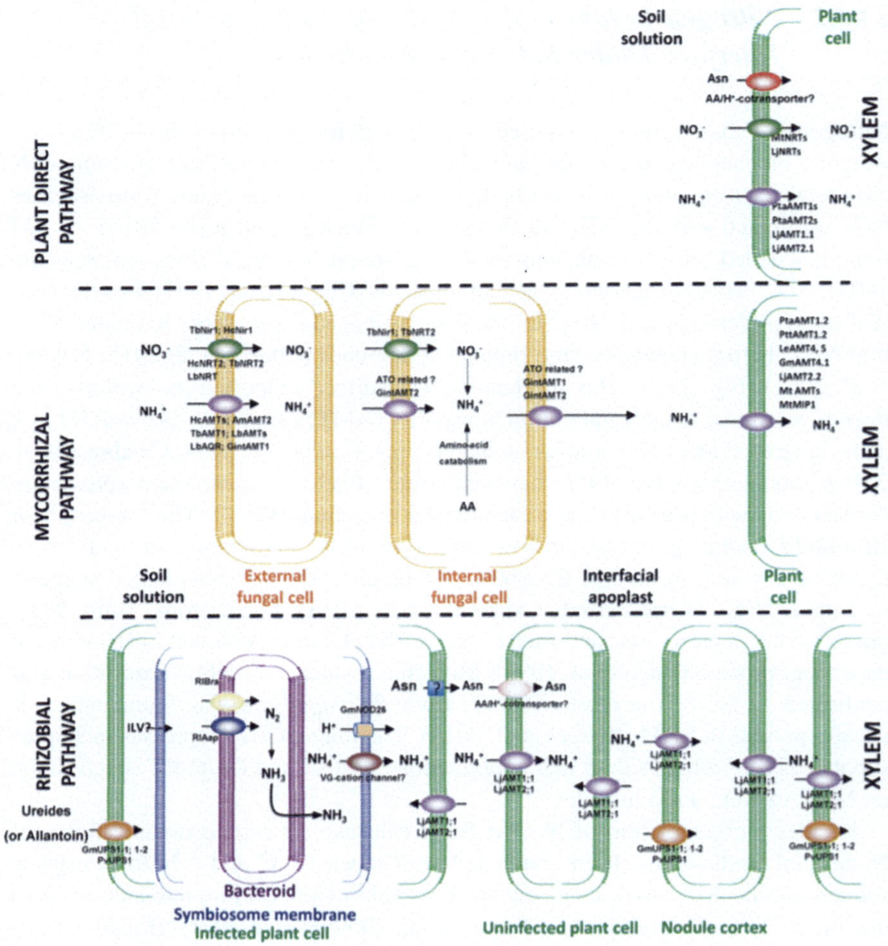

Fig. 23.2 Current knowledge about inorganic N transfer mechanisms in arbuscular and ecto-mycorrhiza and the rhizobium-legume symbiosis. Ammonium and nitrate are exchanged across the different compartments (soil, fungal/bacterial cells, apoplast, plant cells) by several membrane transporters that are not yet fully characterized. The different abbreviations are: NRT, nitrate transporter; AMT, ammonium transporter; LIV, leucine, isoleucine, valine; AA; amino acid; Asn, Asparagine; UPS, ureide permease; VG, voltage gated; Am, *Amanita muscaria*; Gint, *Glomus intraradices* (recently renamed *Rhizophagus irregularis*); Gm, *Glycine max*; Hc, *Hebeloma cylindrosporum*; Lb: *Laccaria bicolor*; Le, *Lycopersicum esculentum*; Lj, *Lotus japonicus*; Mt, *Medicago truncatula*; Pta, *Populus tremula* x alba; Ptt, *Populus trichocarpa*; Pp, *Pinus pinaster*; Pv, *Phaseolus vulgaris*; Rl, *Rhizobium leguminosarum*; Tb, *Tuber borchii*. Adapted from Courty et al. (2015)

improvement in water status through AM fungi linked to enhanced water uptake mainly because of AM fungal hyphae penetration (Jiang and Huang 2003). As water status is severely disturbed by the salinity stress, key role of AM fungi is to save the host plant from dehydration and thus aid the host plant to enable maximum absorption of water via deep fungal hyphal network system (Porcel et al. 2012). But the K concentration and its accumulation in plant under the saline environment did not affect significantly the AM fungi colonization. Minimum uptake of K was reported in tomato plant by AM fungi root colonization under salt stress (Al-Karaki 2000). Reports showed that maximum salt stress-tolerant mycorrhizal plants showed the highest concentration of K in their shoot which is correlated with the stomatal K regulation (Porras-Soriano et al. 2009b). The root colonizing fungi *R. intraradices* in olive plants under the medium saline condition showed the K acquisition by 3.4 fold as compared to nonmycorrhizal root colonized plants, and in turn reaching up to 6.4 fold in plants colonized by *R. mosseae* (Thomas et al. 2003). Under salt stress conditions, maximum effect on the K uptake was only noted upon the inoculation of AM fungi and thus both shoot and root accumulated more K in AM fungi inoculated plant in comparison to uninoculated plant under salt stress condition. In this way, *A. nilotica* inoculated with mycorrhiza showed high amount of K in shoot and root under all of the salt treatments (Giri et al. 2007). Some previous reports showed similar increase in K concentration and hence noted that high K^+:Na^+ was maintained by the mycorrhizal plant under the salinity stress condition by accumulating more K^+ (Mohammad et al. 2003). Furthermore, at medium and high levels of salinity stress; there is a minor effect of AM fungi root colonization on shoot K concentration (Mardukhi et al. 2011). Similarly, AM fungi symbiotic association has the potential to improve the plant tolerance against water deficit condition thereby maintaining the plant water relation (Stevens et al. 2011).

Both under water stress and well water conditions the water relations were prominent for mycorrhizal plants (Asrar et al. 2012). The adaptation of AM fungi to drought stress condition is quick and hence deliberate the fruitful effects on the host plant under water deficit stress conditions (Nasim 2010). The uptake and transportation of water and nutrients from the soil to host plant roots is among the key functions of extraradical hyphae of AM fungi (Peterson and Massicotte 2004). As compared to saturated soil conditions, the hyphal network transport more water under the dried soils. Similarly, the leaf water potential also enhanced by symbiotic association under drought stress condition (Gholamhoseini et al. 2013). The uptake of water is due to hydrophilic nature of hyphal tips, which absorb water from soil and transport via either cytoplasmic pathway or inner layer of wall from a single soil pore, alongside the AM hypha toward the cells of cortex (Allen 2007). When the soil dehydrates severely, the extraradical AM hyphae with a diameter of 2–5 m penetrate through soil pores and provide the mycorrhizal root more access to water zone (Wu et al. 2013). Some previous reports also showed that AM fungal colonization improved plant water relationship (Zhang et al. 2011). The AM symbiosis improved K tissue concentration which is an important physiological attribute to regulate the root water uptake potential of plant (Benlloch-González et al. 2009). The high potassium uptake by mycorrhizal application under drought stress conditions showed that AM symbiosis has the potential of reducing drought stress and reconstruction of ecosystems (Wu et al. 2011).

23.5.4 *Mycorrhizal Fungi: Effects on Macro and Micronutrients Uptake Under Salt and Drought Stress Conditions*

Generally, under salinity stress conditions, due to imbalance in nutrient composition such as excessive Na^+ and Cl^- ions uptake caused the excessive toxicity thereby leading to a reduction in osmotic potential of plants, disruption of cell organelles and their metabolism and ultimately salt affect plant growth and reduce the yield. The nutrient imbalance in the plant cells caused by nutrient uptake and/or transport to the shoot leading to ion deficiencies (Adiku et al. 2001). The AM fungi inoculation significantly increased the content of P and ascorbic acid of the pepper plant and also salinity and AM fungi had a significant effect on the concentration of phenols and ascorbic acid in the fruit (Grimaldo-Pantoja et al. 2017).

The AM fungi have the potential to assist plant for nutrients uptake under different stresses including salinity and drought (Al-Karaki 2000). The performance of mycorrhizal resistant species under salinity condition is determined by the uptake of nutrients (Daei et al. 2009). In mycorrhizal plants, the uptake of N, P, K, Ca, and Mg was significantly enhanced as compared to non-mycorrhizal plants (Heidari and Karami 2014; Ortaş and Rafique 2017). The prominent species of AM fungi such as *R. mosseae* and *R. intraradices* increased Zn and Mn concentration significantly in pistachio plants, regardless of the soil moisture conditions (Bagheri et al. 2012). Similarly Al-Karaki et al. (2001) demonstrated that the uptake of Cu, Fe, and Zn significantly increased for mycorrhizal plants with respect to non-mycorrhizal plants under salinity stress (Al-Karaki et al. 2001). According to Daei et al. (2009), mycorrhizae absorbed maximum amount of Zn and produced high quantity of root and yield under saline stress condition. Various nutrients such as N, Ca, Mg, Fe, Cu, and Mn were absorbed by different cultivars of wheat under saline soil inoculated with AM fungi species including *R. etunicatum*, *R. mosseae*, and *R. intraradices* (Mardukhi et al. 2011). Plants inoculated with AM fungi showed significant uptake of micronutrients as compared to uninoculated control in lettuce plant whereas other studies showed that rhizospheric bacteria with AM fungi inoculated wheat plants had more efficiency against Se (Durán et al. 2016). The study of Mohammad et al. (2003) showed that under salinity stress conditions, the uptake of micronutrients improved upon the inoculation of AM fungi, similarly the indigenouos AM fungi inoculation enhanced the uptake of Fe, Mn, and Cu. While the Zn uptake was promoted by AM fungi under all the treatment of salinity stress.

There is an increase in AM fungi response by decreasing fertility of soil and furthermore along with increasing the severity of drought stress (García et al. 2008). Reports showed that AM fungi application significantly reduced the drought stress and drought-induced deficient nutrients such as Fe and Zn (Gholamhoseini et al. 2013). The drought stress was reduced in pistachio cultivars assisted by the inoculation of AM fungi which improve the uptake of slowly diffusing mineral ions, i.e., PO_4^{2-} and Zn^{2+} (Bagheri et al. 2012). The higher mineral nutrients content such as Fe and Zn were observed in plants inoculated with AM fungi as compared to non-mycorrhizal plants under drought and well water conditions (Amiri et al. 2017). Furthermore, the inoculation of *F. mosseae* and *R. intraradices* incremented the Zn content under severe drought conditions (Shirani et al. 2018). The inoculation of AM fungi species such as *R. mosseae* and *R. intraradices* enhanced the Zn uptake up to 4.14% and 3.95%, Cu concentration up to 12.72% and 11.72% with respect to uninoculated treatment under drought stress (Askari et al. 2018).

23.6 Effects of Mycorrhizal Inoculation and Biochar Application to Reduce the Salt Effects on Nutrient Uptake and Plant Growth

Conversion of the barren saline land into cultivated field is a solution to meet current challenges of world global food security (Biswas and Biswas 2014). The emerging knowledge of biochar addition has shown that it improves the physiochemical and biological properties of salt-affected soils (Dahlawi et al. 2018; Rafique et al. 2017). The increased uptake of P in salt-affected soil by biochar is due to its direct action of P source facility and indirectly biochar promotes the growth medium condition particularly soil organic carbon (Lashari et al. 2013). The higher K concentration in salt-affected soil induced by biochar addition is considered as one of the most significant mechanisms for biochar to enhanced the growth of plants under salt stress (Abbas et al. 2018). The nutrient status of plants affected by biochar ability of increasing nutrient retention, decreased leaching, and gaseous losses through improving surface properties of soil (Mukherjee and Zimmerman 2013). By promoting the root biochar interaction, the application of biochar increment nutrients procurement potential of plants in saline soils (Olmo et al. 2016). The addition of biochar in field and Laboratory condition in saline soil might mitigate its adverse effects and plant growth is augmented by releasing essential macro- and micronutrients including Ca, K, N, P, and Zn (Kim et al. 2016). The combined application of biochar and AM fungi improve plant yield under saline soil with respect to individual application of AM fungi and biochar. The enhanced yield was credited by increasing P and Mn and moreover the Na^+/K^+ ratio in plant grown under salt stress condition inoculated with AM fungi and biochar (Hammer et al. 2015). The green house study showed a significant synergistic application of biochar and AM fungi on the growth parameters and nutrient uptake in seedlings (Budi and Setyaningsih 2013). A generally better plant nutrition status may help to overcome salinity stress. Not only Na^+ levels were reduced, but also K and Mg concentrations and total K and Mg uptake were clearly increased in plants that received biochar and/or AM fungi addition (Hammer et al. 2015).

23.7 Effects of Mycorrhizal Inoculation on Water Uptake

The influence of AM fungi mainly depends on the uptake and transportation of water and nutrients, which enhances the hydration of plant tissue for securing physiological sustainability and improvement in growth (Abdel-Salam et al. 2018). The AM fungi inoculation significantly enhanced plant water relation under drought stress conditions (Borde et al. 2012). Therefore, improved water relation in plant inoculated with mycorrhiza also augmented the nutrient status which in turn exclusively extract the moisture contents of soil (Subramanian et al. 2006). The application of AM fungi regulates activity and expression of aquaporin both in fungal species and host plant in order to tolerate the drought stress conditions (Li et al. 2013). By activating the antioxidant defense system and stabilizing water table of soil aggravates the AM fungi boost the plant tolerance to water deficit conditions (Bedini et al. 2009). The AM fungi symbiosis significantly influenced the water uptake from soil. Regarding water uptake from the soil, effectiveness of different fungal species vary among themselves. This effectiveness in plant water uptake by AM fungi is mainly related to the external mycelium produced by individual AM fungi species and also related with the root colonization rate for alive and active structure (Baum et al. 2015). The increase in water and several macro- and micronutrients uptake was due to AM fungal hyphae that can extend to explorable surface area almost up to 50 times (Berruti et al. 2014). Therefore the host plant along with AM fungal species inoculation showed positive impact on water use efficiency (Bernardo et al. 2019). The AM fungal species alter the rate of translocation of water, hydration of tissue, and thus improved the physiological and water status of host plants (Liu et al. 2015). Through the plant symbiosis with AM fungi, the plant can be beneficial for transportation of water and thus helping the plant to possess stomata opening (Zhu et al. 2012). Mycorrhizal seedling enhanced more uptake of water than non-mycorrhizal seedling under water deficit conditions (Gong et al. 2013). This is due to the expansion of fungal hyphae in the absorption region in host plant thereby enhancing the water absorption through root (Liu et al. 2015).

23.8 Mechanisms of Mycorrhizae on Salt Tolerance in Soil and Inside the Host Plant

Salinity is a major problem for plant growth and yield (Porcel et al. 2012). The toxicity of Na^+ and Cl^- ions causes an imbalance in the nutrient composition and a decrease in plants osmotic potential, therefore salt stress induces physiological drought in plants (Evelin et al. 2009). Under salinity, AM fungi can improve water content and enhance beneficial nutrient uptake such as P, N, Mg, and Ca. The AM fungi lead various biochemical, physiological, and molecular changes in plants. The AM fungi induce photosynthetic efficiency, nitrogen fixation, and the accumulation of various osmolytes, polyamines, prolines, betaines, and carbohydrates and enhance leaf, stem, and root biomass and K^+:Na^+ ratios in leaves. It is also strongly mentioned that antioxidant system is enhanced by AM fungi (Porcel et al. 2012; Evelin et al. 2009; Cekic et al. 2012; Chang et al. 2018). Therefore, AM fungi symbiosis is suggested as a promising method for utilization of salt-alkaline lands (Chang et al. 2018). And mycorrhizal inoculation could be used in order to develop salt-tolerant crop plants (Porcel et al. 2012; Evelin et al. 2009).

The AM fungi can induce the expression of aquaporin and stress-related genes like absciisic acid (*Lsnced*), late embryogenesis abundant protein (*LsLea*), Δ1-pyrroline-carboxylate synthetase (*LsP5CS*) and PIP, Na^+/H^+ antiporters. The expression of these genes protects mycorrhizal plants from the detrimental effects of salt stress(Porcel et al. 2012; Evelin et al. 2009). The AM fungi can also maintain water use efficiency and stomatal behavior by regulating the key genes in the ABA pathway (14-3-3 genes), therefore AM fungi can improve drought tolerance (Xu et al. 2018). Under water deficiency it was reported that AM fungi and N-fixing bacteria could cause an increase in grain protein content and benefits to agricultural production (Oliveira et al. 2017).

Mycorrhizal fungi enhance plant growth, nutrition acquisition, antioxidant system, and siderophore production under adverse conditions. Therefore, they are suggested instead of the use of pesticides and inorganic fertilizers in agricultural applications and for developing sustainable and safer agricultural productions. As it is a biological process, it is essential for sustainable agriculture and it can be replaced by conventional agriculture applications (Kumar and Verma 2018). Hence, AM symbiosis can be a potential answer for conservation of some plant species in their natural ecosystems (Zarik et al. 2016). However, further studies should be done focusing on AM fungal salt-tolerant strains, cyclic nucleotide-gated channels, and cation proton antiporters to develop of salt-tolerant inoculums and for successful environmental and agricultural managements (Kumar et al. 2015).

23.8.1 Mycorrhizal Effectiveness for Hormonal Process and Signaling Under Salt Stress

The symbiosis is older than 450 million years, and it is environmentally friendly. The alleviation of detrimental effects of stress conditions is known to be related to phytohormones, secondary metabolites, and signaling molecules (Lopez-Raez 2016). Phytohormones have vital roles in plant metabolism. They act as stimulators in plant defense response under various environmental stresses. Phytohormones can also be produced by AM fungi which can be used for inducing the host tolerance under various conditions such as salinity, drought, heavy metal stress, and nutrient deficiency (Egamberdieva et al. 2017). The AM fungi can mediate with the phytohormone balance in the host plants, therefore AM fungi have important impact on the plant development by influencing as bioregulator and enhancing tolerance against environmental stresses as bio protector. By selecting the appropriate combinations of plant and fungus, maximum benefits can be achieved for farming. In addition, AM fungi can lead to a reduction in biocides and chemical fertilizers (Rouphael et al. 2015). Therefore, identification of interactions between host, microbe, and stress should be well evaluated (Egamberdieva et al. 2017).

Plant-associated microbes have beneficial impacts on the stimulation of phytohormones such as cytokinins, gibberellins, auxins, ABA, and salicylic acid in plant tissues (Lopez-Raez 2016). Among the phytohormones, AM fungi and salicylic acid mediate carbohydrate metabolism and ion homeostasis in crop plants and they can eliminate the deleterious effects of salt stress. Salicylic acid can enhance number of arbuscules and vesicles and cause an increase in the root colonization under salt stress. Therefore, seed priming with salicylic acid improves AM symbiosis and it is suggested as a potential approach in sustainable agriculture production under salinity (Liu et al. 2018b). Salicylic acid can also modulate water conductivity by regulating the root aquaporins. In addition, there is a strong network between aquaporins and phytohormones, especially salicylic acid, abscisic acid, and jasmonic acid in the controlling of the water transport in the roots (Quiroga et al. 2018). In addition, AM fungi can induce auxin synthesis and lead to high root hair growth under drought stress, therefore AM fungi can help to stimulate the deleterious effects of osmotic stress (Liu et al. 2018b).

The ABA normally regulates stomatal closure under drought stress, however in mycorrhizal plants (Ouledali et al. 2019) demonstrated that ABA was not the key factor in controlling the stomata behavior, AM fungi also control the stomata regulation. The AM fungi can also cause an increase in expressing the jasmonic acid gene in the roots under drought stress and this increase could help to respond to water stress (Duc et al. 2018). Under nutrient deficiency strigolactones (SLs), a plant hormone which modulates the coordinated plant development, can act as signals for the establishment of AM fungal symbiosis. The SLs can help host plant to alleviate the symptoms of stresses. Because of the beneficial effects of strigolactones, they are suggested as sustainable and innovative strategy for modern agricultural processes (Lopez-Raez 2016). The AM symbiosis can mediate various plant hormones and plant growth regulators; therefore, this symbiosis can have beneficial effects on plant metabolism under normal and stress conditions.

23.9 Alleviation of Salt and Drought Stresses by Arbuscular Mycorrhizal (AM) Fungi

Soil salinity and drought are among the most harmful stresses which affect the plant growth by reducing water uptake and cause osmotic stress (Santander et al. 2017). Both salinity and drought effect negatively on the distribution of mineral nutrients balance. Salinity causes imbalances of nutrients because of the deficiencies or the competitions of Na^+ and Cl^- ions with the essential nutrients. In this response, K^+ concentration is important in order to maintain turgor pressure under stress conditions. It is mentioned that high ratios of K^+:Na^+ is important for the improvement of the plant resistance to salinity. The Ca^{2+} can also regulate the plant resistance as a signal under salinity and drought (Hu and Schmidhalter 2005). Sun et al. (2017) mentioned that AM fungi formation could alleviate the deleterious effects of drought stress and eliminate growth retardation, therefore AM fungi increase plants yield in semiarid and/or arid environments. In a previous study, it was reported that AM fungi caused an increase of P uptake in dry soil (Neumann and George 2004). The AM fungal species such as *Septoglomus constrictum* can have a positive effect on the plant tolerance to drought stress by expression of some genes in the roots and it was mentioned that inoculation with *S. constrictum* could have higher stress tolerance to drought than non-mycorrhizal plants (Duc et al. 2018). In another study, it was reported that AM fungi improved photosynthetic efficiency, and proline, protein concentrations, and leaf gas exchanges. The symbiosis enhanced C sequestration in drought and salinization affected regions and increased the resistance of plants to drought and salinity (Zhang et al. 2018). Combination of AM fungal species could also increase the tolerance to drought stress in addition to abiotic stress tolerance (Oyewole et al. 2017). Therefore, AM fungi can be suggested as a promising biological application for the alleviation of salt and drought stresses (Zhang et al. 2018).

23.9.1 Arbuscular Mycorrhizal Fungi Increase Tolerance to Salinity in Plant Species

The AM fungi can make symbiosis with various vegetable crops. This symbiosis can lead to profitable and commercial agricultural and horticultural products. The success of the inoculation is strongly related to the properties of soil and genotypes of AM fungi and host plants. Moreover, environmental conditions such as water supply and nutrient content affect significantly AM fungi efficiency on their host plants. Optimum combinations of AM fungi and crop plants should be well evaluated according to soil properties and inoculation methods (Baum et al. 2015). The AM fungi can have a potential to enhance sustainability and profitability of salt tolerance of plants (Ashok Aggarwal et al. 2012). Hashem et al. (2018) indicated that AM fungi inoculation could ameliorate the deleterious effects of salt stress by enhancing biomass and pigment, phenols, proline contents, jasmonic acid, salicylic acid contents and antioxidant enzyme activities (Ashok Aggarwal et al. 2012).

Some families such as Apiaceae, Amaryllidaceae, Cucurbitaceae, Asteraceae, Solanaceae, and Fabaceae have high mycorrhizal dependency. The AM fungi can induce growth promotion and product quality of these host plants (Baum et al. 2015). The AM fungi affect secondary metabolites and enhance the nutraceutical compounds in the host plants. In order to supply the global food demand, high sustainable horticultural products should be well developed. In this response, the AM fungi is a promising environment friendly strategy (Garg and Bharti 2018). In addition to the applications of biofertilizers and biopesticides, AM fungi inoculation is a promising strategy for future applications (Baum et al. 2015). Amanifar et al. (2019) mentioned that AM fungi induced genes expression which have important role in triterpene saponin glycyrrhizin biosynthesis and pharmaceutical contents quality under salt stress and also enhanced growth, P and K uptake, higher K^+/Na^+ ratio, proline content, and membrane integrity. Therefore, AM fungi can eliminate the deleterious effects of salt stress, and it can be used as a practical application for exploiting the salinity in soils.

23.9.2 Crop Tolerance to Salinity and Drought and Relation with Mycorrhizal Dependency

Drought and salt stress are usually developed at the same time and in the same area. It is well known that plants use some root-rhizosphere mechanisms against drought and salt stress. For instance, mycorrhizal fungi cooperation with the plant roots to reduce the severity of salt and drought stresses. Especially plant genotypes which depend on mycorrhizae have more tolerance than other genotypes. Under pot culture, soil culture 94 bread wheat genotypes tolerance to water stress was tested and it has been shown that drought stress tolerance of wheat was significantly increased in the presence of mycorrhizae compared to drought stress tolerance in the absence of mycorrhizae (Lehnert et al. 2018). Zrnic and Siric (2017) reported that mycorrhiza inoculated plants are more tolerant to nutrients and water stress, soil salinity, and heavy metals concentrations. It has been indicated that plant drought tolerance is under the genes control. Fan and Liu (2011) reported that mRNA abundance of four genes involved in reactive oxygen species homeostasis and oxidative stress battling was higher in the AM plants when compared with the non-AM plants. They indicated that possible drought-induced genes may enhance the tolerance of AM plants to water deficit (Fan and Liu 2011).

23.9.2.1 Selective Interactions Between Different Species of Mycorrhizal Fungi and Plant for Salt and Drought Tolerance

Salinity stress also causes water deficiency of plant tissue, and under low water potential reduces growth by inhibiting cell division and cell expansion (Hasegawa et al. 2000a). Under salinity stress plants can develop several mechanisms such as (1) increasing the plant membrane thickness and enhancing the cell wall thickness (Miransari 2016). (2) increasing the number of vesicles in plant cells. (3) mycorrhizal infected unit can increase against salt damage to reduce the deleterious effects. (4) Plant roots and mycorrhizae increase water efficiency and uptake. Miransari (2016) indicates that especially under stress, the right combination of AM fungi and the host plant may result in the highest level of efficiency. The AM fungi can alleviate salt stress in mycorrhizal inoculated plant species through several mechanisms. Evelin et al. (2009) indicated the mechanisms of AM fungi which can employ to enhance the salt tolerance of host plants such as enhanced nutrient acquisition (P, N, Mg, and Ca), maintenance of the K^+:Na^+ ratio, biochemical changes (accumulation of proline, betaines, polyamines, carbohydrates, and antioxidants), physiological changes (photosynthetic efficiency, relative permeability, water status, ABA accumulation, nodulation, and nitrogen fixation), molecular changes (the expression of genes: PIP, Na^+/H^+ antiporters, Lsnced, Lslea, and LsP5CS) and ultrastructural changes. Mycorrhizal infection seems that significantly control many plant physiological and biochemical mechanisms. Under water stress conditions Wu and Xia (2006) shown that *Citrus tangerine* leaves and root parts have higher K^+ and Ca^{2+} than non-inoculated plants. Ortuno et al. (2018) reported that when the substrate (silt and compost) was well-watered mycorrhizal inoculation reduced the Na and increased phosphorus uptake of *Cistus albidus* plants.

It seems that mycorrhiza species have different effects on reducing salt effects on plant growth. Estrada et al. (2013) treated three native AM fungi inoculation on maize plant and the results showed a significant increase of K^+ and reduced Na^+ accumulation as compared to non-mycorrhizal plants, concomitantly with higher K^+/Na^+ ratios in their tissues. The work of Estrada et al. (2013) has shown that when native AM fungi isolates are used mycorrhizal benefits could be enhanced. One pot experiment was conducted under drought stress conditions by Liu et al. (2018c) and their results indicated that mycorrhizal inoculation stimulated greater root hair growth of trifoliate orange that is independent on AM fungi species related with mycorrhiza-modulated auxin synthesis and transport, which benefits the host plant to enhance drought tolerance.

The study of (2006) suggests that *R. versiforme* mycorrhizal inoculation helps in increments of enzymatic and non-enzymatic antioxidant productions which in turn help plants to enhance drought tolerance. The AM fungi inoculated plants activities of SOD, guaiacol peroxidase (G-POD) and glutathione reductase (GR), catalase (CAT) and ascorbate peroxidase (APX) were significantly higher than in non-AM roots and those higher enzymatic and non-enzymatic antioxidant productions would partly alleviate oxidative damage (Wu et al. 2006). In another work, Lehnert et al. (2018) have shown that genotypes differed in their response to mycorrhizae under drought stress conditions. In many work AM fungi and PGPB dual inoculation have a significant role in stimulation of plants growth, induce tolerance to drought, and salinity (Tobar et al. 1994b). The PGPB (*Pseudomonas aurantiaca*, *P. extremorientalis*) and *R. irregularis* inoculated wheat seed germination is better, seedling growth and root elongation is better, salinity tolerance is high (Egamberdieva and Kucharova 2009). Ruiz-Lozano et al. (2018) reported that a symbiotic association with AM fungi resulted in salinity tolerance, CO_2 utilization, and enhanced growth in rice. Elhindi et al. (2017) have also documented the role of AM fungi in reducing salt stress in sweet basil.

The AM fungi inoculation help pistachio growth in several ways such as growth, success of grifting and water and nutrient uptake (Abbaspour et al. 2012). Also Abbaspour et al. (2012)'results have shown that AM formation enhanced the drought tolerance of pistachio plants. On the other hand, the results of Bagheri et al. (2011) have shown that the adverse effects of water stress were significantly reduced by AM inoculation. Ferguson et al. (1998) indicated that mycorrhizal growth promotion is generally observed in more stressful conditions. The results of Shamshiri and Fattahi (2016) showed the depressing effect of salt stress on mycorrhization extent and showed that the effect of salinity on colonization rate is completely under the influence of host plant of pistachio (*Pistacia vera*) rootstocks.

Mycorrhiza species especially indigenous species have significant effects on salt tolerance. The results of Paymaneh et al. (2019) shown that one of the indigenous AMF communities from low-salinity soils conferred a significant tolerance of pistachio to salinity in terms of maintaining its phosphorus acquisition upon the stress.

23.9.3 Effects of AMF-Colonization on Survival Rate of Horticultural Plants After Transplantation to the Field Conditions

According to Rouphael et al. (2015) the AM fungi interfere with the phytohormone balance of host plants, thereby influencing plant development (bioregulators) and inducing tolerance to soil and environmental stresses (bioprotector) factors. In general, under salinity and water stress conditions plant photosynthesis rate decreases. Salinity also indirectly affect plant growth by affecting photosynthesis, turgor, and enzyme activities of plant (Hasegawa et al. 2000b). Mycorrhizal inoculation is expected to stimulate the photosynthesis. Irrigation with saline water and mycorrhizal inoculation in cucumber plants increased fresh and dry weight, proline, electrolyte leakage, photosynthesis, and stomatal conductance (Haghighi et al. 2017). In many works, it has been shown that mycorrhiza inoculation increases citrus, pistachio, maize, tomato, wheat, clover, lettuce plants tolerances to salinity stress (Al-Karaki et al. 1998; Al-Khaliel 2010; Paymaneh et al. 2019; Feng et al. 2002; Satir et al. 2016). Mycorrhizal inoculated trifoliate orange seedlings displayed significantly lower polyamine oxidase activity and diamine oxidase activity in leaves and roots, irrespective of NaCl concentration (Wang et al. 2016). Also, in that work, they have reported that mycorrhizal inoculated seedlings showed significantly higher soluble protein concentration, ornithine decarboxylase, arginine decarboxylase, and superoxide dismutase activity in leaves and roots.

23.9.4 *Effect of Biochar and Mycorrhizae on Alleviation of Salt and Drought*

Biochar is an important multifunctional carbon material that can have effect biological, chemical, and physical properties of soil and improve the soil quality (Yu et al. 2019). Biochar is a pyrolyzed organic material as a soil amendment. The effects of biochar on soil properties mainly depend on feedstock and pyrolysis conditions, pH, nutrient contents, and ion exchange capacities of biochar (Hammer et al. 2015). The longevity of biochar in the soil presents can act as bioremediation in comparison to other organic materials such as compost and animal manure that more quickly break down. It has shown that the application of biochar has a potential improvement on the soil's physical and chemical properties and also on plant growth under abiotic stress factors such as heat, drought, and salinity. Under salt-affected and water-stressed soil conditions biochar addition usually ameliorated the soil physicochemical and biological properties and also enhanced the plant physiological performance as well as plant growth, yield, and nutrients uptake (Ali et al. 2017). Through this way plants' growth is better than non-biochar amendment control treatments (Farhangi-Abriz and Torabian 2017). Farhangi-Abriz and Torabian (2017) conducted, a pot experiment, under salt added soil conditions biochar amendment reduced osmotic substances and oxidative stress of common bean plant.

It is a potential source of nutrient especially P recycling from the agricultural wastes to enrich the soil fertility and quality. Biochar production and its soil application as an amendment achieved promising results for crop production (Dickinson et al. 2015; Ortas 2016), soil chemical and physical properties improvement and biochemical properties enhancement to facilitate soil biota (Puga et al. 2015), mitigation of climate change effects in a long-term experiment (Smith 2016) and disposal of large scale waste biomass such as sludge wastes (Jeffery et al. 2015). Also, biochar can be used for reducing the abiotic factors such as salt effects. Biochar addition widely increases soil porosity and accordingly which can enhance the potential of soil to boost plant growth (Mollinedo et al. 2016). Co-application of biochar and mycorrhizae that promote plant growth and reduce the salt and drought stress. In a pot experiment it has been demonstrated that biochar application together with AM fungal inoculation resulted in an additional yield increase in *Lactuca sativa* compared to each alone under nonsaline conditions (Hammer et al. 2015). A field experiment was conducted by Thomas et al. (2013) that their results showed that hardwood sawdust biochar addition significantly adsorbed great amounts of added salt from the soil. And they reported that biochar application alone increased biomass of *P. vulgaris*, with a ~50% increase relative to untreated control plants. And their results also indicated that biochar can ameliorate salt stress effects on plants through salt absorption. Through biochar large surface and mycorrhizal hyphae can be novel dual applications to mitigate the effects of salinization in agricultural, urban, and contaminated soils.

Drought stress can cause a decline in colonization of AM fungi; however, biochar amendment can enhance nutrient uptake, chlorophyll content, and photosynthesis efficiency and ameliorate significantly the deleterious effects of drought stress by enhancing mycelium, arbuscule, and spore numbers, therefore the colonization can increase under water deficiency (Hashem et al. 2019). Biochar treatments in the soils can increase AM fungi colonization and improve the interaction between roots and AM fungi (Yu et al. 2019). Biochar application can have beneficial effects on the nutrient especially P uptake and enhance plant growth (Shen et al. 2016). In saline soils biochar treatment may have benefits, the application can cause an increase in AM fungi growth and inhibit the negative effects of salt stress (Hammer et al. 2015). Zhang et al. (2019b) reported that combination of biochar amendment and AM fungi inoculation could have positive effects on both nutrient uptake and decrease the deleterious effects of heavy metal stress in polluted soils. The AM fungal spores were isolated from different sodic soils in Central Europe results are indicating that up to 80 % of all spores from the different sites gave one single PCR-pattern which closely matched that of *R. geosporum* (Bothe 2012).

23.10 Conclusion

All literature records show that AM colonization may alleviate and compensate the growth limitations imposed by salt and water deficiency stress conditions. It seems that AM symbiosis improve plant nutrition by allowing the cells to regulate ions more effectively. Also, inoculation can improve mineral nutrient uptake by availability or transport of mycorrhizal hyphae, thus enhancing salt tolerance. Under AM inoculation conditions higher absorption of P, Zn, Cu, K, Ca, and Mg in plants under saline conditions may improve growth rate, salt tolerance and suppress the adverse effects of the salinity stress. Under saline conditions, there may be a displacement of membrane-associated Ca by Na in roots membranes. Potassium also can have a similar displacement with excess Na. Since AM inoculated plants have high Ca and Mg, Ca may help to keep membrane integrity and protect host plants against salt damage (Läuchli and Epstein 1990). Under salinity conditions, lower Na uptake and the higher Mg absorption by mycorrhizal inoculation might be an important salt-alleviation mechanism for salt tolerance of plant species. In salinity affected soils, AM inoculation may also alleviate some of physiological mechanisms of plants. Mycorrhiza inoculated plants may exudate more carbohydrate like cytokinin to rhizosphere to enhance chloroplast development and increase the chlorophyll levels in order to increase the photosynthesis values.

PNW 601-E • November 2007. D.A. Horneck, J.W. Ellsworth, B.G. Hopkins, D.M. Sullivan, and R.G. Stevens. Managing salt-affected soils for crop production (Horneck et al. 2007)

References

Abbas T, Rizwan M, Ali S, Adrees M, Zia-ur-Rehman M, Qayyum MF et al (2018) Effect of biochar on alleviation of cadmium toxicity in wheat (Triticum aestivum L.) grown on Cd-contaminated saline soil. Environ Sci Pollut Res 25(26):25668–25680

Abbaspour H, Saeidi-Sar S, Afshari H, Abdel-Wahhab MA (2012) Tolerance of mycorrhiza infected Pistachio (Pistacia vera L.) seedling to drought stress under glasshouse conditions. J Plant Physiol 169(7):704–709

Abdel-Salam E, Alatar A, El-Sheikh MA (2018) Inoculation with arbuscular mycorrhizal fungi alleviates harmful effects of drought stress on damask rose. Saudi J Biol Sci 25(8):1772–1780

Abu-Elsaoud AM, Nafady NA, Abdel-Azeem AM (2017) Arbuscular mycorrhizal strategy for zinc mycoremediation and diminished translocation to shoots and grains in wheat. PLoS One 12(11): e0188220

Adiku S, Renger M, Wessolek G, Facklam M, Hecht-Bucholtz C (2001) Simulation of the dry matter production and seed yield of common beans under varying soil water and salinity conditions. Agric Water Manag 47(1):55–68

Al-Karaki GN (1998) Benefit, cost and water-use efficiency of arbuscular mycorrhizal durum wheat grown under drought stress. Mycorrhiza 8(1):41–45

Al-Karaki GN (2000) Growth of mycorrhizal tomato and mineral acquisition under salt stress. Mycorrhiza 10(2):51–54

Al-Karaki GN, Clark RB (1999) Mycorrhizal influence on protein and lipid of durum wheat grown at different soil phosphorus levels. Mycorrhiza 9(2):97–101

Al-Karaki GN, Al-Raddad A, Clark RB (1998) Water stress and mycorrhizal isolate effects on growth and nutrient acquisition of wheat. J Plant Nutr 21(5):891–902

Al-Karaki GN, Hammad R, Rusan M (2001) Response of two tomato cultivars differing in salt tolerance to inoculation with mycorrhizal fungi under salt stress. Mycorrhiza 11(1):43–47

Al-Karaki G, McMichael B, Zak J (2004) Field response of wheat to arbuscular mycorrhizal fungi and drought stress. Mycorrhiza 14(4):263–269

Al-Khaliel AS (2010) Effect of salinity stress on mycorrhizal association and growth response of peanut infected by Glomus mosseae. Plant Soil Envir 56(7):318–324

Ali S, Rizwan M, Qayyum MF, Ok YS, Ibrahim M, Riaz M, Arif MS, Hafeez F, Al-Wabel MI, Shahzad AN (2017) Biochar soil amendment on alleviation of drought and salt stress in plants: a critical review. Environ Sci Pollut Res 24(14):12700–12712

Allen MF (2007) Mycorrhizal fungi: highways for water and nutrients in arid soils. Vadose Zone J 6 (2):291–297

Amanifar S, Khodabandeloo M, Fard EM, Askari MS, Ashrafi M (2019) Alleviation of salt stress and changes in glycyrrhizin accumulation by arbuscular mycorrhiza in liquorice (Glycyrrhiza glabra) grown under salinity stress. Environ Exp Bot 160:25–34

Amiri R, Nikbakht A, Etemadi N, Sabzalian MR (2017) Nutritional status, essential oil changes and water-use efficiency of rose geranium in response to arbuscular mycorrhizal fungi and water deficiency stress. Symbiosis 73(1):15–25

Aroca R, Ruiz-Lozano JM, Zamarreño ÁM, Paz JA, García-Mina JM, Pozo MJ et al (2013) Arbuscular mycorrhizal symbiosis influences strigolactone production under salinity and alleviates salt stress in lettuce plants. J Plant Physiol 170(1):47–55

Aroca R, Abu-Elsaoud AM, Nafady NA, Abdel-Azeem AM (2017) Arbuscular mycorrhizal strategy for zinc mycoremediation and diminished translocation to shoots and grains in wheat. PLoS One 12(11):e0188220

Ashok Aggarwal NK, Karishma, Neetu, Tanwar A, Gupta KK (2012) Arbuscular mycorrhizal symbiosis and alleviation of salinity stress. J Appl Natl Sci 4(1):144–155

Askari A, Ardakani M, Vazan S, Paknejad F, Hosseini Y (2018) The effect of mycorrhizal symbiosis and seed priming on the amount of chlorophyll index and absorption of nutrients under drought stress in sesame plant under field conditions. Appl Ecol Environ Res 16 (1):335–357

Asrar A, Abdel-Fattah G, Elhindi K (2012) Improving growth, flower yield, and water relations of snapdragon (Antirrhinum majus L.) plants grown under well-watered and water-stress conditions using arbuscular mycorrhizal fungi. Photosynthetica 50(2):305–316

Atul-Nayyar A, Hamel C, Hanson K, Germida J (2009) The arbuscular mycorrhizal symbiosis links N mineralization to plant demand. Mycorrhiza 19(4):239–246

Auge RM (2004) Arbuscular mycorrhizae and soil/plant water relations. Can J Soil Sci 84 (4):373–381

Bagheri V, Shamshiri MH, Shirani H, Roosta HR (2011) Effect of mycorrhizal inoculation on ecophysiological responses of pistachio plants grown under different water regimes. Photosynthetica 49(4):531–538

Bagheri V, Shamshiri M, Shirani H, Roosta H (2012) Nutrient uptake and distribution in mycorrhizal pistachio seedlings under drought stress. J Agric Sci Technol 14:1591–1604

Bago B, Vierheilig H, Piché Y, Azcón-Aguilar C (1996) Nitrate depletion and pH changes induced by the extraradical mycelium of the arbuscular mycorrhizal fungus Glomus intraradices grown in monoxenic culture. New Phytol 133(2):273–280

Barea JM, Calvet C, Estaun V, Camprubi A (1996) Biological control as a key component in sustainable agriculture. Plant Soil 185(2):171–172

Barun Kumar Manjhi SP, Meena SK, Yadav RS, Alvina F, Singh HB, Rakshit A (2016) Mycorrhizoremediation of nickel and cadmium: a promising technology. Nat Environ Pollut Technol 15:647–652

Baum C, El-Tohamy W, Gruda N (2015) Increasing the productivity and product quality of vegetable crops using arbuscular mycorrhizal fungi: a review. Sci Hortic 187:131–141

Bedini S, Ramasamy C, Giovannetti M (2009) Glomalin-related soil protein as an indicator of the impact of different agronomical practices on soil quality. Asp Appl Biol 98:75–80

Belmondo S, Fiorilli V, Pérez-Tienda J, Ferrol N, Marmeisse R, Lanfranco L (2014) A dipeptide transporter from the arbuscular mycorrhizal fungus Rhizophagus irregularis is upregulated in the intraradical phase. Front Plant Sci 5:436

Bencherif K, Boutekrabt A, Fontaine J, Laruelle F, Dalpè Y, Lounès-Hadj SA (2015) Impact of soil salinity on arbuscular mycorrhizal fungi biodiversity and microflora biomass associated with Tamarix articulata Vahll rhizosphere in arid and semi-arid Algerian areas. Sci Total Environ 533:488–494

Benlloch-González M, Fournier JM, Benlloch M (2009) K+ deprivation induces xylem water and K + transport in sunflower: evidence for a co-ordinated control. J Exp Bot 61(1):157–164

Bernardo L, Carletti P, Badeck FW, Rizza F, Morcia C, Ghizzoni R et al (2019) Metabolomic responses triggered by arbuscular mycorrhiza enhance tolerance to water stress in wheat cultivars. Plant Physiol Biochem 137:203–212

Berruti A, Borriello R, Orgiazzi A, Barbera AC, Lumini E, Bianciotto V (2014) Arbuscular mycorrhizal fungi and their value for ecosystem management. In: Biodiversity: the dynamic balance of the planet. InTech, Rijeta, Croacia, pp 159–191

Biswas A, Biswas A (2014) Comprehensive approaches in rehabilitating salt affected soils: a review on Indian perspective. Open Trans Geosci 1(1):13–24

Bitterlich M, Rouphael Y, Graefe J, Franken P (2018) Arbuscular mycorrhizas: a promising component of plant production systems provided favorable conditions for their growth. Front Plant Sci 9:1329

Boldt-Burisch K, Naeth MA, Schneider U, Schneider B, Huttl RF (2018) Plant growth and arbuscular mycorrhizae development in oil sands processing by-products. Sci Total Environ 621:30–39

Borde M, Dudhane M, Jite P (2012) Growth, water use efficiency and antioxidant defense responses of mycorrhizal and non mycorrhizal Allium sativum L. under drought stress condition. Ann Plant Sci 1(1):6–11

Bothe H (2012) Arbuscular mycorrhiza and salt tolerance of plants. Symbiosis 58(1–3):7–16

Brundrett M (1991) Mycorrhizas in natural ecosystems. Adv Ecol Res 21:171–313

Bucher M (2007) Functional biology of plant phosphate uptake at root and mycorrhiza interfaces. New Phytol 173(1):11–26

Bücking H, Kafle A (2015) Role of arbuscular mycorrhizal fungi in the nitrogen uptake of plants: current knowledge and research gaps. Agronomy 5(4):587

Budi SW, Setyaningsih L (2013) Arbuscular mycorrhizal fungi and biochar improved early growth of Neem (Melia azedarach Linn.) seedling under greenhouse conditions. Jurnal Manajemen Hutan Tropika 19(2):103–110

Calabrese S, Pérez-Tienda J, Ellerbeck M, Arnould C, Chatagnier O, Boller T et al (2016) GintAMT3–a low-affinity ammonium transporter of the arbuscular mycorrhizal Rhizophagus irregularis. Front Plant Sci 7:679

Cappellazzo G, Lanfranco L, Fitz M, Wipf D, Bonfante P (2008) Characterization of an amino acid permease from the endomycorrhizal fungus Glomus mosseae. Plant Physiol 147(1):429–437

Carvalho LM, Correia PM, Martins-Loucao MA (2004) Arbuscular mycorrhizal fungal propagules in a salt marsh. Mycorrhiza 14(3):165–170

Carvalho LM, Caçador I, Martins-Loução MA (2006) Arbuscular mycorrhizal fungi enhance root cadmium and copper accumulation in the roots of the salt marsh plant Aster tripolium L. Plant Soil 285(1–2):161–169

Cekic FO, Unyayar S, Ortas I (2012) Effects of arbuscular mycorrhizal inoculation on biochemical parameters in Capsicum annuum grown under long term salt stress. Turk J Bot 36(1):63–72

Chang W, Sui X, Fan X-X, Jia T-T, Song F-Q (2018) Arbuscular mycorrhizal symbiosis modulates antioxidant response and ion distribution in salt-stressed Elaeagnus angustifolia seedlings. Front Microbiol 9:652

Chen BD, Zhu YG, Duan J, Xiao XY, Smith SE (2007) Effects of the arbuscular mycorrhizal fungus Glomus mosseae on growth and metal uptake by four plant species in copper mine tailings. Environ Pollut 147(2):374–380

Chen M, Arato M, Borghi L, Nouri E, Reinhardt D (2018) Beneficial services of arbuscular mycorrhizal fungi – from ecology to application. Front Plant Sci 9:1270

Chi G-G, Srivastava AK, Wu Q-S (2018) Exogenous easily extractable glomalin-related soil protein improves drought tolerance of trifoliate orange. Arch Agron Soil Sci 64(10):1341–1350

Chitarra W, Pagliarani C, Maserti B, Lumini E, Siciliano I, Cascone P et al (2016) Insights on the impact of arbuscular mycorrhizal symbiosis on tomato tolerance to water stress. Plant Physiol 2016:00307

Cimen B, Yesiloglu T (2016) Rootstock breeding for abiotic stress tolerance in citrus. In: Abiotic and biotic stress in plants-recent advances and future perspectives. InTech, London

Cliquet JB, Murray PJ, Boucaud J (1997) Effect of the arbuscular mycorrhizal fungus Glomus fasciculatum on the uptake of amino nitrogen by Lolium perenne. New Phytol 137(2):345–349

Courty PE, Smith P, Koegel S, Redecker D, Wipf D (2015) Inorganic nitrogen uptake and transport in beneficial plant root-microbe interactions. Crit Rev Plant Sci 34(1–3):4–16

Coutinho ES, Barbosa M, Beiroz W, Mescolotti DLC, Bonfim JA, Louro Berbara RL et al (2019) Soil constraints for arbuscular mycorrhizal fungi spore community in degraded sites of rupestrian grassland: implications for restoration. Eur J Soil Biol 90:51–57

Daei G, Ardekani M, Rejali F, Teimuri S, Miransari M (2009) Alleviation of salinity stress on wheat yield, yield components, and nutrient uptake using arbuscular mycorrhizal fungi under field conditions. J Plant Physiol 166(6):617–625

Dahlawi S, Naeem A, Rengel Z, Naidu R (2018) Biochar application for the remediation of salt-affected soils: challenges and opportunities. Sci Total Environ 625:320–335

Dickinson D, Balduccio L, Buysse J, Ronsse F, Huylenbroeck G, Prins W (2015) Cost-benefit analysis of using biochar to improve cereals agriculture. GCB Bioenergy 7(4):850–864

Duc NH, Csintalan Z, Posta K (2018) Arbuscular mycorrhizal fungi mitigate negative effects of combined drought and heat stress on tomato plants. Plant Physiol Biochem 132:297–307

Durán P, Acuña J, Armada E, López-Castillo O, Cornejo P, Mora M et al (2016) Inoculation with selenobacteria and arbuscular mycorrhizal fungi to enhance selenium content in lettuce plants and improve tolerance against drought stress. J Soil Sci Plant Nutr 16(1):211–225

Egamberdieva D, Kucharova Z (2009) Selection for root colonising bacteria stimulating wheat growth in saline soils. Biol Fertil Soils 45(6):563–571

Egamberdieva D, Wirth S, Behrendt U, Ahmad P, Berg G (2017) Antimicrobial activity of medicinal plants correlates with the proportion of antagonistic endophytes. Front Microbiol 8:199

Elhindi KM, El-Din AS, Elgorban AM (2017) The impact of arbuscular mycorrhizal fungi in mitigating salt-induced adverse effects in sweet basil (Ocimum basilicum L.). Saudi J Biol Sci 24(1):170–179

Estrada B, Aroca R, Maathuis FJM, Barea JM, Ruiz-Lozano JM (2013) Arbuscular mycorrhizal fungi native from a Mediterranean saline area enhance maize tolerance to salinity through improved ion homeostasis. Plant Cell Environ 36(10):1771–1782

Evelin H, Kapoor R, Giri B (2009) Arbuscular mycorrhizal fungi in alleviation of salt stress: a review. Ann Bot 104(7):1263–1280

Evelin H, Giri B, Kapoor R (2012) Contribution of Glomus intraradices inoculation to nutrient acquisition and mitigation of ionic imbalance in NaCl-stressed Trigonella foenum-graecum. Mycorrhiza 22(3):203–217

Fan Q-J, Liu J-H (2011) Colonization with arbuscular mycorrhizal fungus affects growth, drought tolerance and expression of stress-responsive genes in Poncirus trifoliata. Acta Physiol Plant 33 (4):1533

Farhangi-Abriz S, Torabian S (2017) Antioxidant enzyme and osmotic adjustment changes in bean seedlings as affected by biochar under salt stress. Ecotox Environ Safe 137:64–70
Feng G, Zhang FS, Li XL, Tian CY, Tang C, Rengel Z (2002) Improved tolerance of maize plants to salt stress by arbuscular mycorrhiza is related to higher accumulation of soluble sugars in roots. Mycorrhiza 12(4):185–190
Ferguson L, Kaur S, Epstein L (1998) Arbuscular mycorrhizal fungi on pistachio rootstocks in California. In: Ferguson L, Kester D (eds) Second international symposium on pistachios and almonds. Acta Horticulturae. International Society Horticultural Science, Leuven 1, pp 211–218
Frey B, Schüepp H (1993) Acquisition of nitrogen by external hyphae of arbuscular mycorrhizal fungi associated with Zea mays L. New Phytol 124(2):221–230
Füzy A, Biró B, Tóth T, Hildebrandt U, Bothe H (2008) Drought, but not salinity, determines the apparent effectiveness of halophytes colonized by arbuscular mycorrhizal fungi. J Plant Physiol 165(11):1181–1192
Gadkar V, David-Schwartz R, Kunik T, Kapulnik Y (2001) Arbuscular mycorrhizal fungal colonization. Factors involved in host recognition. Plant Physiol 127(4):1493–1499
Gai JP, Fan JQ, Zhang SB, Mi NN, Christie P, Li XL et al (2018) Direct effects of soil cadmium on the growth and activity of arbuscular mycorrhizal fungi. Rhizosphere 7:43–48
García I, Mendoza R, Pomar MC (2008) Deficit and excess of soil water impact on plant growth of Lotus tenuis by affecting nutrient uptake and arbuscular mycorrhizal symbiosis. Plant Soil 304 (1–2):117–131
Garg N, Bhandari P (2016) Silicon nutrition and mycorrhizal inoculations improve growth, nutrient status, K+/Na+ ratio and yield of Cicer arietinum L. genotypes under salinity stress. Plant Growth Regul 78(3):371–387
Garg N, Bharti A (2018) Salicylic acid improves arbuscular mycorrhizal symbiosis, and chickpea growth and yield by modulating carbohydrate metabolism under salt stress. Mycorrhiza 28 (8):727–746
Garg N, Manchanda G (2009) Role of arbuscular mycorrhizae in the alleviation of ionic, osmotic and oxidative stresses induced by salinity in Cajanus cajan (L.) Millsp.(pigeonpea). J Agron Crop Sci 195(2):110–123
Gholamhoseini M, Ghalavand A, Dolatabadian A, Jamshidi E, Khodaei-Joghan A (2013) Effects of arbuscular mycorrhizal inoculation on growth, yield, nutrient uptake and irrigation water productivity of sunflowers grown under drought stress. Agric Water Manag 117:106–114
Giasson P, Jaouich A, Cayer P, Gagné S, Moutoglis P, Massicotte L (2006) Enhanced phytoremediation: a study of mycorrhizoremediation of heavy metal–contaminated soil. Remediat J 17(1):97–110
Giri B, Kapoor R, Mukerji K (2007) Improved tolerance of Acacia nilotica to salt stress by arbuscular mycorrhiza, Glomus fasciculatum may be partly related to elevated K/Na ratios in root and shoot tissues. Microb Ecol 54(4):753–760
Gong M, Tang M, Chen H, Zhang Q, Feng X (2013) Effects of two Glomus species on the growth and physiological performance of Sophora davidii seedlings under water stress. New For 44 (3):399–408
Grattan S, Grieve C (1998) Salinity–mineral nutrient relations in horticultural crops. Sci Hortic 78 (1–4):127–157
Grimaldo-Pantoja GL, Niu GH, Sun YP, Castro-Rocha A, Alvarez-Parrilla E, Flores-Margez JR et al (2017) Negative effect of saline irrigation on yield components and phytochemicals of pepper (Capsicum annuum) inoculated with arbuscular mycorrhizal fungi. Rev Fitotec Mex 40 (2):141–149
Haghighi M, Mohammadnia S, Attai Z, Pessarakli M (2017) Effects of mycorrhiza inoculation on cucumber growth irrigated with saline water. J Plant Nutr 40(1):128–137
Hamilton EW III, Frank DA (2001) Can plants stimulate soil microbes and their own nutrient supply? Evidence from a grazing tolerant grass. Ecology 82(9):2397–2402
Hammer EC, Nasr H, Pallon J, Olsson PA, Wallander H (2011) Elemental composition of arbuscular mycorrhizal fungi at high salinity. Mycorrhiza 21(2):117–129

Hammer EC, Forstreuter M, Rillig MC, Kohler J (2015) Biochar increases arbuscular mycorrhizal plant growth enhancement and ameliorates salinity stress. Appl Soil Ecol 96:114–121

Hasegawa PM, Bressan RA, Zhu J-K, Bohnert HJ (2000a) Plant cellular and molecular responses to high salinity. Annu Rev Plant Biol 51(1):463–499

Hasegawa PM, Bressan RA, Zhu JK, Bohnert HJ (2000b) Plant cellular and molecular responses to high salinity. Annu Rev Plant Physiol Plant Mol Biol 51:463–499

Hashem A, Alqarawi AA, Radhakrishnan R, Al-Arjani A-BF, Aldehaish HA, Egamberdieva D et al (2018) Arbuscular mycorrhizal fungi regulate the oxidative system, hormones and ionic equilibrium to trigger salt stress tolerance in Cucumis sativus L. Saudi J Biol Sci 25(6):1102–1114

Hashem A, Kumar A, Al-Dbass AM, Alqarawi AA, Al-Arjani A-BF, Singh G et al (2019) Arbuscular mycorrhizal fungi and biochar improves drought tolerance in chickpea. Saudi J Biol Sci 26(3):614–624

Heidari M, Karami V (2014) Effects of different mycorrhiza species on grain yield, nutrient uptake and oil content of sunflower under water stress. J Saudi Soc Agric Sci 13(1):9–13

Helander M, Saloniemi I, Omacini M, Druille M, Salminen J-P, Saikkonen K (2018) Glyphosate decreases mycorrhizal colonization and affects plant-soil feedback. Sci Total Environ 642:285–291

Hodge A (2001) Arbuscular mycorrhizal fungi influence decomposition of, but not plant nutrient capture from, glycine patches in soil. New Phytol 151(3):725–734

Hodge A, Campbell CD, Fitter AH (2001) An arbuscular mycorrhizal fungus accelerates decomposition and acquires nitrogen directly from organic material. Nature 413(6853):297

Hokmabadi H, Arzani K, Grierson P (2005) Growth, chemical composition, and carbon isotope discrimination of pistachio (Pistacia vera L.) rootstock seedlings in response to salinity. Aust J Agric Res 56(2):135–144

Horneck DA, Ellsworth JW, Hopkins BG, Sullivan DM, Stevens RG (2007) Managing salt-affected soils for crop production.

Hu YC, Schmidhalter U (2005) Drought and salinity: a comparison of their effects on mineral nutrition of plants. J Plant Nutr Soil Sci 168(4):541–549

Jeffery S, Bezemer TM, Cornelissen G, Kuyper TW, Lehmann J, Mommer L et al (2015) The way forward in biochar research: targeting trade-offs between the potential wins. GCB Bioenergy 7 (1):1–13

Ji L, Tan W, Chen X (2019) Arbuscular mycorrhizal mycelial networks and glomalin-related soil protein increase soil aggregation in Calcaric Regosol under well-watered and drought stress conditions. Soil Tillage Res 185:1–8

Jiang X, Huang Y (2003) Mechanism of contribution of mycorrhizal fungi to plant saline-alkali tolerance. Ecol Environ 12(3):353–356

Jones CG, Lawton JH, Shachak M (1997) Positive and negative effects of organisms as physical ecosystem engineers. Ecology 78(7):1946

Juniper S, Abbott LK (2006) Soil salinity delays germination and limits growth of hyphae from propagules of arbuscular mycorrhizal fungi. Mycorrhiza 16(5):371–379

Kaya C, Ashraf M, Sonmez O, Aydemir S, Tuna AL, Cullu MA (2009) The influence of arbuscular mycorrhizal colonisation on key growth parameters and fruit yield of pepper plants grown at high salinity. Sci Hortic 121(1):1–6

Khan AG (2005) Role of soil microbes in the rhizospheres of plants growing on trace metal contaminated soils in phytoremediation. J Trace Elem Med Biol 18(4):355–364

Khan AG (2006) Mycorrhizoremediation—an enhanced form of phytoremediation. J Zhejiang Univ Sci B 7(7):503–514

Khan A, Sharif M, Ali A, Shah SNM, Mian IA, Wahid F et al (2014) Potential of AM fungi in phytoremediation of heavy metals and effect on yield of wheat crop. Am J Plant Sci 05 (11):1578–1586

Kim H-S, Kim K-R, Yang JE, Ok YS, Owens G, Nehls T et al (2016) Effect of biochar on reclaimed tidal land soil properties and maize (Zea mays L.) response. Chemosphere 142:153–159

Koegel S, Brulé D, Wiemken A, Boller T, Courty P-E (2015) The effect of different nitrogen sources on the symbiotic interaction between Sorghum bicolor and Glomus intraradices: expression of plant and fungal genes involved in nitrogen assimilation. Soil Biol Biochem 86:159–163

Kumar A, Verma JP (2018) Does plant—microbe interaction confer stress tolerance in plants: a review? Microbiol Res 207:41–52

Kumar A, Dames JF, Gupta A, Sharma S, Gilbert JA, Ahmad P (2015) Current developments in arbuscular mycorrhizal fungi research and its role in salinity stress alleviation: a biotechnological perspective. Crit Rev Biotechnol 35(4):461–474

Lal R (2002) Carbon sequestration in dryland ecosystems of West Asia and North Africa. Land Degrad Dev 13(1):45–59

Lashari MS, Liu Y, Li L, Pan W, Fu J, Pan G et al (2013) Effects of amendment of biochar-manure compost in conjunction with pyroligneous solution on soil quality and wheat yield of a salt-stressed cropland from Central China great plain. Field Crop Res 144:113–118

Latef AAHA, Chaoxing H (2011) Effect of arbuscular mycorrhizal fungi on growth, mineral nutrition, antioxidant enzymes activity and fruit yield of tomato grown under salinity stress. Sci Hortic 127(3):228–233

Latef A, He CX (2011) Effect of arbuscular mycorrhizal fungi on growth, mineral nutrition, antioxidant enzymes activity and fruit yield of tomato grown under salinity stress. Sci Hortic 127(3):228–233

Läuchli A, Epstein E (1990) Plant responses to saline and sodic conditions. Agric Salinity Assess Manag 71:113–137

Le Pioufle O, Declerck S (2018) Reducing water availability impacts the development of the arbuscular mycorrhizal fungus Rhizophagus irregularis MUCL 41833 and its ability to take up and transport phosphorus under in vitro conditions. Front Microbiol 9:1254

Lehnert H, Serfling A, Friedt W, Ordon F (2018) Genome wide association studies reveal genomic regions associated with the response of wheat (Triticum aestivum L.) to mycorrhizae under drought stress conditions. Front Plant Sci 9:1728

Lenoir I, Fontaine J, Sahraoui ALH (2016) Arbuscular mycorrhizal fungal responses to abiotic stresses: a review. Phytochemistry 123:4–15

Li T, Hu YJ, Hao ZP, Li H, Wang YS, Chen BD (2013) First cloning and characterization of two functional aquaporin genes from an arbuscular mycorrhizal fungus Glomus intraradices. New Phytol 197(2):617–630

Li J-T, Yang Y, He C, Huang L, Ban Y, Tang M (2017) The effects of arbuscular mycorrhizal fungi on glomalin-related soil protein distribution, aggregate stability and their relationships with soil properties at different soil depths in lead-zinc contaminated area. PLoS One 12(8):e0182264

Liang C-C, Li T, Xiao Y-P, Liu M-J, Zhang H-B, Zhao Z-W (2009) Effects of inoculation with arbuscular mycorrhizal fungi on maize grown in multi-metal contaminated soils. Int J Phytoremediation 11(8):692–703

Liu T, Sheng M, Wang C, Chen H, Li Z, Tang M (2015) Impact of arbuscular mycorrhizal fungi on the growth, water status, and photosynthesis of hybrid poplar under drought stress and recovery. Photosynthetica 53(2):250–258

Liu C, Dai Z, Cui M, Lu W, Sun H (2018a) Arbuscular mycorrhizal fungi alleviate boron toxicity in Puccinellia tenuiflora under the combined stresses of salt and drought. Environ Pollut 240:557–565

Liu C-Y, Zhang F, Zhang D-J, Srivastava AK, Wu Q-S, Zou Y-N (2018b) Mycorrhiza stimulates root-hair growth and IAA synthesis and transport in trifoliate orange under drought stress. Sci Rep 8(1):1978

Liu CY, Zhang F, Zhang DJ, Srivastava AK, Wu QS, Zou YN (2018c) Mycorrhiza stimulates root-hair growth and IAA synthesis and transport in trifoliate orange under drought stress. Sci Rep 8:9

Loha A, Kashyap AK, Sharma P (2018) A putative cyclin, SiPHO80 from root endophytic fungus Serendipita indica regulates phosphate homeostasis, salinity and heavy metal toxicity tolerance. Biochem Biophys Res Commun 507(1–4):414–419

López-Pedrosa A, González-Guerrero M, Valderas A, Azcón-Aguilar C, Ferrol N (2006) GintAMT1 encodes a functional high-affinity ammonium transporter that is expressed in the extraradical mycelium of Glomus intraradices. Fungal Genet Biol 43(2):102–110

Lopez-Raez JA (2016) How drought and salinity affect arbuscular mycorrhizal symbiosis and strigolactone biosynthesis? Planta 243(6):1375–1385

Lovelock CE, Wright SF, Clark DA, Ruess RW (2004) Soil stocks of glomalin produced by arbuscular mycorrhizal fungi across a tropical rain forest landscape. J Ecol 92(2):278–287

Maggio A, Reddy MP, Joly RJ (2000) Leaf gas exchange and solute accumulation in the halophyte Salvadora persica grown at moderate salinity. Environ Exp Bot 44(1):31–38

Mardukhi B, Rejali F, Daei G, Ardakani MR, Malakouti MJ, Miransari M (2011) Arbuscular mycorrhizas enhance nutrient uptake in different wheat genotypes at high salinity levels under field and greenhouse conditions. C R Biol 334(7):564–571

Marschner H, Dell B (1994) Nutrient uptake in mycorrhizal symbiosis. Plant Soil 159(1):89–102

Miransari M (2014) Use of microbes for the alleviation of soil stresses. Springer, Dordrecht

Miransari M (2016) Stress and mycorrhizal plant. In: Recent Advances on Mycorrhizal Fungi. Springer, Cham, pp 63–79

Mirzaei J, Moradi M (2017) Relationships between flora biodiversity, soil physiochemical properties, and arbuscular mycorrhizal fungi (AMF) diversity in a semi-arid forest. Plant Ecol Evol 150 (2):151–159

Mohammad MJ, Malkawi HI, Shibli R (2003) Effects of arbuscular mycorrhizal fungi and phosphorus fertilization on growth and nutrient uptake of barley grown on soils with different levels of salts. J Plant Nutr 26(1):125–137

Mollinedo J, Schumacher TE, Chintala R (2016) Biochar effects on phenotypic characteristics of "wild" and "sickle" Medicago truncatula genotypes. Plant Soil 400(1–2):1–14

Mosse B, Stribley D, Le Tacon F (1981) Ecology of mycorrhizae and mycorrhizal fungi. Adv Microbial Ecol 5:137–210

Mukherjee A, Zimmerman AR (2013) Organic carbon and nutrient release from a range of laboratory-produced biochars and biochar–soil mixtures. Geoderma 193:122–130

Naher UA, Othman R, Panhwar QA (2013) Beneficial effects of Mycorrhizal Association for Crop Production in the tropics—a review. Int J Agric Biol 15(5):1021–1028

Nakano A, Takahashi K, Koide RT, Kimura M (2001) Determination of the nitrogen source for arbuscular mycorrhizal fungi by 15 N application to soil and plants. Mycorrhiza 10(6):267–273

Nasim G (2010) The role of arbuscular mycorrhizae in inducing resistance to drought and salinity stress in crops. In: Plant adaptation and phytoremediation. Springer, Berlin, pp 119–141

Navarro JM, Pérez-Tornero O, Morte A (2014) Alleviation of salt stress in citrus seedlings inoculated with arbuscular mycorrhizal fungi depends on the rootstock salt tolerance. J Plant Physiol 171(1):76–85

Neumann E, George E (2004) Colonisation with the arbuscular mycorrhizal fungus Glomus mosseae (Nicol. & Gerd.) enhanced phosphorus uptake from dry soil in Sorghum bicolor (L.). Plant Soil 261:245–255

Oliveira RS, Carvalho P, Marques G, Ferreira L, Nunes M, Rocha I et al (2017) Increased protein content of chickpea (Cicer arietinum L.) inoculated with arbuscular mycorrhizal fungi and nitrogen-fixing bacteria under water deficit conditions. J Sci Food Agric 97(13):4379–4385

Olmo M, Villar R, Salazar P, Alburquerque JA (2016) Changes in soil nutrient availability explain biochar's impact on wheat root development. Plant Soil 399(1–2):333–343

Ortas I (2016) Role of mycorrhizae and biochar on plant growth and soil quality. In: Bruckman VJ, Varol EA, Uzun BB, Liu JF (eds) Biochar, a Regıonal supply chain approach in view of climate change mitigation. Cambridge Universitey Press, Cambridge, UK, p 398

Ortaş I, Rafique M (2017) The mechanisms of nutrient uptake by arbuscular mycorrhizae. In: Varma A, Prasad R, Tuteja N (eds) Mycorrhiza—nutrient uptake, biocontrol, ecorestoration. Springer, Cham, pp 1–19
Ortuno MF, Lorente B, Hernandez JA, Sanchez-Blanco MJ (2018) Mycorrhizal inoculation on compost substrate affects nutritional balance, water uptake and photosynthetic efficiency in Cistus albidus plants submitted to water stress. Braz J Bot 41(2):299–310
Osakabe Y, Osakabe K, Shinozaki K, Tran L-SP (2014) Response of plants to water stress. Front Plant Sci 5:86
Ouledali S, Ennajeh M, Ferrandino A, Khemira H, Schubert A, Secchi F (2019) Influence of arbuscular mycorrhizal fungi inoculation on the control of stomata functioning by abscisic acid (ABA) in drought-stressed olive plants. S Afr J Bot 121:152–158
Oyewole BO, Olawuyi OJ, Odebode AC, Abiala MA (2017) Influence of Arbuscular mycorrhiza fungi (AMF) on drought tolerance and charcoal rot disease of cowpea. Biotechnol Rep 14:8–15
Pavithra D, Yapa N (2018) Arbuscular mycorrhizal fungi inoculation enhances drought stress tolerance of plants. Groundw Sustain Dev 7:490–494
Paymaneh Z, Sarcheshmehpour M, Bukovská P, Jansa J (2019) Could indigenous arbuscular mycorrhizal communities be used to improve tolerance of pistachio to salinity and/or drought? Symbiosis:1–15
Pérez-Tienda J, Testillano PS, Balestrini R, Fiorilli V, Azcón-Aguilar C, Ferrol N (2011) GintAMT2, a new member of the ammonium transporter family in the arbuscular mycorrhizal fungus Glomus intraradices. Fungal Genet Biol 48(11):1044–1055
Peterson RL, Massicotte HB (2004) Exploring structural definitions of mycorrhizas, with emphasis on nutrient-exchange interfaces. Can J Bot 82(8):1074–1088
Porcel R, Aroca R, Ruiz-Lozano JM (2012) Salinity stress alleviation using arbuscular mycorrhizal fungi. A review. Agron Sustain Dev 32(1):181–200
Porcel R, Aroca R, Azcon R, Ruiz-Lozano JM (2016) Regulation of cation transporter genes by the arbuscular mycorrhizal symbiosis in rice plants subjected to salinity suggests improved salt tolerance due to reduced Na+ root-to-shoot distribution. Mycorrhiza 26(7):673–684
Porras-Soriano A, Luisa Soriano-Martin M, Porras-Piedra A, Azcon R (2009a) Arbuscular mycorrhizal fungi increased growth, nutrient uptake and tolerance to salinity in olive trees under nursery conditions. J Plant Physiol 166(13):1350–1359
Porras-Soriano A, Soriano-Martín ML, Porras-Piedra A, Azcón R (2009b) Arbuscular mycorrhizal fungi increased growth, nutrient uptake and tolerance to salinity in olive trees under nursery conditions. J Plant Physiol 166(13):1350–1359
Puga A, Abreu C, Melo L, Beesley L (2015) Biochar application to a contaminated soil reduces the availability and plant uptake of zinc, lead and cadmium. J Environ Manag 159:86–93
Quiroga G, Erice G, Aroca R, Zamarreño ÁM, García-Mina JM, Ruiz-Lozano JM (2018) Arbuscular mycorrhizal symbiosis and salicylic acid regulate aquaporins and root hydraulic properties in maize plants subjected to drought. Agric Water Manag 202:271–284
Rafique M, Sultan T, Ortas I, Chaudhary HJ (2017) Enhancement of maize plant growth with inoculation of phosphate-solubilizing bacteria and biochar amendment in soil. Soil Sci Plant Nutr 63:460–469
Redon P-O, Béguiristain T, Leyval C (2008) Influence of Glomus intraradices on Cd partitioning in a pot experiment with Medicago truncatula in four contaminated soils. Soil Biol Biochem 40 (10):2710–2712
Rillig MC, Mummey DL (2006) Mycorrhizas and soil structure. New Phytol 171(1):41–53
Rivero J, Alvarez D, Flors V, Azcon-Aguilar C, Pozo MJ (2018) Root metabolic plasticity underlies functional diversity in mycorrhiza-enhanced stress tolerance in tomato. New Phytol 220 (4):1322–1336
Rouphael Y, Franken P, Schneider C, Schwarz D, Giovannetti M, Agnolucci M et al (2015) Arbuscular mycorrhizal fungi act as biostimulants in horticultural crops. Sci Hortic 196:91–108
Roy M, Roychowdhury R, Mukherjee P (2018) Remediation of fly ash dumpsites through bioenergy crop plantation and generation: a review. Pedosphere 28(4):561–580

Ruiz-Lozano JM, Collados C, Barea JM, Azcón R (2001) Arbuscular mycorrhizal symbiosis can alleviate drought-induced nodule senescence in soybean plants. New Phytol 151(2):493–502
Ruiz-Lozano JM, Porcel R, Aroca R (2006) Does the enhanced tolerance of arbuscular mycorrhizal plants to water deficit involve modulation of drought-induced plant genes? New Phytol 171 (4):693–698
Ruiz-Lozano JM, Porcel R, Azcón C, Aroca R (2012) Regulation by arbuscular mycorrhizae of the integrated physiological response to salinity in plants: new challenges in physiological and molecular studies. J Exp Bot 63(11):4033–4044
Ruiz-Lozano JM, Porcel R, Calvo-Polanco M, Aroca R (2018) Improvement of salt tolerance in rice plants by arbuscular mycorrhizal symbiosis. In: Root biology. Springer, Cham, pp 259–279
Ruíz-Sánchez M, Armada E, Muñoz Y, de Salamone IEG, Aroca R, Ruíz-Lozano JM et al (2011) Azospirillum and arbuscular mycorrhizal colonization enhance rice growth and physiological traits under well-watered and drought conditions. J Plant Physiol 168(10):1031–1037
Santander C, Aroca R, Ruiz-Lozano JM, Olave J, Cartes P, Borie F et al (2017) Arbuscular mycorrhiza effects on plant performance under osmotic stress. Mycorrhiza 27(7):639–657
Satir NY, Ortas I, Satir O (2016) The influence of mycorrhizal species on sour orange (Citrus aurantium L.) growth under saline soil conditions. Pak J Agric Sci 53(2):399–406
Shabani L, Sabzalian MR (2016) Mostafavi pour S. Arbuscular mycorrhiza affects nickel translocation and expression of ABC transporter and metallothionein genes in Festuca arundinacea. Mycorrhiza 26(1):67–76
Shamshiri MH, Fattahi M (2016) Effects of arbuscular mycorrhizal fungi on photosystem II activity of three pistachio rootstocks under salt stress as probed by the OJIP-test. Russ J Plant Physiol 63 (1):101–110
Shen Q, Hedley M, Arbestain MC, Kirschbaum MUF (2016) Can biochar increase the bioavailability of phosphorus? J Soil Sci Plant Nut 16(2):268–286
Sheng M, Tang M, Chen H, Yang B, Zhang F, Huang Y (2008) Influence of arbuscular mycorrhizae on photosynthesis and water status of maize plants under salt stress. Mycorrhiza 18 (6–7):287–296
Shirani H, Shamshiri M, Bagheri V, Roosta H (2018) Nutrient uptake and distribution in mycorrhizal pistachio seedlings under drought stress. J Agric Sci Technol 14:1591–1604
Smith P (2016) Soil carbon sequestration and biochar as negative emission technologies. Glob Chang Biol 22(3):1315–1324
Smith SE, Read DJ (2008) Mycorrhizal symbiosis, Third edn. Academic, San Diego, CA
Soka GE, Ritchie ME (2018) Arbuscular mycorrhizal spore composition and diversity associated with different land uses in a tropical savanna landscape. Tanzania Appl Soil Ecol 125:222–232
Stevens KJ, Wall CB, Janssen JA (2011) Effects of arbuscular mycorrhizal fungi on seedling growth and development of two wetland plants, Bidens frondosa L., and Eclipta prostrata (L.) L., grown under three levels of water availability. Mycorrhiza 21(4):279–288
Subramanian K, Santhanakrishnan P, Balasubramanian P (2006) Responses of field grown tomato plants to arbuscular mycorrhizal fungal colonization under varying intensities of drought stress. Sci Hortic 107(3):245–253
Sudhir P, Murthy S (2004) Effects of salt stress on basic processes of photosynthesis. Photosynthetica 42(2):481–486
Sun X, Shi J, Ding G (2017) Combined effects of arbuscular mycorrhiza and drought stress on plant growth and mortality of forage sorghum. Appl Soil Ecol 119:384–391
Swaty RL, Deckert RJ, Whitham TG, Gehring CA (2004) Ectomycorrhizal abundance and community composition shifts with drought: predictions from tree rings. Ecology 85(4):1072–1084
Thomas H, Morgan W, Humphreys M (2003) Designing grasses with a future–combining the attributes of Lolium and Festuca. Euphytica 133(1):19–26
Thomas SC, Frye S, Gale N, Garmon M, Launchbury R, Machado N et al (2013) Biochar mitigates negative effects of salt additions on two herbaceous plant species. J Environ Manag 129:62–68

Tobar R, Azcón R, Barea J (1994a) Improved nitrogen uptake and transport from 15N-labelled nitrate by external hyphae of arbuscular mycorrhiza under water-stressed conditions. New Phytol 126(1):119–122

Tobar R, Azcón R, Barea J (1994b) The improvement of plant N acquisition from an ammonium-treated, drought-stressed soil by the fungal symbiont in arbuscular mycorrhizae. Mycorrhiza 4 (3):105–108

Treu R, Falandysz J (2017) Mycoremediation of hydrocarbons with basidiomycetes-a review. J Environ Sci Health B 52(3):148–155

Trouvelot S, Heloir MC, Poinssot B, Gauthier A, Paris F, Guillier C et al (2014) Carbohydrates in plant immunity and plant protection: roles and potential application as foliar sprays. Front Plant Sci 5:592

Wang W, Vinocur B, Altman A (2003) Plant responses to drought, salinity and extreme temperatures: towards genetic engineering for stress tolerance. Planta 218(1):1–14

Wang F, Lin X, Yin R (2005) Heavy metal uptake by arbuscular mycorrhizas of Elsholtzia splendens and the potential for phytoremediation of contaminated soil. Plant Soil 269 (1–2):225–232

Wang P, Srivastava AK, Zhang YC, Wu QS (2016) Inoculation response of mycorrhizas on morphology and physiological behaviour of trifoliate orange (Poncirus trifoliata) roots under salt stress. Indian J Agric Sci 86(11):1438–1442

Wu QS, Xia RX (2006) Arbuscular mycorrhizal fungi influence growth, osmotic adjustment and photosynthesis of citrus under well-watered and water stress conditions. J Plant Physiol 163 (4):417–425

Wu QS, Zou YN, Xia RX (2006) Effects of water stress and arbuscular mycorrhizal fungi on reactive oxygen metabolism and antioxidant production by citrus (citrus tangerine) roots. Eur J Soil Biol 42(3):166–172

Wu QS, Zou YN, Liu W, Ye X, Zai H, Zhao L (2010) Alleviation of salt stress in citrus seedlings inoculated with mycorrhiza: changes in leaf antioxidant defense systems. Plant Soil Environ 56 (10):470–475

Wu Q-S, Zou Y-N, He X-H (2011) Differences of hyphal and soil phosphatase activities in drought-stressed mycorrhizal trifoliate orange (Poncirus trifoliata) seedlings. Sci Hortic 129(2):294–298

Wu QS, Srivastava AK, Zou YN (2013) AMF-induced tolerance to drought stress in citrus: a review. Sci Hortic 164:77–87

Xu L, Li T, Wu Z, Feng H, Yu M, Zhang X et al (2018) Arbuscular mycorrhiza enhances drought tolerance of tomato plants by regulating the 14-3-3 genes in the ABA signaling pathway. Appl Soil Ecol 125:213–221

Yamato M, Ikeda S, Iwase K (2008) Community of arbuscular mycorrhizal fungi in a coastal vegetation on Okinawa island and effect of the isolated fungi on growth of sorghum under salt-treated conditions. Mycorrhiza 18(5):241–249

Yang SJ, Zhang ZL, Xue YX, Zhang ZF, Shi SY (2014) Arbuscular mycorrhizal fungi increase salt tolerance of apple seedlings. Bot Stud 55:7

Yang YR, He CJ, Huang L, Ban YH, Tang M (2017) The effects of arbuscular mycorrhizal fungi on glomalin-related soil protein distribution, aggregate stability and their relationships with soil properties at different soil depths in lead-zinc contaminated area. PLoS One 12(8):e0182264

Yu H, Zou W, Chen J, Chen H, Yu Z, Huang J et al (2019) Biochar amendment improves crop production in problem soils: a review. J Environ Manag 232:8–21

Zaefarian F, Rezvani M, Ardakani MR, Rejali F, Miransari M (2013) Impact of mycorrhizae formation on the phosphorus and heavy-metal uptake of alfalfa. Commun Soil Sci Plant Anal 44(8):1340–1352

Zarik L, Meddich A, Hijri M, Hafidi M, Ouhammou A, Ouahmane L et al (2016) Use of arbuscular mycorrhizal fungi to improve the drought tolerance of Cupressus atlantica G. C R Biol 339 (5–6):185–196

Zhang XH, Zhu YG, Chen BD, Lin AJ, Smith SE, Smith FA (2005) Arbuscular mycorrhizal fungi contribute to resistance of upland rice to combined metal contamination of soil. J Plant Nutr 28 (12):2065–2077

Zhang Q, Xu L, Tang J, Bai M, Chen X (2011) Arbuscular mycorrhizal mediation of biomass–density relationship of Medicago sativa L. under two water conditions in a field experiment. Mycorrhiza 21(4):269–277

Zhang Z, Wang Q, Wang H, Nie S, Liang Z (2017) Effects of soil salinity on the content, composition, and ion binding capacity of glomalin-related soil protein (GRSP). Sci Total Environ 581–582:657–665

Zhang T, Hu Y, Zhang K, Tian C, Guo J (2018) Arbuscular mycorrhizal fungi improve plant growth of Ricinus communis by altering photosynthetic properties and increasing pigments under drought and salt stress. Ind Crop Prod 117:13–19

Zhang W, Wang C, Liu M, Yu Y (2019a) Integrated reclamation of saline soil nitrogen transformation in the hyphosphere by earthworms and arbuscular mycorrhizal fungus. Appl Soil Ecol 135:137–146

Zhang F, Liu M, Li Y, Che Y, Xiao Y (2019b) Effects of arbuscular mycorrhizal fungi, biochar and cadmium on the yield and element uptake of Medicago sativa. Sci Total Environ 655:1150–1158

Zhu X, Song F, Liu S, Liu T, Zhou X (2012) Arbuscular mycorrhizae improves photosynthesis and water status of Zea mays L. under drought stress. Plant Soil Environ 58(4):186–191

Zrnic M, Siric I (2017) The application of mycorrhiza in horticulture Primjena mikorize u hortikulturi. J Cent Eur Agric 18(3):706–732

Chapter 24
Combined Use of Beneficial Bacteria and Arbuscular Mycorrhizal Fungi for the Biocontrol of Plant Cryptogamic Diseases: Evidence, Methodology, and Limits

Yuko Krzyzaniak, Maryline Magnin-Robert, Béatrice Randoux, Joël Fontaine, and Anissa Lounès-Hadj Sahraoui

Abstract Keeping up agricultural production, while reducing chemical inputs, under biotic and abiotic stresses, falls within a considerable challenge. Among the solutions, developing more sustainable strategies to protect crops by using biocontrol agents is a promising alternative. Several beneficial microorganisms can limit pathogen development by direct (antibiosis, competition) and/or indirect effects (plant growth promotion and resistance induction) thanks to nutritional and hormonal balance regulations. A great deal of research articles have already reported the beneficial roles of rhizobacteria or arbuscular mycorrhizal fungi (AMF) on reducing the development of fungal pathogens when applied alone, or on plant nutrition when co-inoculated; while fewer of them reviewed the co-inoculation as a mean to protect against fungal pathogens. Hence, this review aims to present recent developments on the effectiveness of the co-inoculation with beneficial bacteria and AMF for fungal disease management. Firstly, main mechanisms beneath the beneficial effects of bacterial or AMF inoculation alone on plants are recalled. Secondly, results relating their combined inoculation on disease severity are presented and sorted according to their outcomes (synergistic, neutral, or antagonistic interactions), along with the possible underlying mechanisms. Finally, we gathered the main methodologies employed in tripartite interaction experiment and brought to light future challenges for practical use.

Keywords Arbuscular mycorrhizal fungi · Plant growth-promoting rhizobacteria · Mycorrhiza helper bacteria · Tripartite interactions · Fungal diseases · Biocontrol

---

Y. Krzyzaniak · M. Magnin-Robert · B. Randoux · J. Fontaine · A. Lounès-Hadj Sahraoui (✉)
Université du Littoral Côte d'Opale, Unité de Chimie Environnementale et Interactions sur le Vivant (UCEIV) UR 4492, Calais Cedex, France
e-mail: lounes@univ-littoral.fr

N. Shrivastava et al. (eds.), *Symbiotic Soil Microorganisms*, Soil Biology 60,
https://doi.org/10.1007/978-3-030-51916-2_24

24.1 Introduction

Protecting agricultural crops from cryptogamic diseases generally requires a massive use of phytosanitary products. However, unintentional negative impacts on the environment and human health (Gill and Garg 2014) have been building up, along with the increasing number of fungicide resistant strains of pathogens (Rupp et al. 2017), making crop management more and more challengeable (Brauer et al. 2019). This led several actors to think out complementary strategies to keep on protecting crops by creating new forms of agroecosystems that would both respect biological processes, while maintaining a sufficient income to the farmers (Gianinazzi et al. 2010). Along with a more restrictive regulatory frame such as the European Union directive 2009/128/EC on the sustainable use of pesticide in pest management, Member States are now compelled to emphasize the use of alternative (non-synthetic) products in pest control, by launching national programs which implement the principles of Integrated Pest Management (IPM) for instance. Indeed, according to the Food and Agriculture Organization of the United Nations : "IPM means the careful consideration of all available pest control techniques and subsequent integration of appropriate measures that limit the development of pest populations and keep pesticides and other technical interventions to levels that are economically justified and reduce or minimize risks to human health and the environment. IPM emphasizes the growth of a healthy crop with the least possible disruption to agro-ecosystems and encourages natural pest control mechanisms" (FAO 2016).

One means among many, is the use of biological control agents. According to Heimpel and Mill, biological control (also called biocontrol) is "simultaneously a natural phenomenon, a pest management strategy and a scientific discipline" (Heimpel and Mills 2017). It allows controlling pests and diseases caused by fungi, bacteria, or viruses, with the help of other organisms or their derivative products. Biocontrol relies on natural mechanisms occurring during plant-microbe interactions (Flint et al. 1998; Alabouvette et al. 2006; Barratt et al. 2018). Based on predation, parasitism, herbivory, antibiosis, competition, or other natural mechanisms, it can maintain the pest population below the economic threshold level (Benjamin and Wesseler 2016). More specifically in the case of fungal disease management, biocontrol consists in using living fungal (Perazzolli et al. 2011; Mustafa et al. 2017) or bacterial (Bach et al. 2016) microorganisms; but also microorganism- (Li et al. 2019) or macroorganism-derived compounds such as polysaccharides from plants, algae, or crustaceans (Trouvelot et al. 2014; El Guilli et al. 2016; De Bona et al. 2019). Among soil microorganisms, Arbuscular Mycorrhiza Fungi (AMF) and the beneficial rhizospheric bacteria such as the Plant Growth-Promoting Rhizobacteria (PGPR) attract much interest as typical models used in laboratories for this past century, with the hope to improve crop health.

As several articles report successful disease severity reduction in response to a singly inoculated microorganism, the combination with two of them is also expected to be more effective than individual application, *via* a synergistic or additive effect

(Saldajeno and Hyakumachi 2011a). Indeed, on the contrary to sterilized substrates used in laboratories, plant roots associated with mycorrhizal fungi in field conditions or their natural environments encounter many other organisms, such as bacteria. These latter can induce neutral, positive, or negative interactions toward mycorrhizal fungi (Bonfante and Anca 2009). However, it was pointed out that mycorrhizal fungi would rather attract more bacterial allies than enemies (Frey-Klett and Garbaye 2005), comforting the idea that synergistic interaction is likely to happen. Evidence reporting positive outcomes from the combined inoculation of AMF and beneficial rhizosphere microbiome are growing, and have been well reviewed. For instance, some bacteria are able to stimulate AMF growth and mycorrhizal colonization rate: this is the case of some bacteria called Mycorrhiza Helper Bacteria, or MHB (Garbaye 1994; Frey-Klett et al. 2007).

Thus, assessing the effects of a tripartite interaction, between AMF, bacteria, and host plant, is a recent but a very interest-drawing issue. The underlying mechanisms in this triangle remain still uncovered, yet some elements have been brought to light (Deveau and Labbé 2016). The combined inoculation with AMF and bacteria was reported to have numerous positive effects, leading to enhanced plant nutritional status and growth as described in many reviews (Artursson et al. 2006; Nadeem et al. 2014); and also on disease protection, but to a lesser extent (Baysal and Silme 2019). However, reviews that focus essentially on the effects on plant disease protection in response to such interaction are still scarce. Thus, in the present review, we will target the bibliography on two types of soil beneficial microorganisms, the AMF, and the beneficial rhizospheric bacteria, and report the state-of-the-art knowledge about the potentials of the combined use of these latter in plant protection, with recent results concerning aerial or soil borne fungal diseases. More particularly, we aimed to provide a comparative study of the methodology employed to set up an experimental design, with the view to make available a practical "starters' guide" for those who initiate research projects on tripartite interactions. A more critical section is dedicated to highlight difficulties and future perspectives linked to the development of such a strategy in field experiments.

24.2 Beneficial Microorganisms in Plant Health

Plants are complex hubs of microscopic life and may suffer from pathogen attacks, as well as host several communities without any harm nor benefits, by actively gathering and selecting their own microbiome (Jones et al. 2019). It is now largely accepted that the plant physiology and performance can be affected by the host-associated microbial community; and more widely, plant ecology cannot be considered on its own anymore, but rather understood as a whole (with its associated organisms), called the holobiont (Vandenkoornhuyse et al. 2015; Agler et al. 2016). Among the microorganisms living in, or on the different compartments of the host, some maintain a mutualistic relationship, providing benefits for the plant. These include a wide variety of microorganisms- bacteria or fungi- and special attention was given for those living in the rhizosphere in the past century (Desai et al. 2016). They are referred as Plant Growth-Promoting Microorganisms/Rhizobacteria/Fungi, or PGPM/R/F (Lugtenberg and Kamilova 2009; Kundan et al. 2015). Yet the denomination is not always adapted, because, over time and discoveries, some of them turn out to both be able to stimulate plant growth and protect them from diseases; therefore, the term Plant-Health Promoting Microorganism (PHPM) is also interestingly used by some authors (Hayat et al. 2010).

Among the fungi, the upmost studied ones for their protective effect are, for example, some species of *Trichoderma, Aspergillus, Pythium, Fusarium, Coniothyrium* (Singh et al. 2018); but also the ectomycorrhizal fungi (Mohan et al. 2015), or the AMF (Comby et al. 2017), which will be detailed in this review. Among the bacteria, the most reported in the literature are mainly the strains of *Bacillus, Pseudomonas*, or symbiotic N_2 fixing bacteria (*Rhizobia* species), the free-living nitrogen-fixing PGPR such as some species of *Azospirillum, Azotobacter*, or *Burkholderia* (Desai et al. 2016).

24.2.1 Arbuscular Mycorrhizal Fungi (AMF)

24.2.1.1 Main Beneficial Effects of AMF on Plants

Among beneficial associations existing between plants and microorganisms, the arbuscular mycorrhizal symbiosis is probably the oldest relationship, back to more than 400 millions years ago (Karandashov and Bucher 2005), and the most widespread spatially and phylogenetically, throughout the plant kingdom (Wang and Qiu 2006). AMF are obligate biotrophic fungi from the phylum Glomeromycota, which are able to establish a symbiotic relation with over 80% of land plant species (Gianinazzi et al. 2010). This mutualistic interaction is based on nutrient exchanges between the plant and the fungus. While the host plant provides carbohydrates up to 20% of its fixed carbon (Parniske 2008), the AMF facilitate the uptake of water, and some major elements such as nitrogen (N), phosphorus (P), potassium (K), sulfur (S), or trace elements like iron, copper, and zinc (Smith and Read 2008; Lehmann and Rillig 2015; Wipf et al. 2019).

This interaction results in various beneficial outcomes at different scales and these latter were already widely reviewed by many authors (Miransari 2010; Gianinazzi et al. 2010; Rouphael et al. 2015). Briefly, these include, among many others: a better plant growth and development, a better stabilization of soil aggregates against erosion (Rillig and Mummey 2006; Bedini et al. 2009), an increased tolerance to abiotic stress (Lenoir et al. 2016; Begum et al. 2019), such as salt (Sannazzaro et al. 2006), drought (Doubková et al. 2013), or pollutants (Hildebrandt et al. 2007). AMF inoculation can also confer a better resistance to biotic stresses (Comby et al. 2017). Indeed, more specifically, the effectiveness of plant protection in response to AMF inoculation against fungal pathogens has been increasingly more reported, along with the interests to develop new alternatives to reduce chemical inputs.

In some cases, the inoculation with AMF alone can improve host tolerance to pathogens (Wehner et al. 2010), but studies about plant protection by AMF target predominantly soil-borne fungal pathogens. Classically, several direct or indirect modes of action are reported in many studies (Azcón-Aguilar et al. 2002; Vierheilig et al. 2008).

24.2.1.2 Underlying Mechanisms in Plant Protection by AMF

24.2.1.2.1 Direct Modes of Action Involved in Reducing Pathogen Development

There are several modes of action described in the literature, and some of them are specific for soil-borne pathogens. Indeed, in that case, the two protagonists may share the same space around the rhizosphere, and can therefore directly interact. Concerning direct mechanisms, competition for infection sites can occur. For example, a *Phytophthora* was unable to penetrate into the roots that were already containing *Funneliformis mosseae* arbuscules, seat of nutrient exchanges between plant and fungal partners, in the cortex cells, leading to a reduced disease severity. Authors concluded that effects on numbers of infection loci are one of the mechanisms leading to the protection of tomato against this pathogen (Vigo et al. 2000). On the contrary, it was shown that *Rhizophagus irregularis* inoculation could not reduce the development of *Ilyonectria liriodendra,* as AMF inoculation increased the abundance of the pathogen within grapevine rootstocks. Reciprocally, the presence of the pathogen resulted in increased AMF abundance compared to AMF inoculation alone (Holland et al. 2019). Authors hypothesized that plants with AMF alone may not have been placed in stressful conditions enough for mycorrhization, while the infection with the pathogen could have triggered a sufficient stress level to allow AMF colonization.

Competition for nutrient sources such as plant photosynthates can also be involved (Lerat et al. 2003), as both AMF and soil-borne pathogenic fungi may depend on the same carbon sources; however, limited data support this mechanism (Smith and Read 2008).

Other studies typically mention direct inhibition as a potential mode of action, but reports on the production of antimicrobial compounds by AMF are not common. Some works report unidentified antimicrobial substances produced by the extraradical mycelium of an *R. irregularis*, reducing conidial germination of *Fusarium oxysporum* f. sp. *chrysanthemi* (Filion et al. 1999). Antibiosis is often cited, but more supposed than demonstrated in literature (Cameron et al. 2013; Schouteden et al. 2015).

24.2.1.2.2 Indirect Modes of Action Involved in Reducing Pathogen Development

Still in the context of soil-borne pathogen—AMF interaction, indirect modes of action that are mediated *via* the host plant can also explain the reduction of pathogen development. Morphological changes of the root in response to AMF are indeed likely to be involved, but with contrasting results. On the one hand, authors observed that an increased root lignification induced by AMF could slow down the penetration by the pathogen (Dugassa et al. 1996); while on the other hand, increased root-branching induced by AMF provided new infection sites for the pathogen (Norman et al. 1996).

Also, main reviews suggest that a better protection is correlated with a compensation of damages by an increased biomass, thanks to an improved nutrient status (Azcón-Aguilar et al. 2002). Research articles putting specifically this mechanism under light in plant–fungal pathogens context are actually scarce, on the contrary to those in the context of herbivory. Indeed, compensation regrowth was, for example, evidenced especially for mycorrhizal tallgrass prairie plants, in response to grasshopper defoliation (Kula et al. 2005). However, this hypothesis was invalidated in some cases, where the enhancement of *Astragalus adsurgens* biomass by AMF was also shown to increase susceptibility against powdery mildew (Liu et al. 2018).

Another indirect mechanism could occur through the activation of plant defenses, which can interestingly lead to resistance against foliar fungal pathogens. The induction of plant defense mechanisms upon AMF colonization causes significant transcriptional and hormonal changes (Pieterse et al. 2014), leading to Mycorrhiza-Induced Resistance or MIR (Jung et al. 2012). This establishes a local or systemic resistance, in the roots and/or on the aerial parts (Liu et al. 2007). As an example, an inoculation of *F. mosseae* on tomato caused a local accumulation of phenolic compounds in roots in response to *Phytophthora nicotianae* var. *parasitica* attack revealed by a strong autofluorescence; leading to a reduced disease severity compared to control (Cordier et al. 1996). Song and collaborators showed that mycorrhizal inoculation alone did not impact the accumulation of transcripts of most defense-related genes tested in tomato plants. However, upon *Alternaria solani* attack on AMF-inoculated plants, authors noted a strong induction of three genes encoding pathogenesis-related proteins, PR1, PR2, and PR3, as well as other defense-related genes encoding lipoxygenases (LOX), allene oxide cyclase (AOC), and phenylalanine ammonia-lyase (PAL) in leaves. The induction of defense responses in AMF pre-inoculated plants was much higher and more rapid than that in un-inoculated plants in response to fungal challenge (Song et al. 2015), implying that systemic resistance was associated with a priming of plant defenses (Conrath et al. 2015). Another study also reported a systemic resistance of wheat against powdery mildew in AMF-inoculated plants, along with the upregulation of genes encoding a PAL, a chitinase, and a peroxidase, however only in the absence of the pathogen, suggesting that it was not related to a primed state of wheat leaves (Mustafa et al. 2017).

Most importantly, AMF-mediated protection can result from either one or a combination of different mechanisms cited above (Whipps 2004; Pozo et al. 2013). Multiple modes of actions can be an advantage in terms of crop protection, since if one mode of action is ineffective under specific conditions, the others can take over and maintain host protection. However, it also implies that protection efficacy can be inconsistent over time, since complex factors affect the pathosystem (Vierheilig et al. 2008).

24.2.2 Beneficial Bacteria: Definition Boundaries and Modes of Action

24.2.2.1 How to Define a Beneficial Bacteria?

On the contrary to AMF, there are no phylogenetic levels to gather all beneficial bacterial in one group; and terms to define them are prolific. The different names given for special bacteria aim to emphasize one feature of their life, such as (1) the spheres where they live, which is the case for free-living bacteria, root-associated ecto- or endophytes bacteria (Ryan et al. 2008), or even for AMF spore-associated bacteria (Bonfante 2003); (2) their beneficial functions that are known so far, such as the PGPR, able to stimulate plant growth (Compant et al. 2010), or Induced Systemic Resistance (ISR) bacteria able to stimulate plant resistance against pathogens and pests (Pieterse et al. 2014), or MHB able to facilitate mycorrhiza establishment (Deveau and Labbé 2016); or even bacteria with very specific abilities, like the case of phosphate-solubilizing bacteria (Kalayu 2019). Therefore, the more research is carried out, the more features are discovered, making sometimes overlay between them.

In the light of these facts, amalgams can rapidly be made, for example, by automatically associating endophytes as beneficial bacteria, however, the definition of endophytes does not specify their functional relationship, and they can exist also as latent pathogens, saprotrophs, or mutualistic associations (Fesel and Zuccaro 2016; Khare et al. 2018). Hence, in order to be as inclusive as possible within this review, we will define as "beneficial bacteria," not only bacteria positively impacting plant growth, but rather all those who have positive outcomes in any aspects within the tripartite interaction, i.e., on plant (growth and defense), and AMF (growth and function).

24.2.2.2 Underlying Mechanisms in Plant Protection by Beneficial Bacteria

It is possible to class bacteria by their positive effects observed on plants (Lugtenberg and Kamilova 2009). Authors distinguish "direct" PGPR, which directly promotes growth in noninfectious context; and the "indirect" PGPR, which can result in growth promotion because they are able to reduce harm caused by phytopathogens, acting therefore as biocontrol agents. Among direct PGPR, many categories of bacteria were then defined, including "Biofertilizers," "Phytostimulators," "Stress Controllers," "Rhizoremediators," etc. (Lugtenberg and Kamilova 2009). Nonetheless, some bacteria can have multiple roles (Martínez-Viveros et al. 2010). Indeed, some bacteria such as *Azotobacter paspali* are well known to both fix N_2 and to produce plant growth factors like indole-3-acetic acid (IAA) (Abbass and Okon 1993); while others are able to both inhibit soil-borne pathogens development and to stimulate plant growth as it was shown for *Pseudomonas oryzihabitans* in rice seedlings (Verma et al. 2018). Therefore, from one bacterial activity (such as metabolite production or nutrient uptake), many effects can be observed; and conversely, one effect can result from an association of many activities, making the classification difficult. Regarding plant protection issues against fungal pathogens only, we will focus on the mechanisms to direct and indirect effects toward the pathogen.

24.2.2.2.1 Direct Modes of Action Involved in Reducing Pathogen Development

Bacteria are able to produce metabolites, which are small molecular weight molecules (Davies 2013). There is a vast chemical diversity, including amino acids, vitamins, pigments, hormones, organic acids, volatile organic compounds, etc. (Kanchiswamy et al. 2015; Singh et al. 2017). Among these metabolites, some turn out to have bioactive properties with direct antimicrobial activities and therefore can have antagonistic effects towards a pathogen. Classical compounds cited in the literature are, for example, hydrogen cyanide, phenazines, or pyrrolnitrin (Compant et al. 2010; Hayat et al. 2010). More recently, bacterial rhamnolipids or cyclic lipopeptides attracted interest because they were found to have multiple roles, such as both antimicrobial and plant defense eliciting effects (Borah et al. 2016; Mejri et al. 2018). Some bacteria also synthesize enzymes such as cellulases, chitinases or protein and lipid degrading enzymes that have the potential to lyse the cell walls of fungal pathogens (Kundan et al. 2015).

Another direct effect involves competition for nutrient sources and ecological niches, similarly to AMF modes of action cited above. Indeed, as bacteria can act as biofertilizers, they can directly provide readily available nutrients for the plant (Fuentes-Ramirez and Caballero-Mellado 2006). This is, for example, the case of Rhizobia nodules in leguminous plant roots such as pea, peanut, alfalfa. Atmospheric N_2 is converted by bacterial nitrogenases into ammonia, which can be used by the plant as a nitrogen source, in contrast to N_2 (Van Rhijn and Vanderleyden 1995). Also, low levels of phosphate can limit plant development, but some bacteria are able to solubilize phosphate from inorganic or organic bound phosphate with bacterial phosphatases and phytases, releasing an assimilable form of phosphorus for the plant (Rodríguez and Fraga 1999; Kalayu 2019). Therefore, uptake of nutrients of great biological interest can constitute a source of competition toward pathogens, in particular for those which require exogenous nutrients to germinate, such as amino acids for *Botrytis cinerea* (Yoder and Whalen 1975). The competition is not only for nitrogen but also for transition metals such as iron, zinc, or manganese (Fones and Preston 2013). Strains of *Pseudomonas* can produce siderophores like pyoverdine or pseudobactin that bind to iron and facilitate its assimilation. It may therefore limit iron availability for pathogens such as *F. oxysporum* sp., resulting in pathogen growth reduction (Duijff et al. 1994; Arya et al. 2018). In parallel, bacteria engage physical interactions with their close environment, which have several consequences in the context of pathogen growth limitation. Along the competition for nutrients, competition for niches also occurs around root colonization sites, and is globally known as CNN, as for "competition for nutrients and niches" (Latha et al. 2019). Pliego and collaborators demonstrated that two similar *Pseudomonas alcaligenes* strains, selected by their ability to colonize avocado roots, use strongly different root colonization strategies. This suggested that in addition to the total bacterial root colonization level, the sites occupied on the root are important for their protective level against avocado root rot caused by *Rosellinia necatrix* (Pliego et al. 2008).

Physical interactions also include modes of action such as hyperparasitism. It is the case of a special direct interaction between two organisms, in which one of them is gaining nutrients from the other. If the latter is considered as a parasite or a pathogen, the interaction is then defined as hyperparasitism or predation (Köhl et al. 2019). Bacterial hyperparasitism has not been widely documented. Indeed, references report mainly confrontation against pathogenic bacteria (McNeely et al. 2017) or parasitic nematodes, but it is not excluded that pathogenic fungi can also be targets.

24.2.2.2.2 Indirect Modes of Action Involved in Reducing Pathogen Development

Bacterial inoculation can also limit pathogen development by other indirect mechanisms, mediated by the host plant defense regulation. In some cases, interaction of bacteria or their metabolites with plant roots can lead to ISR (Pieterse et al. 2014). ISR-inducing bacteria produce elicitors, or Microbial-Associated Molecular Patterns (MAMPs), that involve a vast chemical diversity, including lipopolysaccharides, flagellins; but also metabolites such as 2,4-diacetylphlotoglucinol, pyocyanin (also having antimicrobial effects), N-acyl homoserine lactones, iron-regulated siderophores and biosurfactants (as reviewed in De Vleesschauwer and Höfte 2009). Moreover, volatile organic compounds are also able to elicit ISR, as it was previously reported for the volatiles 2R,3R-butanediol produced by a *Bacillus subtilis,* and a C13 volatile liberated by *Paenibacillus polymyxa* in *Arabidopsis* (Ryu et al. 2004; Lee et al. 2012).

Pathogen development can also be limited by other indirect mechanisms, involving plant growth and metabolism regulation. Many rhizospheric bacteria are known to produce hormones, such as auxins (in particular, IAA), ethylene, cytokinins, gibberellins, abscisic acid, salicylic acid (SA) and jasmonic acid (JA). These are either excreted for root uptake or hormonal balance manipulation in the host plant, to stimulate growth and alleviate stress responses (Tsukanova et al. 2017). Many of those hormones are involved in root/shoot growth, production of root exudates, but also in defense-related pathways (Rosier et al. 2018). As an example, *Paenibacillus lentimorbus* B-30488 was shown to produce aminocyclopropane-1-carboxylase (ACC) deaminase -an enzyme involved in reducing ethylene production in plants leading to an enhanced tolerance against southern blight disease in tomato caused by *Scelerotium rolfsii*. The inoculated plants showed a modulation of the ethylene pathway and antioxidant enzyme activities; while systemic tolerance could be confirmed by overexpression of defense-related genes such as *PR1*, *PR2*, *PR4*, or *PR7* (Dixit et al. 2016). However, hormonal changes induced by PGPR is not consistently correlated to an increased resistance toward a pathogen, as antagonistic signaling molecules such as JA, SA, and IAA are part of multiple crossroads. Although not a PGPR, but rather in the context of hormonal change inductors, the biotrophic pathogen *Pseudomonas syringae* pv. *tomato*, which induces SA-mediated defense, made *Arabidopsis* more susceptible to the necrotrophic pathogen, *Alternaria brassicicola*, by suppressing the JA signaling pathway (Spoel et al. 2007). These data suggest the possibility that a prior inoculation with a microorganism with one lifestyle might modulate, or even compromise the ability of the plant to defend itself against pathogens with another lifestyle, due to hormonal antagonism (Walters et al. 2013).

Finally, another indirect mechanism would be for the case of interactions between soil microorganisms and their impact on plant health. There is increasingly more evidence for bacterial strains reported as promoters of AMF symbiosis, called MHB (Garbaye 1994; Duponnois 2006; Frey-Klett et al. 2007). MHB promote the establishment of mycorrhizal symbiosis by several means, including: (1) the improvement of root receptiveness of the fungus; (2) the interference in plant-fungus recognition and symbiosis establishment; (3) the promotion of AMF propagule germination, as well as spore survival and mycelium growth; (4) the modification of soil chemical properties for better conduciveness to the fungus (for review: Deveau and Labbé 2016). So, MHB can have two levels of impacts: on the AMF itself, and on its biological activities, such as its protection abilities against plant diseases.

24.3 Plant Protection Against Fungal Diseases Using AMF and Bacteria Co-Inoculation: Several Scenarios and Possible Mechanisms

The concept of co-inoculation of plants with AMF and bacteria was first reviewed about 30 years ago (Garbaye 1994), and some authors have overviewed these experiments since then (Frey-Klett et al. 2007; Deveau and Labbé 2016). The cited literature was mostly focused on evaluating the impact of the co-inoculation on (1) the main stages of AMF life cycle in vitro (spore germination, hyphal elongation, root colonization) and (2) plant growth and development parameters such as aerial and root biomass, yield, mineral content (Saldajeno and Hyakumachi 2011b). Indeed, the literature showing results of protection efficacy against diseases in response to the combined inoculation of AMF and bacteria is still in short supply, and the underlying mechanisms are often suggested more than brought out, undoubtedly due to the complexity to lead these kinds of multifactorial complex experiments.

24.3.1 Common Base Grounds Between Studies

After gathering data from the last 30 years, studies mostly report protection assays of co-inoculation against soil-borne diseases. They represent more than 80% of research content cited so far in this review, against less than 20% focused on aerial-borne pathogens. Among these, more than half of the content is reporting efficacies against root rots or wilts, caused essentially by *Fusarium* or *Rhizoctonia* species, and to a lesser extent, against *Verticillium, Pythium,* or *Phytophthora* species. The AMF species frequently tested are *F. mosseae* (previously known as *Glomus mosseae*), *R. irregularis* (previously known as *Glomus intraradices*), *Glomus fasciculatum,* and some *Gigaspora* sp., as commercial inoculants or produced by the research laboratories themself. The bacterial inoculants are either sampled from the corresponding AMF spores, host plant or simply formulated as a commercial inoculant. The main bacteria which are used as inoculant belong to *Pseudomonas, Rhizobium,* or *Bacillus* genus. The main studies were gathered in Table 24.1 and were reported according to the efficacy of protection obtained in response to (1) the combined inoculation; compared to (2) the AMF alone and (3) to the bacteria alone. In this way, it is possible to understand which microorganism contributes the most to the resulting protection; or if the co-inoculation leads to a synergistic, or perhaps antagonistic type of interaction.

First, all of the articles found have in common that the dual inoculation protects significantly against the tested pathogen, compared to the double negative control (without any beneficial microorganism). Few cases of "negative" interactions are found in the literature, maybe and especially because these data are not accepted, or unfortunately not submitted due to complexity and variability of these interactions. Also, these articles have also in common that authors used at least already one well-protective microorganism (AMF or bacteria) in the interaction. But their combined level of protection resulted in different ways. Hence, with these different possibilities, we could categorize the studies according to three types of outcomes so far. For further sections, we defined as a protection "gain" when the co-inoculation resulted in better protection as compared to one of a single inoculant (AMF or bacteria).

Table 24.1 Table summarizing the main studies reported in the literature concerning plant co-inoculation with arbuscular mycorrhizal fungi (AMF) and beneficial bacteria for the control of fungal pathogens

Scenario	Disease (causal agent)	Host plant	AMF	Bacteria	AMF	Bacteria	AMF + Bacteria	References
"Full-gain"	Red crown rot (*Cylindrocladium parasiticum*)	Soybean (*Glycine max* L.)	*Funneliformis mosseae*	*Bradyrhizobium* sp. BXYD3	+	+	+++	Gao et al. (2012)
	Root rot disease (*Rhizoctonia solani*)	Common bean (*Phaseolus vulgaris* L.)	*Glomus sinuosum, Gigaspora albida*	*Pseudomonas fluorescens*	+	+	+++	Neeraj and Singh (2011)
	Root rot and wilt (*Fusarium chlamydosporum* and *Ralstonia solanacearum*)	Coleus (*Coleus forskohlii* Briq.)	*Glomus fasciculatum*	*Pseudomonas monteilii*	+	+	+++	Singh et al. (2013)
	Fusarium wilt (*Fusarium oxysporum*) root rot disease (*R. solani*)	Mung bean (*Vigna radiata* L. cv. Wilczek VA02)	*G. fasciculatum*	*P. fluorescens* strain VuPf1	+	+	+++	Basu and Santhaguru (2009)
	Fusarium wilt (*F. oxysporum* f. sp. Cubense)	Banana (*Musa acuminata* Colla cv. Grand Naine)	25 different isolates	29 isolates of AMF spore-associated bacteria	+	+	+++	Sumathi and Thangavelu (2016)
	Fusarium wilt (*F. oxysporum* FOPV001)	Papaya (*Carica papaya* L. cv. Maradol)	AMF complex (MTZ01) : *Rhizophagus irregularis, F. mosseae, Glomus etunicatum, G. albida*	*Pseudomonas* sp. (PPV3)	+	++	+++	Hernández-Montiel et al. (2013)
	Fusarium root rot (*Fusarium solani*)	Geranium (*Pelargonium graveolens* L'hér)	*F. mosseae*	*Bacillus subtilis*	+	++	+++	Haggag et al. (2001)

"Full-gain" (continued)	Charcoal rot (*Macrophomina phaseolina*)	Geranium (*P. graveolens* L'hér.)	*F. mosseae*	*B. subtilis*	+	++	+++	Haggag et al. (2001)
	Buckeye rot (*Phytophthora parasitica*)	Tomato (*Lycopersicon esculentum* Mill.)	*F.mosseae*	*Paenibacillus* sp. strain B2	++	+	+++	Budi et al. (1999)
	Fusarium wilt (*F. oxysporum* f. sp. *lycopersici*)	Tomato (*Lycopersicon esculentum* Mill. cv Pant Tomato-3)	*R. irregularis*	*P.fluorescens*	++	+	+++	Srivastava et al. (2010)
	Spring black stem and leaf spot (*Phoma medicaginis*)	Lucerne (*Medicago sativa* L.)	*F.mosseae*	*Sinorhizobium medicae*	++	+	+++	Gao et al. (2018)
	Basal stem rot (*Ganoderma boninense*)	Oil palm (*Elaeis guineensis* Jacq.)	*R. irregularis* UT126, *Glomus clarum* BR152B	*Pseudomonas aeruginosa* U	+	Not tested	+++	Sundram et al. (2015)
"No gain, no loss"	Pythium root damping-off (*Pythium ultimum*)	Cucumber (*Cucumeris sativus* L.)	*R. irregularis*	*Burkholderia cepacia*	+	+	+	Larsen et al. (2003)
	Verticillium wilt (*Verticillium dahliae*)	Strawberry (*Fragaria x ananassa* cv. Selva)	Vaminoc (commercial mix) : *Glomus caledonium, G. fasciculatum, F. mossae*	Commercial inoculum : *B. subtilis* FZB24	+	+	+	Tahmatsidou et al. (2006)
	Fusarium root rot (*F. solani*)	Common bean (*Phaseolus vulgaris* L.)	*F. mosseae*	*Rhizobium leguminosarum* pv. *phaseoli*	++	+	++	Hassan Dar et al. (1997)

(continued)

Table 24.1 (continued)

Scenario	Disease (causal agent)	Host plant	AMF	Bacteria	AMF	Bacteria	AMF + Bacteria	References
"Partial loss"	Fusarium wilt (*F. oxysporum* f. sp. *lycopersici*)	Tomato (*Lycopersicon esculentum* Mill.)	*R. irregularis*	*P. fluorescens, Pseudomonas putida, Enterobacter cloacae*	+	++	+	Akköprü and Demir (2005)
	Powdery mildew (*Sphaerotheca macularis*)	Strawberry (*Fragaria vesca* L. cv Elvira)	Vaminoc®:*G. caledonium, G. fasciculatum, F. mosseae*	Commercial inoculum : *B. subtilis* FZB24	+	++	+	Lowe et al. (2012)

The studies are classified according to several protection gain scenarios: "full-gain," "no loss, no gain," "partial loss." Protection rates assigned with different signs "+, ++ or +++" should be compared within the same study. A variable number of "+" signs suggests that protection rates are different between the conditions tested within the same case study: AMF alone or bacteria alone and AMF/bacteria co-inoculation. For the co-inoculation column, protection rates are assigned according to the gain over those obtained with simple inoculants: "+++" means a net gain in protection over those obtained compared to simple inoculation; "++" means a comparable level of protection, as the most protective of the two inoculants; "+" means a comparable level of protection, as the least protective of the two inoculants

24.3.2 "Full-Gain" Scenario

24.3.2.1 Evidence for Protection Gain with a Dual Inoculation

In this case, AMF or bacteria alone protects partially the plant against the disease, at the same or different levels, and the combined inoculation leads to a better protection level (slight or strong). Many authors used *F. mosseae* as AMF inoculant: (1) combined with a *Bacillus subtilis*, the protection was better against *Fusarium solani* or *Macrophomina phaseolina* of geraniums than with single inoculants (Haggag et al. 2001); (2) with a *Paenibacillus sp.* strain B2, the protection level against *Phytophtora parasitica* on tomato plants was improved (Budi et al. 1999); (3) with *Sinorhizobium medicae*, *Phoma medicaginis* development was better reduced on *Medicago sativa* compared to AMF or bacteria alone (Gao et al. 2018).

The combination of *P. fluorescens* with *R. irregularis* leads to a stronger protection of tomato against *F. oxysporum* f. sp. *lycopersici* than the microorganisms used alone (Srivastava et al. 2010). Similarly, Singh and collaborators observed that the combination of *Pseudomonas monteilii* with *Glomus fasciculatum* reduced the severity of root rot and wilt (*Fusarium chlamydosporum* and *Ralstonia solanacearum*) by 63% on *Coleus forskohlii*, while the AMF alone could already reduce by 56% (Singh et al. 2013). The same AMF in combination with *Pseudomonas fluorescens* strain VuPf1 also reduced disease severity caused by *F. oxysporum* or *Rhizoctonia solani* on mung bean plants, by increasing the plant vigour index by two times higher than the AMF alone (Basu and Santhaguru 2009).

Other studies tested a mix of different AMF inoculants with a bacteria. Co-inoculation of *Glomus sinuosum* and *Gigaspora albida* with *P. fluorescens* could better reduce root rot disease caused by *R. solani* on common bean plants, compared to microbial agents alone (Neeraj and Singh 2011). An AMF complex (MTZ01) composed of three species of *Glomus* and one *Gigaspora* combined with a *Pseudomonas sp.* (PPV3) could better protect against fusarium wilt on papaya plants (*F. oxysporum*), than AMF or bacteria used singly (Hernández-Montiel et al. 2013).

Another extensive work studied 25 different isolates of AMF and 29 AMF spore-associated bacteria against fusarium wilt of banana (*F. oxysporum* f. sp. *cubense*): to give an insight of their work, they observed for instance that *Glomus etunicatum* and *Pseudomonas aeruginosa* together could reduce disease severity by 60%, while inoculants alone reduce by 40 and 45%, respectively (Sumathi and Thangavelu 2016).

It is noteworthy that the protection efficacy is not always visible in any circumstances, because many factors may influence the interaction, such as fertirrigation and thus, nutrient bioavailability. Indeed, co-inoculation of a *Rhizobium* with *F. mosseae* in soybean in low P condition could reduce red crown rot (*Cylindrocladium parasiticum*) incidence by 89% compared to the double negative control, while the AMF or bacteria reduced it by 50% (Gao et al. 2012), thus representing more than 40% gain in protection level. Interestingly, this differential was not as strong in higher P condition. The authors suggest that high P availability has been demonstrated to play opposite roles in nodulation and mycorrhization as indicated by enhanced nodulation but suppressed mycorrhization with increasing P availability. They also found that P addition significantly decreased AMF colonization rate, implying that the increased disease incidence might be partly due to the decrease of AMF colonization rate. However, the link between AMF colonization rate and protection level might not always be correlated.

24.3.2.2 Possible Mechanisms Explaining a "Full-Gain" Scenario

In the scenario where protection rate is higher with co-inoculation than with one microorganism alone, it can be explained by these following hypotheses. On the one hand, bacteria and AMF may work independently from each other and the plant. In this case, bacteria can protect the plant in its way and AMF can add up supplemental protection rate in parallel. As it was shown in Sects. 24.2.1.2.1 and 24.2.2.2.1 beneficial microorganisms can exhibit direct and indirect effects against pathogens, and may, therefore, work independently from the presence of each other (Fig. 24.1). It is possible that two direct effects are involved (e.g., antimicrobial compounds of bacteria, combined with the competition for infections sites and nutrients by the AMF, against the soil-borne pathogen, see mechanisms in red Fig. 24.1); or that two indirect effects are additive (e.g., ISR and MIR building up stronger plant defense responses, see mechanisms in blue Fig. 24.1). However, few studies pointed out specifically independent and adding modes of action between microbial inoculants to explain a protection gain when co-inoculated, and these explanations still remain speculative and need further research to support this hypothesis. On the other hand, beneficial microorganisms can act in synergy with each other. As mentioned in Sect. 24.2.2.1, some bacteria are able to facilitate mycorrhizal establishment as MHB (see mechanism in purple, Fig. 24.1) and therefore, set up mutual beneficial interactions. The study mentioned earlier (Gao et al. 2018) showed that improved plant growth of lucerne resulted from a higher P and N uptake *via* the symbiotic *F. mosseae* and *S. medicae*, and a mutual promotion of arbuscular mycorrhizal fungal and rhizobial colonizations. AMF or rhizobium inoculation in the roots reduced the adverse effects of the pathogen *Phoma medicaginis*, which was correlated with enhanced defense-related enzymatic activities such as chitinase, β-1,3-glucanase or PAL, and with higher lignin or JA contents.

Alternatively, the co-inoculation with *Glomus* species and *P. aeruginosa* were reported to reduce the disease severity of basal stem rot disease (*Ganoderma boninense*) of palm seedlings by 80% (compared to negative control), while *Glomus* alone reduced it by 68%; but no difference in mycorrhizal colonization rate was observed with or without the bacteria (Sundram et al. 2015). So on the contrary to the previous study case, AMF colonization rate and the resulting protection are not always correlated, and alternative mechanisms can be involved.

In the context of possible indirect synergistic effects *via* plant metabolic regulations (see mechanisms in green Fig. 24.1), it is well known that the role of nitrogen (N) and its metabolism in plant immune defenses are fundamental (Fagard et al. 2014). It was reported that N starvation of tomato plants increased plant susceptibility to the necrotrophic pathogen *B. cinerea*, by partially impairing MIR

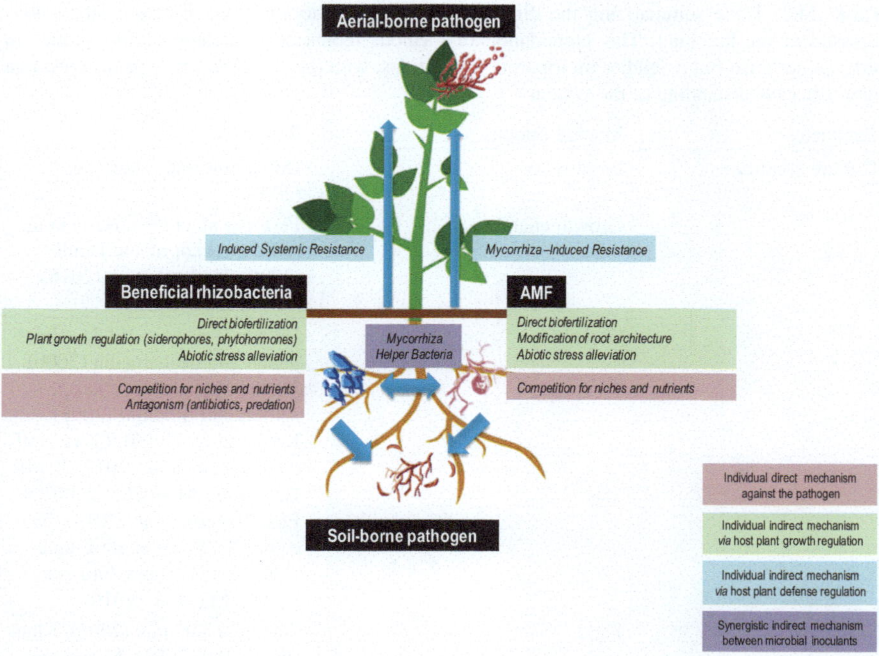

Fig. 24.1 Main mechanisms involved in reducing aerial- or soil-borne fungal pathogen developments, in response to a single or a combined inoculation of plants with arbuscular mycorrhizal fungi (AMF) and beneficial bacteria. For soil-borne pathogens, beneficial microorganisms can reduce pathogen development, thanks to their direct effect (mechanisms identified in red), such as competition for niches and nutrients, or antagonism. They also limit pathogen development *via* indirect mechanisms, that are mediated by the plant: either, through the regulation of the host plant growth (identified in green), or through the activation of plant defenses (identified in blue). In this latter case, mechanisms can be effective for both aerial- or soil-borne pathogens. Microbial inoculants can also interact with each other, as the example of bacteria called Mycorrhiza Helper Bacteria, able to facilitate mycorrhization (identified in purple). Co-inoculation of both beneficial microorganisms may result in controlling pathogen development *via* additive effect of direct or indirect individual mechanisms of each beneficial agent or synergistic effect of indirect mechanisms mediated by both agents

(Sanchez-Bel et al. 2016). Therefore, the combination with a N_2 fixing bacteria or a *Rhizobium*, may also explain how an enhanced nitrogen uptake would contribute to stronger MIR, leading to an overall improved resistance to pathogens.

Finally, root colonization by AMF is known to change the quality and the quantity of root exudates released in soil. And since these latter affect microbial populations in soil (Hage-Ahmed et al. 2013), it is not excluded that it can both inhibit the pathogen and also profit to beneficial bacteria at the same time. For example, on the one hand, it was shown that AMF-inoculated corn releases a benzoxazinoid, the 2,4-dihydroxy-7-methoxy-2 H-1,4-benzoxazin-3(4 H)-one (DI MBOA) in root exudates, enhancing the resistance to sheath blight caused by *R. solani* (Song et al. 2011). In parallel, this secondary metabolite with antimicrobial

Table 24.2 Table summarizing the different methods used to study the tripartite interactions reported in the literature. This compilation of methods represent a "methodological guide" in order to optimize future studies on tripartite interactions, with the aim to improve plant protection by a better understanding of the system

Parameters		Method options	References
Culture conditions		In vitro	Duponnois and Plenchette (2003)
		Growth chamber	Fester et al. (1999), Berta et al. (2005), Akköprü and Demir (2005), Hashem et al. (2016), Battini et al. (2017)
		Greenhouse	Hassan Dar et al. (1997), Abdel-Fattah and Mohamedin (2000), Jaizme-Vega et al. (2006), Akhtar and Siddiqui (2008), Kohler et al. (2009), Lowe et al. (2012), Gao et al. (2012, 2018), Hernández-Montiel et al. (2013), Flor-Peregrín et al. (2014), Xun et al. (2015), Pérez-de-Luque et al. (2017), Todeschini et al. (2018), Ma et al. (2019)
		Field	Tahmatsidou et al. (2006), Khan and ZAIDI (2007), Neeraj and Singh (2011), Singh et al. (2013), Cely et al. (2016), Bona et al. (2017), Espidkar et al. (2017), Aalipour et al. (2019), Jaffuel et al. (2019)
AMF	Inoculation method	Adding solid inoculum in the sowing/transplanting hole	Gao et al. (2018), Hernández-Montiel et al. (2013), Singh et al. (2013), Tahmatsidou et al. (2006), Abdel-Fattah and Mohamedin (2000), Akköprü and Demir (2005), Cely et al. (2016), Duponnois and Plenchette (2003), Khan and Zaidi (2007), Hassan Dar et al. (1997), Lowe et al. (2012), Neeraj and Singh (2011), Akhtar and Siddiqui (2008), Espidkar et al. (2017)
		Adding solid inoculum and mixing to the whole substrate	Gao et al. (2012), Sumathi and Thangavelu (2016), Battini et al. (2017), Fester et al. (1999), Hashem et al. (2016), Jaizme-Vega et al. (2006), Kohler et al. (2009), Todeschini et al. (2018), Xun et al. (2015), Berta et al. (2005), Aalipour et al. (2019)
		Adding liquid mycorrhizal spore suspension in sowing/planting spot	Pérez-de-Luque et al. (2017), Jaffuel et al. (2019)

(continued)

Table 24.2 (continued)

Parameters		Method options	References
		Adding liquid mycorrhizal spore suspension at planting + at transplanting to field	Bona et al. (2017)
		Adding inoculum with a seed coating material	Ma et al. (2019)
		Adding solid inoculum in holes at sowing + at transplanting	Flor-Peregrin et al. (2014)
	Control	Adding autoclaved AMF+ corresponding washing filtrate of the inoculum	Abdel-Fattah and Mohamedin (2000), Battini et al. (2017), Kohler et al. (2009)
		Adding washing filtrate of inoculum	Aalipour et al. (2019)
		Adding autoclaved AMF	Xun et al. (2015), Jaffuel et al. (2019)
		Adding matching volume of distilled water	Pérez-de-Luque et al. (2017)
		Adding non-mycorrhizal pieces of roots of host plant used for propagation	Duponnois and Plenchette (2003)
		Adding matching volume of substrate	Flor-Peregrin et al. (2014), Jaffuel et al. (2019)
		Adding the seed coating material without inoculum	Ma et al. (2019)
		No specific addition or unprecised	Khan and Zaidi (2007), Hassan Dar et al. (1997), Lowe et al. (2012), Neeraj and Singh (2011), Akhtar and Siddiqui (2008), Fester et al. (1999), Hashem et al. (2016), Jaizme-Vega et al. (2006), Todeschini et al. (2018), Berta et al. (2005), Espidkar et al. (2017), Bona et al. (2017)
	Detection	Trypan blue staining (Phillips and Hayman 1970) or chitin staining (Koske and Gemma 1989) + observations according to Trouvelot et al. (1986), Giovanetti and Mosse (1980) or McGonigle et al. (1990).	
Bacteria	Inoculation method	Adding once the bacterial suspension in the transplanting spot	Gao et al. (2018, 2012), Hernández-Montiel et al. (2013), Sumathi and Thangavelu (2016), Duponnois and Plenchette (2003), Hassan Dar et al. (1997), Pérez-de-Luque et al. (2017), Akhtar and Siddiqui (2008), Flor-Peregrin et al. (2014), Aalipour et al. (2019), Jaffuel et al. (2019), Bona et al. (2017), Ma et al. (2019)

(continued)

Table 24.2 (continued)

Parameters		Method options	References
		Immersing rooted seedlings in bacterial suspension before transplanting	Akköpü and Demir (2005), Tahmatsidou et al. (2006), Hashem et al. (2016), Lowe et al. (2012), Berta et al. (2005)
		Adding bacterial suspension at multiple times	Battini et al. (2017); Jaizme-Vega et al. (2006); Kohler et al. (2009); Todeschini et al. (2018)
		Incubating seeds in bacterial suspension before sowing	Fester et al. (1999), Khan and Zaidi (2007), Xun et al. (2015), Espidkar et al. (2017)
		Immersing stem cuttings in bacterial suspension before planting	Singh et al. (2013)
		Incubating seeds in bacterial suspension before sowing + adding suspension after planting	Cely et al. (2016)
		Mixing thoroughly a volume of bacterial suspension with substrate	Neeraj and Singh (2011)
	Resuspension solution for inoculation	Saline solution	Cely et al. (2016), Duponnois and Plenchette (2003), Hashem et al. (2016), Jaizme-Vega et al. (2006), Pérez-de-Luque et al. (2017), Todeschini et al. (2018), Berta et al. (2005), Bona et al. (2017), Ma et al. (2019)
		Culture medium	Akköprü and Demir (2005), Sumathi and Thangavelu (2016), Khan and Zaidi (2007), Akhtar and Siddiqui (2008), Flor-Peregrin et al. (2014)
		Distilled water	Tahmatsidou et al. (2006), Kohler et al. (2009), Lowe et al. (2012), Xun et al. (2015), Jaffuel et al. (2019), Espidkar et al. (2017)
		Unprecised	Fester et al. (1999), Aalipour et al. (2019)
	Control	Distilled water	Abdel-Fattah and Mohamedin (2000), Akhtar and Siddiqui (2008), Aalipour et al. (2019), Jaffuel et al. (2019)
		Culture medium	Khan and Zaidi (2007), Neeraj and Singh (2011), Flor-Peregrin et al. (2014)
		Saline solution	Duponnois and Plenchette (2003), Pérez-de-Luque et al. (2017), Todeschini et al. (2018), Ma et al. (2019)

(continued)

Table 24.2 (continued)

Parameters		Method options	References
		No specific addition or unprecised	Hassan Dar et al. (1997), Kohler et al. (2009), Lowe et al. (2012), Neeraj and Singh (2011), Fester et al. (1999), Hashem et al. (2016), Jaizme-Vega et al. (2006), Xun et al. (2015), Berta et al. (2005), Espidkar et al. (2017), Bona et al. (2017), Cely et al. (2016)
	Detection	Counting CFUs after plating on antibiotics supplemented media	Berta et al. (2005), Hernandez-montiel et al. (2013), Abdel-Fattah and Mohamedin (2000)
		P-solubilizing strains : Enrichment culture on NBRIP growth medium + counting CFUs with clear halo of P solubilization	Khan and Zaidi (2007)
		Rhizobium strains : Counting nodules numbers	Gao et al. (2018, 2012)
		GFP-tagged/fluorescent strains: CFUs counting after plating on media under UV	Pérez-de-Luque et al. (2017)
		DNA extraction + gel migration with PCR products (targeting 16S rRNA)	Ma et al. (2019)
		No assessment or unprecised	Xun et al. (2015), Cely et al. (2016), Jaizme-Vega et al. (2006), Kohler et al. (2009), Duponnois and Plenchette (2003), Hashem et al. (2016), Lowe et al. (2012), Todeschini et al. (2018), Akhtar and Siddiqui (2008), Flor-Peregrin et al. (2014)

properties was also shown to be involved in semiochemical recruitment of a beneficial bacteria *Pseudomonas putida* (Neal et al. 2012). Taken all together, these studies support the potential synergies between AMF and attraction of beneficial bacteria through changes in root exudates.

24.3.3 "Partial-Loss" Scenario

24.3.3.1 Evidence for a Partial Protection Loss with a Dual Inoculation

In this case, AMF or bacteria protects at same or different levels, but the protection level of their combination becomes lower than the greatest of the two (although still better than the double negative control). A commercial bacterial (*B. subtilis* FZB24) inoculation alone reduced powdery mildew (*Sphaerotheca aphanis*) severity by 56% on strawberry plants, while the commercial mix of *Glomus* (Vaminoc®) inoculation alone reduced disease severity by 13%. Co-inoculation did not improve protection rate and was similar to AMF alone (Lowe et al. 2012). Similarly, Akköpru and Demir observed that *R. irregularis* inoculation alone reduced fusarium wilt (*F. oxysporum* f. sp. *lycopersici*) on tomato plants by 17%, and *P. putida* by 58%, while the combination resulted in 30% of protection efficacy (Akköprü and Demir 2005). Apart from the practical viewpoint of losing disease suppression potential, this is a worthy outcome to know, as it might reveal antagonistic relationships between the AMF and the bacteria, which are usually less readily published.

24.3.3.2 Possible Mechanisms Explaining a "Partial-Loss" Scenario

A partial loss in protection with the co-inoculation, compared to the most effective microbial inoculant, can be explained by several ways. Firstly, in the case where a partial loss in protection is observed along with a partial decrease in AMF colonization rate in dual inoculation compared to AMF alone, this would probably suggest that a direct or indirect antagonistic effect from the bacteria against the AMF can be suspected. If a bacteria produces an antimicrobial compound able to limit a fungal pathogen development or is more competitive for root niches, it is not excluded that it can also have antagonistic effect against AMF, *via* its diffusible or volatile metabolites (Xavier and Germida 2003). Indeed, Akköpru and Demir (2005) observed that the combination of a *P. putida* with *R. irregularis* ended up in partial loss of protection efficacy compared to single inoculations, as it was correlated to a decrease of AMF root colonization rate in presence of this bacteria, with a 12% less colonization rate compared to AMF alone. Authors suggested that the inhibiting effects are thought to be related to a hypothetic secretion of antimicrobial substances toward AMF, although no analysis revealing the presence of such substances could unfortunately be pursued. However, they did demonstrated that their tested bacterial strain had a potential direct inhibitory effect against *F. oxysporum* in vitro, that was not associated to a siderophore effect: hence, we could question that, if this strain could exhibit an antagonistic activity toward a fungal pathogen, would it also be possible to exert it to another fungus, such as an AMF?

Another explanation is that some PGPR can lead to changes in the chemical composition and therefore structural properties of root cell walls (Vacheron et al. 2013). For example, the biocontrol agent *Bacillus pumilus* INR-7 is able to enhance lignin deposition in pearl millet epidermal tissues. These cell wall modifications induced by PGPR have been reported to limit pathogen development, by stimulating ISR plant defense responses, such as enhanced lignin synthesis and callose apposition (García-Gutiérrez et al. 2013). Therefore, if the bacteria colonizes the root cells faster than the AMF, it may induce root structural changes that would limit the penetration of other microorganisms, AMF included. However few data are available to support this hypothesis, but further studies would be worth carried out to examine such cases of antagonism. Conversely, if a partial loss in protection is observed with dual inoculation, but no decrease in AMF colonization rate is observed, it would either mean that (1) AMF limits the bacteria to exert its mode of action, thus lessening its efficacy against the pathogen; and/or (2) protection rate is not linearly correlated with AMF colonization rate. As stated above for bacteria, causes of such incompatibility may also find the source in root exudates or structural changes, as induced by AMF, but negatively impacting the inoculated bacteria.

Otherwise, microbial inoculants may act indirectly, *via* the metabolic regulation of the host plant (see mechanisms listed in green, Fig. 24.1). If the two microbial inoculants induce contradictory metabolic pathways, the biochemical antagonism may trouble defense signalization and response toward the pathogen. Lowe (2012) suggested that inoculation of *Glomus* species with *B. subtilis* may induce conflicting hormonal pathways (ex: SA *vs.* JA) in controlling powdery mildew on strawberry plant, because defense responses to biotrophic pathogens are generally associated with SA-hormonal pathway (Glazebrook 2005).

Indeed, as PGPR may promote plant growth on the one hand, while AMF can activate plant defenses (or vice-versa) on the other hand, this raises a fundamental question regarding the plant reactions to these stimuli. If a plant allocates its energy to its defense activation in response to a microbial inoculation, how can it ensure the supply of other biological functions, such as growth? This balance, or "trade-off," between growth and defense is an ongoing issue in plant immunity and physiology (Heil 2001; Bolton 2009). Many studies more focused on plant–pathogen interactions, then evidenced that the allocation of energetic resources for defense reactions are taking place at the expense of their "fitness". This latter, also called selective value, is described in evolutionary biology as the ability of an individual of a certain genotype to reproduce; and can be measured in many ways such as in number of fruits, or growth rate. Visible or not, systematic or not, the impacts of defense activation (by MIR or ISR) exist, as there are common metabolic pathways between growth and defense, mediated by several phytohormones such as SA, JA or IAA (Pieterse et al. 2009). And since some AMF can transiently lead to the accumulation of SA (García-Garrido and Ocampo 2002), it might act against the action of a bacteria, that was meant to stimulate IAA accumulation (Vacheron et al. 2013). Indeed, one typical example of hormonal crosstalks is between SA and IAA signaling (Wang and Wang 2014). It was shown in Arabidopsis plants, that SA does not directly affect auxin synthesis, but instead, inhibits plant responses regulated by auxin-dependant gene expression. Without SA, auxin is supposed to be perceived and linked to the auxin receptor TIR1 F-box protein. This complex is able to fix and degrade a class of repressors, called the auxin/indole-3-acetic-acid proteins

(AUX/IAA), that is meant to repress the expression of auxin-dependant genes. Therefore, its degradation frees the expression of auxin-dependant genes, leading to typical auxin responses. However with SA, the expression of TIR1 F-box is repressed: auxin cannot bind to its receptor, and no complex can degrade Aux/IAA proteins, which are then able to exert their role of repressor of auxin-dependant genetic transcription. At the end, no auxin-like responses can be observed in the presence of SA, independently from auxin concentration (Wang et al. 2007; Wang and Wang 2014; Huot et al. 2014).

24.3.4 "No-Gain, No-Loss" Scenario

24.3.4.1 Evidence for No Protection Gain with a Dual Inoculation

In that case, AMF or bacteria alone protects the plant against the disease at the same or different level, but the combined inoculation leads neither to an improvement, neither to a lower protection than the greatest of the two inoculants. The protection efficacy of a dual inoculation is therefore at the same level compared to the best obtained by one of the two microbial inoculants. This case was observed by Larsen and collaborators, where *R. irregularis* or *Burkholderia cepacia* could singly reduce *Pythium ultimum* population density at similar efficacy level, but the combined use did not significantly improve the pathogen control on cucumber plants (Larsen et al. 2003). Similarly, *F. solani* propagules number was reduced in response to *F. mosseae* inoculation, but was not more lowered in the presence of *Rhizobium leguminosarum* bv. *phaseoli* on common beans (Hassan Dar et al. 1997) and disease index was barely improved (0,04 unit less) by the addition of the bacteria.

Tahmatsidou and collaborators (2006) demonstrated that the protection level against *Verticillium* wilt on strawberry runners was so strong that, in any case, it was not possible to improve it any higher, as no pathogen could develop, in presence of the commercial inoculants Vaminoc® (AMF), *B. subtilis* FZB24, alone or in combination (Tahmatsidou et al. 2006).

24.3.4.2 Possible Mechanisms Explaining a "No-Gain, No-Loss" Scenario

No changes in protection level by adding a supplemental microbial inoculant is a complex scenario to explain. First of all, the above-cited studies reported that one of the microbial inoculants, particularly AMF, was already very effective in reducing pathogen development (Tahmatsidou et al. 2006), placing, therefore, the pathosystem in a context where there was only a tight scope for improvement. Indeed a total protection ensured by one of the microorganisms would mask a potential gain conferrable by the other one. In this context, it would be interesting to carry out preliminary tests to evaluate which conditions are optimal to obtain an intermediate protection rate (by exposing the plants to more stressful conditions, such as abiotic stress or higher pathogen pressure), and test an additional inoculation thereafter.

If no or slight changes are observed in terms of protection while an increased colonization rate of AMF or bacterial population is reported in co-inoculation, as it was reported by Hassan Dar and collaborators, it supports again that protection rate may not be always positively correlated to a quantitative presence of the microorganism (Hassan Dar et al. 1997). Similarly, within a tripartite interaction context, but apart from a plant–pathogen context-, no changes in date palm seedlings physiological parameters (root and leaf dry weight) was measured with a phosphate-solubilizing bacteria strain such as *Pseudomonas oryzihabitans* with *F. mosseae*, while the P content or the mycorrhizal colonization rate were enhanced compared to AMF alone (Boutheina et al. 2019). Another study also reported that papaya plant roots were better mycorrhized with *Glomus* species in presence of *B. coagulans* than *Glomus* alone, but no improvement was observed on plant height, fruits per plant, density of fruits (Mamatha et al. 2002). Then, it is not because an MHB increases mycorrhizal colonization rate, that the repercussion at phenotype scale will be an obligatory issue in response to mutualistic interactions at microscopic scale.

Finally, it also occurs that no change in microbial population density or colonization rates is observed, along without any gain in protection rates. In that case, one can suspect whether the bacterial inoculation was effective or not: is the mode of inoculation or the culture system suitable for the beneficial microorganism lifestyle? Did the latter survive along the experiment? If so, are the conditions favorable enough to reveal the best potential of microbial biological activity? Or on the contrary, are the conditions stressful enough to reveal differentials, as suggested by Nadeem and collaborators (Nadeem et al. 2014)? Indeed, several studies show that both individual biological activity and microbial interactions are submitted to the environmental factors such as pH or soil characteristics (Rousk et al. 2009; Ratzke and Gore 2018). It legitimately raises the issues on the factors influencing such interactions, and how to cope with them by testing the most suitable experimental design and methods.

24.4 Methodological Guide: How to Start Working on Tripartite Interactions ?

The aim of this following synthesis is to produce an overview of the main methods employed to study tripartite interactions, in order to facilitate the first step of experiment designing that would best answer to working hypotheses. This does not aim to compare the methods between them in order to rule out one method among others, which would surely ensure a high protection efficacy, as it is believed that there is no “ideal” method because every system is unique in such complexity.

We rather aim to compile what is possible to do, but also where are the lines of improvement, which would be interesting to explore for future research.

The different methods are summarized in Table 24.2 and are organized according to three main entries: cultivation design, AMF, and bacterial inoculation methods. For the last two sections, we decided to emphasize on the different reported modes of inoculation (frequency, time, mode of application), as well as which negative controls were used, and which protocols were applied to check microbial colonization or their presence at the end of the experiment.

Among the thirty references dealing with the combined inoculation of AMF and bacteria, half of them carried out the experiments in greenhouse conditions, one third in field conditions, one-sixth in growth chamber, and only one in vitro conditions, as far as we found.

24.4.1 Microbial Inoculation Methods

Concerning AMF inoculation, six principal means to bring the AMF inoculum to the system were identified. Equivalent number of publications reported adding solid inoculum, either by putting it right in the sowing/transplanting hole or by mixing it to the whole substrate. More recently, some research teams brought the inoculum by pouring spore suspension in the sowing or transplanting spot (Pérez-de-Luque et al. 2017; Jaffuel et al. 2019) or even by applying a specific coating material to the seeds in order to fix AMF inoculum thereafter (Ma et al. 2019). The volumes of inoculum brought to the soil are very variable between studies, and the concentration in propagule numbers constituting it are not always precise.

Concerning bacteria-related methods, seven principal means were identified. The vast majority reported adding a bacterial suspension in the transplanting or sowing hole at the beginning of the experiment (Hernández-Montiel et al. 2013; Gao et al. 2018). Other methods reported also immersing rooted seedlings in bacterial suspension before transplanting (Lowe et al. 2012; Hashem et al. 2016); incubating seeds in bacterial suspension before sowing (Xun et al. 2015; Espidkar et al. 2017); immersing stem cuttings in bacterial suspension before planting (Singh et al. 2013); or mixing thoroughly a volume of bacterial suspension with the substrate (Neeraj and Singh 2011). Finally, others bring bacterial suspension twice during the experiment, with a combination of some of the methods cited above, such as incubating the seeds in bacterial suspension and adding again at planting time (Cely et al. 2016).

In both cases, for AMF or bacterial inoculation, most of the studies reported the addition of the inoculum only once during the whole experiment (at sowing or transplanting), while a minority add it twice (at sowing, then at transplanting, for example).

24.4.2 *Choosing the "Mock-Inoculum"*

Consequently, the technique of microbial inoculation questions which negative control (or "mock"-inoculum) to use. There does not seem to have one kind of control used universally, especially for AMF mock inoculation.

Concerning AMF-associated controls, several types of controls were found in the framework of this review. Most predominantly, authors either (1) do mention that their negative control conditions consist of adding matching volume of substrate; or (2) do not mention anything specifically. Other kinds of more specific "mock-inoculum" were used: authors mentioned the addition of autoclaved AMF propagules with the view to provide same amount of organic matter, but with the inactivated AMF (Jaffuel et al. 2019), or of the washing filtrate of the inoculum in order to simulate equal microflora (Aalipour et al. 2019), or both (Battini et al. 2017). Authors also added a matching volume of distilled water, in correspondence to the AMF spore suspension prepared in water (Pérez-de-Luque et al. 2017), or by adding non-mycorrhizal pieces of roots of the host plant used for propagation, in correspondence to in vitro-produced mycorrhized roots placed in sowing holes (Duponnois and Plenchette 2003).

Concerning bacterial-associated controls, the options are less varied. Indeed, it is associated with which kind of solution was used to prepare bacterial suspension. In this way, if bacterial cells were resuspended in saline solution (usually 0.85% NaCl or 0.1 M $MgSO_4$) or fresh nutrient liquid medium, then mock-inoculated plants received the same volume of the corresponding saline solution (Todeschini et al. 2018) or liquid medium (Flor-Peregrín et al. 2014). However, it sometimes occurs that even if the resuspension method is described, the corresponding mock inoculation may not be explicitly mentioned.

It is possible that multiple controls (substrate only, AMF, and bacterial mocks) cannot be all carried out for obvious material reasons. However, preliminary experiments can be carried out in order to check whether the disease develops at the same level in every mock-condition. In that case, controls could be considered as equivalent in terms of disease protection, and the following experiments may include fewer "control" conditions for saving reasons.

24.4.3 *Colonization or Bacterial Viability Assessment Methods*

In some cases, protection rates are not always improved with a dual inoculation, compared to AMF or bacteria alone, especially results exhibit great variability between biological repetitions. The first checks made concern the microbial inoculation effectiveness, i.e., are AMF or bacteria alive, and did they successfully colonize the roots (or other compartments in cases of migration by bacteria) at the end of the experiment?

To check AMF colonization in roots, protocols are basically using a staining procedure, in the articles studying tripartite interactions such as Trypan blue staining (Phillips and Hayman 1970), followed by microscopic observations of fungal stuctures and counting (Giovanetti and Mosse 1980; McGonigle et al. 1990). However, apart from tripartite interactions, there are several other options to track and quantify AMF root colonization. A few of the following methods have been developed: AMF species-specific isoenzymes (Tisserant et al. 1998), genus or species-specific antibodies (Treseder and Allen 2002) and various nuclear (Van Tuinen et al. 1998; Lee et al. 2008) or mitochondrial DNA-based (Sarma et al. 2017) or RNA-based molecular methods such as the fragments of a RNA polymerase II gene (Thioye et al. 2019).

However, tracking and monitoring the presence of the bacteria in the culture system (plant organs or rhizospheric soil) is more difficult and challenging. Tracking refers to phenotypic or genotypic detection, quantification, and localization of inoculated PGPB strains, whereas monitoring includes tracking as well as determining the physiological activity of the inoculated PGPB strains, such as N-fixation, P-solubilization, and phytohormone production, among others.

Antibiotic-resistant strains have the advantage to be identified by dilution-plating of root buffer extracts, onto solid media supplemented with an antibiotic like rifampicin (Berta et al., 2005). Also, rhizobium strains on leguminous plants have the advantage to establish nodules, so that a simple visual counting gives a good information about successful bacterial colonization and the set up of the symbiosis (Gao et al. 2018, 2012). An extensive review was recently published to gather and sort the main methods for tracking and monitoring of bacteria into three main categories: (1) reporter genes-based methods, (2) immunological methods, and (3) nucleic acid-based methods (Rilling et al. 2019). The reporter gene is a gene attached to a regulatory sequence of a target gene that can be used to detect the presence and/or expression of the target. The reporter genes are usually identified by color, luminescence, or fluorescence, such as green fluorescent protein marking (Zhao et al. 2011).

24.5 Future Challenges Regarding Current Limits for Practical Use

Biocontrol, and more specifically, the use of microbial inoculants could constitute one of the ways to set up a more sustainable agriculture. They possess several advantages when compared with chemical agricultural inputs (mineral fertilizers, conventional pesticides). Indeed, along with a pre-selection of the most efficient microorganism, there is a potential reduced risk on environmental and human health. They are safer to apply, with potential multiple modes of actions, and also are able to multiply if inoculated in appropriate conditions and may survive to the next season (Berg 2009). There is strong evidence that microorganisms play a central role in plant health. However, studying several living organisms at the same time, which can be affected each differently by a slight change, may be challenging, even in controlled conditions. In this section, we will identify the different factors that can influence tripartite interactions: this will therefore reveal the limits of their use, but more importantly, list out the potential research axis to work on for their better use.

Microbial inoculant efficacy in plant protection depends on quite a number of factors, including: (1) biotic factors such as disease pressure, crop type/plant species/ genotype (Zhao et al. 2016), trophic lifestyle of the pathogen (Ojiambo and Scherm 2006), AMF species in relation to their aggressiveness toward plant colonization (Martinez-Medina et al. 2009); (2) abiotic factors, including climatic parameters such as temperature and relative humidity (De Curtis et al. 2012), seasonal variations (Escudero and Mendoza 2005), soil types or nutrient bioavailability (Baar et al. 2011); (3) agronomical or application-related factors included, for example, crop management, formulations of inoculants, mode and timing of applications of the tested strains (Fedele et al. 2020; Tabassum et al. 2017); (4) and finally industrial and socioeconomic situations, such as cost and profitability balance, biocontrol market, national or EU regulation for authorizations (Nicot et al. 2012).

Indeed, we should note that most of the studies were carried out in pots under controlled conditions, on artificial substrates in order to simplify the model with the view to unravel the mechanisms and the part of protection rate conferred by each of the microorganisms. But in natural field conditions, roots are confronted with a plethora of microorganisms, varying with the seasons and geographic locations. The applied inoculants may compete with naturally occurring microbial populations and be placed under complete unfavorable conditions to thrive and deploy all the mechanisms that could have been observed in laboratory. Therefore, it is legitimate to wonder how to fill the gap between the lab and the field, as under certain cases, the results obtained in field are not similar to those of laboratory (Smyth et al. 2011). The probability of working in compatible conditions can be very thin, to obtain a favorable context that matches for each component of the system (plant, inoculant 1, inoculant 2, pathogen, physicochemical characteristics of soil).

To cope with these issues, it would be interesting to isolate from the targeted soil, indigenous microbial strains which would be the most adapted to pedoclimatic conditions, or naturally the most competitive ones as compared to other native strains, or to exogenous strains (Khalid et al. 2004; Mahmoudi et al. 2017). Then, after isolation and study of their potential beneficial effects, authors underline the importance of the development process for further marketable products. Indeed, some Gram-positive bacteria are potential efficient biocontrol strains, but they are difficult to formulate because they do not produce spores of long shelf life: the process of dessication for the formulation is then compromising their practical use. Due to their persistence, Gram-negative bacteria can be preferred over and can last and stabilize more efficiently within a formulation (Tabassum et al. 2017).

As a conclusion, co-inoculation of AMF and beneficial bacteria is visibly worth using in some studies. Taken all the cited literature, one of the most studied or successful cases involve AMF and Rhizobia associations on legume plants, through their mutual biological stimulation in their interactions. Main technical locks reside in how to deliver viable, easily formulable microbial inoculants, that are both compatible with themselves, and with the environmental conditions in which there are brought to. Hence, robustness of results and transferability of the tests to the field can be hoped to improve, by a better understanding of the specific needs of each inoculant, alone and combined, with the plant. In the context of the upcoming of more and more variable and extreme abiotic stresses linked to climate change, priority in crop managements should be to promote plant health, by restoring soil health. Helping plants to cope with the stresses by the use of beneficial microorganisms is then, one of the many drivers for a more sustainable agriculture, along with integrated pest and diseases management strategies.

Acknowledgments The authors wish to thank the "Université du Littoral Côte d'Opale (ULCO)" for providing financial supports for Y. Krzyzaniak post-doctoral fellowship. This work has been carried out in the framework of TRIPLET project which was supported by the partnership A2U (Artois, UPJV, ULCO) and in the framework of CPER ALIBIOTECH project which was financed by European Union, French State and the French Region of Hauts-de-France.

Conflict of Interests The authors declare no conflict of interest.

References

Aalipour H, Nikbakht A, Etemadi N et al (2019) Biochemical response and interactions between arbuscular mycorrhizal fungi and plant growth promoting rhizobacteria during establishment and stimulating growth of Arizona cypress (Cupressus arizonica G.) under drought stress. Sci Hortic (Amsterdam) 261:108923. https://doi.org/10.1016/j.scienta.2019.108923

Abbass Z, Okon Y (1993) Plant growth promotion by Azotobacter paspali in the rhizosphere. Soil Biol Biochem 25:1075–1083. https://doi.org/10.1016/0038-0717(93)90156-6

Abdel-Fattah GM, Mohamedin AH (2000) Interactions between a vesicular-arbuscular mycorrhizal fungus (*Glomus intraradices*) and *Streptomyces coelicolor* and their effects on sorghum plants grown in soil amended with chitin of brawn scales. Biol Fertil Soils 32(5):401–409. https://doi.org/10.1007/s003740000269

Agler MT, Ruhe J, Kroll S et al (2016) Microbial hub taxa link host and abiotic factors to plant microbiome variation. PLoS Biol 14:e1002352. https://doi.org/10.1371/JOURNAL.PBIO.1002352

Akhtar MS, Siddiqui ZA (2008) Glomus intraradices, Pseudomonas alcaligenes, and Bacillus pumilus: effective agents for the control of root-rot disease complex of chickpea (Cicer arietinum L.). J Gen Plant Pathol 74:53–60. https://doi.org/10.1007/s10327-007-0062-4

Akköprü A, Demir S (2005) Biological control of Fusarium wilt in tomato caused by Fusarium oxysporum f. sp. lycopersici by AMF Glomus intraradices and some rhizobacteria. J Phytopathol 153:544–550. https://doi.org/10.1111/j.1439-0434.2005.01018.x

Alabouvette C, Olivain C, Steinberg C (2006) Biological control of plant diseases: the European situation. Eur J Plant Pathol 114:329–341

Artursson V, Finlay RD, Jansson JK (2006) Interactions between arbuscular mycorrhizal fungi and bacteria and their potential for stimulating plant growth. Environ Microbiol 8:1–10

Arya N, Rana A, Rajwar A et al (2018) Biocontrol efficacy of siderophore producing indigenous pseudomonas strains against Fusarium wilt in tomato. Natl Acad Sci Lett 41:133–136. https://doi.org/10.1007/s40009-018-0630-5

Azcón-Aguilar C, Jaizme-Vega MC, Calvet C (2002) The contribution of arbuscular mycorrhizal fungi to the control of soil-borne plant pathogens. In: Mycorrhizal technology in agriculture. Birkhäuser, Basel, pp 187–197

Baar J, Paradi I, Lucassen ECHET, Hudson-Edwards KA, Redecker D, Roelofs JGM, Smolders AJP (2011) Molecular analysis of AMF diversity in aquatic macrophytes: a comparison of oligotrophic and utraoligotrophic lakes. Aquat Bot 94(2):53–61. https://doi.org/10.1016/j.aquabot.2010.09.006

Bach E, Dos Santos Seger GD, de Carvalho Fernandes G et al (2016) Evaluation of biological control and rhizosphere competence of plant growth promoting bacteria. Appl Soil Ecol 99:141–149. https://doi.org/10.1016/j.apsoil.2015.11.002

Barratt BIP, Moran VC, Bigler F, van Lenteren JC (2018) The status of biological control and recommendations for improving uptake for the future. BioControl 63:155–167. https://doi.org/10.1007/s10526-017-9831-y

Basu M, Santhaguru K (2009) Impact of *Glomus fasciculatum* and fluorescent pseudomonas on growth performance of *Vigna radiata* (l.) wilczek challenged with phytopathogens. Journal of Plant Protection Research 49(2):190–194. https://doi.org/10.2478/v10045-009-0028-y

Battini F, Grønlund M, Agnolucci M et al (2017) Facilitation of phosphorus uptake in maize plants by mycorrhizosphere bacteria. Sci Rep 7:4686. https://doi.org/10.1038/s41598-017-04959-0

Baysal Ö, Silme RS (2019) The beneficial influence of microbial interactions on plant diseases and plant growth promoting effect. In: Mycorrhizosphere and pedogenesis. Springer, Singapore, pp 151–166

Bedini S, Pellegrino E, Avio L et al (2009) Changes in soil aggregation and glomalin-related soil protein content as affected by the arbuscular mycorrhizal fungal species Glomus mosseae and Glomus intraradices. Soil Biol Biochem 41:1491–1496. https://doi.org/10.1016/j.soilbio.2009.04.005

Begum N, Qin C, Ahanger MA et al (2019) Role of arbuscular mycorrhizal fungi in plant growth regulation: implications in abiotic stress tolerance. Front Plant Sci 10:1068. https://doi.org/10.3389/fpls.2019.01068

Benjamin EO, Wesseler JHH (2016) A socioeconomic analysis of biocontrol in integrated pest management: a review of the effects of uncertainty, irreversibility and flexibility. NJAS-Wageningen J Life Sci 77:53–60. https://doi.org/10.1016/j.njas.2016.03.002

Berg G (2009) Plant–microbe interactions promoting plant growth and health: perspectives for controlled use of microorganisms in agriculture. Appl Microbiol Biotechnol 84(1):11–18

Berta G, Sampo S, Gamalero E et al (2005) Suppression of Rhizoctonia root-rot of tomato by Glomus mossae BEG12 and Pseudomonas fluorescens A6RI is associated with their effect on the pathogen growth and on the root morphogenesis. Eur J Plant Pathol 111:279–288. https://doi.org/10.1007/s10658-004-4585-7

Bolton MD (2009) Primary metabolism and plant defense–fuel for the fire. Mol Plant-Microbe Interact 22:487–497. https://doi.org/10.1094/MPMI-22-5-0487

Bona E, Cantamessa S, Massa N et al (2017) Arbuscular mycorrhizal fungi and plant growth-promoting pseudomonads improve yield, quality and nutritional value of tomato: a field study. Mycorrhiza 27:1–11. https://doi.org/10.1007/s00572-016-0727-y

Bonfante P (2003) Plants, mycorrhizal fungi and endobacteria: a dialog among cells and genomes. Biol Bull 204:215–220. https://doi.org/10.2307/1543562

Bonfante P, Anca I-A (2009) Plants, mycorrhizal fungi, and bacteria: a network of interactions. Annu Rev Microbiol 63:363–383. https://doi.org/10.1146/annurev.micro.091208.073504

Borah SN, Goswami D, Sarma HK et al (2016) Rhamnolipid biosurfactant against fusarium verticillioides to control stalk and ear rot disease of maize. Front Microbiol 7:1505. https://doi.org/10.3389/fmicb.2016.01505

Boutheina Z-E, Aya H, Naima B, Ahmed N (2019) Responses of date palm seedling to co-inoculation with phosphate solubilizing bacteria and mycorrhizal arbuscular fungi. Int J Environ Agric Biotechnol 4:581–588. https://doi.org/10.22161/ijeab/4.2.43

Brauer VS, Rezende CP, Pessoni AM et al (2019) Antifungal agents in agriculture: friends and foes of public health. Biomol Ther 9:521. https://doi.org/10.3390/biom9100521

Budi SW, Van Tuinen D, Martinotti G, Gianinazzi S (1999) Isolation from the Sorghum bicolor mycorrhizosphere of a bacterium compatible with arbuscular mycorrhiza development and antagonistic towards soilborne fungal pathogens. Appl Environ Microbiol 65:5148–5150

Cameron DD, Neal AL, van Wees SCM, Ton J (2013) Mycorrhiza-induced resistance: more than the sum of its parts? Trends Plant Sci 18:539–545

Cely MVT, Siviero MA, Emiliano J et al (2016) Inoculation of Schizolobium parahyba with mycorrhizal fungi and plant growth-promoting Rhizobacteria increases wood yield under field conditions. Front Plant Sci 7:1708. https://doi.org/10.3389/fpls.2016.01708

Comby M, Mustafa G, Magnin-Robert M et al (2017) Arbuscular mycorrhizal fungi as potential bioprotectants against aerial phytopathogens and pests. In: Arbuscular mycorrhizas and stress tolerance of plants. Springer, Singapore, pp 195–223

Compant S, Clément C, Sessitsch A (2010) Plant growth-promoting bacteria in the rhizo- and endosphere of plants: their role, colonization, mechanisms involved and prospects for utilization. Soil Biol Biochem 42:669–678. https://doi.org/10.1016/J.SOILBIO.2009.11.024

Conrath U, Beckers GJM, Langenbach CJG, Jaskiewicz MR (2015) Priming for enhanced defense. Annu Rev Phytopathol 53:97–119. https://doi.org/10.1146/annurev-phyto-080614-120132

Cordier C, Gianinazzi S, Gianinazzi-Pearson V (1996) Colonisation patterns of root tissues by Phytophthora nicotianae var. Parasitica related to reduced disease in mycorrhizal tomato. Plant Soil 185:223–232

Davies J (2013) Specialized microbial metabolites: functions and origins. J Antibiot (Tokyo) 66:361–364

De Bona GS, Adrian M, Negrel J et al (2019) Dual mode of action of grape cane extracts against *Botrytis cinerea*. J Agric Food Chem 67:5512–5520. https://doi.org/10.1021/acs.jafc.8b07098

De Curtis F, de Felice DV, Ianiri G, De Cicco V, Castoria R (2012) Environmental factors affect the activity of biocontrol agents against ochratoxigenic Aspergillus carbonarius on wine grape. Int J Food Microbiol 159(1):17–24. https://doi.org/10.1016/j.ijfoodmicro.2012.07.023

De Vleesschauwer D, Höfte M (2009) Rhizobacteria-induced systemic resistance. In: Van Loon L (ed) Advances in botanical research, vol 51, pp 223–281

del Jaizme-Vega CM, Rodríguez-Romero AS, Antonio L, Núñez B (2006) Effect of the combined inoculation of arbuscular mycorrhizal fungi and plant growth-promoting rhizobacteria on papaya (Carica papaya L.) infected with the root-knot nematode Meloidogyne incognita. Effect of the combined inoculation of arbuscular mycorrhiz. Fruits 61:162. https://doi.org/10.1051/fruits:2006013

Desai S, Kumar GP, Daniel Amalraj L et al (2016) Exploiting PGPR and AMF biodiversity for plant health management. In: Microbial inoculants in sustainable agricultural productivity, Research perspectives, vol 1. Springer, India, New Delhi, pp 145–160

Deveau A, Labbé J (2016) Mycorrhiza helper bacteria. In: Martin F (ed) Molecular mycorrhizal symbiosis. John Wiley, New York, pp 437–450

Dixit R, Agrawal L, Gupta S et al (2016) Southern blight disease of tomato control by 1-aminocyclopropane-1-carboxylate (ACC) deaminase producing *Paenibacillus lentimorbus* B-30488. Plant Signal Behav 11:e1113363. https://doi.org/10.1080/15592324.2015.1113363

Doubková P, Vlasáková E, Sudová R (2013) Arbuscular mycorrhizal symbiosis alleviates drought stress imposed on Knautia arvensis plants in serpentine soil. Plant Soil 370:149–161. https://doi.org/10.1007/s11104-013-1610-7

Dugassa GD, Von Alten H, Schönbeck F (1996) Effects of arbuscular mycorrhiza (AM) on health of Linum usitatissimum L. infected by fungal pathogens. Plant Soil 185:173–182

Duijff BJ, Bakker PAHM, Schippers B (1994) Suppression of fusarium wilt of carnation by pseudomonas putida WCS358 at different levels of disease incidence and iron availability. Biocontrol Sci Tech 4:279–288. https://doi.org/10.1080/09583159409355336

Duponnois R (2006) Bacteria Helping Mycorrhiza Development. In: Mukerji KG, Manoharachary C, Singh J (eds) Microbial Activity in the Rhizoshere, Soil Biology, vol 7. Springer, Heidelberg, Berlin, pp 297–310

Duponnois R, Plenchette C (2003) A mycorrhiza helper bacterium enhances ectomycorrhizal and endomycorrhizal symbiosis of Australian Acacia species. Mycorrhiza 13:85–91. https://doi.org/10.1007/s00572-002-0204-7

El Guilli M, Hamza A, Clément C et al (2016) Effectiveness of postharvest treatment with chitosan to control citrus green mold. Agriculture 6:12. https://doi.org/10.3390/agriculture6020012

Escudero V, Mendoza R (2005) Seasonal variation of arbuscular mycorrhizal fungi in temperate grasslands along a wide hydrologic gradient. Mycorrhiza 15(4):291–299. https://doi.org/10.1007/s00572-004-0332-3

Espidkar Z, Yarnia M, Ansari MH et al (2017) Differences in nitrogen and phosphorus uptake and yield components between barley cultivars grown under arbuscular mycorrhizal fungus and pseudomonas strains Co-inoculation in rainfed condition. Appl Ecol Environ Res 15:195–216. https://doi.org/10.15666/aeer/1504_195216

Fagard M, Launay A, Clément G et al (2014) Nitrogen metabolism meets phytopathology. J Exp Bot 65:5643–5656. https://doi.org/10.1093/jxb/eru323

FAO (2016) Integrated Pest Management (IPM). EU Com

Fedele G, Bove F, González-Domínguez E, Rossi V (2020) A Generic Model Accounting for the Interactions among pathogens, host plants, biocontrol agents, and the environment, with parametrization for Botrytis cinerea on Grapevines. Agronomy 10(2):222. https://doi.org/10.3390/agronomy10020222

Fesel PH, Zuccaro A (2016) Dissecting endophytic lifestyle along the parasitism/mutualism continuum in Arabidopsis. Curr Opin Microbiol 32:103–112

Fester T, Maier W, Strack D (1999) Accumulation of secondary compounds in barley and wheat roots in response to inoculation with an arbuscular mycorrhizal fungus and co-inoculation with rhizosphere bacteria. Mycorrhiza 8:241–246. https://doi.org/10.1007/s005720050240

Filion M, St-Arnaud M, Fortin JA (1999) Direct interaction between the arbuscular mycorrhizal fungus Glomus intraradices and different rhizosphere microorganisms. New Phytol 141:525–533. https://doi.org/10.1046/j.1469-8137.1999.00366.x

Flint ML, Dreistadt SH, Clark JK, University of California Integrated Pest Management Program (1998) Natural enemies handbook : the illustrated guide to biological pest control. UC Division of Agriculture and Natural Sciences

Flor-Peregrín E, Azcón R, Martos V et al (2014) Effects of dual inoculation of mycorrhiza and endophytic, rhizospheric or parasitic bacteria on the root-knot nematode disease of tomato. Biocontrol Sci Tech 24:1122–1136. https://doi.org/10.1080/09583157.2014.925091

Fones H, Preston GM (2013) The impact of transition metals on bacterial plant disease. FEMS Microbiol Rev 37:495–519. https://doi.org/10.1111/1574-6976.12004

Frey-Klett P, Garbaye J (2005) Mycorrhiza helper bacteria: a promising model for the genomic analysis of fungal-bacterial interactions. New Phytol 168:4–8. https://doi.org/10.1111/j.1469-8137.2005.01553.x

Frey-Klett P, Garbaye J, Tarkka M (2007) The mycorrhiza helper bacteria revisited. New Phytol 176:22–36. https://doi.org/10.1111/j.1469-8137.2007.02191.x

Fuentes-Ramirez LE, Caballero-Mellado J (2006) Bacterial biofertilizers. In: PGPR: biocontrol and biofertilization. Springer, Netherlands, pp 143–172

Gao X, Lu X, Wu M et al (2012) Co-inoculation with rhizobia and AMF inhibited soybean red crown rot: from field study to plant defense-related gene expression analysis. PLoS One 7: e33977. https://doi.org/10.1371/journal.pone.0033977

Gao P, Li Y, Guo Y, Duan T (2018) Co-inoculation of lucerne (Medicago sativa) with an AM fungus and a rhizobium reduces occurrence of spring black stem and leaf spot caused by Phoma medicaginis. Crop Pasture Sci 69:933. https://doi.org/10.1071/CP18135

Garbaye J (1994) Helper bacteria: a new dimension to the mycorrhizal symbiosis. New Phytol 128:197–210. https://doi.org/10.1111/j.1469-8137.1994.tb04003.x

García-Garrido JM, Ocampo JA (2002) Regulation of the plant defence response in arbuscular mycorrhizal symbiosis. J Exp Bot 53:1377–1386. https://doi.org/10.1093/jexbot/53.373.1377

García-Gutiérrez L, Zeriouh H, Romero D et al (2013) The antagonistic strain Bacillus subtilisUMAF6639 also confers protection to melon plants against cucurbit powdery mildew by activation of jasmonate- and salicylic acid-dependent defence responses. Microb Biotechnol 6:264–274. https://doi.org/10.1111/1751-7915.12028

Gianinazzi S, Gollotte A, Binet MN et al (2010) Agroecology: the key role of arbuscular mycorrhizas in ecosystem services. Mycorrhiza 20:519–530

Gill KH, Garg H (2014) Pesticides: environmental impacts and management strategies. In: Pesticides-toxic aspects. InTech, Croatia, pp 187–230

Giovanetti M, Mosse B (1980) An evaluation of techniques for measuring vesicular arbuscular mycorrhizal infection in roots. New Phytol 84(3):489–500. https://doi.org/10.1111/j.1469-8137.1980.tb04556.x

Glazebrook J (2005) Contrasting mechanisms of defense against biotrophic and necrotrophic pathogens. Annu Rev Phytopathol 43:205–227. https://doi.org/10.1146/annurev.phyto.43.040204.135923

Hage-Ahmed K, Moyses A, Voglgruber A et al (2013) Alterations in root exudation of intercropped tomato mediated by the Arbuscular Mycorrhizal fungus *Glomus mosseae* and the soilborne pathogen *Fusarium oxysporum* f.sp. *lycopersici*. J Phytopathol 161:763–773. https://doi.org/10.1111/jph.12130

Haggag WM, Faten M, Faten A (2001) Interaction between vesicular arbuscular mycorrhizae and antagonistic biocontrol micro-organisms on controlling root-rot disease incidence of geranium plants. J Biol Sci 1:1147–1153. https://doi.org/10.3923/jbs.2001.1147.1153

Hashem A, Abd_Allah EF, Alqarawi AA et al (2016) The interaction between arbuscular mycorrhizal fungi and endophytic bacteria enhances plant growth of Acacia gerrardii under salt stress. Front Microbiol 7:1089. https://doi.org/10.3389/fmicb.2016.01089

Hassan Dar G, Zargar MY, Beigh GM (1997) Biocontrol of Fusarium root rot in the common bean (Phaseolus vulgaris L.) by using symbiotic Glomus mosseae and rhizobium leguminosarum. Microb Ecol 34:74–80. https://doi.org/10.1007/s002489900036

Hayat R, Ali S, Amara U et al (2010) Soil beneficial bacteria and their role in plant growth promotion: a review. Ann Microbiol 60:579–598

Heil M (2001) The ecological concept of costs of induced systemic resistance (ISR). Eur J Plant Pathol 107:137–146. https://doi.org/10.1023/A:1008793009517

Heimpel GE, Mills NJ (2017) Biological control: ecology and applications. Cambridge University Press, Cambridge

Hernández-Montiel LG, Rueda-Puente EO, Cordoba-Matson MV, Holguín-Peña JR, Zulueta-Rodríguez R (2013) Mutualistic interaction of rhizobacteria with arbuscular mycorrhizal fungi and its antagonistic effect on *Fusarium oxysporum* in Carica papaya seedlings. Crop Prot 47:61–66. https://doi.org/10.1016/J.CROPRO.2013.01.008

Hildebrandt U, Regvar M, Bothe H (2007) Arbuscular mycorrhiza and heavy metal tolerance. Phytochemistry 68:139–146

Holland T, Bowen P, Kokkoris V et al (2019) Does inoculation with arbuscular mycorrhizal fungi reduce trunk disease in grapevine rootstocks? Horticulturae 5:61. https://doi.org/10.3390/horticulturae5030061

Huot B, Yao J, Montgomery BL, He SY (2014) Growth-defense tradeoffs in plants: a balancing act to optimize fitness. Mol Plant 7:1267–1287

Jaffuel G, Imperiali N, Shelby K et al (2019) Protecting maize from rootworm damage with the combined application of arbuscular mycorrhizal fungi, Pseudomonas bacteria and entomopathogenic nematodes. Sci Rep 9:3127. https://doi.org/10.1038/s41598-019-39753-7

Jones P, Garcia BJ, Furches A et al (2019) Plant host-associated mechanisms for microbial selection. Front Plant Sci 10:862

Jung SC, Martinez-Medina A, Lopez-Raez JA, Pozo MJ (2012) Mycorrhiza-induced resistance and priming of plant defenses. J Chem Ecol 38:651–664. https://doi.org/10.1007/s10886-012-0134-6

Kalayu G (2019) Phosphate solubilizing microorganisms: promising approach as biofertilizers. Int J Agron 2019:1–7

Kanchiswamy CN, Malnoy M, Maffei ME (2015) Chemical diversity of microbial volatiles and their potential for plant growth and productivity. Front Plant Sci 6:151

Karandashov V, Bucher M (2005) Symbiotic phosphate transport in arbuscular mycorrhizas. Trends Plant Sci 10:22–29. https://doi.org/10.1016/j.tplants.2004.12.003

Khalid A, Arshad M, Zahir ZA (2004) Screening plant growth-promoting rhizobacteria for improving growth and yield of wheat. J Appl Microbiol 96:473–480. https://doi.org/10.1046/j.1365-2672.2003.02161.x

Khan MS, ZAIDI A (2007) Synergistic effects of the inoculation with plant growth-promoting Rhizobacteria and an arbuscular mycorrhizal fungus on the performance of wheat. Turkish J Agric For 31:355–362

Khare E, Mishra J, Arora NK (2018) Multifaceted interactions between endophytes and plant: developments and prospects. Front Microbiol 9:2732

Köhl J, Kolnaar R, Ravensberg WJ (2019) Mode of action of microbial biological control agents against plant diseases: relevance beyond efficacy. Front Plant Sci 10:845

Kohler J, Hernández JA, Caravaca F, Roldán A (2009) Induction of antioxidant enzymes is involved in the greater effectiveness of a PGPR versus AM fungi with respect to increasing the tolerance of lettuce to severe salt stress. Environ Exp Bot 65:245–252. https://doi.org/10.1016/J.ENVEXPBOT.2008.09.008

Koske RE, Gemma JN (1989) A modified procedure for staining roots to detect VA mycorrhizas. Mycol Res 92(4):486–488

Kula AAR, Hartnett DC, Wilson GWT (2005) Effects of mycorrhizal symbiosis on tallgrass prairie plant-herbivore interactions. Ecol Lett 8:61–69. https://doi.org/10.1111/j.1461-0248.2004.00690.x

Kundan R, Pant G, Jadon N, Agrawal PK (2015) Plant growth promoting rhizobacteria: mechanism and current prospective. J Fertil Pestic 6:155. https://doi.org/10.4172/2471-2728.1000155

Larsen J, Ravnskov S, Jakobsen I (2003) Combined effect of an arbuscular mycorrhizal fungus and a biocontrol bacterium against Pythium ultimum in soil. Folia Geobot 38:145–154. https://doi.org/10.1007/BF02803147

Latha P, Karthikeyan M, Rajeswari E (2019) Endophytic bacteria: prospects and applications for the plant disease management. In: Plant health under biotic stress. Springer, Singapore, pp 1–50

Lee J, Lee S, Young JPW (2008) Improved PCR primers for the detection and identification of arbuscular mycorrhizal fungi. FEMS Microbiol Ecol 65:339–349. https://doi.org/10.1111/j.1574-6941.2008.00531.x

Lee B, Farag MA, Park HB et al (2012) Induced resistance by a long-chain bacterial volatile: elicitation of plant systemic defense by a C13 volatile produced by Paenibacillus polymyxa. PLoS One 7:e48744. https://doi.org/10.1371/journal.pone.0048744

Lehmann A, Rillig MC (2015) Arbuscular mycorrhizal contribution to copper, manganese and iron nutrient concentrations in crops - a meta-analysis. Soil Biol Biochem 81:147–158. https://doi.org/10.1016/j.soilbio.2014.11.013

Lenoir I, Fontaine J, Lounès-Hadj Sahraoui A (2016) Arbuscular mycorrhizal fungal responses to abiotic stresses: a review. Phytochemistry 123:4–15

Lerat S, Lapointe L, Gutjahr S et al (2003) Carbon partitioning in a split-root system of arbuscular mycorrhizal plants is fungal and plant species dependent. New Phytol 157:589–595. https://doi.org/10.1046/j.1469-8137.2003.00691.x

Li Y, Héloir M, Zhang X et al (2019) Surfactin and fengycin contribute to the protection of a *Bacillus subtilis* strain against grape downy mildew by both direct effect and defence stimulation. Mol Plant Pathol 20:1037. https://doi.org/10.1111/mpp.12809

Liu J, Maldonado-Mendoza I, Lopez-Meyer M et al (2007) Arbuscular mycorrhizal symbiosis is accompanied by local and systemic alterations in gene expression and an increase in disease resistance in the shoots. Plant J 50:529–544. https://doi.org/10.1111/j.1365-313X.2007.03069.x

Liu Y, Feng X, Gao P et al (2018) Arbuscular mycorrhiza fungi increased the susceptibility of astragalus adsurgens to powdery mildew caused by erysiphe pisi. Mycology 9:223–232. https://doi.org/10.1080/21501203.2018.1477849

Lowe A, Rafferty-McArdle SM, Cassells AC (2012) Effects of AMF- and PGPR-root inoculation and a foliar chitosan spray in single and combined treatments on powdery mildew disease in strawberry. Agric Food Sci 21:28–38. https://doi.org/10.23986/afsci.4997

Lugtenberg B, Kamilova F (2009) Plant-growth-promoting Rhizobacteria. Annu Rev Microbiol 63:541–556. https://doi.org/10.1146/annurev.micro.62.081307.162918

Ma Y, Látr A, Rocha I et al (2019) Delivery of inoculum of Rhizophagus irregularis via seed coating in combination with Pseudomonas libanensis for cowpea production. Agronomy 9:33. https://doi.org/10.3390/agronomy9010033

Mahmoudi N, Mahdhi M, Abdedaiem R et al (2017) Growth response and drought tolerance of lens culinaris inoculated with arbuscular mycorrhizal fungi. Int J Pure App Biosci 5:674–685. https://doi.org/10.18782/2320-7051.6063

Mamatha G, Bagyaraj DJ, Jaganath S (2002) Inoculation of field-established mulberry and papaya with arbuscular mycorrhizal fungi and a mycorrhiza helper bacterium. Mycorrhiza 12:313–316. https://doi.org/10.1007/s00572-002-0200-y

Martínez-Medina A, Pascual JA, Lloret E, Roldán A (2009) Interactions between arbuscular mycorrhizal fungi and Trichoderma harzianum and their effects on Fusarium wilt in melon plants grown in seedling nurseries. J Sci Food Agric 89(11):1843–1850. https://doi.org/10.1002/jsfa.3660

Martínez-Viveros O, Jorquera MA, Crowley DE et al (2010) Mechanisms and practical considerations involved in plant growth promotion by Rhizobacteria. J Soil Sci Plant Nutr 10:293–319. https://doi.org/10.4067/S0718-95162010000100006

McGonigle TP, Miller MH, Evans DG, Fairchild GL, Swan JA (1990) A new method which gives an objective measure of colonization of roots by vesicular-arbuscular mycorrhizal fungi. New Phytol 115(3):495–501. https://doi.org/10.1111/j.1469-8137.1990.tb00476.x

McNeely D, Chanyi RM, Dooley JS et al (2017) Biocontrol of Burkholderia cepacia complex bacteria and bacterial phytopathogens by Bdellovibrio bacteriovorus. Can J Microbiol 63:350–358

Mejri S, Siah A, Coutte F et al (2018) Biocontrol of the wheat pathogen Zymoseptoria tritici using cyclic lipopeptides from Bacillus subtilis. Environ Sci Pollut Res 25:29822–29833. https://doi.org/10.1007/s11356-017-9241-9

Miransari M (2010) Contribution of arbuscular mycorrhizal symbiosis to plant growth under different types of soil stress. Plant Biol 12:563–569

Mohan V, Nivea R, Menon S (2015) Evaluation of ectomycorrhizal fungi as potential bio-control agents against selected plant pathogenic fungi. J Acad Ind Res 3:408

Mustafa G, Khong NG, Tisserant B et al (2017) Defence mechanisms associated with mycorrhiza-induced resistance in wheat against powdery mildew. Funct Plant Biol 44:443–454. https://doi.org/10.1071/FP16206

Nadeem SM, Ahmad M, Zahir ZA et al (2014) The role of mycorrhizae and plant growth promoting rhizobacteria (PGPR) in improving crop productivity under stressful environments. Biotechnol Adv 32:429–448. https://doi.org/10.1016/J.BIOTECHADV.2013.12.005

Neal AL, Ahmad S, Gordon-Weeks R, Ton J (2012) Benzoxazinoids in root exudates of maize attract pseudomonas putida to the rhizosphere. PLoS One 7:e35498. https://doi.org/10.1371/journal.pone.0035498

Neeraj, Singh K (2011) Organic amendments to soil inoculated arbuscular mycorrhizal fungi and Pseudomonas fluorescens treatments reduce the development of root-rot disease and enhance the yield of Phaseolus vulgaris L. Eur J Soil Biol 47:288–295. https://doi.org/10.1016/J.EJSOBI.2011.07.002

Nicot PC, Alabouvette C, Bardin M, Blum B, Köhl J, Ruocco M (2012) Review of factors influencing the success or failure of biocontrol: technical, industrial and socio-economic perspectives. IOBC-WPRS Bulletin 78:95–98. http://ec.europa.eu/sanco_pesticides/

Norman JR, Atkinson D, Hooker JE (1996) Arbuscular mycorrhizal fungal-induced alteration to root architecture in strawberry and induced resistance to the root pathogen Phytophthora fragariae. Plant Soil 185:191–198

Ojiambo PS, Scherm H (2006) Mini-review biological and application-oriented factors influencing plant disease suppression by biological control: a meta-analytical review. Phytopathology 96 (11):1168–1174. https://doi.org/10.1094/PHYTO-96-1168

Parniske M (2008) Arbuscular mycorrhiza: the mother of plant root endosymbioses. Nat Rev Microbiol 6:763–775. https://doi.org/10.1038/nrmicro1987

Perazzolli M, Roatti B, Bozza E, Pertot I (2011) Trichoderma harzianum T39 induces resistance against downy mildew by priming for defense without costs for grapevine. Biol Control 58:74–82. https://doi.org/10.1016/j.biocontrol.2011.04.006

Pérez-de-Luque A, Tille S, Johnson I et al (2017) The interactive effects of arbuscular mycorrhiza and plant growth-promoting rhizobacteria synergistically enhance host plant defences against pathogens. Sci Rep 7:16409. https://doi.org/10.1038/s41598-017-16697-4

Phillips JM, Hayman DS (1970) Improved procedures for clearing roots and staining parasitic and vesicular-arbuscular mycorrhizal fungi for rapid assessment of infection. Trans Br Mycol Soc 55 (1):158–IN18. https://doi.org/10.1016/S0007-1536(70)80110-3

Pieterse CMJ, Leon-Reyes A, Van Der Ent S, Van Wees SCM (2009) Networking by small-molecule hormones in plant immunity. Nat Chem Biol 5:308–316

Pieterse CMJ, Zamioudis C, Berendsen RL et al (2014) Induced systemic resistance by beneficial microbes. Annu Rev Phytopathol 52:347–375. https://doi.org/10.1146/annurev-phyto-082712-102340

Pliego C, De Weert S, Lamers G et al (2008) Two similar enhanced root-colonizing pseudomonas strains differ largely in their colonization strategies of avocado roots and Rosellinia necatrix hyphae. Environ Microbiol 10:3295–3304. https://doi.org/10.1111/j.1462-2920.2008.01721.x

Pozo MJ, Jung SC, Martínez-Medina A et al (2013) Root allies: Arbuscular mycorrhizal fungi help plants to cope with biotic stresses. In: Aroca R (ed) Soil biology. Springer, Berlin, pp 289–307

Ratzke C, Gore J (2018) Modifying and reacting to the environmental pH can drive bacterial interactions. PLoS Biol 16:e2004248. https://doi.org/10.1371/journal.pbio.2004248

Rillig MC, Mummey DL (2006) Mycorrhizas and soil structure. New Phytol 171:41–53

Rilling JI, Acuña JJ, Nannipieri P et al (2019) Current opinion and perspectives on the methods for tracking and monitoring plant growth–promoting bacteria. Soil Biol Biochem 130:205–219. https://doi.org/10.1016/J.SOILBIO.2018.12.012

Rodríguez H, Fraga R (1999) Phosphate solubilizing bacteria and their role in plant growth promotion. Biotechnol Adv 17:319–339. https://doi.org/10.1016/S0734-9750(99)00014-2

Rosier A, Medeiros FHV, Bais HP (2018) Defining plant growth promoting rhizobacteria molecular and biochemical networks in beneficial plant-microbe interactions. Plant Soil 428:35–55

Rouphael Y, Franken P, Schneider C et al (2015) Arbuscular mycorrhizal fungi act as biostimulants in horticultural crops. Sci Hortic (Amsterdam) 196:91–108

Rousk J, Brookes PC, Bååth E (2009) Contrasting soil pH effects on fungal and bacterial growth suggest functional redundancy in carbon mineralization. Appl Environ Microbiol 75:1589–1596. https://doi.org/10.1128/AEM.02775-08

Rupp S, Weber RWS, Rieger D et al (2017) Spread of Botrytis cinerea strains with multiple fungicide resistance in German horticulture. Front Microbiol 7:2075. https://doi.org/10.3389/fmicb.2016.02075

Ryan RP, Germaine K, Franks A et al (2008) Bacterial endophytes: recent developments and applications. FEMS Microbiol Lett 278:1–9

Ryu CM, Farag MA, Hu CH et al (2004) Bacterial volatiles induce systemic resistance in Arabidopsis. Plant Physiol 134:1017–1026. https://doi.org/10.1104/pp.103.026583

Saldajeno MGB, Hyakumachi M (2011a) Arbuscular mycorrhizal interactions with Rhizobacteria or saprotrophic fungi and its implications to biological control of plant diseases. In: Fulton SM (ed) Mycorrhizal fungi: soil, agriculture and environmental implications. Nova Science Publishers, New York, pp 1–28

Saldajeno MGB, Hyakumachi M (2011b) The plant growth-promoting fungus Fusarium equiseti and the arbuscular mycorrhizal fungus Glomus mosseae stimulate plant growth and reduce severity of anthracnose and damping-off diseases in cucumber (Cucumis sativus) seedlings. Ann Appl Biol 159:28–40. https://doi.org/10.1111/j.1744-7348.2011.00471.x

Sanchez-Bel P, Troncho P, Gamir J et al (2016) The nitrogen availability interferes with mycorrhiza-induced resistance against Botrytis cinerea in tomato. Front Microbiol 7:1598. https://doi.org/10.3389/FMICB.2016.01598

Sannazzaro AI, Ruiz OA, Albertó EO, Menéndez AB (2006) Alleviation of salt stress in lotus glaber by Glomus intraradices. Plant Soil 285:279–287. https://doi.org/10.1007/s11104-006-9015-5

Sarma RK, Saikia R, Talukdar NC (2017) Mitochondrial DNA based molecular markers in arbuscular mycorrhizal fungi (amf) research. Springer, Cham, pp 243–250

Schouteden N, De Waele D, Panis B, Vos CM (2015) Arbuscular mycorrhizal fungi for the biocontrol of plant-parasitic nematodes: a review of the mechanisms involved. Front Microbiol 6:1280

Singh R, Soni SK, Kalra A (2013) Synergy between Glomus fasciculatum and a beneficial Pseudomonas in reducing root diseases and improving yield and forskolin content in Coleus forskohlii Briq. Under organic field conditions. Mycorrhiza 23:35–44. https://doi.org/10.1007/s00572-012-0447-x

Singh R, Kumar M, Mittal A, Mehta PK (2017) Microbial metabolites in nutrition, healthcare and agriculture. 3 Biotech 7(1):15

Singh S, Bhatnagar S, Choudhary S et al (2018) Fungi as biocontrol agent: an alternate to chemicals. In: Fungi and their role in sustainable development: current perspective. Springer, Singapore, pp 23–33

Smith S, Read D (2008) Mycorrhizal Symbiosis. Elsevier Ltd, Amsterdam

Smyth EM, McCarthy J, Nevin R, Khan MR, Dow JM, O'Gara F, Doohan FM (2011) *In vitro* analyses are not reliable predictors of the plant growth promotion capability of bacteria; a *Pseudomonas fluorescens* strain that promotes the growth and yield of wheat. J Appl Microbiol 111(3):683–692. https://doi.org/10.1111/j.1365-2672.2011.05079.x

Song YY, Cao M, Xie LJ et al (2011) Induction of DIMBOA accumulation and systemic defense responses as a mechanism of enhanced resistance of mycorrhizal corn (Zea mays L.) to sheath blight. Mycorrhiza 21:721–731. https://doi.org/10.1007/s00572-011-0380-4

Song Y, Chen D, Lu K et al (2015) Enhanced tomato disease resistance primed by arbuscular mycorrhizal fungus. Front Plant Sci 6:786. https://doi.org/10.3389/fpls.2015.00786

Spoel SH, Johnson JS, Dong X (2007) Regulation of tradeoffs between plant defenses against pathogens with different lifestyles. Proc Natl Acad Sci U S A 104:18842–18847. https://doi.org/10.1073/pnas.0708139104

Srivastava R, Khalid A, Singh USS, Sharma AKK (2010) Evaluation of arbuscular mycorrhizal fungus, fluorescent Pseudomonas and Trichoderma harzianum formulation against Fusarium oxysporum f. sp. lycopersici for the management of tomato wilt. Biol Control 53:24–31. https://doi.org/10.1016/j.biocontrol.2009.11.012

Sumathi S, Thangavelu R (2016) Co-inoculation of arbuscular mycorrhizal fungi (AMF) and their mycorrhizae helper bacteria (MHB) effectively suppresses fusarium wilt in banana. Plant Archives 16(1):365–375

Sundram S, Meon S, Seman IA, Othman R (2015) Application of arbuscular mycorrhizal fungi with Pseudomonas aeruginosa UPMP3 reduces the development of Ganoderma basal stem rot disease in oil palm seedlings. Mycorrhiza 25(5):387–397. https://doi.org/10.1007/s00572-014-0620-5

Tabassum B, Khan A, Tariq M, Ramzan M, Iqbal Khan MS, Shahid N, Aaliya K (2017) Bottlenecks in commercialisation and future prospects of PGPR. In: Applied Soil Ecology Vol. 121, Elsevier, pp 102–117. https://doi.org/10.1016/j.apsoil.2017.09.030

Tahmatsidou V, O'Sullivan J, Cassells AC et al (2006) Comparison of AMF and PGPR inoculants for the suppression of Verticillium wilt of strawberry (Fragaria×ananassa cv. Selva). Appl Soil Ecol 32:316–324. https://doi.org/10.1016/j.apsoil.2005.07.008

Thioye B, van Tuinen D, Kane A, de Faria SM, Ndiaye C, Duponnois R, Sylla SN, Bâ AM (2019) Tracing *Rhizophagus irregularis* isolate IR27 in *Ziziphus mauritiana* roots under field conditions. Mycorrhiza 29(1):77–83. https://doi.org/10.1007/s00572-018-0875-3

Tisserant B, Brenac V, Requena N et al (1998) The detection of Glomus spp. (arbuscular mycorrhizal fungi) forming mycorrhizas in three plants, at different stages of seedling development, using mycorrhiza-specific isozymes. New Phytol 138:225–239. https://doi.org/10.1046/j.1469-8137.1998.00112.x

Todeschini V, Ait Lahmidi N, Mazzucco E et al (2018) Impact of beneficial microorganisms on strawberry growth, fruit production, nutritional quality, and volatilome. Front Plant Sci 9:1611. https://doi.org/10.3389/fpls.2018.01611

Treseder KK, Allen MF (2002) Direct nitrogen and phosphorus limitation of arbuscular mycorrhizal fungi: a model and field test. New Phytol 155:507–515. https://doi.org/10.1046/j.1469-8137.2002.00470.x

Trouvelot A, Kough J, Gianinazzi-Pearson V (1986) Mesure du taux de mycorrhization VA d'un système radiculaire. Recherche de méthodes d'estimation ayant une signification functionnelle. In: Gianninazzi-Pearson V, Gianinazzi S (eds) Physiol Genet Asp mycorrhizae. INRA, Paris, pp 217–221

Trouvelot S, Héloir M-C, Poinssot B et al (2014) Carbohydrates in plant immunity and plant protection: roles and potential application as foliar sprays. Front Plant Sci 5:1–14. https://doi.org/10.3389/fpls.2014.00592

Tsukanova KA, Chebotar V, Meyer JJM, Bibikova TN (2017) Effect of plant growth-promoting Rhizobacteria on plant hormone homeostasis. S Afr J Bot 113:91–102

Vacheron J, Desbrosses G, Bouffaud ML et al (2013) Plant growth-promoting rhizobacteria and root system functioning. Front Plant Sci 4:356. https://doi.org/10.3389/fpls.2013.00356

Van Rhijn P, Vanderleyden J (1995) The rhizobium-plant symbiosis. Microbiol Rev 59:124–142

Van Tuinen D, Jacquot E, Zhao B et al (1998) Characterization of root colonization profiles by a microcosm community of arbuscular mycorrhizal fungi using 25s rDNA-targeted nested PCR. Mol Ecol 7:879–887. https://doi.org/10.1046/j.1365-294x.1998.00410.x

Vandenkoornhuyse P, Quaiser A, Duhamel M et al (2015) The importance of the microbiome of the plant holobiont. New Phytol 206:1196–1206

Verma SK, Kingsley K, Bergen M et al (2018) Bacterial endophytes from rice cut grass (Leersia oryzoides L.) increase growth, promote root gravitropic response, stimulate root hair formation, and protect rice seedlings from disease. Plant Soil 422:223–238. https://doi.org/10.1007/s11104-017-3339-1

Vierheilig H, Steinkellner S, Khaosaad T, Garcia-Garrido JM (2008) The biocontrol effect of mycorrhization on soilborne fungal pathogens and the autoregulation of the AM symbiosis: one mechanism, two effects? In: Mycorrhiza: state of the art, genetics and molecular biology, eco-function, biotechnology, eco-physiology, structure and systematics, 3rd edn. Springer, Berlin, pp 307–320

Vigo C, Norman JR, Hooker JE (2000) Biocontrol of the pathogen Phytophthora parasitica by arbuscular mycorrhizal fungi is a consequence of effects on infection loci. Plant Pathol 49:509–514. https://doi.org/10.1046/j.1365-3059.2000.00473.x

Walters DR, Ratsep J, Havis ND (2013) Controlling crop diseases using induced resistance: challenges for the future. J Exp Bot 64:1263–1280. https://doi.org/10.1093/jxb/ert026

Wang B, Qiu YL (2006) Phylogenetic distribution and evolution of mycorrhizas in land plants. Mycorrhiza 16:299–363

Wang W, Wang Z-Y (2014) At the intersection of plant growth and immunity. Cell Host Microbe 15:400–402. https://doi.org/10.1016/j.chom.2014.03.014

Wang D, Pajerowska-Mukhtar K, Culler AH, Dong X (2007) Salicylic acid inhibits pathogen growth in plants through repression of the auxin signaling pathway. Curr Biol 17:1784–1790. https://doi.org/10.1016/j.cub.2007.09.025

Wehner J, Antunes PM, Powell JR et al (2010) Plant pathogen protection by arbuscular mycorrhizas: a role for fungal diversity? Pedobiologia (Jena) 53:197–201. https://doi.org/10.1016/j.pedobi.2009.10.002

Whipps JM (2004) Prospects and limitations for mycorrhizas in biocontrol of root pathogens. In: Canadian Journal of Botany. NRC Research Press, Ottawa, Canada, pp 1198–1227

Wipf D, Krajinski F, Tuinen D et al (2019) Trading on the arbuscular mycorrhiza market: from arbuscules to common mycorrhizal networks. New Phytol 223:1127–1142. https://doi.org/10.1111/nph.15775

Xavier LJC, Germida JJ (2003) Bacteria associated with Glomus clarum spores influence mycorrhizal activity. Soil Biol Biochem 35:471–478. https://doi.org/10.1016/S0038-0717(03)00003-8

Xun F, Xie B, Liu S, Guo C (2015) Effect of plant growth-promoting bacteria (PGPR) and arbuscular mycorrhizal fungi (AMF) inoculation on oats in saline-alkali soil contaminated by petroleum to enhance phytoremediation. Environ Sci Pollut Res 22:598–608. https://doi.org/10.1007/s11356-014-3396-4

Yoder OC, Whalen ML (1975) Factors affecting postharvest infection of stored cabbage tissue by Botrytis cinerea. Can J Bot 53:691–699. https://doi.org/10.1139/b75-085

Zhao Q, Shen Q, Ran W et al (2011) Inoculation of soil by Bacillus subtilis Y-IVI improves plant growth and colonization of the rhizosphere and interior tissues of muskmelon (Cucumis melo L.). Biol Fertil Soils 47:507–514. https://doi.org/10.1007/s00374-011-0558-0

Zhao X, Yuan S, Song H, Su X, Mao H, Shen W, Qu X, Dong J (2016) Arbuscular mycorrhizal and dark septate fungal associations in riparian plants of the three gorges reservoir region, Southwest China. Aquat Bot 133:28–37. https://doi.org/10.1016/j.aquabot.2016.05.003

Chapter 25
Remediation of Toxic Metal-Contaminated Soil and Its Revitalisation with Arbuscular Mycorrhizal Fungi

Irena Maček

Abstract Soil is a finite resource and its preservation is essential for food security and our sustainable future. Soil contamination caused by the large amounts of pollutants is becoming a major problem on a global scale. In particular, toxic (heavy) metals are a big concern as they are non-degradable, persist in soil and accumulate in food webs. Soil that is heavily contaminated with toxic metals is referred to as a hazardous waste. As the world human population rise also urban agriculture is increasing and fulfilling diverse functions including food production. However, samples from urban areas are frequently exceeding safety standards for toxic metals concentration, which is particularly concerning for children as the most sensitive group to metal toxicity impacts. Thus it is of great importance to find good solutions for efficient restoration of degraded areas and toxic metal-contaminated soil. Recently, a novel and efficient chelant-based (EDTA—ethylenediamine tetraacetate) soil washing technology for remediation of toxic metal-contaminated soils has been developed. The innovative procedure is well suited for cleaning contaminated soils and has a potential to return the valuable resource—fertile soil—back into function. However, the harsh treatment of soil EDTA washing eliminates most of the biota in the remediated soil and significantly reduces its biological activity. It has been shown to be particularly harmful to soil filamentous fungi, including plant symbiotic arbuscular mycorrhizal fungi. After the treatment practically no arbuscular mycorrhizal fungal colonisation has been detectable in plant roots growing in remediated soil several months after the soil remediation treatment procedure, however latter functional mycorrhiza is established. The same was true for the molecular signal (arbuscular mycorrhizal fungal DNA) in the plants growing in remediated soil immediately after the treatment. Currently, there is very little knowledge on the composition, succession, function, and dynamics of

I. Maček (✉)
Department of Agronomy, Biotechnical Faculty, University of Ljubljana, Ljubljana, Slovenia

Department of Biodiversity, Faculty of Mathematics, Natural Sciences and Information Technologies (FAMNIT), University of Primorska, Koper, Slovenia
e-mail: Irena.Macek@bf.uni-lj.si

N. Shrivastava et al. (eds.), *Symbiotic Soil Microorganisms*, Soil Biology 60,
https://doi.org/10.1007/978-3-030-51916-2_25

microbial communities in the soil after the EDTA based remediation treatment. This chapter focuses on the importance of the integration of the role of microbial ecology in environmental sciences along with the application of the modern molecular techniques to improve our understanding of the succession of soil microbial communities and determinants of the survival of introduced microbial inocula in the remediated soil. Particular stress is put on the ubiquitous arbuscular mycorrhizal fungi, potential for the development of functional microbial communities, and following plant–soil feedbacks, including mycorrhiza, after the remediation procedure.

Keywords Abiotic stress · Arbuscular mycorrhizal fungi · Biodiversity · Contaminated soil · Heavy metals · Pollution · Remediation · Revitalisation · Soil · Soil ecology · Toxic metals

25.1 Introduction

Contamination of soils with toxic metals is a serious issue with soil being a finite resource, meaning its loss and degradation is not recoverable within a human lifespan. New soil formation is a slow process taking millennia for a few cm to form, thus it is of great importance to find good solutions for efficient restoration of degraded areas, including toxic (heavy) metals contaminated soils. In most industrialised countries as well as developing countries land contamination exists and is a growing problem. Waste disposal and treatment, together with industrial and commercial activities often result in local soil contamination. Toxic metals (As, Cd, Cr, Cu, Ni, Pb, Zn) are one of the major causes of concern as they are difficult to remove, are non-degradable, persist in environment for a long time and accumulate in food webs. Toxic metals are known to cause severe toxicity and represent a health hazard for people and animals, with children being the most endangered group (Jez and Lestan 2015). In the recent document (EEA 2020, The EU Environment, State and Outlook) EU Environmental Agency reports that in the EU-28[1], potentially polluting activities took place on an estimated 2.8 million sites, but only 24% of the sites are inventoried. Currently, only 28% of the registered sites are investigated, a prerequisite to deciding whether remediation is needed or not (Payá Pérez and Rodríguez Eugenio 2018). Considering the estimated extent of past and current pollution, and the uncertainties of reliable estimates, little progress has been made in the assessment and management of contaminated sites (EEA 2020, The EU Environment, State and Outlook). In a large part of these sites, toxic metals are the most important contaminants (EU Environmental Agency Report No. 1/2007). Soil

[1]EU-28: The European Union (EU) currently counts 27 EU countries. The UK withdrew from the European Union on 31 January 2020. The EEA (EU Environmental Agency) 2020 report still includes contaminated sites in the United Kingdom.

containing dangerous substances—including high concentrations of toxic metals—is considered a hazardous waste (EU Waste Catalogue & Hazardous Waste List, EPA 2002). EU is implementing the Landfill Directive (1999/31/EC) with heavy restriction of hazardous waste disposal which must be subject to remediation treatments.

Thirty of the 39 countries surveyed in the *'Progress in management of contaminated sites'* (European Environment Agency 2014) maintain comprehensive inventories for contaminated sites. Contaminated soil continues to be commonly managed using 'traditional' techniques, e.g. excavation and off-site disposal, which accounts for about one-third of management practices, however, also new technologies are evolving (e.g. Lestan 2017). Due to the severity of the heavy metal contamination problem several soil remediation processes have been developed. One possibility is EDTA (ethylenediaminetetraacetic acid) chelating agent extraction of the contaminants. The recently developed and modified procedure of EDTA soil washing has shown good results in metal removal from contaminated soil (e.g. Finzgar and Lestan 2007; Pociecha and Lestan 2012; Voglar and Lestan 2013) and therefore could be a cost-efficient and sustainable strategy for reclamation of contaminated urban gardens and farmlands. This procedure however has strong negative impact on soil microbial communities, with the filamentous arbuscular mycorrhizal fungi appearing to represent one of the most sensitive groups of the soil microbiota to both, chemical and mechanical disturbance during the remediation procedure (Maček et al. 2016b).

The objective of this chapter is to summarise the current knowledge on the subject of soil fungal diversity and ecology in metal-polluted and EDTA-washed soils before and after remediation. The emphasis is on a ubiquitous soil fungal group that forms arbuscular mycorrhiza, an ancient symbiosis between plants and arbuscular mycorrhizal fungi. Arbuscular mycorrhiza is present in roots of the large majority of terrestrial plant species in a wide range of ecosystems, including urban soils and agroecosystems. Within the chapter soil remediation using EDTA washing is presented along with its negative impacts on soil biota (Maček et al. 2016b). The importance of the new methodological development in the field of molecular ecology with its possible applications in monitoring of the succession and development of microbial communities and revitalisation of the soil after the remediation treatment is emphasised. Finally, future directions in microbial ecology research in these specific systems and advances in the use of the new high-throughput molecular techniques are presented, with an acknowledgement of the possibilities for widening the scope of research, which could include bioprospecting toxic metal-contaminated sites for industrially important microbes and research into potential toxic metal-tolerant fungal taxa (Maček et al. 2016a).

25.2 Soil Remediation Using EDTA (Ethylenediamine Tetraacetate) Soil Washing

In selecting appropriate remediation methods for a specific polluted site, characteristics of the soil and contaminants need to be considered (Lestan et al. 2008; Lestan 2017). The method of soil washing by EDTA (ethylenediamine tetraacetic acid) chelating agent and extraction of the contaminants has shown to result in high multimetal (Pb, Zn, Cd) removal efficiency (e.g. Finzgar and Lestan 2007; Pociecha and Lestan 2012; Voglar and Lestan 2013) especially from bioavailable and labile fractions (Jelusic and Lestan 2014), see Fig. 25.1. This innovative procedure promises a cost-efficient and sustainable strategy for the reclamation of contaminated urban areas and farmlands, also on a larger scale (Lestan 2017). No wastewater is generated and solid wastes are efficiently bitumen stabilised before disposal (Voglar and Lestan 2013, 2014).

The procedure successfully removes available forms of toxic metals and thus lowers the human and environmental hazards of the remediated soil, however, it also significantly diminishes soil microbial activity (Jelusic and Lestan 2014; Maček et al. 2016b). This has been shown by enzymatic tests (Jelusic and Lestan 2014; Kaurin et al. 2018; Kaurin and Lestan 2018), has been studied on soil fauna (Tica et al. 2013; Udovic and Lestan 2010), and in a pot experiment also on plant symbiotic arbuscular mycorrhizal fungi (Maček et al. 2016b). Thus far no trends (succession) of the remediated soil recovering its microbiological properties have been followed for a longer time-period and mainly short-term experiments have been done using remediated substrate (Maček et al. 2016b; Kaurin et al. 2018; Kaurin and Lestan 2018). It is known, however, that in addition to physical and chemical properties ecosystem services of soils largely depend on the diversity and activity of soil microbes (e.g. Jeffries et al. 2003), thus further research of this aspect is needed.

25.3 Arbuscular Mycorrhizal Fungi in Metal Contaminated Soil

Microbes represent the largest portion of biodiversity and biomass in soils. Healthy soils have high biodiversity that regulates ecosystem functions and processes, which lead to a great variety of ecosystem services. Understanding the mechanisms regulating the diversity and structure of microbial communities is urgently required for predicting the ecological impacts of rapidly changing environments. To be able to observe the effect remediation treatment has on the microbial communities, we first need to characterise the communities in the contaminated sites (original soil). Those communities however represent an interesting research object themselves. It is important to know how soil microorganisms respond to disturbance or environmental change (Griffiths and Philippot 2013) and at the extreme end of that is also soil that has been exposed to a long-term toxic metals contamination. Soils that are

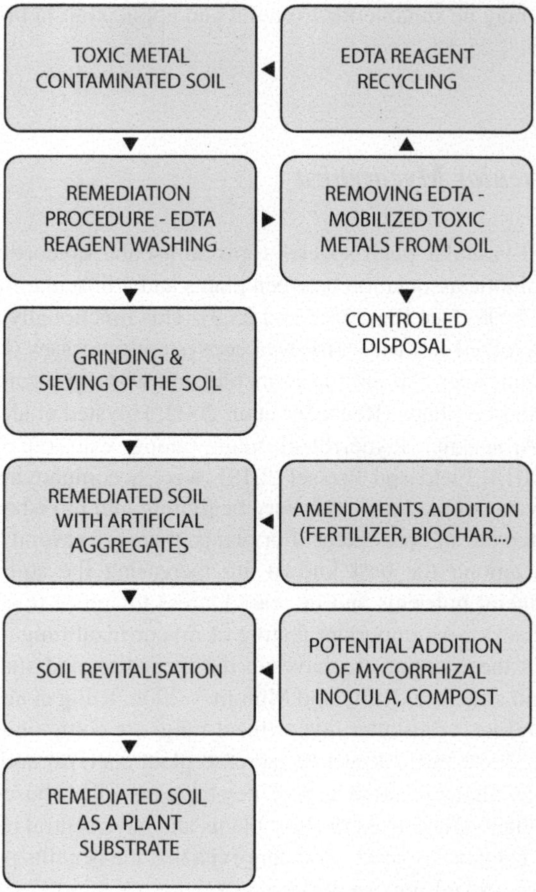

Fig. 25.1 A simplified scheme of the toxic metal-contaminated soil remediation procedure using EDTA (ethylenediamine tetraacetic acid) chelating agent soil washing and extraction of the contaminants. EDTA recycling loop is shown, returning used EDTA reagent solution in the remediation process. The soil remediation procedure has been fully described in Lestan (2017). Different amendments (e.g. compost, fertilizer, soil, mycorrhizal inocula, biochar) have been tested for better performance of the remediated soil to serve as a substrate for plant growth and for its revitalisation with microbiota (e.g. Jelusic et al. 2014; Maček et al. 2016b; Kaurin et al. 2018). Revitalisation of the EDTA-washed remediated soil with arbuscular mycorrhizal fungi has been tested for commercial inocula [Symbivit Remed, Symbiom Ltd., Czech Republic (See footnote 1)] and indigenous inoculum (rhizosphere soil with root particles) in Maček et al. (2016b), however further research on succession of the microbial communities in the newly formed soil substrate is needed

severely contaminated with toxic (heavy) metals still harbour very diverse, microbial communities including organisms that can support plant growth and development like symbiotic arbuscular mycorrhizal fungi. Some of the organisms inhabiting those soils may also have specific adaptations to high levels of contaminants in the environment and may be suitable for isolation and application in biotechnology and agriculture (see Fig. 25.4).

25.3.1 Arbuscular Mycorrhiza

About 80% of all vascular plant species form arbuscular mycorrhiza, which is an underground symbiotic association between plants and arbuscular mycorrhizal fungi (Smith and Read 2008) (see Figs. 25.2 and 25.3). This functionally important group of soil fungi is involved in many terrestrial ecosystem processes (Fitter 2005). The symbiosis is ancient, over 450 million years old, and was significant in enabling the colonisation of land by plants (Redecker et al. 2002; Hoysted et al. 2018; Field and Pressel 2018). Arbuscular mycorrhizal fungi, along with other fungal groups (Hoysted et al. 2018; Field and Pressel 2018), were accompanying plants in their transition from water to land from the very beginning and have been evolving in a range of diverse terrestrial ecosystems. For plants, there are several benefits of being mycorrhizal and among the best known are increasing the soil volume for the acquisition of mineral nutrients and increased stress tolerance (e.g. drought, pollution, pathogen attacks). An important feature of mycorrhizal fungal presence in soil is also the impact these organisms have on the formation and stabilisation of soil aggregates and soil structure (Rillig and Mummey 2006; Rillig et al. 2017; Lehmann et al. 2020). Therefore, arbuscular mycorrhizal fungi are common in many stressed environments and have been shown to increase plant survival and vitality in such ecosystems along with the positive impact they have on soil. Arbuscular mycorrhizal fungi acquire all their carbon from the host plants and have central roles (e.g. nutrient cycling) in many habitats. Several indicators exist that the benefits provided to plants by arbuscular mycorrhizal fungi will become even more important due to increased

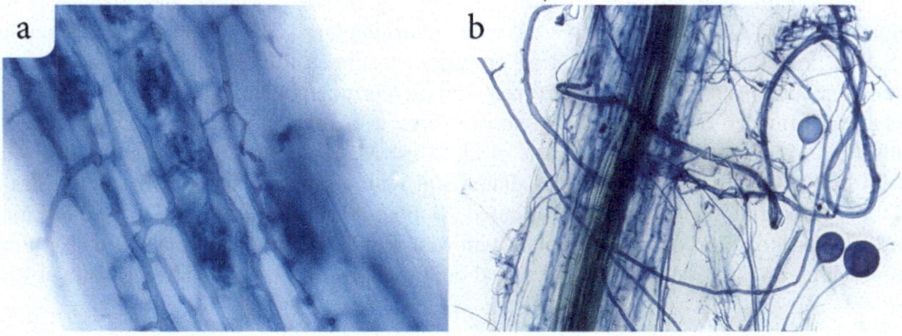

Fig. 25.2 Arbuscular mycorrhizal fungal colonisation of carrot roots (*Daucus carota* L.). Arbuscules can be seen as darker spots connected with intraradicular hyphae (**a**) in the root cortex. A longitudinal root view with arbuscular mycorrhizal fungal hyphae and spores visible around the root and fungal colonisation with arbuscules in the root cortex (**b**). Root associated fungal structures were stained with trypan blue dye. Photos were taken with Olympus Provis AX70 microscope and digital camera Olympus DP70

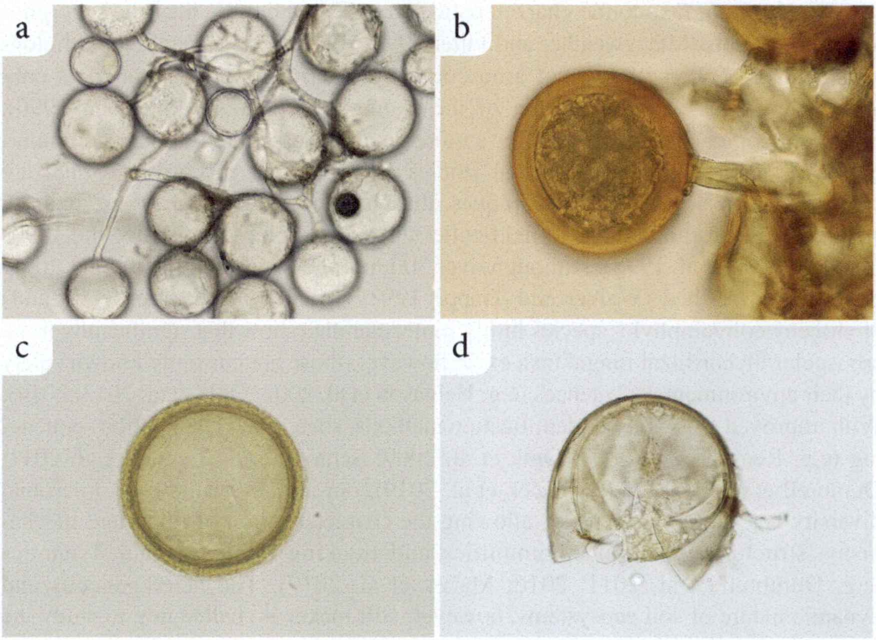

Fig. 25.3 Arbuscular mycorrhizal fungal spores of several species, representing three families Glomeraceae (**a**, **b**), Acaulosporaceae (**c**), and Gigasporaceae (**d**). A crushed spore, mounted in PVLG (polyvinyl alcohol-lactic acid-glycerol) with visible cell wall layers and lipid reserves—oil droplets (**d**). Morphology of the cell wall is used for taxonomical identification and differentiation among the arbuscular mycorrhizal fungal species. Photos were taken by Olympus Provis AX70 microscope and a digital camera

abiotic stresses caused by global change in the future (e.g. Hanson and Welzin 2000).

25.3.2 Use of Molecular Methods in Community Ecology of Arbuscular Mycorrhizal Fungi

Understanding arbuscular mycorrhizal fungal ecology and identification of the main predictors of their community-level processes is still difficult and it applies to a wide range of habitats. By delivering to the plant a range of benefits, arbuscular mycorrhizal fungi have a profound effect on plant community dynamics and diversity, highlighting the central role they have in terrestrial ecosystem processes (Fitter 2005; Rosendahl 2008). Recent molecular studies have shown that communities of arbuscular mycorrhizal fungi in nature are more diverse than originally thought on the basis of spore morphology (Fig. 25.3). Arbuscular mycorrhizal fungal spores still serve as the main taxonomic characteristic since they cannot be identified morphologically in roots (Merryweather and Fitter 1998) (Fig. 25.2). Since the morphological features of the structures of arbuscular mycorrhizal fungi in plant roots only allow low levels of identification of these fungi (Fitter and Moyersoen 1996), molecular approaches are needed for a more detailed description of their communities. The majority of the ecological studies on arbuscular mycorrhizal fungi are constructed on DNA-based techniques that have been developed to quantify arbuscular mycorrhizal fungi in field-collected soil and plant roots since the 1990s (e.g. Helgason et al. 1998). Although over 300 morpho-species of this fungal group have been described (Walker and Trappe 1993; Schüßler 2008, http://www.amf-phylogeny.com/amphylo_species.html), molecular data show that significantly more arbuscular mycorrhizal fungal taxa exist, however, those are currently known solely by their environmental sequences (e.g. Helgason et al. 2002; Öpik et al. 2013, 2014). With improved DNA-based identification methods, such as next generation sequencing (e.g. Roesch et al. 2007; Öpik et al. 2009; Schloss 2009; Lemos et al. 2011; Dumbrell et al. 2011, 2016; Maček et al. 2019), our ability to study soil microbial diversity has started to increase, allowing the characterisation of important mechanisms structuring natural communities and tracking their seasonal dynamics (e.g. Dumbrell et al. 2011, 2016; Maček et al. 2019). The heterogeneous and dynamic nature of soil ecosystems, however, still makes it challenging to study the effect of the soil environment on natural microbial communities in situ.

In the last decade, the next generations of sequencing approaches (e.g. Dumbrell et al. 2011, 2016; Maček et al. 2019) have revealed a much higher diversity than reported before the wide application of these technologies, mainly by reported more rare taxa, which are usually found only with the use of high-throughput technology. The latter allows also more intense sampling and is increasing the number of the analysed sequences and sequencing depth for each analysed sample. In the sampling sites, a combination of different approaches and expertise is needed. In particular, uniting species- (taxonomy) (Figs. 25.3 and 25.4) and community-oriented (ecology) approaches would be a major advantage in studying arbuscular mycorrhizal fungal community ecology and global diversity patterns (Öpik and Davison 2016). Moreover, extreme environments (including toxic metal-contaminated soils) can serve as systems to examine how long-term abiotic selection pressures drive natural communities and their evolution and possibly result in new specialists and extremophilic taxa. Potential new taxa could be isolated, characterised, and stored in international collections (Maček 2017) (Fig. 25.4).

Typically in temperate ecosystems with well-developed arbuscular mycorrhizal associations like semi-natural grasslands >50 arbuscular mycorrhizal fungal taxa (Operational Taxonomic Units - OTUs) from a range of different arbuscular mycorrhizal fungal families are reported using next generation of sequencing methods and sampling over several seasons, based on 18S rRNA SSU (small subunit ribosomal ribonucleic acid) marker genes. For example, in an old-growth (>100 years) semi-natural grassland ecosystem (Giessen free-air carbon dioxide enrichment experiment, Germany) a total of 55 arbuscular mycorrhizal fungal taxa (OTUs) from eight arbuscular mycorrhizal fungal families has been reported colonising mixed root

Fig. 25.4 Technique of establishing single-spore arbuscular mycorrhizal fungal cultures in pipette tips (**a**) with the goal to grow monospecific cultures of arbuscular mycorrhizal fungi established from a single spore isolated from environment (toxic metal-polluted site). Each individual plant-spore combination is later transferred to a bigger pot (**b**) where additional plant host of different species (e.g. *Plantago lanceolata* L., *Lolium perenne* L.) can be planted in order to initiate trap cultures and grow mycorrhizal fungi with their plant hosts to produce enough arbuscular mycorrhizal fungal spores of sufficient quality to enable taxonomical and functional studies on the potentially new (adapted) species of fungi

samples within one year vegetation season (Maček et al. 2019). This estimate of arbuscular mycorrhizal fungal OTU richness is as expected from grassland systems, given both the number and nature (i.e. mixed roots) of samples examined, and the next generation of sequencing methods used (Dumbrell et al. 2011; Hiiesalu et al. 2014; Moora et al. 2014).

Moreover, recent molecular studies on arbuscular fungal ecology from environmental samples (e.g. semi-natural temperate grassland) show that ecological studies should avoid relying only on broad-scale community-level responses of soil microbes including mycorrhizal fungi (for example, like commonly used community composition visualised using nonmetric multidimensional scaling—NMDS) (Maček et al. 2019). Likely, at a local scale, subtle changes in the relative abundances of specific arbuscular fungal populations are often driven by environmental factors and also stochastic processes (e.g. Dumbrell et al. 2010; Maček et al. 2019) and this is not necessarily reflected in the presence of a novel community that is entirely compositionally distinct from some other investigated or control community. A novel approach includes a more detailed examination of specific taxa. For example, methods, relying on modelling taxa (OTU—operational taxonomic units) abundances using multivariate generalised linear models (MV-GLMs) (Wang et al. 2012) or bias reduction binomial generalised linear models (BR-GLMs, Firth 1993), give a good insight into details of the population-level dynamics of specific taxa as shown in studies on the arbuscular mycorrhizal fungal community dynamics from environmental samples (e.g. Maček et al. 2019). In addition, snapshot data observed on single sampling time-point are not sufficient for a full investigation on the soil community composition. Therefore, temporal sampling throughout several seasons and at several time-points and years should be done. Such approaches can show us the susceptibility of the functionally important soil microbes to global change, including land-use change and soil pollution, with responses evident across both population and community levels. This raises critical questions and highlights the need for far greater research into the temporal response of soil microbes including arbuscular mycorrhizal fungi to environmental change; the functional differentiation observed across arbuscular mycorrhizal fungal taxa means any changes in their temporal dynamics has the potential to resonate throughout associated plant communities, changing aboveground competition dynamics and thus future ecosystem productivity in currently unpredictable ways (Maček et al. 2019).

25.3.3 Arbuscular Mycorrhizal Fungal Diversity in Toxic Metal Contaminated Soil

By using molecular methods, fungal communities have been described also in toxic (heavy) metal-contaminated sites (e.g. Zarei et al. 2008, 2010; Hassan et al. 2011; Maček et al. 2016b). Among other abiotic factors toxic metal concentration and biological availability in soils has been shown to impact the composition of arbuscular mycorrhizal fungal communities (e.g. Zarei et al. 2008, 2010, Hassan et al. 2011). Many reports indicate a reduction of arbuscular mycorrhizal fungal diversity in heavy metal-contaminated areas, based both on spore morphology (e.g. Griffioen 1994; Pawlowska et al. 1996; Leyval et al. 1997; del Val et al. 1999) and molecular data (e.g. Zarei et al. 2008, 2010; Hassan et al. 2011). A predominance of the taxa within the genus *Glomus* (old nomenclature before the major modifications published in the years 2010 by Schüßler and Walker, and 2011 by Oehl et al., see also Öpik et al. 2013) has been reported in most of the studied areas with severe toxic metal disturbance (e.g. Whitfield et al. 2004; Vallino et al. 2006; Zarei et al. 2008; Sonjak et al. 2009; Hassan et al. 2011), as well as other anthropogenic environments, such as agricultural sites, phosphate-contaminated sites (Daniell et al. 2001; Renker et al. 2005), and sites with fungicide treatments (Helgason et al. 2007). Zarei et al. (2008) analysed the diversity of arbuscular mycorrhizal fungal associated to *Veronica rechingeri* growing in the heavy metal-contaminated soil of the Anguran Zn and Pb mining region in Iran. Three species could be separated morphologically, while phylogenetic analyses revealed seven different arbuscular mycorrhizal fungal MOTUs (molecular operational taxonomic units) in plant roots, all within the genus *Glomus*. Some MOTUs were only found at sites with the highest and lowest soil toxic metal concentrations and some in both, which is a pattern also observed in other studies (e.g. Zarei et al. 2010; Hassan et al. 2011). Thus, the patterns of new taxa identifications are also showing in extreme environments that entirely originate from human-impacted pollution.

In most of the reported cases, arbuscular mycorrhizal fungal communities have not been sampled to saturation, and more intensive sampling and a higher depth of sequencing (for example, using next generation of sequencing) might result in detecting additional taxa and would allow for a more realistic description of the patterns in the community ecology of this group of organisms. Importantly, questions on long-term (press) related changes in soil microbial communities are relevant to many human drivers with long-term nature, including climate change, nutrient input, land-use change, soil contamination, and others (Maček et al. 2016a). However, in particular questions about the stability against press perturbations have received relatively little attention so far (Ives and Carpenter 2007). The introduction of the next generation of sequencing into ecology can largely change this. These tools are allowing us to obtain an exceptional amount of data (DNA sequences for different molecular markers) in a short time. This is a key condition to be fulfilled in order to understand the complex temporal and spatial patterns in soil microbial communities along with their environmental drivers, including soil pollutants. Moreover, the same molecular techniques and data analyses protocols could also be used for following development and succession of the microbial populations and communities in soil and rhizosphere after the contaminated soil remediation treatment.

25.4 Microbial Communities in Soils After Soil Remediation with EDTA

Remediation technologies often concentrate only on metal removal efficiency and overlook the treated soil's overall health, functioning, and potential use after remediation. The negative side of EDTA soil washing is its impact on soil biota and in addition other soil properties (e.g. soil structure) (Jelusic et al. 2014; Jelusic and Lestan 2014). It significantly diminishes soil microbial activity (Jelusic and Lestan 2014) and diversity (Maček et al. 2016b). This has been shown by using enzymatic tests (Jelusic and Lestan 2014), lately also by Kaurin et al. (2018), Kaurin and Lestan (2018), and has partially been studied on soil fauna (Tica et al. 2013; Udovic and Lestan 2010). Jelusic et al. (2014) have shown that soil remediation reduces the toxic metal concentrations in plants grown in experimental plots but the biomass of tested plants significantly diminished. Presumably, micronutrients were removed along with the toxic metals due to the nonselective nature of EDTA chelation. Effective means of revitalisation are needed to restore health and reclaim the remediated soil as a fertile plant substrate, including fertilisation to restore the soil nutrient pool (Jelusic et al. 2014) (see also Fig. 25.1). In addition, different amendments (e.g. hydrogel, vermiculite), each carrying a specific function, may help to improve the properties of the soil (e.g. Tica et al. 2013). Moreover, the addition of microbial inocula might help the soil to restore functional microbial communities (Fig. 25.1).

The harsh remediation conditions destroy the majority of life in the remediated soil, though direct measurements of biodiversity and community structure change have only been done for plant symbiotic arbuscular mycorrhizal fungi (Maček et al. 2016b). The results from this study show that after revitalisation of soil that was subjected to remediation treatment functional arbuscular mycorrhizal symbiosis can establish, either by commercial or indigenous inoculum addition to the soil substrate with higher diversity of arbuscular mycorrhizal fungi resulting from the soil inoculation with the indigenous inoculum (Maček et al. 2016b). Soil diversity is important for ecosystem stability and function thus following diversity loss after soil remediation treatment should be a part of the procedure integrated into the remediation plan (see Fig. 25.1 for the EDTA washing soil remediation procedure scheme).

25.4.1 Importance of Soil Biodiversity for Ecosystem Stability

According to the insurance hypothesis (Stuart and Pimm 1984), one of the proposed consequences of biodiversity loss is a reduction in the ecosystem stability in the sense of resistance (ability of the system to withstand the disturbance) and resilience (the speed by which the system returns to its pre-disturbance state) (e.g. Stuart and Pimm 1984). This hypothesis is based on the idea that the probability of finding species able to adapt to changing conditions and allowing ecosystem functioning is greater in a more diverse ecosystem. Several attempts have been made to test this hypothesis, for example, one of the approaches is to experimentally build up different levels of biodiversity. Assemblages of up to 43 species of fungi did show evidence of increasing stability with increasing biodiversity (Setälä and McLean 2004). As did communities containing up to 72 species of bacteria (Bell et al. 2005).

However, soils contain far more species than generally used in such community assembly experiments and effects of biodiversity are more evident in systems with low diversity (Nielsen et al. 2011). Arbuscular mycorrhizal fungi and other plant growth-promoting microorganisms have been used as inocula for biofertilisation and phytostimulation or different soil types (Rana et al. 2012). In addition to using indigenous inocula a growing number of inocula are being marketed which may help to restore the functionality of remediated soil. The commercial inocula, however, typically include only a limited taxa richness, which is usually also not very well defined (typically no molecular data on the taxa identity is available on the product specification). Therefore, a range of diversity that has been present in original soil before the remediation treatment is difficult to establish only with using commercial inocula (Maček et al. 2016b). This has also been shown by the study of Maček et al. (2016b) where both, commercial and indigenous inocula were tested (see the next section for details). Nevertheless, the potential for the arbuscular mycorrhizal fungi to establish in the remediated substrate has been shown also in this soil type and long-term experiments are urgently needed in order to follow the succession of the soil microbial and mycorrhizal fungal communities in such substrates (see Fig. 25.5 for the details on the mesocosm experiment with remediated soil at Biotechnical Faculty, University of Ljubljana, Slovenia).

The use of commercial fungal inocula is still connected to a lot of uncertainties, including persistence of inoculated organisms in the new environment, introduction of invasive species, competitive exclusion etc. In a perspective paper Rodrigues and Sanders (2015) address the role of community and population ecology in applying arbuscular mycorrhizal fungi for improved food security. The authors argue that despite the huge potential of the use of symbionts of plants for improving yields of

Fig. 25.5 A mesocosm experiment set as a full factorial experiment with remediated (EDTA washed) and original (toxic metal-contaminated) soil at the laboratory field of Department of Agronomy, Biotechnical Faculty, University of Ljubljana, Slovenia. Contaminated soil from two locations with different soil types has been selected for the experiment; acidic soil (pH < 6.1) from Stossau, Arnoldstein (Austria) and calcareous soil (pH > 7.1) from Meža Valley (Slovenia). Factors include soil type (acidic, calcareous), remediation treatment (original, remediated soil), plant (with *Lolium perenne* L., no plants), inoculum (grassland rhizosphere soil with roots, no inoculum). Temporal dynamics of the soil and plant root microbial community development (succession) and function (archaea, bacteria and fungi), mycorrhiza development, soil chemical parameters and plant responses have been investigated with sampling in regular intervals during the experiment, including the winter season. The experiment has been set as part of the activities within the Slovenian Research Agency (ARRS) research project J4-7052 with I. Maček as the principal investigator

globally important crops, the application of arbuscular mycorrhizal fungi in agriculture is too simplistic and ignores basic ecological principals. Thus interdisciplinary work with ecologists could significantly improve our understanding of the determinants of the survival, and ecological roles of the introduced microbes, and the effect this has on plant production and soil health.

25.4.2 Diversity of Arbuscular Mycorrhizal Fungal Communities in Remediated Soil

Establishing diverse communities of soil microorganisms, and importantly also arbuscular mycorrhizal fungi are urgent for the restoration of biological activity of the remediated soil. The first attempt to revitalise the EDTA-washed remediated soil with arbuscular mycorrhizal fungi was made in a pot experiment, where the newly established diversity of the fungal community was evaluated using molecular tools, at the time by using cloning and Sanger sequencing (Maček et al. 2016b).

The contaminated soil used in this experiment was collected from the upper 30 cm layer of a managed vegetable garden near an abandoned lead smelter in the Meža Valley, Slovenia (Maček et al. 2016b). More than 300 years of active lead mining and smelting was present in the area until 1990, leaving behind 6600 ha of agricultural land polluted primarily with Pb, Zn, and Cd (Jelusic and Lestan 2014). The site is home to more than 6000 inhabitants, with children being the most endangered group regarding toxic metals impact on their health (Jez and Lestan 2015). Despite the adoption of the action programme for improving the quality of the environment and reducing the health hazard for people in the Meža Valley in 2007, the results of annual surveys of blood Pb concentration indicated that the number of children with levels higher than 10 μg dL (toxicity threshold under Slovenian legislation) did not drop over the first 6-year duration of the programme, thus additional measures are immediately needed (Jez and Lestan 2015). After the remediation of soil collected in Meža Valley using EDTA washing technique, the initial metal concentrations of 1585 mg kg^{-1} Pb, 525 mg kg^{-1} Zn, and 8.8 mg kg^{-1} Cd were reduced to 313, 378, and 2.52 mg kg^{-1} for Pb, Zn, and Cd, respectively (Jelusic and Lestan 2014). The effect of EDTA washing of the soil used also in the study of Maček et al. (2016b), on soil properties, toxicity hazards, and plant biomass and fitness has been described before (Jelusic and Lestan 2014; Jelusic et al. 2014).

Maček et al. (2016b) have quantified arbuscular mycorrhizal fungal colonisation in plant roots as an indicator of mycorrhizal potential in toxic metal-contaminated (original) vs. remediated soil, and after inoculation with commercial (Symbivit Remed, Symbiom Ltd., Czech Republic[2]) and indigenous mycorrhizal inoculum (rhizosphere soil and root particles of *Plantago lanceolata* L. roots sampled at the original soil sampling location in Meža Valley, Slovenia). After soil remediation, the mycorrhizal potential and capability of fungi to form arbuscules in roots of several plant hosts (*Plantago lanceolata* L., *Lolium perenne* L., and *Sorghum bicolor* (L.) Moench) were severely reduced to practically hardly detectable root colonisation with fungal hyphae and no arbuscules present. This indicates that immediately after remediation mycorrhizal potential of remediated soil is very low and that the addition of inoculum into such substrate could accelerate the mycorrhiza establishment (Maček et al. 2016b). Further tests have shown that functional symbiosis could be established in the remediated soil by the use of indigenous inoculum (rhizosphere soil with root particles) as a good source of diverse fungal propagules for revitalisation of remediated soil, which has in this experiment performed better

[2]Commercial inoculum—Symbivit Remed, Symbiom Ltd., Czech Republic, that—according to the producer's specification—contains the following arbuscular mycorrhizal fungal isolates; *Rhizophagus irregularis*, *Funneliformis mosseae*, *Claroideoglomus claroideum*, and *Funneliformis constrictus*, isolated from temperate anthropogenic sites. Using molecular methods in our experiment, only *Funneliformis mosseae* could be amplified from the collected root particles that were an integrative part of the Symbivit Remed inoculum (Maček et al. 2016b).

(higher diversity of AM fungi in the final substrate), when compared to the commercial inoculum used (Maček et al. 2016b).

The evaluation of plant root colonisation however cannot give sufficient information on arbuscular mycorrhizal fungal diversity and community composition in the plant roots. Thus, molecular methods based on the analyses of arbuscular mycorrhizal fungal 18S rRNA SSU (small subunit ribosomal ribonucleic acid) marker genes were used in order to explore whether the arbuscular mycorrhizal fungal taxa present were distinct in toxic metal-polluted habitats in Meža Valley (the metal-contaminated site where the original soil has been collected), and in inoculated remediated soil (Maček et al. 2016b). In addition, arbuscular mycorrhizal fungal diversity in plant roots growing in remediated and original soil has been analysed, dependent on different inoculation methods (commercial vs. indigenous inoculum). The experiment has confirmed functional mycorrhizal symbiosis with plants was established either by commercial or indigenous inoculum addition. This was followed in two time-windows, 3- and 5-months after experiment initiation (soil remediation treatment), with progression in the development of mycorrhiza with time, which has been confirmed also as a molecular signal (arbuscular mycorrhizal fungal specific PCR (polymerase chain reaction) products of the 18S rRNA SSU marker genes) 5-months after the initiation of the experiment. The results clearly show that the use of the indigenous inoculum (grassland roots and rhizosphere soil) resulted in higher arbuscular mycorrhizal fungal taxa richness in the roots of plants growing in the remediated soil compared to the use of the commercial inoculum and that the use of the indigenous inocula (rhizosphere soil) resembles the diversity of the environmental (field) mycorrhizal communities to a greater extent. The only taxon that has been detected in the commercial inoculum and later in the remediated soil that has been inoculated by this inoculum type was *Funneliformis mosseae*. No arbuscular mycorrhizal fungal specific PCR products were yielded from the roots of *Plantago lanceolata* plants growing in non-inoculated remediated soil with a very low level of fungal root colonisation intensity.

However, the question on how diverse arbuscular mycorrhizal fungal communities develop in remediated soil still stays as in the same study of Maček et al. (2016b) only a limited number of arbuscular mycorrhizal taxa (OTUs) have been confirmed. Most of the arbuscular mycorrhizal fungal sequences (45 sequences), both from the indigenous inoculum samples and the community developed in the remediated soil could be assigned to the genus *Glomus*, followed by the genera *Rhizophagus*, and *Funneliformis* (both latter previously also known as *Glomus*, see Öpik et al. 2013 for Glomeromycota taxa nomenclature including modifications) (Maček et al. 2016b). A predominance of the taxa within the genus *Glomus* (old nomenclature before the major modifications published in the years 2010 by Schüßler and Walker, and 2011 by Oehl et al., see also Öpik et al. 2013) has been reported from most of the studied areas with severe toxic metal disturbance (e.g. Whitfield et al. 2004; Vallino et al. 2006; Zarei et al. 2008; Sonjak et al. 2009; Hassan et al. 2011), as well as other disturbed sites like agricultural sites, phosphate-contaminated sites (Daniell et al. 2001; Renker et al. 2005), and sites with fungicide treatments (Helgason et al. 2007).

This could however partially be conditioned also with the methodology used at the time of these explorations into fungal communities.

Representatives of *Funneliformis mosseae* were the most frequently reported as toxic metal-tolerant fungal taxa in several independent studies (Turnau et al. 2008; Vallino et al. 2006; Zarei et al. 2008; Hassan et al. 2011), suggesting adaptation of specific ecotypes of this species to toxic metal contamination (e.g. Turnau et al. 2008; Vallino et al. 2006, Zarei et al. 2008; Zarei et al. 2010; Hassan et al. 2011). This group includes arbuscular mycorrhizal fungi with a ruderal strategy as reported for *Funneliformis mosseae*, which is encountered as spores in natural soils and in trap cultures but infrequently in natural plant roots (Öpik et al. 2014). Indeed, *Funneliformis mosseae* has been reported to increase in abundance in many disturbed environments and competitive release has been suggested by Helgason et al. (2007) as the mechanism behind the high abundance of *Funneliformis mosseae* after using the fungicide benomyl to alter the community of arbuscular mycorrhizal fungi in undisturbed monoliths of soil in a natural community. Sequences of *Funneliformis mosseae* (VT67—virtual taxon, according to Öpik et al. 2013) represented also the only arbuscular mycorrhizal fungal taxon that was detected using molecular markers in the dry roots from the commercial inoculum (Symbivit Remed, Symbiom Ltd., Czech Republic) also used in the study with the remediated soil (Maček et al. 2016b).

Since in the study of Maček et al. (2016b) a low-throughput technique (cloning and Sanger sequencing) was used it is possible that only the most abundant arbuscular mycorrhizal fungal taxa have been reported. Therefore, more intensive sampling could result in detecting additional taxa in the samples since the arbuscular mycorrhizal fungal communities in this study were not sampled to saturation due to methodological limitations. Nevertheless, the results of this study clearly show that the most abundant taxa in the newly established communities were among the fungi that are well known for their ruderal (opportunistic) strategy and are tolerant to anthropogenic disturbance (e.g. *Funneliformis mosseae*).

25.5 Future Prospects

The study of Maček et al. (2016b) represented the first preliminary experiment with the focus on arbuscular mycorrhizal fungal communities development in remediated soil based on EDTA washing procedure. More work will be needed in the future to follow the dynamics, stability, and succession of soil microbial communities in remediated soil, along with the development of the soil functional diversity and functional traits. An effort for further investigation of the arbuscular mycorrhizal fungal community composition in the remediated soils on a larger scale (field experiments) is necessary, including monitoring successional stages and seasonal dynamics of the newly established communities, in order to thoroughly evaluate to what extent they can resemble natural communities in healthy soils (Maček et al. 2016b). This is the goal of the project and experiment set with the remediated soil at

University of Ljubljana, Biotechnical Faculty (Fig. 25.5) with the aim to follow temporal dynamics of the soil and plant root microbial community throughout several years. This goes along with evaluating the development (succession) of the communities and functional traits of soil archaea, bacteria, and fungi, mycorrhiza development, soil chemical parameters, and plant–soil responses. The remediated substrate enables studies of succession of the microbial communities in the soil after the treatment and depending on the revitalisation (inoculation) with a different range of microbial inocula. The idea of this experiment (see Fig. 25.5 for details) is to study the temporal dynamics, stability, and the extent of the recovery (resilience) of the microbial communities in two remediated soil types (calcareous and acidic soil). Moreover, studies of plant performance (plant–soil feedbacks) in the new substrate will be included to target the most widely recognised function of soil, which is its support of plant (food) production. The newly developed microbial community of a known composition could also serve as a soil type specific inoculum that could be used to inoculate larger amounts of remediated substrate for faster and more efficient revitalisation.

25.6 Conclusion

Fertile soil is a valuable, limited resource, often contaminated with substances that have negative impacts on life. Soil, heavily contaminated with toxic metals is referred as a hazardous waste. Soil remediation is a solution, but the harsh procedures result in soil that has heavily reduced soil diversity. There still is little knowledge of the establishment, succession, function, and dynamics of soil microbial communities in remediated soil. Moreover, the stability of soil microbial community composition is another question getting increasing attention especially in the light of global change. Therefore, in order to tackle the problem of soil toxic metal contamination from different angles, interdisciplinary efforts should be taken to combine the innovative soil remediation techniques with the integration of the rapidly increasing knowledge on soil microbial ecology, driven by the latest development of molecular tools (e.g. next generation sequencing techniques). This will give a better insight into the biological component of the remediated soil, including soil microbial communities and populations. Application of indigenous inocula (rhizosphere soil) has already been shown to result in a faster development of a more diverse microbial community compared to the no inocula treatment (Maček et al. 2016b), while commercial inocula have been shown to have only a limited range of fungal diversity, often limited to only few or even a single generalist taxon.

Acknowledgements This work was supported by the Slovenian Research Agency (ARRS) funding: basic research project J4-7052 and research programme P4-0085. All of the support given is gratefully acknowledged by the author.

References

Bell T, Newman JA, Silverman et al (2005) The contribution of species richness and composition to bacterial services. Nature 436:1157–1160

Daniell TJ, Husband R, Fitter AH, Young JPW (2001) Molecular diversity of arbuscular mycorrhizal fungi colonizing arable crops. FEMS Microbiol Ecol 36:203–209

del Val C, Barea JM, Azon-Aguilar C (1999) Diversity of arbuscular mycorrhizal fungus populations in heavy-metal-contaminated soils. Appl Environ Microbiol 65:718–723

Dumbrell AJ, Nelson M, Helgason T et al (2010) Relative roles of niche and neutral processes in structuring a soil microbial community. ISME J 4:337–345

Dumbrell AJ, Ashton PD, Aziz N et al (2011) Distinct seasonal assemblages of arbuscular mycorrhizal fungi revealed by massively parallel pyrosequencing. New Phytol 190:794–804

Dumbrell AJ, Ferguson RMW, Clark DR (2016) Microbial community analysis by single-amplicon high-throughput next generation sequencing: data analysis – from raw output to ecology. In: McGenity TJ, Timmis KN, Nogales B (eds) Hydrocarbon and lipid microbiology protocols: microbial quantitation, community profiling and array approaches. Springer, Heidelberg, pp 155–206

EEA (2020) The EU Environment, State & Outlook, https://www.eea.europa.eu/publications/soer-2020

EU Waste Catalogue & Hazardous Waste List, EPA (2002)

European Environment Agency. State of the environment No 1/2007 Chapter 2 (2007). Office for official publications of the European communities.

European Environment Agency (2014). https://www.eea.europa.eu/publications/eea-annual-report-2014

Field KJ, Pressel S (2018) Unity in diversity: structural and functional insights into the ancient partnerships between plants and fungi. New Phytol 220:996–1011

Finzgar N, Lestan D (2007) Multi-step leaching of Pb and Zn contaminated soils with EDTA. Chemosphere 66(5):824–832

Firth D (1993) Bias reduction of maximum likelihood estimates. Biometrika 80:27–38. https://doi.org/10.1093/biomet/80.1.27

Fitter AH (2005) Darkness visible: reflections on underground ecology. J Ecol 93:231–243

Fitter AH, Moyersoen B (1996) Evolutionary trends in root-microbe symbioses. Philos Trans R Soc B Biol Sci 351:1367–1375

Griffioen WAJ (1994) Characterization of a heavy metal-tolerant endomycorrhizal fungus from the surroundings of a zinc refinery. Mycorrhiza 4:197–200

Griffiths BS, Philippot L (2013) Insights into the resistance and resilience of the soil microbial community. FEMS Microbiol Rev 37:112–129

Hanson, Welzin (2000) Drought disturbance from climate change: response of United States forests. ·. Sci Total Environ 262(3):205–220

Hassan SED, Boon E, St-Arnaud M, Hijri M (2011) Molecular biodiversity of arbuscular mycorrhizal fungi in trace metal-polluted soils. Mol Ecol 20:3469–3483

Helgason T, Daniell TJ, Husband R et al (1998) Ploughing up the wood-wide web? Nature 394:431

Helgason T, Merryweather JW, Denison J et al (2002) Selectivity and functional diversity in arbuscular mycorrhizas of co-occurring fungi and plants from a temperate deciduous woodland. J Ecol 90:371–384

Helgason T, Merryweather JW, Young JPW, Fitter AH (2007) Specificity and resilience in the arbuscular mycorrhizal fungi of a natural woodland community. J Ecol 95:623–630

Hiiesalu I, Pärtel M, Davison J et al (2014) Species richness of arbuscular mycorrhizal fungi: associations with grassland plant richness and biomass. New Phytol 203(1):233–244

Hoysted GA, Kowal J, Jacob A et al (2018) A mycorrhizal revolution. Curr Opin Plant Biol 44:1–6

Ives AR, Carpenter SR (2007) Stability and diversity of ecosystems. Science 317:58–62

Jeffries P, Gianinazzi S, Perotto S et al (2003) The contribution of arbuscular mycorrhizal fungi in sustainable maintenance of plant health and soil fertility. Biol Fertil Soils 37:1–16

Jelusic M, Lestan D (2014) Effect of EDTA washing of metal polluted garden soils. Part I: toxicity hazards and impact on soil properties. Sci Tot Environ 475:132–141

Jelusic M, Vodnik D, Maček I, Lestan D (2014) Effect of EDTA washing of metal polluted garden soils. Part II: can remediated soil be used as a plant substrate? Sci Tot Environ 475:142–152

Jez E, Lestan D (2015) Prediction of blood lead levels in children before and after remediation of soil samples in the upper Meza Valley, Slovenia. J Hazard Mater 296:138–146

Kaurin A, Lestan D (2018) Multi-substrate induced microbial respiration, nitrification potential and enzyme activities in metal-polluted, EDTA-washed soils. Environ Pollut 243:238–245

Kaurin A, Cernilogar Z, Lestan D (2018) Revitalisation of metal-contaminated, EDTA-washed soil by addition of unpolluted soil, compost and biochar: effects on soil enzyme activity, microbial community composition and abundance. Chemosphere 193:726–736

Landfill Directive (1999/31/EC) https://www.eea.europa.eu/themes/waste/links/landfill-directive-1999-31-ec

Lehmann A, Zheng W, Ryo M, Soutschek K et al (2020) Fungal traits important for soil aggregation. Front Microbiol 10:2904. https://doi.org/10.3389/fmicb.2019.02904

Lemos LN, Fulthorpe RR, Triplett EW, Roesch LFW (2011) Rethinking microbial diversity analysis in the high throughput sequencing era. J Microbiol Methods 86:42–51

Lestan D (2017) Novel chelant-based washing method for soil contaminated with Pb and other metals—a pilot-scale study. Land Degrad Dev 28:2585–2595

Lestan D, Luo C, Li X (2008) The use of chelating agents in the remediation of metal-contaminated soils: a review. Environ Pollut 153:3–13

Leyval C, Turnau K, Haselwandter K (1997) Effect of heavy metal pollution on mycorrhizal colonization and function: physiological, ecological and applied aspects. Mycorrhiza 7:139–153

Maček I (2017) Arbuscular mycorrhizal fungal communities pushed over the edge—lessons from extreme ecosystems. In: Lukac M, Grenni P, Gamboni M (eds) Soil biological communities and ecosystem resilience, sustainability in plant and crop protection. Springer, Cham, pp 157–172

Maček I, Vodnik D, Pfanz H et al (2016a) Locally extreme environments as natural long-term experiments in ecology. In: Dumbrell AJ, Kordas R, Woodward G (eds) Advances in ecological research, large scale ecology: model systems to global perspectives, vol 55. Academic, Oxford, UK, pp 283–323

Maček I, Šibanc N, Kavšček M, Leštan D (2016b) Diversity of arbuscular mycorrhizal fungi in metal polluted and EDTA washed garden soils before and after soil revitalization with commercial and indigenous fungal inoculum. Ecol Eng 95:330–339

Maček I, Clark DR, Šibanc N et al (2019) Impacts of long-term elevated atmospheric CO_2 concentrations on communities of arbuscular mycorrhizal fungi. Mol Ecol 28:3445–3458

Merryweather JW, Fitter AH (1998) The arbuscular mycorrhizal fungi of *Hyacinthoides non-scripta* I. diversity of fungal taxa. New Phytol 138:117–129

Moora M, Davison J, Öpik M et al (2014) Anthropogenic land use shapes the composition and phylogenetic structure of soil arbuscular mycorrhizal fungal communities. FEMS Microbiol Ecol 90(3):609–621

Nielsen UN, Ayres E, Wall DH, Bardgett RD (2011) Soil biodiversity and carbon cycling: a review and synthesis of studies examining diversity-function relationships. Eur J Soil Sci 62:105–116

Oehl F, Sieverding E, Palenzuela J et al (2011) Advances in Glomeromycota taxonomy and classification. IMA Fungus 2:191–199

Öpik M, Davison J (2016) Uniting species- and community-oriented approaches to understand arbuscular mycorrhizal fungal diversity. Fungal Ecol 24B:106–113

Öpik M, Metsis M, Daniell TJ et al (2009) Large-scale parallel 454 sequencing reveals host ecological group specificity of arbuscular mycorrhizal fungi in a boreonemoral forest. New Phytol 184:424–437

Öpik M, Zobel M, Cantero JJ et al (2013) Global sampling of plant roots expands the described molecular diversity of arbuscular mycorrhizal fungi. Mycorrhiza 23:411–430

Öpik M, Davison J, Moora M, Zobel M (2014) DNA-based detection and identification of Glomeromycota: the virtual taxonomy of environmental sequences. Bot Rev 147:135–147

Pawlowska TE, Blaszkowski J, Rühling A (1996) The mycorrhizal status of plants colonizing a calamine spoil mound in southern Poland. Mycorrhiza 6:499–505

Payá Pérez A, Rodríguez Eugenio N (2018) Status of local soil contamination in Europe: Revision of the indicator "Progress in the management Contaminated Sites in Europe", EUR 29124 EN. Publications Office of the European Union, Luxembourg. https://doi.org/10.2760/093804. isbn:978-92-79-80072-6

Pociecha M, Lestan D (2012) Washing of metal contaminated soil with EDTA and process water recycling. J Hazard Mater 235–236:384–387

Rana A, Joshi M, Prasanna R et al (2012) Biofortification of wheat through inoculation of plant growth promoting rhizobacteria and cyanobacteria. Eur J Soil Biol 50:118–126

Redecker D, Kodner R, Graham LE (2002) *Palaeoglonius grayi* from the Ordovician. Mycotaxon 84:33–37

Renker C, Blanke V, Buscot F (2005) Diversity of arbuscular mycorrhizal fungi in grassland spontaneously developed on area polluted by a fertilizer plant. Environ Pollut 135:255–266

Rillig MC, Mummey DL (2006) Mycorrhizas and soil structure. New Phytol 171:41–53

Rillig M, Muller L, Lehmann A (2017) Soil aggregates as massively concurrent evolutionary incubators. ISME J 11:1943–1948

Rodrigues A, Sanders IR (2015) The role of community and population ecology in applying mycorrhizal fungi for improved food security. ISME J 9:1053–1061

Roesch LFW, Fulthorpe RR, Riva A et al (2007) Pyrosequencing enumerates and contrasts soil microbial diversity. ISME J 1:283–290

Rosendahl S (2008) The first glance into the *Glomus* genome: an ancient asexual scandal with meiosis? New Phytol 193:546–548

Schloss PD (2009) A high-throughput DNA sequence aligner for microbial ecology studies. PLoS One 4:e8230

Schüßler A (2008) Glomeromycota species list. http://www.amf-phylogeny.com/amphylo_species.html

Schüßler A, Walker C (2010) The Glomeromycota: a species list with new families and new genera http://www.amf-phylogeny.com. Published in December 2010 in libraries at The Royal Botanic Garden Edinburgh The Royal Botanic Garden Kew, Botanische Staatssammlung Munich, and Oregon State University, Gloucester 2010; Electronic copy freely available online at http://www.amf-phylogeny.com. Printed copy of available under ISBN-13: 978-1466388048, ISBN-10: 1466388048

Setälä H, McLean MA (2004) Decomposition rate of organic substrates in relation to the species diversity of soil saprophytic fungi. Oecologia 139:98–107

Smith SE, Read DJ (2008) Mycorrhizal symbiosis, 3rd edn. Academic, London, pp 787

Sonjak S, Beguiristain T, Leyval C, Regvar M (2009) Temporal temperature gradient gel electrophoresis (TTGE) analysis of arbuscular mycorrhizal fungi associated with selected plants from saline and metal polluted environments. Plant Soil 314:25–34

Stuart L, Pimm SL (1984) The complexity and stability of ecosystems. Nature 307:321–326

Tica D, Udovic M, Lestan D (2013) Long-term efficiency of soil stabilization with apatite and Slovakite: the impact of two earthworm species (Lumbricus terrestris and Dendrobaena Veneta) on lead bioaccessibility and soil functioning. Chemosphere 91:1–6

Turnau K, Anielska T, Ryszka P et al (2008) Establishment of arbuscular mycorrhizal plants originating from xerothermic grasslands on heavy metal rich industrial wastes – new solution for waste revegetation. Plant Soil 305:267–280

Udovic M, Lestan D (2010) Fractionation and bioavailability of Cu in soil remediated by EDTA leaching and processed by earthworms (Lumbricus terrestris L.). Environ Sci Pollut Res 17:561570

Vallino M, Massa N, Lumini E et al (2006) Assessment of arbuscular mycorrhizal fungal diversity in roots of *Solidago gigantea* growing in a polluted soil in Northern Italy. Environ Microbiol 8:971–983

Voglar D, Lestan D (2013) Pilot scale washing of Pb, Zn and Cd contaminated soil using EDTA and process water recycling. Chemosphere 91:76–82

Voglar D, Lestan D (2014) Chelant soil washing technology for metal contaminated soil. Environ Technol 35:1389–1400

Walker C, Trappe JM (1993) Names and epithets in the Glomales and Endogonales. Mycol Res 97:339–344

Wang Y, Naumann U, Wright ST, Warton DI (2012) Mvabund—an R package for model-based analysis of multivariate abundance data. Methods Ecol Evol 3:471–474

Whitfield L, Richards AJ, Rimmer DL (2004) Relationships between soil heavy metal concentration and mycorrhizal colonization in *Thymus polytrichus* in northern England. Mycorrhiza 14:55–62

Zarei M, König S, Hempel S et al (2008) Community structure of arbuscular mycorrhizal fungi associated to *Veronica rechingeri* at the Anguran zinc and lead mining region. Environ Pollut 156:1277–1283

Zarei M, Hempel S, Wubet T et al (2010) Molecular diversity of arbuscular mycorrhizal fungi in relation to soil chemical properties and heavy metal contamination. Environ Pollut 158:2757–2765

The manufacturer's authorised representative in the EU is Springer Nature Customer Service Centre GmbH, Europaplatz 3, 69115 Heidelberg, Germany. If you have any concerns regarding our products, please contact ProductSafety@springernature.com

Printed and bound by CPI Group (UK) Ltd, Croydon, CR0 4YY
30/09/2026
02225689-0017